Seventh Edition

An Introduction to the Biology of

Marine Life

James L. Sumich

Grossmont College

CONTRIBUTING AUTHORS
Sneed Collard, *University of West Florida*
John Morrissey, *Hofstra University*
Valerie Pennington, *Southwestern College*
Douglas Tupper, *Southwestern College*

Boston   Burr Ridge, IL   Dubuque, IA   Madison, WI   New York   San Francisco   St. Louis
Bangkok   Bogotá   Caracas   Lisbon   London   Madrid
Mexico City   Milan   New Delhi   Seoul   Singapore   Sydney   Taipei   Toronto

WCB/McGraw-Hill

A Division of The **McGraw·Hill** *Companies*

AN INTRODUCTION TO THE BIOLOGY OF MARINE LIFE, SEVENTH EDITION

 This book is printed on recycled, acid-free paper containing 10% postconsumer waste.

1 2 3 4 5 6 7 8 9 0 VNH/VNH 9 3 2 1 0 9 8

ISBN 0-697-34541-6

Vice president and editorial director: *Kevin T. Kane*
Publisher: *Michael D. Lange*
Sponsoring editor: *Margaret J. Kemp*
Developmental editor: *Adora L. Pozolinski*
Marketing manager: *Thomas C. Lyon*
Project manager: *Marilyn M. Sulzer*
Production supervisor: *Mary E. Haas*
Freelance coordinator: *Mary L. Christianson*
Photo research coordinator: *Lori Hancock*
Art editor: *Brenda A. Ernszen*
Compositor: *Shepherd, Inc.*
Typeface: *11/12 Garamond*
Printer: *Von Hoffmann Press, Inc.*

Freelance cover designer: *Kristyn A. Kalnes*
Cover photograph: © *Gary Bell/Masterfile*

Library of Congress Cataloging-in-Publication Data

Sumich, James L.
    An introduction to the biology of marine life / James L. Sumich;
  contributing authors, Sneed Collard . . . [et al.]. — 7th ed.
    p.   cm.
    Includes bibliographical references (p.   ) and indexes.
    ISBN 0-697-34541-6
    1. Marine biology.   I. Collard, Sneed.   II. Title.
  QH91.S95   1999
    578.77—dc21   97-53180
                                                                        CIP

www.mhhe.com

Brief Contents

Contents

Chapter 10
Coral Reefs   265

Chapter 11
Below the Tides   299

Part V
The Pelagic
Realm   319

Chapter 12
The
Zooplankton   321

Chapter 13
Nekton-Distribution,
Locomotion, and
Feeding   338

Chapter 14
Nekton-Migration,
Sensory Reception,
and
Reproduction   374

A*n Introduction to the Biology of Marine Life* was written to satisfy the demand for an introductory college-level text dealing with the biology of marine organisms. Ongoing developments in the field of marine biology have shown that courses in this subject provide an exciting and effective framework for illustrating basic biological principles.

This text is written for the introductory marine biology student. No previous knowledge of marine biology is assumed. However, some exposure to the basic concepts of biology is helpful. Selected groups of marine organisms are used to develop an understanding of biological principles and processes that are basic to all forms of life in the sea. To build on these basics, information dealing with several aspects of taxonomy, evolution, ecology, behavior, and physiology of selected groups of marine organisms is presented. I have intentionally avoided adopting any one of these major subdivisions as the framework of this text. Biology is an inclusive term, and a student's initial venture into this field should provide some flavor of the mix of disciplines that constitutes modern biology.

Scope of Subject Matter

While this text includes more material than covered in a standard semester, instructors can select and mold the material to match their teaching styles and time limitations. The accompanying instructor's manual provides suggestions for use of this text, with judicious use of outside supplementary readings, in a two-quarter or two-semester course.

Sequence of Topics

As in most textbooks, the sequence of topics is somewhat arbitrary and is intended to be flexible. Section I consists of an introduction to the marine environment (Chapter 1) and an examination of the general features of life in the sea (Chapter 2). Section II includes a survey of marine phytoplankton (Chapter 3), benthic plants (Chapter 4) and an examination of the major factors that shape the pattern of marine primary productivity. In Section III, the major marine animal groups are introduced; invertebrates in Chapter 6, and the marine vertebrate classes in Chapter 7. In Section IV, benthic communities are presented; intertidal organisms in Chapter 8, estuarine habitats in Chapter 9, coral reefs in Chapter 10, and subtidal communities in Chapter 11. Adaptations of pelagic animals to their habitat are stressed in Section V, Chapters 12, 13, and 14. The final section develops important perspectives for understanding the effects of human intervention upon marine ecosystems, including fishing (Chapter 15) and pollution (Chapter 16).

New to This Edition

The widespread and positive reception of the six previous editions of this text continues to be encouraging. This edition represents a continuing effort to better meet the needs of those who use the text. The many suggestions and comments from readers and reviewers have been considered in the changes that were made, while retaining the organization of the last edition. Chapter 7 includes expanded coverage of vertebrates, including marine mammals. Boxed current topics, are presented as "Research in Progress" topics, with selected references to lead students to additional background information. New to this edition are "Ecological Perspectives," essays discussing an aspect of the general ecology of the organisms considered in each section. Many new color photographs have been included to better support the text material.

Student Aids

Each chapter contains numerous illustrations, graphs, and charts to assist students in visualizing the concepts presented. End-of-chapter summaries, questions for discussion, and supplementary reading lists encourage further in-depth exploration of covered topics. Additional references, listed at the end of the text, will be useful to students who have the enthusiasm and communication skills necessary to cope with the challenges of reading original literature. Topics in "Research in Progress" boxes are designed to direct interested students to some of the original literature addressing these topics of current research in marine biology.

Supplementary Materials

Instructor's Manual/Test Item File

Prepared by Sneed Collard University of West Florida, the Instructor's Manual offers course schedules, Chapter summaries, objectives, key concepts, review questions, short case studies to be presented as extra lecture material and questions to enhance critical thinking. The Test Item File consists of 25–30 questions/answers for each chapter and was carefully written to test both factual and conceptual information.

Slide Set

A set of 50 additional images not found in the book are available free to instructors.

Transparencies

Fifty acetate transparencies are available with this text. The transparencies are taken from the text and represent figures that merit extra visual review and discussion.

Micro Test III

User-friendly computerized software is available to instructors for testing and grading. It can be ordered in either IBM DOS or Windows format and MacIntosh.

Student Study Guide

Prepared by Larry Lewis, Salem State College, The Student Study Guide will aid students in testing their comprehension of important concepts and principles in marine biology, through the use of informal writing. This unique approach offers study techniques based on the principles of critical thinking developed at the Institute for Writing and Thinking at Bard College.

Laboratory Manual

A companion text, *Laboratory and Field Investigations in Marine Biology,* is available for courses emphasizing field or laboratory experiences. Additionally, several regional identification and field guides are listed in Appendix C.

Videotapes

Several videotapes carefully selected from offerings by Films For the Humanities are available to qualified adopters. Less than 30 minutes in length, these tapes provide an excellent complement to a lecture.

Acknowledgments

Much credit for the development of this text goes to students and instructors who have used previous editions and have offered valuable comments and criticisms. I thank my instructors of the past and colleagues of the present for their contributions to my education and to this book. Four individuals, Dr. John Morrissey of Hofstra University, Dr. Valarie Penning of Southwestern College, Dr. Douglas Tupper of Southwestern College, and Dr. Sneed Collard of University of West Florida, have made valuable contributions to the seventh edition of this text. Special thanks also go to the many individuals and institutions that graciously supplied many of the photographs. Finally and especially, I thank my present and former students for their interest and enthusiasm in discovering rewarding methods of communicating this information.

James L. Sumich
La Mesa, California

List of Reviewers for the Seventh Edition

William G. Ambrose, Bates College

Gil Bane, Kodiak College

Paul A. Billeter, Charles County Community College

Sneed Collard, University of West Florida

Harold N. Cones, Christopher Newport University

Gina Erickson, Highline Community College

Susan Flanagan, Nunez Community College

Dominic Gregorio, Cypress College

Marty L. Harvill, Bowling Green State University

Richard Heard, Gulf Coast Research Lab

Susan Keys, Springfield College

Vicky J. Martin, University of Notre Dame

John F. Morrissey, Hofstra University

Valerie Pennington, Southwestern College

Robert F. Shields, U.S. Coast Guard Academy

Doug Tupper, Southwestern College

Richard Turner, Florida Institute of Technology

Jacqueline F. Webb, Villanova University

Mary Katherine Wicksten, Texas A&M University

I also remain grateful to the following individuals, who reviewed the previous edition and whose guidance and input has been carried forward to this edition:

Brenda Blackwelder, Central Piedmont Community College

James L. Campbell, Los Angeles Valley College

Susan Cormier, University of Louisville

Sheldon Dobkin, Florida Atlantic University

J. Nicholas Ehringer, Hillsborough Community College

Robert T. Galbraith, Crafton Hills College

Hal M. Genger, College of the Redwoods

Lynn Hansen, Modesto Jr. College

Lester Knapp, Palomar College

Matthew Landau, Stockton State College

Cynthia Lewis, San Diego State University

Vicki J. Martin, University of Notre Dame

Donald Munson, Washington College

Joel Ostroff, Brevard Community College

L. Scott Quackenbush, Los Angeles Valley College

Richard A. Roller, University of Wisconsin–Stevens Point

Mary Beth Saffo, University of California–Santa Cruz

Cynthia C. Strong, Bowling Green State University

Jefferson T. Turner, Southeastern Massachusetts University

Richard Turner, Florida Institute of Technology

Jacqueline Webb, New York St. College of Vet. Med., Cornell University

Robert Whitlatch, University of Conneticut

Richard B. Winn, Duke University Marine Laboratory

A synthetic view of North Pacific Ocean Basin   Courtesy National Aeronautics and Space Administration

Introductory Concepts

ECOLOGICAL PERSPECTIVES
Part I Introductory Concepts: The Ocean as a Habitat

The ocean, as a habitat, is substantially different from land. Chapter 1 discusses and quantifies many of the most significant differences. It is useful to place those differences in perspective, and that is the objective of this essay. For example, large thermal oscillations are common on land, whereas the high specific heat of water ensures that marine organisms are rarely exposed to large changes in temperature. In addition, niches are relatively easy to identify on land. Conversely, niches in the open ocean are a challenge to identify. The diversity of natural populations of plankton seems to contradict the competitive exclusion principle in what can be called the "paradox of the plankton." That is to say, although most photosynthetic cells in the sea compete for the same inorganic nutrients, often more than 30 species coexist in small volumes of water. Most producers in the sea are in constant three-dimensional motion. Terrestrial plants are immobile.

The ocean provides a great deal of buoyancy for animals, whereas structural support (i.e., a skeleton) is required on land. Also, these structural materials are quite refractory on land, whereas all is eaten in the sea. The food web of the sea is based on protein. Carbohydrates dominate terrestrial food webs. Marine populations are characteristically different from populations on land. For example, marine populations are often characterized by rapid growth, and terrestrial populations are characterized by slow growth. Populations on land are dominated by adults with low mortality, whereas larvae that suffer high mortality dominate populations in the sea. Marine populations are usually food limited (except during brief periods of blooms; see chapter 3), whereas terrestrial populations are limited by the availability of water. Finally, a "patch" in the sea typically refers to the presence of species, whereas a patch on land usually refers to the absence of species. Nevertheless, classification of the marine environment is analogous to the classification of land, in terms of relative productivity and energy flow. For example, the open ocean is analogous to grasslands, coastal areas are similar to forests, and zones of upwelling and deep estuaries are analogous to rain forests and areas of intensive agricultural activity. By keeping these comparisons and relationships in mind as you read this text, you may achieve a greater understanding of the ocean as a habitat.

The Ocean as a Habitat

T he ocean is home to a tremendous variety of living organisms highly adapted to the special conditions of the sea. The general features of these organisms and the variety of marine life itself are products of the many properties of the ocean habitat. This chapter will provide a survey of the developmental history and present geography of the ocean basins and a general discussion of some properties of seawater and of ocean circulation processes.

We need to develop a special perspective to study the oceans. We naturally tend to see the world from a human viewpoint, with human scales of time and distance. To begin to understand the marine environment of the earth and how it evolved to its present form, we must broaden our perspective to include very different time and distance scales. Terms such as *young* and *old* or *large* and *small* have limited meaning unless placed in some useful context. Figure 1.1 compares size and time scales for a few common oceanic features and inhabitants. Throughout this book, these scales will be revisited and others will be introduced to help you develop a practical sense of the time and space scales experienced by marine organisms.

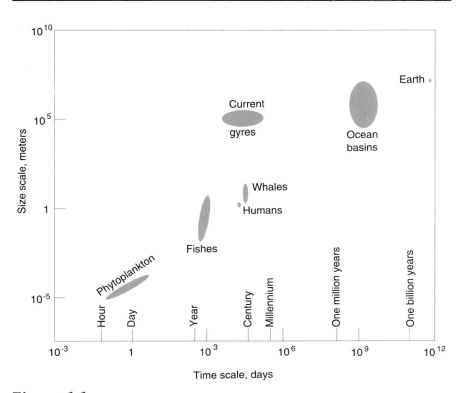

Figure 1.1

Some important marine features with appropriate time and size scales.

The Changing Marine Environment

Our solar system, including the earth, is thought to have been formed approximately 5.0 billion years ago. Modern theories on the origin of the solar system suggest that the planets aggregated from a vast cloud of cold gas and dust particles into clusters of solid matter. These clumps continued to grow as gravity attracted them together. As the earth grew in this manner, pressure from the outer layers compressed and heated the earth's center. Aided by heat from decay of radioactive elements, the interior of the earth melted. Iron, nickel, and other heavy metals settled to the core, while the lighter materials floated to the surface and cooled to form a thin crust (figure 1.2).

Numerous volcanic vents poked through the crust and tapped the upper mantle for liquid material and gases that were then spewed out over the surface of the young earth, and a primitive atmosphere developed. Water vapor was certainly present. As the water vapor condensed, it fell as rain, accumulated in low places on the earth's surface, and formed primitive oceans. Atmospheric gases dissolved into accumulating seawater, and other chemicals, dissolved from rocks and carried to the seas by rivers, added to the mixture, eventually creating seawater.

Since their initial formation, ocean basins have experienced considerable change. New material derived from the earth's mantle has extended the continents so that they are now larger and stand higher than at any time in the past. The oceans have kept pace, getting deeper with accumulations of new water from volcanic gases and of chemical breakdown of rock. Earth's early life forms (represented by fossils older than about 3 billion years) had a significant impact on

Figure 1.2

A section through the earth, showing its density-layered interior structure and the thickness of each layer.

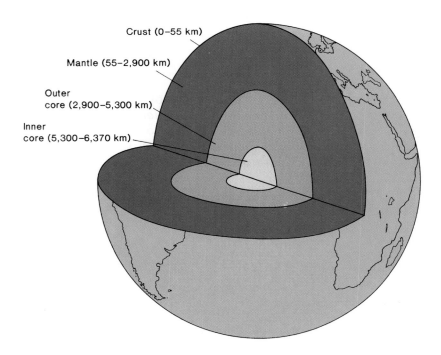

Crust (0–55 km)

Mantle (55–2,900 km)

Outer
core (2,900–5,300 km)

Inner
core (5,300–6,370 km)

the character of their physical environment. Free oxygen (O_2) was produced in increasing amounts by microscopic photosynthetic cells. The O_2 content of the atmosphere 600 million years ago was probably about 1% of its present concentration. It was not much, but it is believed to have been the turning point, the time when organisms utilizing O_2 (in aerobic respiration) became dominant and organisms not utilizing O_2 became less prevalent.

The evolution of more complex life forms that used increasingly efficient methods of energy utilization set the stage for an explosion of marine species. By 500 million years ago, most major groups of marine organisms had made their appearance. Worms, sponges, corals, and the distant ancestors of terrestrial animals and plants were abundant. But life at that time could exist only in the sea, where a protective blanket of seawater shielded it from harmful ultraviolet radiation.

As O_2 became more abundant in the upper atmosphere, some of it was converted to **ozone** (O_3). The process of forming ozone absorbed much of the lethal ultraviolet radiation coming from the sun and prevented the radiation from reaching the earth's surface. The O_2 concentration of the atmosphere 400 million years ago is estimated to have reached 10% of its present level (its current concentration was achieved in the Mesozoic Era). The additional ozone screened out enough ultraviolet radiation to permit a few life forms to abandon their sheltered marine home and colonize the land. Only recently have we become aware that industrialized society's increasing use of aerosols, refrigerants, and other pollutants is gradually depleting this protective layer of ozone. Figure 1.3 summarizes significant events of the origin and early development of life on the earth.

Early in this century, Alfred Wegener proposed that the oceans were changing in other ways. Wegener developed a detailed hypothesis of **continental drift** to explain several global geological features, including the remarkable jigsaw-puzzle fit of some continents (especially Africa and South America). He proposed that our present continental masses had drifted apart after the breakup of a single supercontinent, **Pangaea.** His evidence was ambiguous, and most scientists at the time remained unconvinced. It was not until the early 1960s that new evidence elicited wider endorsement by the scientific community. The early cruises of the *Glomar Challenger* (see Research in Progress: Explorations in Different Dimensions, p. 8) provided the telling evidence that verified Wegener's hypothesis.

The evidence that supports the closely related concepts of **seafloor spreading** and **plate tectonics** indicates that the earth's crust is divided into giant irregular plates (figure 1.4). These rigid plates float on the denser and slightly more plastic mantle material. Each plate is bounded by oceanic trench and ridge systems, and some plates include both oceanic and continental crusts. New oceanic crustal material is formed continually along the axes of oceanic ridges and rises. As the crustal plates grow on either side of the ridge, they move laterally in opposite directions, carrying bottom sediments and attached continental masses with them (figure 1.5).

In 1977, scientists, working in the deep-diving research submersible *Alvin* made a remarkable discovery of new marine animal

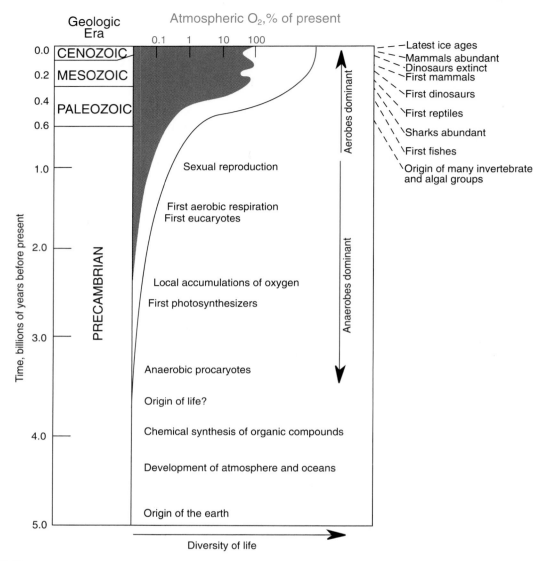

Figure 1.3

A summary of some biological and physical milestones in the early development of life on the earth. The black curve represents the relative diversity of life; the blue curve represents the O_2 concentration of the atmosphere. Several of the terms used here will be defined in chapter 2.

communities associated with seafloor hot water vents (see figure 11.13). These vents are integral parts of some oceanic ridge or rise systems. Members of these and other recently discovered deep-sea communities are discussed in chapter 11.

The changes that seafloor spreading and plate tectonics have wrought on the shapes and sizes of the oceans have been impressive. The African continent is drifting northward on a collision course with Europe, relentlessly closing the Mediterranean Sea. The Atlantic Ocean is becoming wider at the expense of the Pacific Ocean. Australia and India continue to creep northward, slowly changing the shapes of the ocean basins they border. Occasional violent earthquakes are only incidental

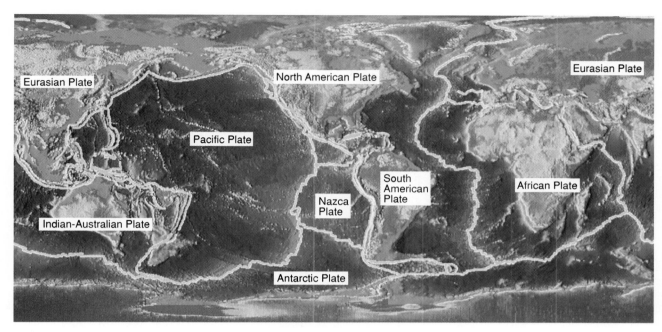

Crustal Plate Boundaries

Figure 1.4

The major plates of the earth's crust. Compare the features of this map with those of figures 1.9 and 11.12.
Courtesy National Geophysical Data Center.

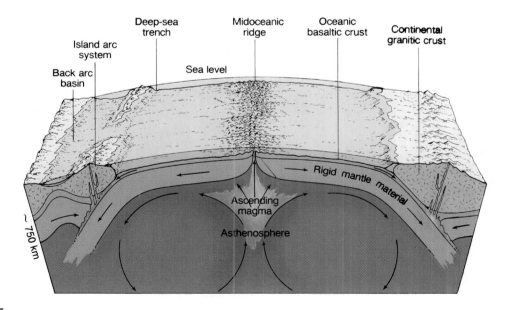

Figure 1.5

Cross section of a spreading ocean floor, illustrating the relative motions of oceanic and continental crusts. New crust is created at the ridge axis, and old crust is lost in trenches.

Research in Progress

Explorations in Different Dimensions

Primitive humans must have explored their local coastal environments very early in their history, but few of their discoveries were recorded. By 325 B.C., Pytheas, a Greek explorer, had sailed to Iceland and developed a method for determining **latitude** (see map). About a century later, Eratosthenes of Alexandria, Egypt, provided the earliest recorded estimate of the earth's size, its first dimension. His calculated circumference of 45,000 km was only about 12% greater than today's accepted value of 40,000 km. During the Middle Ages, Vikings, Arabians, Chinese, and Polynesians sailed over major portions of the earth's oceans. By the fifteenth century, all the major inhabitable land areas were occupied; only Antarctica remained unknown to and untouched by humans. Even so,

precise charting of the ocean basins had to await three key developments, each one associated with its own voyage of discovery.

Between 1768 and 1779, James Cook, an English navigator, conducted three exploratory voyages, mostly in the Southern Hemisphere. He was the first to cross the Antarctic Circle and to understand and conquer scurvy (a disease caused by a deficiency of vitamin C). He is best remembered as the first global explorer to make extensive use of the marine chronometer developed by John Harrison, a British inventor. The chronometer, a very accurate shipboard clock, was necessary to establish the **longitude** of any fixed point on the earth's surface (see map). Together with Pytheas's 2,000-year-old

technique for fixing latitude, accurate positions of geographical features anywhere on the globe could be established for the first time, and our two-dimensional view of the earth's surface was essentially complete. Today, coastal LORAN stations and satellite-based positioning systems enable individuals to determine their position to within a few tens of meters anywhere on the earth.

In 1882, one century after Cook's voyages, the first truly interdisciplinary global voyage for scientific exploration of the seas departed from England. The H.M.S. *Challenger* was converted expressly for this voyage. The voyage lasted over three years, sailed almost 125,000 km in a circumnavigation of the globe, and returned with such a wealth of information that 15 years and

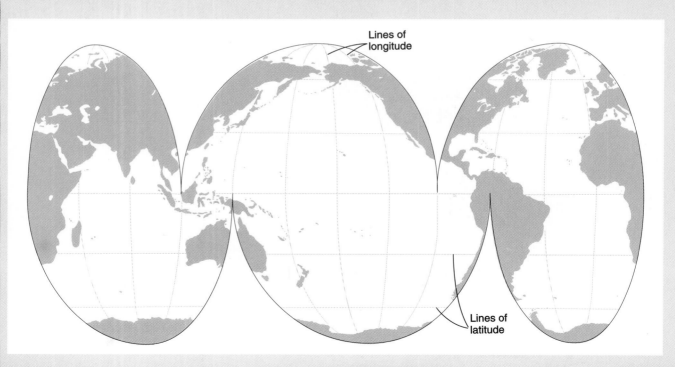

Lines of longitude

Lines of latitude

50 large volumes were required to publish the findings. During the voyage, 492 depth soundings were made. These soundings traced the outlines of the Mid-Atlantic Ridge, plumbed the Mariannas Trench to a depth of 8,185 m, and filled in rough outlines of the third dimension of the world ocean, its depth.

In 1968, a new and unusual ship, the *Glomar Challenger,* was launched to probe time, the fourth dimension of the oceans. Equipped with a deck-mounted drilling rig, the *Glomar Challenger* was capable of drilling into the seafloor in water over 7,000 m deep. Within two years, the *Glomar Challenger* recovered vertical sediment core samples from enough sites on both sides of the Mid-Atlantic Ridge to finally and firmly confirm the concept of seafloor spreading and continental drift. Before being decommissioned in 1983, the *Glomar Challenger* traveled almost 700,000 km and drilled 318,461 m of seafloor in 1,092 drill holes at 624 sites in all ocean basins. Subsequent analyses of microscopic marine fossils recovered from this tremendous store of marine sediment samples have led to refined estimates of the ages and patterns of evolution of all the major ocean basins.

Within two years of her retirement, the *Glomar Challenger* was replaced by the larger and more modern drill ship, the *JOIDES Resolution.*

For more information on current research aboard the JOIDES Resolution, see the following:
Maxwell, A. E. 1993. An abridged history of deep ocean drilling. Oceanus 36(4):8–12.
Other articles in Oceanus 36(4).

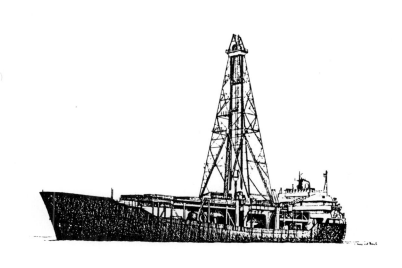

tremors in this monumental collision of crustal plates. The rates of seafloor spreading have been determined for some oceans, and they vary widely. The South Atlantic is widening about 3 cm each year (or approximately your height in your lifetime). The Pacific Ocean is shrinking somewhat faster.

The breakup of the megacontinent, Pangaea, produced ocean basins where none existed before. The seas of 200 million years ago changed size or disappeared altogether. The past positions of the continents and ocean basins, based on our present understanding of the processes involved, are reconstructed in figure 1.6. Excess crust produced by seafloor spreading folds into mountain ranges (the Himalayas is a dramatic example) or slips down into the mantle and remelts (figure 1.5).

Figure 1.6

About 200 million years ago, Pangaea, the megacontinent, separated into two large continental blocks, Laurasia and Gondwana. Since then, they have fragmented into smaller continents and continue to drift. These maps outline the past positions of the changing continents and ocean basins.

Adapted from Dietz and Holden 1970.

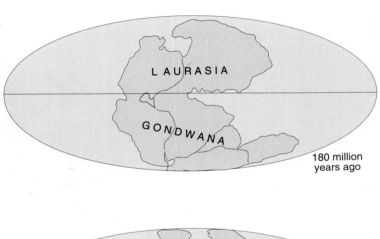

180 million years ago

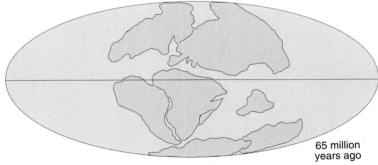

65 million years ago

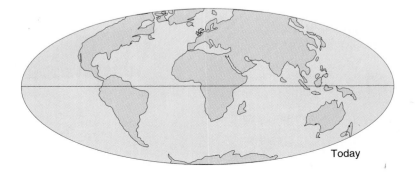

Today

Unfortunately, most marine fossils older than 200 million years can never be studied; they too have been carried to destruction in the mantle by the "conveyor belt" of the seafloor crust. Hence, the only fossil evidence we have for the first 90% of the evolutionary history of marine life is found in landforms that were once ancient seabeds.

On a much shorter time scale, other processes have been at work to alter the shapes and sizes of ocean basins. Only 18,000 years ago, northern reaches of Europe, Asia, and North America were frozen under the grip of the most recent ice age, or the **last glacial maximum (LGM).** The massive amount of water contained in those glaciers lowered sea level about 150 m below its present (and also its preglacial) level. Between 18,000 and 10,000 years ago, the shrinking of these continental glaciers was accompanied by a 150-m rise in sea level and the flooding of land exposed during the LGM. Coral reefs, estuaries, and other shallow coastal habitats were modified extensively during this flooding; these topics will be discussed in chapters 8 through 10.

The World Ocean

At the present time, the world ocean covers approximately 70% of the earth's surface and has an average depth of about 3,800 m. This may seem like a lot of water, but when compared with the earth's diameter of 13,250 km, the ocean is actually a very thin film of water covering the earth's crust. On the scale of this part's front image of the earth from space, the average ocean water depth is represented by a distance of about 0.04 mm, or slightly more than one one-thousandth of an inch.

Conventionally, the world ocean has been separated into four major ocean basins: the Atlantic, Pacific, Indian, and Arctic oceans. A more realistic approach views the marine environment as one large interconnected ocean system. This can be visualized in figure 1.7. The

Figure 1.7

Another view of the world ocean, this one emphasizing the extensive oceanic connections between major ocean basins.

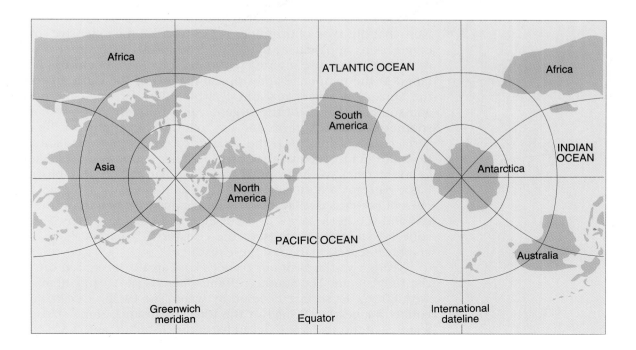

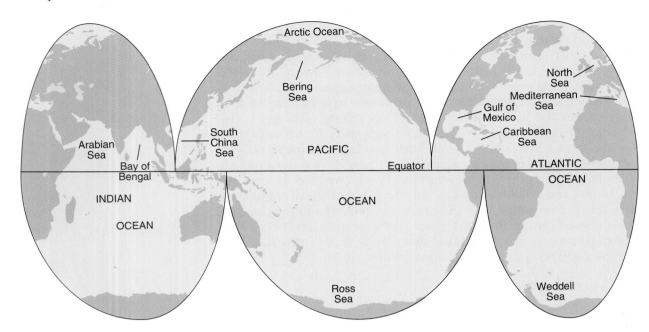

Figure 1.8

An equatorial view of the world ocean.

Antarctic continent is surrounded by a "Southern Ocean," which has three large embayments extending northward. These three oceanic extensions, partially separated by continental barriers, are the Atlantic, Pacific, and Indian oceans. Other smaller oceans and seas, such as the Arctic Ocean and the Mediterranean Sea, project from the margins of the larger ocean basins. Connections between the major ocean basins permit exchange of seawater and marine organisms.

Figure 1.8 presents a more conventional view of the world ocean. Note that this type of map does not emphasize the extensive southern connections apparent in figure 1.7. The format of figure 1.8 is often more useful because interest in the marine environment has been focused in the temperate and tropical regions of the earth. In addition, the equator is a very real physical boundary between the northern and southern halves of the large ocean basins, dividing northern and southern current patterns and life zones. The curvature of the earth's surface causes areas near the equator to receive more radiant energy from the sun than equal-sized areas in polar regions. The resultant heat gradient from warm tropical to cold polar regions establishes the basic pattern of atmospheric and oceanic circulation. Surface ocean current patterns display a nearly mirror-image symmetry in the northern and southern halves of the Pacific and Atlantic oceans. The hemispheric symmetry establishes the equator as a natural focus for the graphical representation of these features.

Nearly two-thirds of our planet's land area is located in the Northern Hemisphere. The Southern Hemisphere is an oceanic hemisphere, with 80% of its surface covered by water. The Pacific Ocean alone accounts for nearly one-half of the total ocean area. Some statistics for features of the six largest ocean basins are listed in table 1.1.

Table 1.1
Some Comparative Features of the Major Ocean Basins

Ocean	Area $\times 10^6$ km^2	Volume $\times 10^6$ km^3	Average Depth, m	Maximum Depth, m
Pacific	165.2	707.6	4,282	11,033
Atlantic	82.4	323.6	3,926	9,200
Indian	73.4	291.0	3,963	7,460
Arctic	14.1	17.0	1,205	4,300
Caribbean	4.3	9.6	2,216	7,200
Mediterranean	3.0	4.2	1,429	4,600
Other	18.7	17.3		
Total	361.1	1,370.3	3,795	

Oceanic depths extend to over 11,000 m, but most of the ocean bottom lies between 3,000 and 6,000 m. A synthetic satellite view of the northern and central parts of the Atlantic Ocean (figure 1.9) illustrates the large-scale features of the ocean bottom.

The **continental shelf,** which extends seaward from the shoreline and is actually a structural part of the continental landmass, would not be considered an oceanic feature if sea level were lowered by as little as 5% of its present average depth. The width of continental shelves varies from almost nonexistent off southern Florida to over 800 km north of Siberia in the Arctic Ocean. Continental shelves account for about 8% of the ocean's surface area; this is equivalent to one-sixth of the earth's total land area.

Most continental shelves are relatively smooth and slope gently seaward. The outer edge of the shelf, sometimes called the **shelf break,** is a vaguely defined feature that usually occurs at depths of 120 to 200 m. Beyond the shelf break, the bottom steepens slightly to become the **continental slope.** The continental slope is the boundary between continental masses and the true ocean basins. The slope is steep, with depths rapidly reaching 3,000 to 4,000 m.

A large portion of the deep ocean basin consists of flat, sediment-covered areas called **abyssal plains.** Most abyssal plains are situated near the margins of the ocean basins at depths between 3,000 and 5,000 m. Oceanic **ridge and rise systems,** such as the Mid-Atlantic Ridge and East Pacific Rise, occupy over 30% of the ocean basin area. The ridge and rise systems are rugged linear features that form a continuous underwater mountain chain that encircles the earth. Isolated peaks of these mountain systems extend above sea level to form islands such as Iceland and Ascension Island in the Atlantic Ocean.

Trenches are distinctive ocean-floor features that are generally deeper than 6,000 m. Most trenches, including the five deepest, are located along the margins of the Pacific Ocean. The Challenger Deep, in the Mariana Trench of the western North Pacific, extends to 11,033 m, the greatest ocean depth found anywhere. (This is as far below sea level as commercial jets fly above sea level.) Trenches account for less than 2% of the ocean-bottom area, but they are interesting because of the high pressure and constant temperature regimes imposed on their inhabitants. Trenches (along with ridge and rise systems) are integral parts in the processes of seafloor spreading and plate tectonics.

Figure 1.9

Large-scale features of the North
Atlantic seafloor.
Courtesy National Geophysical Data Center.

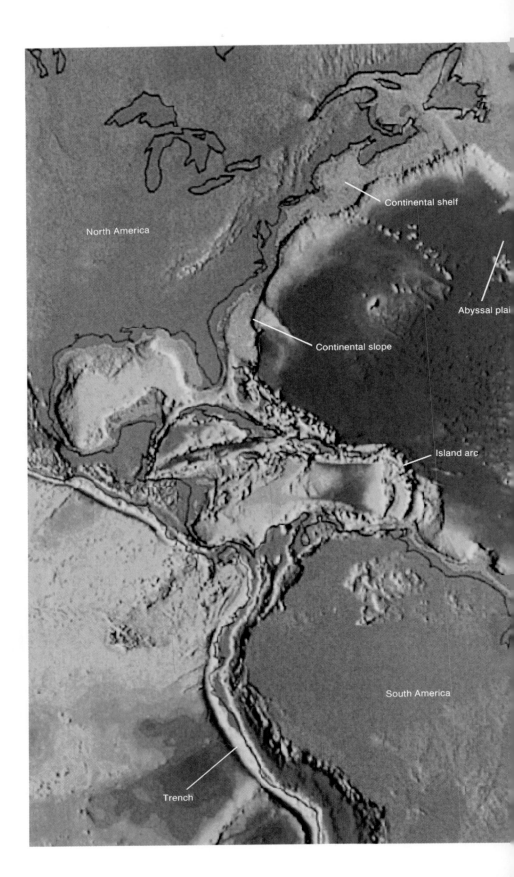

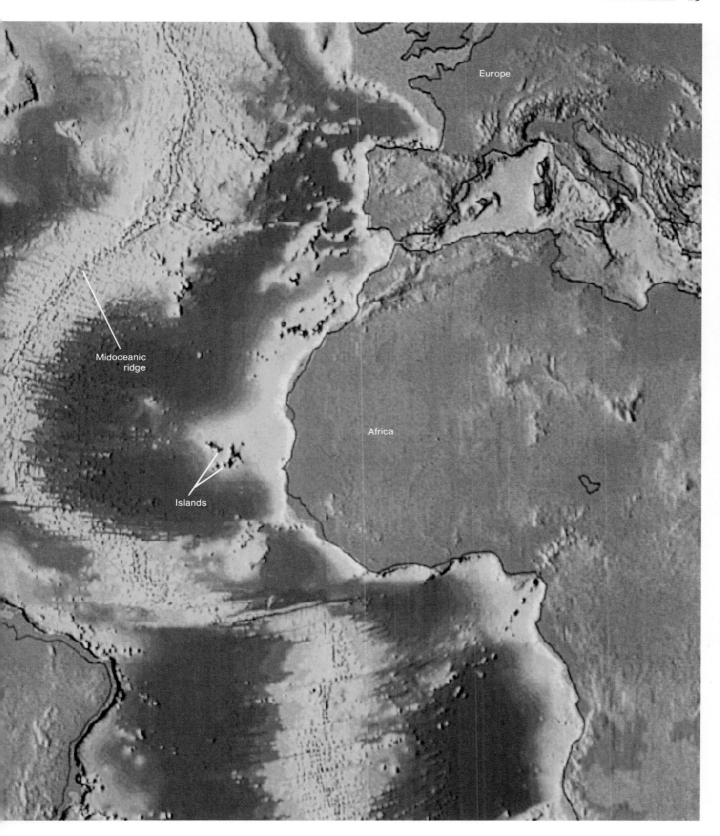

Europe

Midoceanic
ridge

Islands

Africa

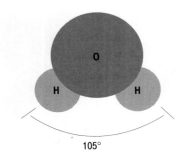

Figure 1.10

The arrangement of H and O atoms in a molecule of water (H_2O). The oxygen end has a slight net negative charge; the hydrogen end has a slight net positive charge.

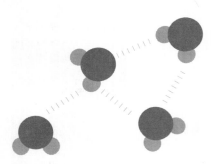

Figure 1.11

Hydrogen bonding between adjacent molecules of liquid water. Blue lines represent hydrogen bonds.

Most oceanic islands and **seamounts** have been formed by volcanic action. Oceanic islands are volcanic mountains that extend above sea level; seamounts are volcanic mountains whose tops remain below the sea surface. The majority of these features are located in the Pacific Ocean. Islands in tropical areas are often submerged and capped by coral atolls or fringed by coral reefs. These reefs, examined in chapter 10, form some of the most beautiful and complex communities found anywhere.

Properties of Seawater

Many properties of seawater are crucial to the survival and well-being of the ocean's inhabitants. Water accounts for 80 to 90% of the volume of most marine organisms. It provides buoyancy and body support for swimming and floating organisms, thereby reducing the need for heavy skeletal structures. Water is also the medium for most chemical reactions needed to sustain life. The life processes of marine organisms in turn alter many fundamental physical and chemical properties of seawater, including its transparency and chemical makeup, making organisms an integral part of the total marine environment. Understanding the interactions between organisms and their marine environment requires a brief study of some of the more important physical and chemical attributes of seawater. The characteristics of pure water and seawater differ in some respects and are similar in other respects. Let us consider first the basic properties of pure water and then study the effects of dissolved substances on those properties.

Pure Water

Water is a common, yet very remarkable, substance on the earth. Although abundant in its liquid form, large quantities of water also exist as a gas in the atmosphere and as a solid in the form of ice and snow. Although individual water molecules have a simple structure, the collective properties of many water molecules together are quite complex. Each water molecule has one atom of oxygen (O) and two atoms of hydrogen (H), which together form water (H_2O). (Some properties of these and other biologically important elements are listed in Appendix B.) The many unusual properties of water stem from its molecular shape: Two hydrogen atoms form an angle of about 105° with the oxygen atom (figure 1.10). This configuration produces an asymmetrical, dipole water molecule, with the oxygen atom dominating one end of the molecule and the hydrogen atoms dominating the other end. The bond between each hydrogen and the oxygen atom is formed by the sharing of two negatively charged electrons. The oxygen atom attracts the electron pair of each bond, causing the oxygen end of the water molecule to assume a slight negative charge. The hydrogen end of the molecule, by giving up part of its electron complement, is left with a small positive charge. The resultant electrical charge separation causes each water molecule to behave like a miniature magnet, one end with a positive charge and the other end with a negative charge. Each end of one water molecule attracts the oppositely charged end of other water molecules. This attractive force creates a weak bond, a **hydrogen bond** or **H-bond**, between adjacent water molecules (figure 1.11). These bonds are much weaker

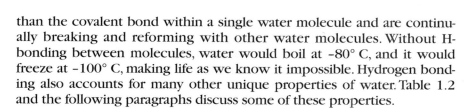

Table 1.2
Some Biologically Important Physical Properties of Water

Properties	Comparison with Other Substances	Importance in Biological Processes
Boiling point	High (100° C) for molecular size	Causes most water to exist as a liquid at earth surface temperatures
Freezing point	High (0° C) for molecular size	Causes most water to exist as a liquid at earth surface temperatures
Surface tension	Highest of all liquids	Critical to position maintenance of sea-surface organisms
Density of solid	Unique among common natural substances	Causes ice to float and inhibits complete freezing of large bodies of water
Latent heat of evaporation	Highest of all common natural substances (540 cal/g)	Moderates sea-surface temperatures by transferring large quantities of heat to the atmosphere through evaporation Inhibits large-scale freezing oceans
Latent heat of freezing	Highest of all common natural substances (80 cal/g)	Maintains a large variety of substances in solution, enhancing a variety of chemical reactions
Solvent power	Dissolves more substances in greater amounts than any other liquid	Moderates daily and seasonal temperature changes
Heat capacity	High (1 cal/g/° C) for molecular size	Stabilizes body temperatures of organisms

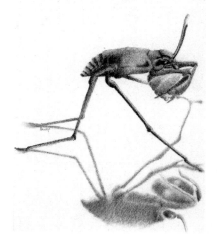

Figure 1.12
A water strider (*Halobates*), one of the few completely marine insects, is supported by the surface tension of seawater.

Redrawn from a photograph by Lana Cheng, Scripps Institution of Oceanography.

than the covalent bond within a single water molecule and are continually breaking and reforming with other water molecules. Without H-bonding between molecules, water would boil at –80° C, and it would freeze at –100° C, making life as we know it impossible. Hydrogen bonding also accounts for many other unique properties of water. Table 1.2 and the following paragraphs discuss some of these properties.

Viscosity and Surface Tension
Hydrogen bonding between adjacent water molecules within the fluid mass tends to resist external forces that would separate these molecules. This property, known as **viscosity,** has a significant effect on all marine organisms. The viscosity of water reduces the sinking tendency of some organisms by increasing the frictional resistance between themselves and nearby water molecules. At the same time, viscosity magnifies problems of drag that actively swimming animals must overcome.

The mutual attraction of water molecules at the surface of a water mass (such as the air-sea boundary) creates a flexible molecular "skin" over the water surface. This, the **surface tension** of water, is sufficiently strong to support the full weight of a water strider (figure 1.12). Both surface tension and viscosity are temperature dependent, increasing with decreasing temperature.

Density-Temperature Relationships
Most liquids contract and become denser as they cool. The solid form of these substances is denser than the liquid form. The structure of

liquid water at low temperatures is not known, but several models have been proposed to explain its behavior. Over most of the temperature range at which pure water is liquid, it behaves like other liquids. At 4° C or above, the **density** increases with decreasing temperature.[1] Below 4° C, the normal density-temperature pattern of pure water reverses. One model suggests that at near-freezing temperatures, less-dense icelike clusters consisting of several water molecules form and disintegrate very rapidly within the body of liquid water. As liquid water continues to cool, more clusters form and the clusters survive longer. Eventually, at 0° C, all the water molecules become locked into the rigid crystal lattice of ice. The ice formed is about 8% less dense than liquid water at the same temperature, so ice always floats on liquid water. This is an unusual, but very fortunate, property of water. Without this unique density-temperature relationship, ice would sink as it formed, and lakes, oceans, and other bodies of water would freeze solid from the bottom up. Winter survival for organisms living in such an environment would be much more difficult.

Heat Capacity

Heat is a form of energy, the energy of molecular motion. The sun is the source of almost all energy entering the earth's heat budget. At the surface of the sea, radiant energy is converted to heat energy. In the sea, heat is transferred from place to place primarily by **convection** (mixing) and secondarily by **conduction** (molecular exchange of heat). Heat energy is measured in **calories.**[2]

Water has the ability to absorb or give up heat without experiencing a large temperature change. To illustrate the high **heat capacity** of water, imagine a 1-g block of ice at –20° C on a heater that provides heat at a constant rate. Heating the ice from –20° C to 0° C requires 10 calories, or 0.5 calories per degree of temperature increase. However, converting 1 g of ice at 0° C to liquid at 0° C requires 80 calories. Conversely, 80 calories of heat must be extracted from 1 g of liquid water at 0° C to convert it to ice at the same temperature. This is referred to as the **latent heat of fusion.** Continued heating of the 1-g water sample from 0° C requires one calorie of heat energy for each one-degree change in temperature until the boiling point (100° C) is reached. At this point, further temperature increase is halted until all the water is converted to water vapor. For this conversion, 540 calories of heat energy are necessary; this is referred to as the **latent heat of vaporization.** Figure 1.13 summarizes the energy requirements for water temperature changes. The high heat capacity and the large amount of heat required for evaporation enable large bodies of water to resist extreme temperature fluctuations. Heat energy is absorbed slowly by water when the air above is warmer and is gradually given up when the air is colder.

---

[1]The maximum density of pure water is used to define the fundamental metric measurement of mass, the gram. The gram is defined as the mass of pure water at 4° C contained in the volume of one cubic centimeter. Thus the density, the ratio of mass to volume, of pure water at 4° C is 1.000 g/cm³.
[2]A calorie is a unit of heat energy, defined as the quantity of heat needed to elevate the temperature of 1 g of pure water 1° C.

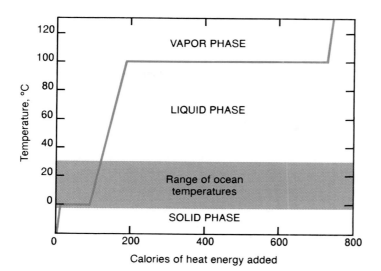

Figure 1.13

The heat energy necessary to cause temperature and phase changes in 1 g of water.

This process provides a temperature-moderating effect for the marine environment and adjacent land areas.

Solvent Action

The small size and polar charges of each water molecule enable it to interact with and dissolve most naturally occurring substances, especially salts, which are composed of atoms or simple molecules, called **ions,** that carry an electrical charge. Salts held together by **ionic bonds** (bonds between oppositely charged adjacent ions) are particularly susceptible to the solvent action of water. Figure 1.14 illustrates the process of a salt crystal dissolving in water. Initially, several water molecules form weak H-bonds with each Na^+ and Cl^- ion, and they eventually overcome the mutual attraction of those ions that previously bound them together in the crystalline structure. As more Na^+ and Cl^- ions are removed in this way, the solid crystal structure disintegrates and the salt dissolves. Nonpolar substances such as oxygen are generally less soluble in water.

Seawater

Seawater has accumulated during billions of years of water's eroding rocks and soil, the breakdown of organisms, and the condensation of rain from the atmosphere. About 3.5% of seawater is composed of dissolved compounds from these sources. The other 96.5% is pure water. Traces of all naturally occurring substances probably exist in the ocean and can be separated into three general categories: (1) inorganic substances, usually referred to as salts, including nutrients necessary for plant growth; (2) dissolved gases; and (3) organic compounds derived from living organisms. Organic compounds dissolved in seawater include fats, oils, carbohydrates, vitamins, amino acids, proteins, and other substances. Scientists think that these compounds are an important source of nutrition for marine bacteria and several other types of organisms. Current research indicates that other organic compounds, especially synthetics such as DDT, PCBs, and other chlorinated

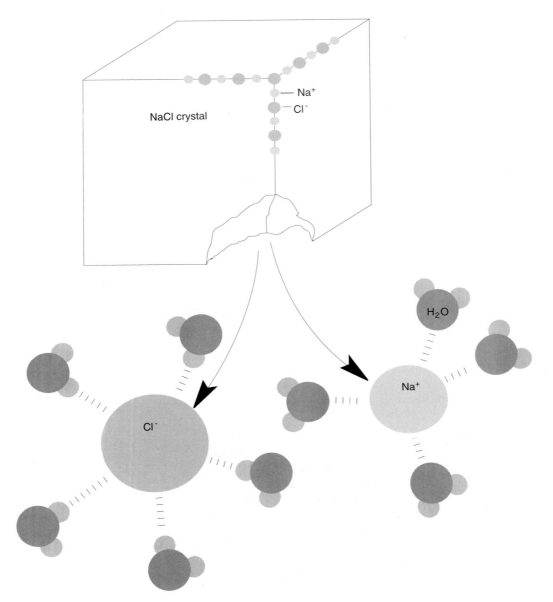

Figure 1.14

A salt crystal (above) and the action of charged water molecules in dissolving the crystal to dissociated Na^+ and Cl^- ions.

hydrocarbons that accumulate in seawater, can have devastating effects on some forms of marine life.

Dissolved Salts

Salts account for the majority of dissolved substances in seawater. The
total amount of dissolved salts in seawater is referred to as its **salinity,**
measured in parts per thousand (‰). Average seawater salinity is approximately 35‰. Salinity values range from nearly zero at river
mouths to over 40‰ in some areas of the Red Sea. Yet, in open-ocean

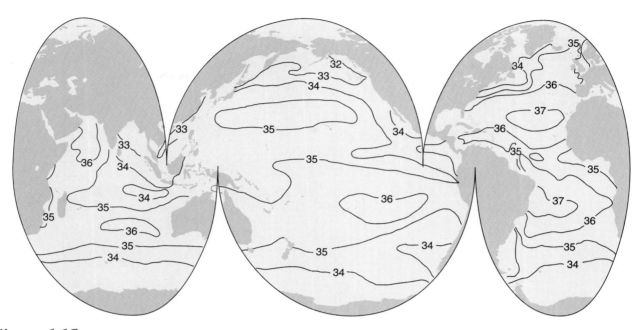

Figure 1.15

Geographical variations of average surface ocean salinities, expressed in ‰.

areas away from coastal influences, the salinity varies only slightly over large distances (figure 1.15).

Salinity is altered by processes that add or remove salts or water from the sea. The primary mechanisms of salt and water addition or removal are evaporation, precipitation, river runoff, and the freezing and thawing of sea ice. When evaporation exceeds precipitation, it removes water from the sea surface, thereby concentrating the remaining salts and increasing the salinity. Excess precipitation decreases salinity by diluting the sea salts. Freshwater runoff from rivers has the same effect. Figure 1.16 illustrates the average annual north-south variation of sea-surface evaporation and precipitation. The areas with greater evaporation than precipitation (hatched portions of figure 1.16) generally correspond to the high-surface-salinity regions shown in figure 1.15. These latitudes also coincide with most of the great land deserts of the world.

When seawater freezes, only the water molecules are incorporated into the developing ice crystal. The dissolved salts are excluded, thus increasing the salinity of the remaining seawater. The process is reversed when ice melts. Freezing and thawing of seawater are usually seasonal phenomena, resulting in little long-term salinity differences.

When dissolved in water, salts dissociate to produce both positively and negatively charged ions. For example, table salt (sodium chloride) dissociates to form positively charged sodium ions (Na^+) and negatively charged chloride ions (Cl^-). The more abundant ions found in seawater are listed in table 1.3 and are grouped as major or minor constituents according to their abundance. The major ions account for over 98% of the total salt concentration in seawater. In relation to each

Figure 1.16

Average north-south variation of sea-surface evaporation and precipitation.

From G. Dietrich, *General Oceanography,* 1963. Copyright © Gebruder Borntraeger, Stuttgart, Germany. Reprinted by permission.

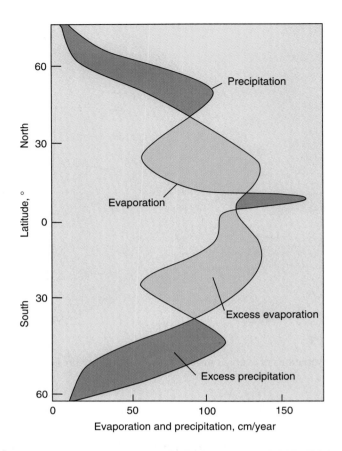

Table 1.3
Major and Minor Ions in Seawater of 35‰ Salinity

Ion	Chemical Symbol	Concentration ‰	
Chloride	Cl⁻	19.3	
Sodium	Na⁺	10.6	
Sulfate	SO_4^{-2}	2.7	
Magnesium	Mg^{+2}	1.3	Major
Calcium	Ca^{+2}	0.4	
Potassium	K⁺	0.4	
Bicarbonate	HCO_3^-	0.1	
Bromide	Br⁻	0.066	
Borate	H_3BO_3	0.027	
Strontium	Sr^{+2}	0.013	Minor
Fluoride	F⁻	0.001	
Silica	$Si(OH)_4$	0.001	

Plus traces of other naturally occurring elements

other, concentrations of the major ions remain remarkably constant even though their total abundance may differ from place to place.

Seawater is a complete chemical medium for life for it provides all the substances necessary for the growth and maintenance of plant tissue. Magnesium, calcium, bicarbonate, and silica are important components of the hard skeletal parts of marine organisms. Plants need

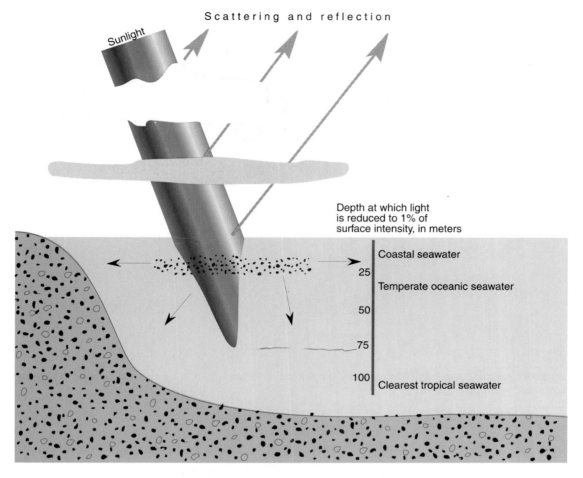

Scattering and reflection

Sunlight

Depth at which light
is reduced to 1% of
surface intensity, in meters

Coastal seawater

25

Temperate oceanic seawater

50

75

100   Clearest tropical seawater

Figure 1.17

Fate of visible spectrum near the sea surface.

nitrate and phosphate for the synthesis of organic material. In addition, a vital similarity exists between the chemical composition of seawater and the composition of the body fluids of marine organisms. Most of the more abundant ions enumerated in table 1.3 are important components of the body fluids of all organisms.

Light and Temperature in the Sea

Most marine organisms living in the upper portions of the sea utilize light energy from the sun for one of two functions, vision or photosynthesis. The amount of energy reaching the sea surface through the atmosphere depends on the presence of dust, clouds, and gases that absorb, reflect, and scatter a portion of the incoming solar radiation (figure 1.17). On an average day, about 65% of the sun's radiation arriving at the outer edge of our atmosphere reaches the earth's surface. The magnitude of incoming solar radiation is reduced when the angle of the sun is low, as in winter or at high latitudes, and a portion of the light that makes it through the atmosphere is reflected back into space by the sea surface.

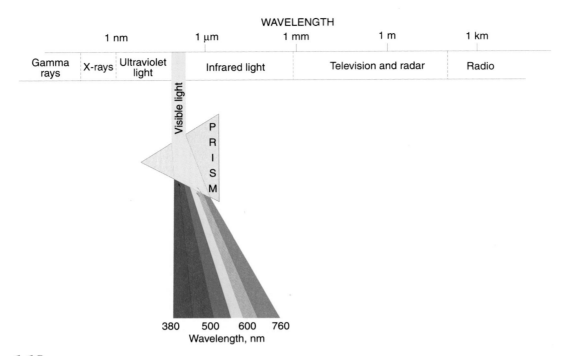

Figure 1.18

The electromagnetic radiation spectrum. The small portion known as visible light is passed through a prism to separate the light into its component colors.

Of the broad spectrum of electromagnetic radiation that is generated by the sun (top of figure 1.18), most marine animals can visually detect only a very narrow band near the center of the spectrum. Our eyes visually respond only to the portion labeled **visible light** in figure 1.18 (violet through red), and most other animals with eyes, whether they see in color or not, respond visually to approximately the same portion of the electromagnetic radiation spectrum.

The band of light energy used by animals for vision is nearly identical to that used in photosynthesis. Photosynthetic organisms must remain in the upper region of the ocean (the **photic zone**) where solar energy sufficient to support photosynthesis will reach them. The depth of the photic zone is determined by how rapidly seawater absorbs light and converts it to heat energy. Dissolved substances, suspended sediments, and even plankton populations diminish the amount of light available for photosynthetic activity and cause the depth of light penetration to differ dramatically between coastal and oceanic water (figure 1.17).

As sunlight travels through our atmosphere and into the sea, its color characteristics are altered as seawater rapidly absorbs the violet and the orange-red portions of the visible spectrum, leaving the green and blue wavelengths to penetrate deeper. Even in the clearest tropical waters, almost all of the red light is absorbed in the upper 10 m. Clear seawater is most transparent to the blue and green portions of the spectrum (450 to 550 nm); 10% of the blue light penetrates to 100 m. However, even this light is eventually absorbed or scattered

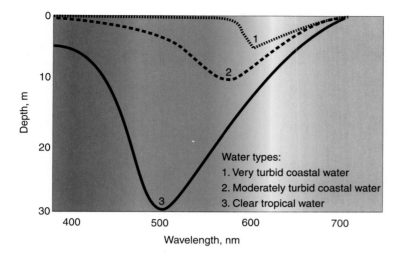

Figure 1.19

Penetration of various wavelengths of light in three water types. Note the shift to shorter wavelengths (bluer light) in clearer water.

(figure 1.19). The deeper penetration and eventual back-scattering of blue light account for the characteristic blue color of clear, tropical seawater. Coastal waters are commonly more turbid, with a greater load of suspended sediments and dissolved substances derived from land runoff. Here, there is a shift in the relative penetration of light energy, with green light penetrating deepest. In many coastal regions, green light is reduced to 1% of its surface intensity in less than 30 m. Adaptations to these different light regimes by photosynthetic organisms are explained in chapter 5.

When sunlight is absorbed by water molecules, it is converted to heat energy, and the motion of the water molecules increases. **Temperature,** commonly recorded as degrees Celsius (° C), is the way we measure that change in molecular motion. Temperature is a universal factor governing the existence and behavior of living organisms. Life processes cease to function above the boiling point of water, when protein structures are irreversibly altered (as when you cook an egg), or at subfreezing temperatures, when the formation of ice crystals disrupts cellular structures. But between these absolute limits, life flourishes.

The high heat capacity of water limits marine temperatures to a much narrower range than land temperatures (figure 1.20). Some marine organisms survive in coastal tropical lagoons at temperatures as high as 40° C. Some bacteria associated with deep-sea hydrothermal vents (figures 11.13 and 11.14) experience water temperatures above 60° C. Other deep-sea animals spend their lives in water perpetually one or two degrees below 0° C. Penguins and a few other birds and mammals well-adapted to extreme cold commonly tolerate air temperatures far below 0° C in polar regions. Penguins even manage to incubate and hatch eggs under these conditions. But these are exceptions; most marine species live at water temperatures between 0° and 30° C.

The distribution of various forms of marine life is closely associated with geographical differences in seawater temperatures. Surface ocean temperatures are highest near the equator and decrease toward both poles. This temperature gradient establishes several east-west–trending marine climatic zones (figure 1.21). The approximate temperature range of each zone is included in figure 1.20.

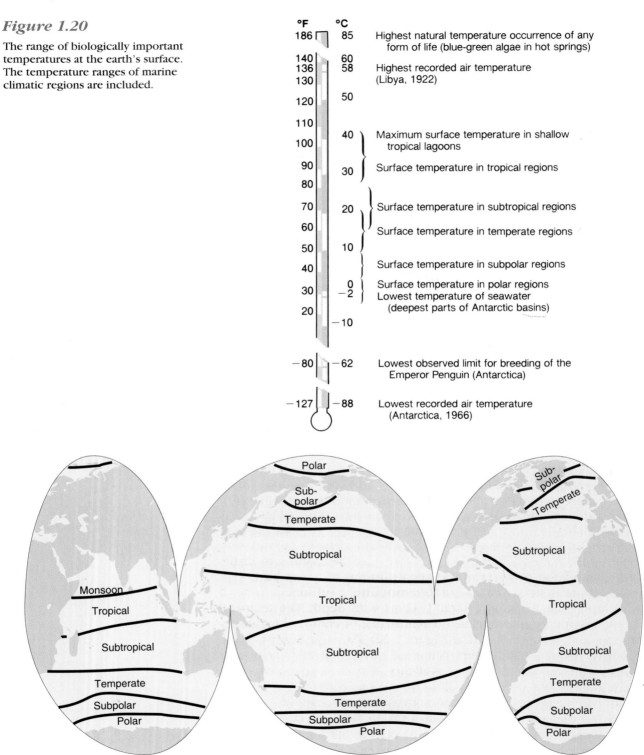

Figure 1.20

The range of biologically important temperatures at the earth's surface. The temperature ranges of marine climatic regions are included.

°F	°C	
186	85	Highest natural temperature occurrence of any form of life (blue-green algae in hot springs)
140	60	Highest recorded air temperature
136	58	(Libya, 1922)
130		
120	50	
110		
100	40	Maximum surface temperature in shallow tropical lagoons
90	30	Surface temperature in tropical regions
80		
70	20	Surface temperature in subtropical regions
60		Surface temperature in temperate regions
50	10	
40		Surface temperature in subpolar regions
30	0	Surface temperature in polar regions
	−2	Lowest temperature of seawater (deepest parts of Antarctic basins)
20	−10	
−80	−62	Lowest observed limit for breeding of the Emperor Penguin (Antarctica)
−127	−88	Lowest recorded air temperature (Antarctica, 1966)

Figure 1.21

Marine climatic zones.
Adapted from Bogdanov 1963.

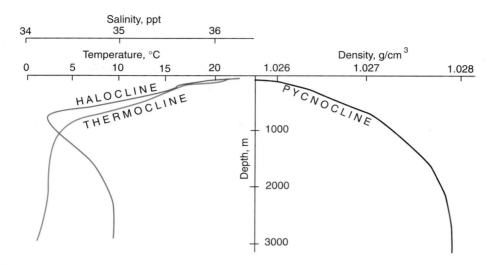

Figure 1.22

Variations in temperature (orange curve) and salinity (blue curve) at a GEOSECS station in the western South Atlantic Ocean. The resulting density profile is shown at right (black curve).

Salinity-Temperature-Density Relationships

Seawater density is a function of both temperature and salinity. The density increases with a temperature decrease or a salinity increase. Under normal oceanic conditions, temperature fluctuations exert a greater influence on seawater density because the range of marine temperature values is much greater (–2° C to 30° C) than the range of open-ocean salinities.

Some generalizations also can be made about the vertical distribution of ocean temperature, salinity, and density. The densest water is found on the bottom; however, the physical processes that create this dense water (evaporation, freezing, cooling) are strictly ocean surface features. Therefore, dense water on the sea bottom must originally sink from the surface. This sinking process is the only mechanism available to drive circulation of water in the deep portions of ocean basins. An obvious feature in most oceans is a **thermocline,** a subsurface zone of rapid temperature decrease (about 1° C/m) with depth. The temperature drop across the thermocline creates a zone of comparable density increase known as a **pycnocline** (figure 1.22). The large density differences on either side of the thermocline effectively separate the oceans into a two-layered system: a thin, well-mixed surface layer above the thermocline overlying a heavier, cold, thick, stable zone below. The pycnocline and thermocline inhibit mixing and the exchange of gases, nutrients, and sometimes even organisms between the two layers. In temperate and polar regions, the thermocline is a seasonal feature. During the winter, the surface water is cooled to the same low temperature as the deeper water. This cooling causes the thermocline to disappear and results in wintertime mixing between the two layers. Warmer marine climates of the tropics and subtropics are more often characterized by well-developed, permanent thermoclines.

Pressure

Organisms living below the sea surface constantly experience the pressure created by the weight of the overlying water. At sea level, pressure from the weight of the earth's envelope of air is about 1 kg/cm^2, or 15 lb/in^2, or 1 **atmosphere** (atm). Pressure in the sea increases another 1 atm for every 10-m increase in depth to more than 1,000 atm in the deepest trenches. Most marine organisms can tolerate the pressure changes that accompany moderate changes in depth, and some even thrive in the constant high-pressure environment of the deep sea. However, animals with gas-filled organs that collapse and expand with depth (and pressure) changes, such as fishes with swim bladders or whales with lungs, have evolved some sophisticated solutions to the problems associated with large and rapid pressure changes. These adaptations are discussed in chapter 13.

Dissolved Gases and Acid/Base Buffering

The solubility of gases in seawater is a function of temperature. Greater solubility occurs at lower temperatures. Nitrogen, carbon dioxide, and oxygen are the most abundant gases dissolved in seawater. Nitrogen (N_2) is comparatively inert and, therefore, is not involved in the basic life processes of most organisms. (Notable exceptions are some N_2-fixing microorganisms and the occasional careless SCUBA diver who dives too deep for too long.) Carbon dioxide and oxygen, on the other hand, are metabolically very active. Carbon dioxide and water are utilized in photosynthesis to produce oxygen and high-energy organic compounds such as carbohydrates and fats. Respiration reverses the results of the photosynthetic process by releasing the usable energy incorporated in the organic components of an organism's food. In contrast to photosynthesis, oxygen is used in respiration, and carbon dioxide is given off.

Carbon dioxide (CO_2) is abundant in most regions of the sea; concentrations too low to support plant growth do not normally occur. Seawater has an unusually large capacity to absorb CO_2 because most dissolved CO_2 does not remain as a gas. Rather, much of the CO_2 combines with water to produce a weak acid, carbonic acid (H_2CO_3). Typically, carbonic acid dissociates to form a hydrogen ion (H^+) and a bicarbonate ion (HCO_3^-) or two H^+ ions and a carbonate ion (CO_3^{-2}). These reactions can be summarized in the following chemical equations:

1. CO_2 + H_2O $\rightleftarrows$ H_2CO_3
   carbon   water   carbonic
   dioxide   acid

2. H_2CO_3 $\rightleftarrows$ H^+ + HCO_3^-
   carbonic   hydrogen   bicarbonate
   acid   ion   ion

3. HCO_3^- $\rightleftarrows$ H^+ + CO_3^{-2}
   bicarbonate   hydrogen   carbonate
   ion   ion   ion

The arrows pointing in both directions indicate that each reaction is reversible, either producing or removing H^+ ions. The abundance of hydrogen ions in water solutions controls the acidity or alkalinity of that solution and is measured on a scale of 0 to 14 pH units

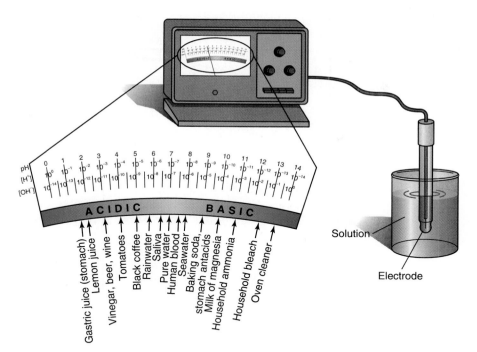

Figure 1.23

The pH scale, showing the concentration of H⁺ and OH⁻ at each pH unit and typical value of open ocean seawater. Note that the concentration scale is exponential.

(figure 1.23). The pH units are a measure of the hydrogen ion concentration. Low pHs are very acidic and represent a high H⁺ ion concentration. A pH of 14 is very basic (or alkaline) and denotes low H⁺ ion concentrations. Neutral pH (the pH of pure water) is 7 on the pH scale. The carbonic acid–bicarbonate-carbonate system in seawater functions to **buffer** or to limit changes of seawater pH. If excess hydrogen ions are present, the reactions above proceed to the left and the excess hydrogen ions are removed from solution. Otherwise, the solution would become more acidic. If too few hydrogen ions are present, more are made available by the conversion of carbonic acid to bicarbonate, and bicarbonate to carbonate. In open-ocean conditions, this buffering system is very effective, limiting ocean water pH values to a range between 7.5 and 8.4. This dynamic system functions as a crucial storage device for accumulating atmospheric CO_2 that is a result of our human activities on land (see Research in Progress: Our Planetary Greenhouse, p. 56).

Oxygen in the form of O_2 is necessary for the survival of most organisms. (The major exceptions are some species of anaerobic microorganisms.) The concentration of O_2 in seawater strongly influences the distribution of marine life. Oxygen is utilized by organisms in all areas of the marine environment, including the deepest trenches. However, the transfer of oxygen from the atmosphere to seawater and the production of excess oxygen by photosynthetic marine organisms are the only methods available to introduce oxygen into seawater. Both of these processes occur only in the near-surface region of the ocean. Oxygen consumed near the bottom can only be replaced by oxygen from the surface. If replenishment is not rapid enough, available oxygen supplies will be reduced or removed completely. Oxygen replenishment occurs by very slow diffusion processes from the oxygen-rich

Figure 1.24

Vertical distribution of dissolved O_2 in the North Pacific (150° W, 47° N) during winter.
Data from Barkley 1968.

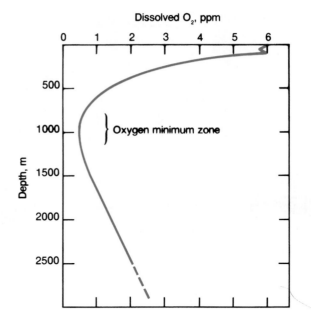

surface layers downward and also by downward vertical water movements that carry oxygen-enriched waters to deep-ocean basins. At intermediate depths, about 1,000 m, animal respiration and bacterial decomposition use O_2 as fast as it is replaced, creating an **oxygen (O_2) minimum zone.** Figure 1.24 illustrates a vertical profile of dissolved oxygen concentration from the surface to the bottom of the sea.

Dissolved Nutrients

Nitrate (NO_3^{-2}) and phosphate (PO_4^{-3}) are the fertilizers of the sea. These and smaller amounts of other nutrients are utilized by photosynthetic organisms living in the near-surface waters and are excreted back into the water at all depths as waste products of the organisms that had consumed and digested the photosynthetic material. This sinking of once-living material eventually removes nutrients from near-surface waters and increases their concentrations in deeper waters. (More details concerning these patterns of nutrient distribution appear in chapter 5.)

The vertical distribution of dissolved nutrients is usually opposite that of dissolved oxygen. The opposing patterns of vertical oxygen and nutrient distribution reflect the contrasting biological processes that influence their concentrations in seawater. Oxygen is normally produced by near-surface photosynthesizers and consumed by animals and bacteria, whereas O_2 is consumed and nutrients are excreted by organisms at all depths.

The Ocean in Motion

Ocean water is constantly in motion, providing a near-uniform medium for living organisms. Such motion enhances mixing and minimizes variations in salinity and temperature characteristics. Oceanic circulation processes also serve to disperse swimming and floating organisms and

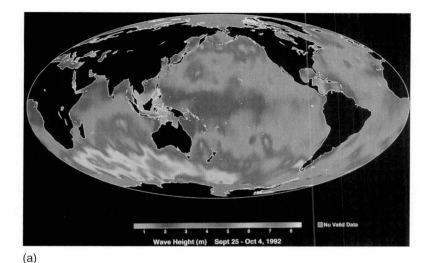

(a)

No Valid Data

Wave Height (m)   Sept 25 - Oct 4, 1992

(b)

Figure 1.25

(*a*) Range of oceanic wave heights, determined over a 10-day period by the TOPEX/Poseidon satellite. (*b*) Wave periods and their corresponding range of heights. Most wave energy is concentrated in waves with periods less than 20 s.
(*a*) Courtesy of NASA.

their eggs, spores, and larvae. Toxic body wastes are carried away, and food, nutrients, and essential elements are replenished. Heat from the sun is the driving force behind oceanic circulation processes. These circulation processes, so beneficial to all forms of marine life, are wave action, tides, currents, and vertical water movements.

Waves

Differential heating of various regions of the earth's atmosphere by the sun produces winds. Winds that blow across the sea surface produce **waves** and **surface currents.** Waves are periodic vertical disturbances of the sea surface. They typically travel in a repeating series of alternating wave crests and troughs. The size and energy of waves are dependent on the wind's velocity, duration, and **fetch** (the distance over which the wind blows in contact with the sea surface). Ocean waves range in height from a few millimeters for very small capillary waves to greater than 30 m for towering storm waves. Waves are commonly characterized by their height (figure 1.25), wavelength, and **period** (the time required for two successive wave crests

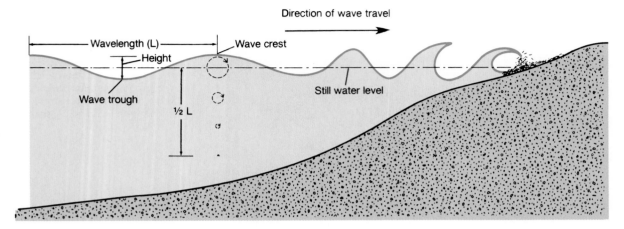

Figure 1.26

Wave form and pattern of water motion in a deep-water wave as it moves toward shore. Circles indicate orbits of water particles diminishing with depth. There is little water motion below the depth of one-half of the wavelength.

to pass a fixed point). Regardless of their size, period, or cause, the general features of waves apply to all ocean waves.

Once generated, waves move away from the area of formation. However, only the wave shape advances, transmitting the energy forward. The water particles themselves do not advance horizontally. Instead, their paths approximate vertical circles with little net forward motion (figure 1.26). Waves provide an important mechanism to mix the near-surface layer of the sea. The depth to which waves produce noticeable motion is about one-half the wavelength. As wavelengths seldom exceed 100 m in any ocean, the depth of effective mixing by wind-driven waves is generally no greater than 50 m.

Waves entering shallow water behave differently than open-ocean waves. When the water depth is less than one-half the wavelength, bottom friction begins to slow the forward speed of the waves. This slowing causes the waves to become higher and steeper. At the point where the wave height/wavelength ratio exceeds 1/7, the wave top becomes unstable as it overruns the bottom, pitches forward, and breaks. The energy released on shorelines (and on the organisms living there) by breaking waves is sometimes tremendous and is a major force in shaping the character of the seashore.

Tides

Tides are ocean surface phenomena familiar to anyone who has spent time on a seashore. They are very-long-period waves that are usually imperceptible in the open ocean and become noticeable only at the shoreline, where they can be observed as a periodic rise and fall of the sea surface. The maximum elevation of the tide, known as **high tide,** is followed by a fall in sea level to a minimum elevation, or **low tide.** On most coastlines, two high tides and two low tides occur each day. The vertical difference between consecutive high tides and low tides is the **tidal range,** which varies from a few centimeters in the

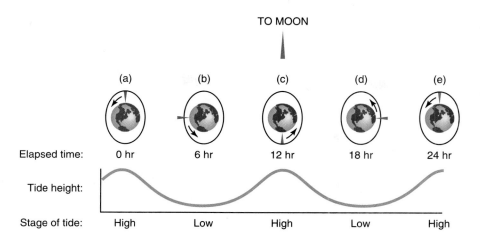

TO MOON

| (a) | (b) | (c) | (d) | (e) |

Elapsed time: 0 hr   6 hr   12 hr   18 hr   24 hr

Tide height:

Stage of tide: High   Low   High   Low   High

Figure 1.27

Each day as the earth rotates, a point on its surface (indicated by the marker) experiences high tides when under tidal bulges and low tides when at right angles to tidal bulges.

Mediterranean Sea to more than 15 m in the long, narrow Bay of Fundy between Nova Scotia and New Brunswick.

In 1687, in his *Principia Mathematica*, Sir Isaac Newton explained ocean tides as the consequence of the gravitational attraction of the moon and sun on the oceans of the earth. According to Newton's law of universal gravitation, *the gravitational attraction between two bodies is directly proportional to their masses and inversely proportional to the square of the distance between the bodies.* Our moon, owing to its closeness to the earth, exerts about twice as much tide-generating force as does the more distant but much larger sun.

The moon completes one orbit around the earth each lunar month (27.5 days). To maintain that orbit, the gravitational attraction between the earth and moon must exactly balance the centrifugal force holding the bodies apart. In concert, these two opposing forces create two tide-producing forces at the earth's surface.

Hypothetically, if the earth were completely covered with water, two bulges of water, or lunar tides, would pile up: one on the side of the earth facing the moon and the other on the opposite side of the globe (figure 1.27). As the earth makes a complete rotation every 24 hours, a point on the earth's surface (indicated by the marker in figure 1.27) would first experience a high tide (a), then a low tide (b), another high tide (c), another low tide (d), and finally another high tide (e). During that rotation, however, the moon advances in its own orbit so that an additional 50 minutes of the earth's rotation is required to bring that point directly in line with the moon again. Thus the reference point experiences only two equal high and two equal low tides every 24 hours and 50 minutes (a lunar day).

In a similar manner, the sun-earth system also generates tide-producing forces that yield a solar tide about one-half as large as the lunar tide. The solar tide is expressed as a variation on the basic lunar tidal pattern, not as a separate set of tides. When the sun, the moon, and the earth are in alignment (at the time of the new moon and full moon, figure 1.28), the solar tide has an additive effect on the lunar tide, creating extra-high high tides and very-low low tides (**spring tides**). One week later, when the sun and moon are at right angles to

Figure 1.28

Weekly tidal variations caused by changes in the relative positions of the earth, moon, and sun.

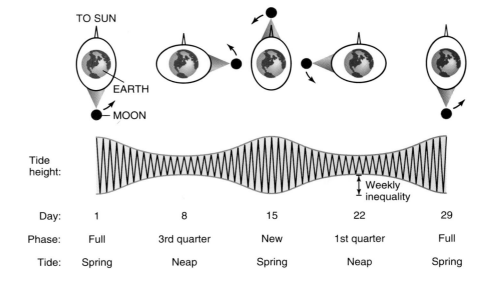

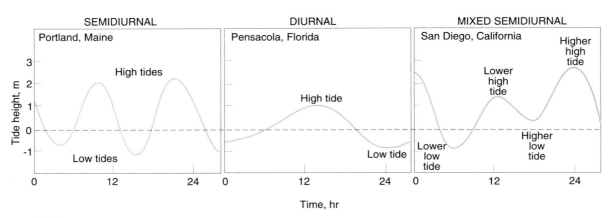

Figure 1.29

Three common types of tides.

each other, the solar tide partially cancels the lunar tide to produce moderate tides known as **neap tides.** During each lunar month, two sets of spring tides and two sets of neap tides occur.

So far, we have considered only the effects of tide-producing forces in a not very realistic ocean covering a hypothetical planet without continents. What happens when continental landmasses are taken into consideration? The continents act to block the westward passage of the tidal bulges as the earth rotates under them. Unable to move freely around the globe, these tidal impulses establish complex patterns within each ocean basin that often differ markedly from the tidal patterns of adjacent ocean basins or other regions of the same ocean basin.

Figure 1.29 shows some regional variations in the daily tidal configuration at three stations along the East and West coasts of North America. Portland, Maine, experiences two high tides and two low tides each lunar day. The two high tides are quite similar to each other, as are the two low tides. Such tidal patterns, referred to as **semidiurnal**

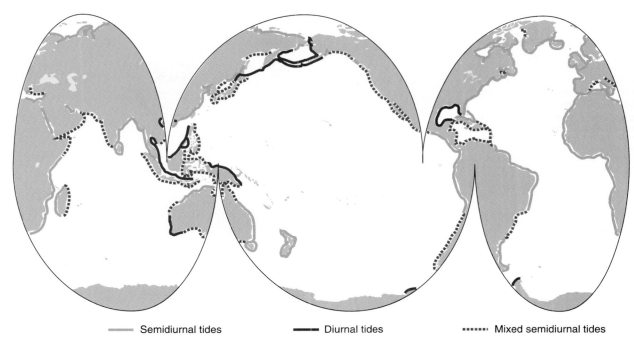

Figure 1.30

The geographical occurrence of the three types of tides described in figure 1.29.

(semidaily) **tides,** are characteristic of much of the East Coast of the United States. The tidal pattern at Pensacola, Florida, on the Gulf Coast, consists of one high tide and one low tide each lunar day. This is a **diurnal,** or daily, **tide.** Different yet is the daily tidal pattern at San Diego, California. There, two high tides and two low tides occur each day, but successive high tides are quite different from each other. This type of tidal pattern, characteristic of the West Coast of North America, is a **mixed semidiurnal tide.** Figure 1.30 outlines the geographical occurrence of diurnal, semidiurnal, and mixed semidiurnal tides.

Tidal conditions for any day on a selected coastline can be predicted because the periodic nature of tides is easily observed and recorded. For the most part, prediction of the timing and amplitude of future tides is based on historical observations of tidal occurrences at tide-recording stations along coastlines and in harbors around the world. The National Ocean Survey of the U.S. Department of Commerce uses information from these records to compile and publish annual "Tide Tables of High and Low Water Predictions" for principal ports along most coastlines of the world.

Surface Currents

Measurable ocean surface currents occur in regions where winds blow over the ocean with a reasonable constancy of direction and velocity. Unlike wave motion, surface currents represent large-scale horizontal movements of water molecules. Three major wind belts occur in the Northern Hemisphere. The **trade winds,** near 15° N latitude,

Figure 1.31

A spiral of current directions, indicating greater deflection to the right (in the Northern Hemisphere), which increases with depth because of the Coriolis effect. Arrow length indicates relative current speed.

Redrawn from H. U. Sverdrup, Martin W. Johnson, and Richard H. Fleming, *The Oceans: Their Physics, Chemistry, and General Biology,* © 1942, renewed 1970. By permission of Prentice Hall, Inc., Englewood Cliffs, New Jersey.

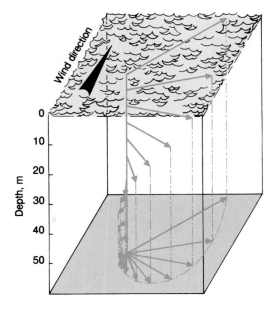

blow from northeast to southwest. The **westerlies,** in the middle latitudes, blow primarily from the west-southwest. And the **polar easterlies,** at very high latitudes, blow from east to west. Each of these wind belts has its mirror-image counterpart in the Southern Hemisphere.

The momentum imparted to the sea by these winds drives regular patterns of broad, slow, relatively shallow ocean surface currents. Some currents transport more than 100 times the volume of water carried by all of the earth's rivers combined. Currents of such magnitude greatly affect the distribution of marine organisms and the rate of heat transport from tropical regions to polar regions.

As the surface layer of water is forced horizontally by the wind, momentum is transferred downward. The speed of the deeper water steadily diminishes as momentum is lost to overcome the viscosity of the water. Eventually, at depths generally less than 200 m, the speed of wind-driven currents becomes negligible.

The surface water moved by the wind does not flow parallel to the wind direction but experiences an appreciable deflection—a deflection to the right in the Northern Hemisphere and a deflection to the left in the Southern Hemisphere. This deflection is known as the **Coriolis effect.** As successively deeper water layers are set into motion by the water above them, they undergo a further Coriolis deflection from the direction of the water just above to produce a spiral of current directions from the surface downward (figure 1.31). The magnitude of the Coriolis deflection of wind-driven currents varies from about 15° in shallow coastal regions to nearly 45° in the open ocean. The net Coriolis deflection from the wind headings creates a pattern of wind-forced ocean-surface currents that flow primarily in an east-west direction.

Continental masses obstruct the continuous east-west flow of currents. Water transported by these currents is moved from one side of the ocean and accumulates on the other side. The surface of the equatorial Pacific Ocean, for example, is higher on the west side than it is on the east side. The opposite is true in the middle latitudes of both

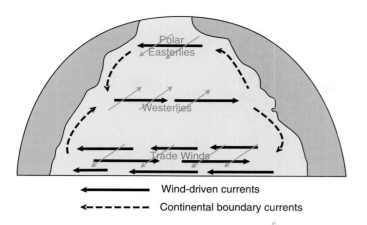

Figure 1.32

Generalized surface-current flow in
the North Pacific Ocean. Blue arrows
indicate general directions of ocean-
surface winds.

hemispheres, where the east side is higher. Eventually, the water must
flow from areas where it has accumulated to regions where it origi-
nated. Either the water flows directly back against the established current, producing a **countercurrent,** or it flows as a **continental
boundary current** in a north-south direction from areas of accumulation to areas where water has been removed. Both these current
patterns exist, but they are especially obvious in the North Pacific
(figure 1.32). An east-flowing Equatorial Countercurrent divides the
west-flowing North Pacific Equatorial Current. The north-south–flowing
continental boundary currents meet the east-west currents to produce
large, circulating currents, or **gyres.** Similar current patterns are found
in the other major ocean basins (figure 1.33).

   Maps similar to figure 1.33 are useful for describing long-term average patterns of surface ocean circulation. However, they tend to hide
the complexity and even the beauty that exist in these currents at any
moment in time. Current maps are analogous to the blurred images
taken of a night freeway scene when the camera shutter is held open
for hours. The pattern of traffic flow is obvious, yet the details of vehicles' slowing, accelerating, and changing lanes are completely lost. The
recent advent of satellite monitoring of ocean surface phenomena (see
Research in Progress: Oceanography from Space, p. 76) has opened a
completely new approach for visualizing and understanding global-scale
surface current patterns. Figure 1.34 is a satellite image of a portion of
the North Atlantic Ocean, including the Gulf Stream. This image empha-
sizes ocean-surface temperature differences and reveals remarkable me-
anders, constrictions, and nearly detached rings of Gulf Stream water as
the current flows north and east along the path shown in figure 1.33.

   Several short-term and dramatic departures from the average current patterns shown in figure 1.33 do occur. One departure, El Niño, is
characterized by a prominent warming of the equatorial Pacific surface
waters. El Niño occurs irregularly every few years, usually around
Christmastime (hence the name El Niño, "The Child"); each occurrence
lasts from several months to over a year. El Niño is associated with the
Southern Oscillation, a transpacific linkage of atmospheric pressure systems. Normally, the trade winds blow around the South Pacific high-
pressure center located near Easter Island and then blow westward to
the large Indonesian low-pressure center. As these winds move water

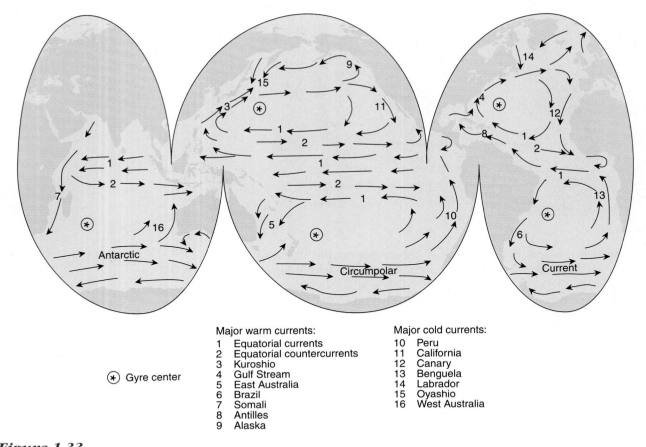

Major warm currents:
1 Equatorial currents
2 Equatorial countercurrents
3 Kuroshio
4 Gulf Stream
5 East Australia
6 Brazil
7 Somali
8 Antilles
9 Alaska

Major cold currents:
10 Peru
11 California
12 Canary
13 Benguela
14 Labrador
15 Oyashio
16 West Australia

⊛ Gyre center

Figure 1.33

The major surface currents of the world ocean. Compare with figure 1.8.
Adapted from Picard and Emery 1982.

westward, the water is warmed and the thermocline is depressed from about 50 m below the surface on the east side of the Pacific to about 200 m deep on the west side. El Niños occur when this pressure difference across the tropical Pacific relaxes (for reasons not yet known), and both surface winds and ocean currents either cease to flow westward or actually reverse themselves. Although the effects of an El Niño-Southern Oscillation (ENSO) event are somewhat variable, they are often global in extent and occasionally severe in impact. The 1982–1983 ENSO event, for example, was associated with heavy flooding on the West Coast of the United States, intensification of the drought in sub-Saharan Africa and Australia, and severe hurricane-force storms in Polynesia. Surface ocean water temperatures from Peru to California soared to as much as 8° C above normal. The impact that such severe short-term departures from normal conditions have on local marine populations is considered in chapters 5 and 15.

Vertical Water Movements

Vertical water movements are produced by sinking and upwelling processes. Such processes tend to break down the vertical stratification

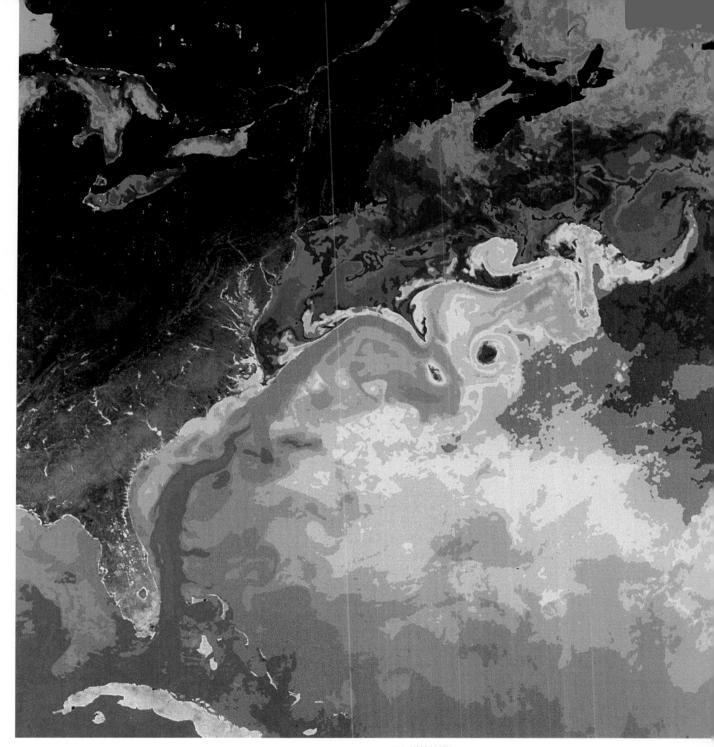

Figure 1.34

NOAA satellite image of the Gulf Stream along the East Coast of the United States during early April 1984. Water temperatures are represented by a range of colors, from purple (cold water 2°-9° C) through blue, green, yellow, orange, and red (very warm water, 26°-28° C). The Gulf Stream appears as a narrow band of warm water off the Florida coast, then cools gradually as it moves northeastward and releases heat to the atmosphere. As it moves offshore from Cape Hatteras, the current begins to meander and forms isolated rings.

Courtesy O. Brown, University of Miami.

Figure 1.35

A stylized illustration of the general pattern of deep-ocean circulation in the major ocean basins. Light blue represents currents at intermediate depths. Dark blue represents deeper currents.
Adapted from Broecker et al. 1985.

established by the thermocline. Seawater sinks when its density increases. The physical processes that increase seawater density are strictly surface features. Thus, dense seawater, which sinks from the surface and usually is highly oxygenated, transports dissolved oxygen to deep areas of the ocean basins, areas which would otherwise be **anoxic** (lacking oxygen). The chief areas of sinking are located in the colder latitudes, where sea surface temperatures are low. Figure 1.35 outlines the general patterns of large-scale deep-ocean circulation. These patterns of water transport are very slow and ill defined. Time spans of a few hundred to a thousand years are required for water that sinks in the North Atlantic to reach the surface again in the Southern Hemisphere.

Rising water masses are produced by **upwelling** processes and are considered in chapter 5. Whatever the cause, all upwelling processes bring deeper nutrient-rich waters to the surface. The continuous availability of deep-water nutrients that can be used by photosynthetic organisms accounts for the high productivity characteristic of regions of upwelling. Several of the world's important fisheries are based in upwelling areas.

In the arid climate of the Mediterranean Sea, evaporation from the sea surface greatly exceeds precipitation and runoff. The resulting high-salinity water sinks and fills the deeper parts of the Mediterranean basin. The sinking of surface water provides substantial mixing and O_2 replenishment for the deep water of the Mediterranean and is similar to the deep circulation of the open ocean. Part of this deep, dense water eventually flows out of the Mediterranean over the shallow sill at Gibraltar and down into the Atlantic Ocean. To compensate

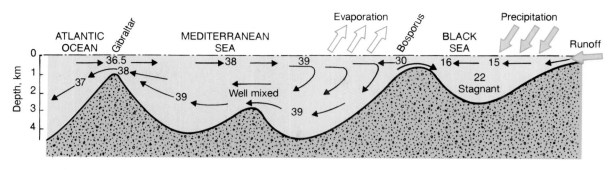

Figure 1.36

A comparison of the deep-ocean circulation patterns of two marginal seas, the Mediterranean Sea and the Black Sea. The numbers represent salinity (‰).

for the outflow and losses due to evaporation, nearly 2 million cubic meters of Atlantic surface water flow into the Mediterranean each second. The currents at Gibraltar can be compared to two large rivers flowing in opposite directions, one above the other (figure 1.36).

Like the Mediterranean, the Black Sea is isolated by a shallow sill (at the Bosporus). In contrast to the Mediterranean Sea, however, the Black Sea is characterized by a large excess of precipitation and river runoff. The dilute surface waters form a shallow, low-density layer that does not mix with the higher salinity, denser water below but, instead, flows into the Mediterranean Sea through the Bosporus (figure 1.36). For all practical purposes, the water below 150 m in the Black Sea is stagnant. Low-salinity, oxygen-rich surface water does not sink, so the more common oxygen-dependent forms of marine life are restricted to the uppermost layer. But the anoxic deep waters of the Black Sea (more than 80% of its volume) are by no means lifeless. The rain of organic material from above accumulates and provides abundant nourishment for numerous types of anaerobic bacteria. These bacteria exist without O_2 and in turn produce hydrogen sulfide (H_2S), which is toxic to other forms of life. Thus, the lack of deep circulation in the Black Sea limits the input of O_2 and allows the buildup of nutrients and H_2S. In this sense, the circulation of the Black Sea resembles that of some semienclosed fjords of Scandinavia and the west coast of Canada.

Classification of the Marine Environment

The size and complexity of the marine environment make it a difficult system to classify. Many systems of classification have been proposed, each reflecting the interest and bias of the classifier. The system presented here is a slightly modified version of a widely accepted scheme proposed by Hedgpeth. The terms used in figure 1.37 designate particular zones of the marine environment; these terms should not be confused with the names of groups of organisms that normally inhabit these zones. The boundaries of these zones are defined on the basis of physical characteristics such as water temperature, water depth, and available light.

The limits of the **intertidal zones** are defined by tidal fluctuations of sea level along the shoreline. These zones and their inhabitants are examined in detail in chapter 8. The splash, intertidal, and inner shelf

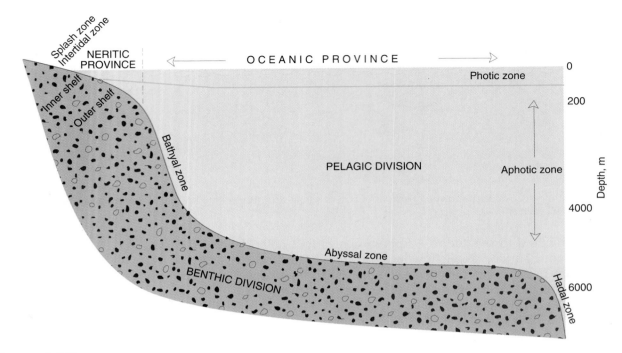

Figure 1.37

A system for classifying the marine environment.
Adapted from Hedgpeth 1957.

zones occur in the **photic** (lighted) **zone,** where the light intensity is great enough to accommodate photosynthesis. The depth of the photic zone depends on conditions that affect light penetration in water, extending much deeper in clear, tropical waters than in murky, coastal waters of temperate areas. The average depth of the photic zone is 50–100 m. The remaining zones are located in the **aphotic** (unlighted) **zone,** where the absence of sunlight prohibits photosynthesis.

The **benthic division** refers to the environment of the sea bottom. The inner shelf includes the seafloor from the low-tide line to the bottom of the photic zone. Beyond that, to the edge of the continental shelf, is the outer shelf. The bathyal zone is approximately equivalent to the continental slope areas. The abyssal zone refers to abyssal plains and other ocean-bottom areas between 3,000 and 6,000 m in depth. The upper boundary of this zone is sometimes defined as the region where the water temperature never exceeds 4° C. The hadal zone is that part of the ocean bottom below 6,000 m, primarily the trench areas.

The **pelagic division** includes the entire water mass of the ocean. For our purposes, it will be sufficient to separate the pelagic region into two provinces: the **neritic province,** which includes the water over the continental shelves, and the **oceanic province,** which includes the water that overlies the deep ocean basins.

Each of these subdivisions of the ocean environment is inhabited by characteristic assemblages of marine organisms. It is these organisms and their interactions with their immediate surroundings that are the subject of this book.

- The world ocean is a large, interconnected body of seawater separated by continents into several ocean basins and marginal seas. The sizes and shapes of the ocean basins as well as their seawater contents are continually changing, consequences of a long history of geological vulcanism, seafloor spreading and plate tectonics, and extensive continental glaciation.
- The unusual characteristics of water itself determine most of the basic properties of seawater. The asymmetrical shape of a water molecule creates an electrical charge separation that initiates hydrogen-bonding interactions with adjacent water molecules. Hydrogen bonding, in turn, affects water's basic properties, including viscosity, surface tension, heat capacity, solvent capability, density-temperature relationships, and its stability as a liquid.
- Seawater contains a variety of dissolved salts, gases, and other substances. These dissolved substances affect the density of seawater, its osmotic properties, buffering capacity, and other biologically significant features. This water is constantly in motion, mixed and moved by winds, waves, tides, currents, sinking water masses, and upwelling.
- Energy from the sun divides most of the world ocean into a two-layered system, with a shallow, well-mixed, warm, sunlit layer overlaying a much deeper, cold, dark, high-pressure layer of slowly moving water below.
- The marine environment can be separated into two broad units, the benthic and pelagic divisions. These in turn may be subdivided into smaller categories based on water depth, light availability, and tidal exposure.

1. List the processes that occur when ice forms on seawater. Explain why these processes, once initiated, tend to establish conditions that resist further freezing.
2. How are the surface currents of the North Atlantic similar to those of the South Atlantic? In what major way do they differ?
3. A bottle is tossed into the ocean off the northern coast of Peru. Three years later the bottle is recovered on a beach in Norway. Describe the bottle's most likely path from Peru to Norway, assuming it was transported solely by ocean surface currents. Do any other reasonable routes exist?
4. List the two most abundant ions and gases dissolved in seawater.
5. List four major causes of salinity variation in seawater.
6. List three properties of seawater that change with salinity variations, and indicate how those properties change as the salinity *increases*.
7. Name two major *cold* surface currents found in the North Pacific Ocean and in the North Atlantic Ocean.

1. Describe why surface ocean-water temperatures vary less from season to season than do air temperatures over nearby landmasses.
2. Explain why both latitude and longitude are necessary to fix the position of any location on the earth's surface.

3. List and describe the major physical and chemical features of seawater that change markedly from the sea surface downward. How do these same features change along the sea surface as one proceeds from the equator north or south to higher latitudes?

4. Draw and label a tide curve representing a typical lunar day on your local coastline. How does it compare with a tide curve for the same day and the same latitude on the opposite side of the continent?

Suggestions for Further Reading

Books

Borgese, E., and N. Ginsburg, eds. 1979-1993. *The ocean yearbooks.* vols. 1-10. Chicago: University of Chicago Press.

Cloud, P. 1989. *Oasis in space: Earth history from the beginning.* New York: W.W. Norton.

Duxbury, A. C., and A. B. Duxbury. 1991. *An introduction to the world's oceans.* Dubuque, IA: Wm. C. Brown Publishing.

Open University. 1989. *The ocean basins: Their structure and evolution.* Oxford, England: Pergamon Press.

———. 1989. *Ocean circulation.* Oxford, England: Pergamon Press.

———. 1989. *Seawater: Its composition, properties, and behavior.* Oxford, England: Pergamon Press.

———. 1989. *Waves, tides, and shallow water processes.* Oxford, England: Pergamon Press.

Rand McNally. 1987. *The Rand McNally atlas of the oceans.* Chicago: Rand McNally.

Articles

Bonati, E. 1987. The rifting of continents. *Scientific American* March 256:97-103.

Broecker, W. S. 1983. The ocean. *Scientific American* (September) 249:146-60.

Bryan, K. 1978. The ocean heat balance. *Oceanus* 21(3):18-26.

Edmond, J. M. 1982. Ocean hot springs. *Oceanus* 25(2):22-27.

Hogg, N. 1992. The Gulf Stream and its recirculation. *Oceanus* 35(2):28-37.

Jenkyns, H. C. 1994. Early history of the oceans. *Oceanus* 36(4):49-52.

McDonald, J. E. 1970. The Coriolis effect. *Scientific American* (May) 222:72-76.

Vink, G. E., W. Morgan, and P. Vogt. 1985. The earth's hot spots. *Scientific American* April 252:50-57.

Webster, P. J. 1981. Monsoons. *Scientific American* (February) 244:108-18.

Weller, R. A., and D. M. Farmer. 1992. Dynamics of the ocean's mixed layer. *Oceanus* 35(2):46-55.

Whitworth III, T. 1988. The Antarctic circumpolar current. *Oceanus* 31(2):53-58.

Internet Addresses

The Salt Marsh
http://www.tip.net.au/~dfry/amanda/saltmars.htm

The Role of Salt Marshes in the Marine Environment
http://www.tip.net.au/~dfry/amanda/smimport.htm

The Deep Blue Sea
http://www.tip.net.au/~dfry/amanda/oceanic.htm

The Open Water
http://www.tip.net.au/~dfry/amanda/dbopen.htm

Visitors' Guide to Marine Ecosystems of Northern New South Wales by Amanda Fry at **http://www.tip.net.au/~dfry/amanda/** Alternate site: **http://www.pcug.org.au/~dfry/amanda/** Web pages' names aptly describe contents. These are introductory concepts to coastal influences on marine biology in northern New South Wales.

International Data Exchange
http://www.lib.noaa.gov/

Click on link for: World data center, Oceanography, cooperation with IOC/IODE and ICES, Temperature-salinity profile, Oceanographic data archeology and rescue.

Laboratory for Satellite Altimetry
http://www.lib.noaa.gov/
Click on link for altimetry data.

Top 10 Questions about the Oceans
http://www.lib.noaa.gov/
Click on link for answers to: bathymetry, tides, organisms, charts, Gulf
   Stream, surface temperature, dive locations and conditions, beach
   weather and water temperatures, careers. National Oceanographic
   Data Center (NODC)

National Association of Marine Laboratories
http://www.mbl.edu/html/NAML/NAML.html
Site has links to: Western Association of Marine Laboratories, Northeast
   Association of Marine and Great Lakes Labs, and Southern
   Association of Marine Labs.

Some Ecological and Biological Concepts

Life is a special phenomenon. We observe, dissect, analyze, and discuss it. We can characterize the attributes of life and can state with a good deal of authority how living systems work. Yet life remains difficult to define. Life forms require energy to break down and resynthesize complex chemical substances. These materials and energy are used by all life forms to accomplish the four *R*s of living:

Respiration: Obtaining, with a series of enzyme-controlled chemical reactions, usable energy from complex high-energy molecules

Reproduction: Transferring copies of genes to subsequent generations

Response: Reacting to and interacting with stimuli from the surroundings

Regulation: Controlling the exchange of materials between internal and external environments while maintaining an organized internal environment

How marine organisms accomplish these life requirements is the theme of this book.

The Cellular Structure of Life

Living organisms are modular: They are composed of a single cell or a complex assemblage of specialized cells. Fossil evidence of complex life forms older than 600 million years is not abundant. Even so, fossil remains of simple cells over 3 billion years old have been reported from scattered sites in Africa and Australia. These microscopic organisms achieved a level of structural organization strikingly similar to some modern bacteria. Like modern bacteria and cyanobacteria (figure 2.1), these early life forms lacked much of the complex subcellular structure found in other modern cells. Yet, they presumably contained the necessary complement of cellular machinery needed to function as living cells. A **cell wall** provides form and mechanical support for the cell. Inside the cell wall, a selectively permeable **plasma membrane** separates the internal fluid environment (the **cytoplasm**) from the exterior environment of the cell and regulates exchange between the cell and its external medium, with limited movement provided by a whiplike **flagellum.** Internally, the genetic information is coded and stored in a single, circular **chromosome.** Small ribosomes use that information to direct the synthesis of enzymes. The enzymes, in turn, control and regulate all other chemical reactions that occur in living cells.

The structurally simple bacteria and cyanobacteria (the **procaryotes**) have been eclipsed in some environments by groups of relatively large organisms, the **eucaryotes.** The complexity and diversity of eucaryotic cells are responsible for most of the immense

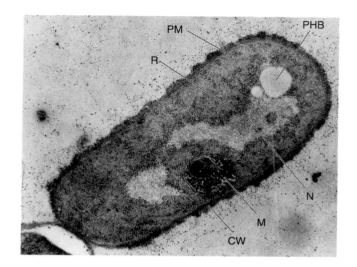

Figure 2.1

Transmission electron micrograph of a procaryotic bacterial cell, *Bacillus* (CW, cell wall; M, the mesosome; N, nucleoid; PHB, poly-β-hydroxybutyrate inclusion body; PM, plasma membrane; and R, ribosomes).

Courtesy © Ralph A. Slepecky/Visuals Unlimited.

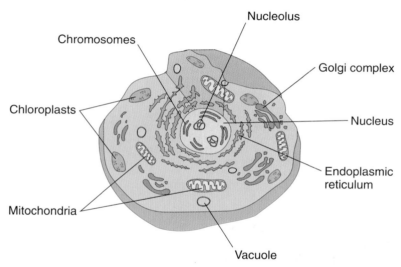

Figure 2.2

Simplified diagram of a eucaryotic cell. In addition to several types of subcellular organelles common to most eucaryotes, photosynthetic cells contain chloroplasts and are typically supported by external cell walls.

variety of species on the earth today. Eucaryotic cells are generally larger than procaryotes and house a variety of membrane-bound structures (figure 2.2) not found in procaryotes (figure 2.1).

The chromosomes and their surrounding **nuclear membrane** form a central structure, the **nucleus.** (Bacteria and cyanobacteria lack a true nucleus.) The enzymes involved in respiration and energy release are associated with numerous small **mitochondria.** Many of the enzyme-synthesizing ribosomes are free in the cytoplasm; others are arranged on a membranous **endoplasmic reticulum.** Food particles are ingested and stored in **vacuoles** within the cell. Other subcellular structures are involved in excretion of wastes, osmotic balance, and other cellular chores. In addition, photosynthetic eucaryotes typically possess two special features: (1) **Chloroplasts** serve as the sites of photosynthesis, and (2) a rigid cell wall provides shape and support in a manner similar to procaryotic cell walls.

From the basic module of the cell, organisms of greater size and organizational complexity have evolved. Yet each level of organization

Table 2.1
Levels of Organization in Living Systems

Level of Organization*	Definition	Examples
Ecosystem	The organisms of a particular type of area and the physical features of the environment in which they live	Coral reef ecosystem
Community	An ecologically integrated group consisting of all the populations living in a given, limited area	Coral reef lagoon
Population	A group of interbreeding organisms coexisting in the same time and place	Barracuda school
Organism	An individual structure of one or more cells capable of reproduction and mutation	Barracuda
Organ	A specific body part consisting of several tissues performing as an identifiable, functional unit	Heart, intestine
Tissue	An aggregation of similar cells, usually with a specific function	Muscle tissue, fatty tissue
Cell	The fundamental organizational unit of living material	Muscle cell, nerve cell
Cellular organelle	A well-defined structure within a cell	Mitochondrion, nucleus
Macromolecule	A very large molecule consisting of numerous simple molecules linked together	Proteins, carbohydrates
Simple molecule	A small chemical unit consisting of two or more atoms bonded together	Amino acids, sugars
Atom	The smallest unit of an element; not divisible by ordinary chemical procedures	C, H, O

*The living systems are listed from the most complex systems to the least complex systems. *Each level is composed of numerous units from one or more levels below it.*

within any living system is derived from one or more components smaller and less complex than itself. Table 2.1 lists and defines some common levels of organization found in living systems.

Adaptations to Life in the Sea

All living organisms exhibit varying capabilities for both **ecological** and **evolutionary adaptations** to changing conditions. Ecological adaptations occur within one's lifetime, are accomplished by individuals, and sometimes show immediate results. Evolutionary adaptations are products of the changing response of a population of individuals over many generations. The ultimate effect of ecological adaptations is the ability of individuals to secure sufficient resources so they might survive until they successfully reproduce. By the simplest of definitions, to reproduce successfully means only that an organism must replace itself with an offspring also capable of reproducing successfully (by the same definition).

Natural populations are characterized by reproductive potentials that exceed those needed to maintain the population size and that are in excess of the number their habitat can support. Eventually, expanding populations outgrow their necessary resources, and competition between individual members of the population intensifies. An individual's ability to survive and reproduce depends on its genetic and physical uniqueness. Natural conditions cause many to perish before reaching sexual maturity. Only those equipped to compete and survive in their current, local environment succeed in passing on their genetic traits to future generations.

The offspring inherit characteristics that, in turn, provide a similar ability to compete and survive. Superior fitness may result from a resistance to disease, starvation, or climatic variations, or it may be a capacity to reproduce quickly. This competition and differential survival are summarized in the overworked phrase "survival of the fittest." However, the rules and conditions for survival change continuously and unpredictably. The selection factor for one generation might be a food shortage, and for the next generation, disease. As a result, "survival of the fitter" might be a more appropriate phrase, for organisms never evolve to perfectly fit their total environment.

The basic biological units of evolutionary adaptation, then, are populations. Evolutionary adaptation occurs only in populations, never in individuals. Individuals perish regardless of whether or not their populations evolve. When populations cease to adapt and change, extinction becomes inevitable. Of the millions of types of organisms that have evolved in more than 3 billion years of life's history on the earth, only a tiny fraction exist today. These are the temporary winners.

In the sea, these winners perpetually confront fluctuations in temperature, salinity, available oxygen, light, and food, as well as attempts by their neighbors to crowd them out or consume them. This chapter examines a few general strategies exhibited by marine organisms for coping with such stresses. In day-to-day ecological time, these stresses mold the structures of communities and ecosystems. Over much longer periods of time, they shape the evolutionary path of the affected populations. Ecological adaptations can be reduced to the following three general categories:

1. Adaptations to accommodate the physical and chemical environment
2. Adaptations to secure food and avoid being eaten
3. Adaptations to ensure successful reproduction

An organism's (or population's) success in the first two categories ultimately reflects success in the third. This section provides a few examples of strategies to deal with the physical environment and with problems related to nutrition. Other chapters will introduce additional strategies. Finally, successful reproduction strategies are so varied and so specific to individual groups that their discussion is placed within the chapters or sections dealing with other aspects of the biology of those groups. However, this chapter will discuss a few important generalizations regarding reproduction.

The Value of Sex

Although we often refer to reproduction as the process by which we replicate ourselves from one generation to the next, it also can be described as the process by which sets of genes are transferred through

Figure 2.3

Four approaches to asexual reproduction.

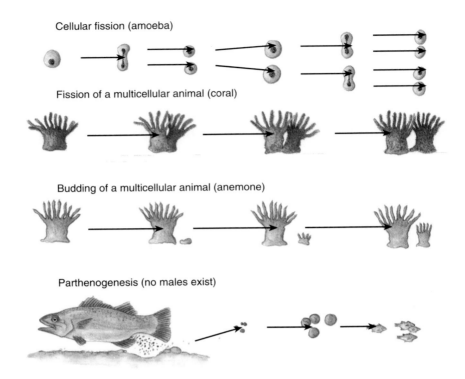

Cellular fission (amoeba)

Fission of a multicellular animal (coral)

Budding of a multicellular animal (anemone)

Parthenogenesis (no males exist)

generations of organisms. Organisms that reproduce **asexually** (figure 2.3) copy themselves; the products of their reproductive efforts are genetically identical to each other and to their parent. However, it is not accurate to say that sexually reproducing organisms produce replicates of themselves. In fact, the whole point of **sexual reproduction** is to provide a mechanism to produce diverse offspring from two parents. Some, but not necessarily all, offspring of sexually reproducing parents may coexist better where resources are variable than will the progeny of asexual reproducers simply because offspring of sexual parents are different. They have different resource needs, and competition between individuals for the resources is less severe than it might be between genetically identical offspring. Therefore, the relative advantage for sexual versus asexual reproduction is a function of the resources available to the reproducing population and the size of the population. In small populations (less than about 1 million individuals), the costs of sexual reproduction begin to outweigh the advantages, and asexual reproduction often becomes the more advantageous method, providing a means of rapid reproduction to exploit temporarily favorable conditions.

Sexual reproduction is not only more complex than asexual reproduction, it is also more costly. The high costs of sexual reproduction are related to the expenses of producing and maintaining males and to the expenses associated with meiosis. Regardless of the approach taken to accomplish reproduction, all sexually reproducing organisms include the same basic elements in the process (figure 2.4). In **hermaphroditic** animals, such as most barnacles and some fishes, all adults function in both female and male gender roles, some at the same time (simultaneous

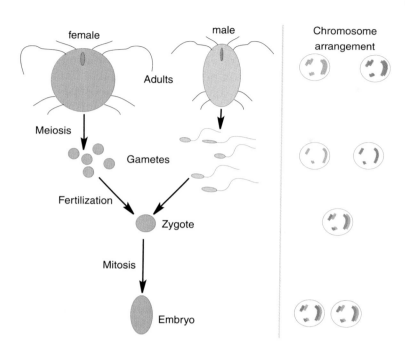

Figure 2.4

The basic components of sexual reproduction. The chromosome arrangement of each cell is shown to the right.

hermaphrodites); others function as one gender and then transform to the other (sequential hermaphrodites). Most other animals retain the same gender role for their entire lives, with approximately equal numbers of males and females. Plants exhibit different life cycle patterns (see chapter 4) but still express the basic attributes of sexual reproduction: meiosis followed by fertilization.

When mature, sexually reproducing adults produce **gametes** (either eggs or sperm) by **meiosis.** Meiosis is a cell division process in which the chromosomes of the gametes produced include one of each of the pairs of chromosomes characteristic of the other cells of the adult individuals (figure 2.4). This is termed a **haploid** chromosome condition. Halving of chromosome numbers in the formation of gametes is a necessary component of sexual reproduction, for gamete production is followed by **fertilization,** the remaining obligatory part of sexual reproduction. In fertilization, the chromosomes carried by the sperm cell are combined with those of the egg to form a **zygote** with double the chromosome number of either of the gametes. This double, or **diploid,** set of chromosomes is carried by all cells in the development to sexual maturity. The process is complex, but it produces variety within a population and within a short time span. Sex in reproduction, then, is the method of choice for most multicellular animals and plants. The fusion of the eggs and sperm from two genetically different individuals promotes genetic diversity, the basis for adaptation in evolutionary time scales.

Salinity Effects

It is essential to the well-being of all living things that they maintain reasonably constant internal environmental conditions. **Homeostasis** is the tendency of living organisms to control or regulate fluctuations of their internal environment. Homeostasis is the result of coordinated biological processes that regulate conditions such as body temperature or blood ion concentrations. Homeostasis is not a static situation; it is a situation that varies within definite and tolerable limits. This section describes those processes that affect the homeostasis of salt and water exchange between the body fluids of an organism and its seawater environment.

The body fluids of marine organisms are separated from seawater by boundary membranes that participate in many vital exchange processes, including absorption of oxygen, intake of nutrients, and excretion of waste materials. Small molecules, such as water, easily pass through some of these membranes, but the passage of larger molecules and the ions abundant in seawater is restricted. Such membranes are **selectively permeable;** they allow only small molecules and ions to pass through while regulating the exchange of larger molecules and ions. When substances are free to move, as they are when dissolved in seawater, they move along a gradient from regions of high concentrations to regions of lower concentrations. This type of molecular or ionic transfer is known as **diffusion.** Diffusion causes both water molecules and dissolved substances to move along concentration gradients within living organisms and across selectively permeable membranes between organisms and surrounding seawater.

To illustrate the basic problem of salt and water balance in marine organisms, let's examine two representative animals: a sea cucumber and a salmon. A sea cucumber avoids problems of salt and water imbalance by maintaining an internal fluid medium chemically similar to seawater (about 35‰ dissolved salts). It can easily maintain this balance as long as the salt concentrations of the fluids on either side of its boundary membranes are equal and no concentration gradient exists. (This is known as an **isosmotic** condition.) A state of equilibrium is maintained as long as water diffuses out of the sea cucumber as rapidly as it enters and the salt content of the internal fluids remains equal to that of the seawater outside (figure 2.5, top).

However, if the sea cucumber is removed from the sea and placed in a freshwater lake, the salt concentration is then greater inside the animal (still 35‰) than outside (the body fluids are now **hyperosmotic** to the lake water), and the internal water concentration (965‰) is correspondingly less than the concentration of the lake water (1,000‰). Water molecules, following their concentration gradient, diffuse across the selectively permeable boundary membranes into the sea cucumber. The movement of water across such a membrane is a special type of diffusion known as **osmosis.** The dissolved salts, now more concentrated within the animal than outside the animal, cannot diffuse out of the sea cucumber because this movement is blocked by the impermeability of the membranes to the salts. The net result is an increase in the amount of water inside the sea cucumber. The additional water creates an internal **osmotic pressure** that is potentially damaging because the animal is incapable of expelling the excess water and so it swells. Most other marine invertebrates and many marine plants have little or no capability for countering such osmotic stress. As a consequence, these organisms are limited to regions where salinity rarely fluctuates.

IN SEAWATER (35‰ S)   **IN FRESH WATER** (0‰ S)

SEA CUCUMBER
(Body fluids = 35‰ S)

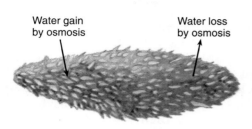

Water gain
by osmosis

Water loss
by osmosis

Isosmotic environment

No osmotic problems arise.

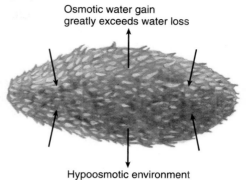

Osmotic water gain
greatly exceeds water loss

Hypoosmotic environment

Excess water that cannot be
excreted causes tissue damage or death.

SALMON
(Body fluids = 18‰ S)

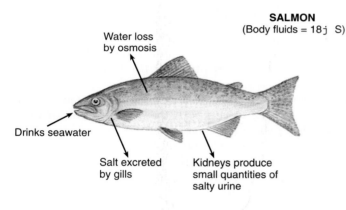

Water loss
by osmosis

Drinks seawater

Salt excreted
by gills

Kidneys produce
small quantities of
salty urine

Hyperosmotic environment

Osmotic water loss is countered by the
drinking of seawater and the excretion of excess salts.

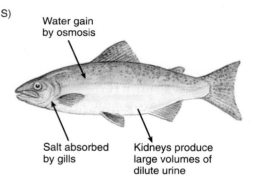

Water gain
by osmosis

Salt absorbed
by gills

Kidneys produce
large volumes of
dilute urine

Hypoosmotic environment

Excess water is excreted by kidneys
to maintain normal fluid balance.

Figure 2.5

A comparison of the osmotic
conditions of a sea cucumber and a
salmon in seawater and fresh water.

In contrast to the lack of control that sea cucumbers have over their osmotic situation, bony fishes and some other marine animals and some marine plants possess well-developed osmoregulatory mechanisms. As a result, some of these organisms are free to move between regions of varying salinities unhindered by osmotic upsets. (Estuarine organisms are discussed in chapter 9.) A salmon (which spends part of its life in seawater and the remainder in fresh water) will serve as an example of how some organisms maintain a homeostatic internal medium regardless of external environmental conditions (figure 2.5, bottom).

The salt concentration of a salmon's body fluids, like those of most other bony fishes, is midway between the concentrations found in fresh water and in seawater (about 18‰). As such, the body fluids are hyperosmotic to fresh water and **hypoosmotic** to seawater. Thus, these fishes never achieve an osmotic balance with their external environment. Instead, they must constantly expend energy to maintain a stable internal osmotic condition different from either river or ocean water. In seawater, salmon lose body water by osmosis and are constantly fighting dehydration. To counter this, the salmon drink large amounts of seawater, which is absorbed through their digestive tracts and into their bloodstreams. The water is retained in the body tissues, and excess salts are actively excreted by special **chloride cells** located in the gills. Because the kidneys of salmon are unable to produce urine with a salt concentration higher than that of their body fluids, their kidneys cannot get rid of the excess salts.

The osmotic problems of salmon are completely reversed when they are in freshwater rivers and lakes. Now the problem is one of osmotic water gain across the gill and digestive membranes and a steady loss of salts to the surrounding water. Now the salmon drink very little fresh water. To balance the inflow of water, the kidneys produce large amounts of dilute urine after effectively recovering most of the salts from that urine. Needed salts also are obtained from food and are actively absorbed from the surrounding water through specialized cells in the gills. Thus, at a considerable expense of energy, salmon maintain a homeostatic internal fluid environment in either river or ocean water.

Marine autotrophs also deal with their osmoregulatory challenges in various ways. Like sea cucumbers, many marine autotrophs maintain cell fluids such that little or no concentration gradient exists across membranes. Unlike animal cells, most plant cells are surrounded by inelastic cell walls that provide structural resistance to the stresses internal osmotic pressures place on fragile cell membranes. Often, cell walls alone are sufficient to deal with pressures generated by ion imbalances across cell membranes.

Temperature Effects

Individual activity, cell growth, oxygen consumption, and other physiological functions collectively termed **metabolism** proceed at temperature-regulated rates. Most animals lack mechanisms for body temperature regulation. These are **poikilotherms** (often inappropriately described as cold-blooded). These organisms are also referred to as **ectotherms;** their body temperatures vary with, and are largely controlled by, outside environmental temperatures. The terms *poikilotherm* and *ectotherm,* often used interchangeably, refer to distinct aspects of body temperature control. Poikilotherms do not regulate their body temperatures; external conditions govern the body temperatures of ectotherms. Most organisms are simultaneously ectothermic and poikilothermic.

For marine ectotherms, water temperature is a principal factor controlling metabolic rates. Marine ectotherms generally have rather narrow optimum temperature ranges (**stenothermal**), bracketed on either side by wider suboptimal, but tolerable, ranges. The temperature-moderating properties of water described in chapter 1 restrict

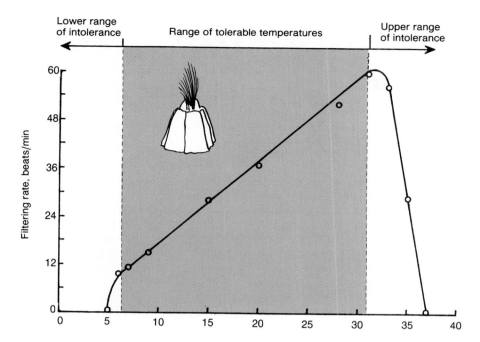

Lower range of intolerance | Range of tolerable temperatures | Upper range of intolerance

Figure 2.6

Filtering rate of a submerged intertidal barnacle as a function of water temperature. Only within the range of tolerable temperatures is the filtering rate proportional to the water temperature.
Adapted from Southward 1964.

fluctuations of temperatures experienced by marine ectotherms. Within these tolerable temperature limits, the metabolic rate of many poikilotherms is roughly doubled by a 10° C temperature increase. This, however, is only a general rule of thumb; some processes may accelerate sixfold with a 10° C temperature increase, whereas other processes may not change at all. The actual effect of water temperature on the feeding rate of a typical marine ectotherm, a submerged barnacle, is shown in figure 2.6.

Only birds and mammals have nearly constant body temperatures. They are known as **homeotherms.** Their normal body temperatures are maintained near 40° C by the release of heat in internal tissues. Thus, they are also **endotherms.** Endothermic homeotherms are less restricted by environmental temperatures than are their poikilothermic neighbors. As a result, they often range widely over all thermal regimes present in the sea (**eurythermal**).

A few large tunas, billfishes, and sharks occupy a thermal position intermediate to the two just discussed. These fishes are poikilothermic, so their body temperatures fluctuate with that of the surrounding seawater. Even so, they are unlike most other poikilotherms as they retain some of the heat produced by their swimming muscles. These animals are endothermic, yet they lack the constant body temperatures characteristic of birds and mammals. Chapter 13 will describe the mechanisms for this heat retention.

Trophic Relationships

Relationships between different organisms can be described by their **trophic** associations. This approach involves determining what an organism eats and what eats it. Living organisms require two fundamental things from their nourishment, matter and energy.

Research in Progress

Our Planetary Greenhouse

Since the 1970s, a consensus has emerged among atmospheric scientists that increasing atmospheric greenhouse gases will cause (or have already initiated) a general global warming of the surface of our planet. The average temperature of the earth's surface is maintained at its present temperature by a finely tuned global heat engine. About half of the solar energy hitting our upper atmosphere penetrates to the earth's surface, where it is converted to heat energy as it is absorbed by water, vegetation, soil, and human-made structures. If the average temperature of the earth's surface is to remain stable, an equal amount of heat energy must be reradiated from the earth's surface back into space.

Heat energy, however, radiates at longer wavelengths than does incoming visible light, and some atmospheric gases are more transparent to visible light than they are to radiated heat. These atmospheric greenhouse gases (especially water vapor, carbon dioxide, methane, and ozone) serve as a natural part of the global heat budget system by trapping heat near the earth's surface and keeping most of our solar-powered planet well above the freezing temperature of water. Why then is the greenhouse effect considered a problem? We have, since the beginning of the Industrial Revolution, begun to enhance the greenhouse effect by substantially increasing the concentrations of natural greenhouse gases in our atmosphere. Fossil fuel combustion and devegetation of land surfaces (especially burning of tropical rain forests, clear-cutting of temperate forests, and urban development) appear to be the main sources of carbon dioxide. Within the next century, a doubling of preindustrial levels of this gas is a virtual certainty. In addition, two other greenhouse gases will make their way into the atmosphere: substantial amounts of industrially produced chlorofluorocarbons (CFCs), used in aerosol products and refrigerants, and agriculturally generated methane.

What is not yet certain is what effect these greenhouse gases will have on planetary temperatures and ultimately on climate. The most likely eventual response of our climate to increased concentrations of greenhouse gases is a general, long-term global warming, causing deserts to expand, continental ice caps to partially melt, and the sea level to rise. However, the magnitude and geographical distribution of such a warming trend cannot be predicted with confidence using current climate prediction models. We cannot know whether a particular climatic event, such as the 1988 drought in the U.S. Midwest, is part of a global warming trend or merely a fluctuation of normal climatic cycles (see Research in Progress: El Niño, p. 136), but these models do suggest that major droughts will occur more frequently and in more areas over the next several decades.

One reason for the uncertainty of predictions from the best available climate models is the role of the world ocean in both global heat and carbon budgets. Seawater has a high heat capacity, and the temperature of the water filling the deep-ocean basins is substantially lower than surface-water temperatures. Can the deep sea serve as an effective sink for excess heat from the surface? One difficulty in understanding the processes governing heat transfer from air to water is the difference in time scales; days or weeks for the atmosphere are years or centuries for the ocean. Any temperature response by the ocean to increased atmospheric temperatures may be too slow to moderate rising earth surface temperatures. Other complications also exist. As water warms, it expands. This thermal expansion of the world ocean could cause average sea level to rise as much as 1 m for each 1° C increase in temperature and cause significant worldwide changes in the present pattern of human occupation of low-lying coastal plains.

Matter is necessary for individual growth and for reproduction. Energy is needed to maintain the ordered chemical state that distinguishes living organisms from nonliving assemblages of similar material. To satisfy their energy needs, all living organisms use **adenosine triphosphate (ATP)** as their fundamental molecule of energy exchange. A molecule of ATP is composed of an adenosine compound with three phosphate groups: adenosine — ℗ ~ ℗ ~ ℗. The symbol ~ represents a slightly unstable chemical bond that,

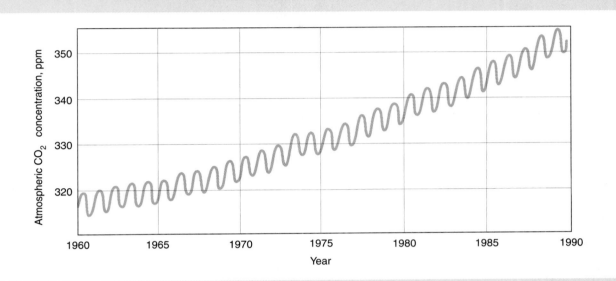

The world ocean is also a carbon dioxide sink; it already contains more than 50 times as much CO_2 as the atmosphere and 20 times as much CO_2 as the earth's total biosphere. Several features of the ocean govern its capacity to absorb and hold CO_2. First, the rate of vertical circulation and mixing limits the direct exchange of CO_2 between the atmosphere and the ocean. Once the CO_2 is in seawater its solubility is much greater than that of other atmospheric gases because it reacts with water to form carbonate and bicarbonate ions. In addition, there exists a biological "carbon pump" of organisms that absorb carbonate in surface waters and make hard skeletons of it; these skeletons rapidly sink to the deep seafloor. These features make it exceptionally difficult to accurately predict the behavior of CO_2 exchange between the atmosphere and the world ocean. It is known that only half as much CO_2 as was expected from present rates of industrial emissions is accumulating in the atmosphere. The oceans may be the sink for the missing CO_2.

To assess the long-term effects of CO_2, methane, and other greenhouse gases, the Acoustic Thermography of Ocean Climate (ATOC) experiment has been designed to measure precisely the temperature-dependent travel time of loud underwater sound pulses across oceanic distances.

For more information on this topic, see the following:
White, R. M. 1990. The great climate debate. Scientific American *(July)* *263:36–43.*

when broken, provides the energy necessary for metabolic work. Usually, only the terminal bond is broken to release energy, a Ⓟ unit, and adenosine — Ⓟ ~ Ⓟ (adenosine diphosphate, or ADP):

$$ATP \xrightarrow{\text{enzyme}} ADP + P + energy$$

Photosynthesis is a biochemical process that uses **chlorophyll** pigments to absorb some of the abundant energy of the sun's rays. In this process, ATP and other high-energy substances are made and then used to synthesize sugars, amino acids, and lipids from CO_2 and H_2O. For the present, photosynthesis can be summarized by the following general equation:

$$6CO_2 + 12H_2O \xrightarrow[\text{chlorophyll}]{\text{sunlight}} C_6H_{12}O_6 + 6H_2O + 6O_2$$

carbon   water   chlorophyll   sugar   water   oxygen
dioxide

Fossil evidence suggests that early procaryotes were the first to capitalize on photosynthesis as a solution to their energy needs. Fossil remains of cyanobacteria nearly 3 billion years old indicate that photosynthesis evolved at an early stage in the development of life on the earth.

Most nonphotosynthetic organisms on the earth rely directly or indirectly on the energy-rich organic substances produced by photosynthetic organisms. In environments with limited amounts of free O_2 (such as anoxic basins or deep-ocean bottom muds) and abundant supplies of organic material, **anaerobic respiration** (respiration without O_2) provides a mechanism to obtain energy for use in cellular processes. Several variations of anaerobic respiration are exhibited by plants and animals, yet all release energy from organic substances without using O_2. In **alcoholic fermentation,** for example, sugar is degraded, or broken down, to alcohol and CO_2. Energy is released in the form of ATP:

$$C_6H_{12}O_6 \xrightarrow[\text{enzymes}]{\text{respiratory}} 2C_2H_5OH + 2CO_2 + \text{energy}$$

sugar   enzymes   alcohol   carbon (equivalent
                                           dioxide   to 2 ATP)

In most eucaryotic organisms, respiratory processes more complex than that of anaerobic respiration completely oxidize high-energy compounds such as sugar to carbon dioxide and water and, in the process, release energy:

$$C_6H_{12}O_6 + 6O_2 \xrightarrow[\text{enzymes}]{\text{respiratory}} 6CO_2 + 6H_2O + \text{energy}$$

sugar   oxygen enzymes carbon   water (equivalent
                         dioxide   to 37 ATP)

This process utilizes oxygen and is called **aerobic respiration.** In aerobic respiration, each molecule of sugar yields 18 times as much energy as it would if used in anaerobic respiration. Organisms that metabolize food and oxygen in this manner secure a tremendous energetic advantage over their anaerobic competitors.

The transfer of matter and energy for use in metabolic processes has resulted in a close interdependence of three major categories of marine organisms: **producers, consumers,** and **decomposers. Autotrophs** are self-nourishing organisms capable of absorbing solar energy and photosynthetically building high-energy organic substances, such as carbohydrates. In the process, autotrophs use inorganic nutrients (primarily nitrate and phosphate), water, and dissolved

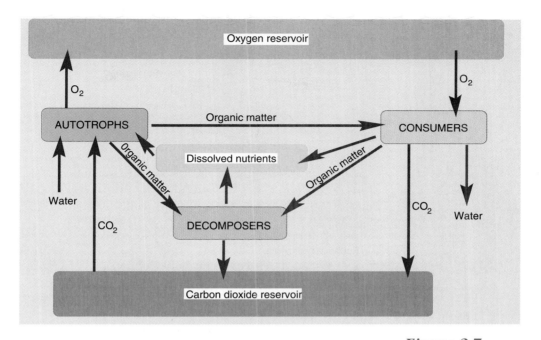

Figure 2.7

Simplified paths of the flow of oxygen and carbon in an idealized marine ecosystem.

gases. They are the **primary producers** of marine ecosystems and are placed in the first **trophic level.** Some bacterial autotrophs extract energy from inorganic compounds to build high-energy organic molecules. These autotrophs are **chemosynthetic,** and they will be discussed more in chapter 11. The consumers and decomposers are unable to synthesize their own food from inorganic substances and must depend on autotrophs for nourishment. These are consumers, or **heterotrophs,** each having some specialization in terms of nutrition. Animals who feed on autotrophs are **herbivores** and occupy the second trophic level, whereas those that prey on other animals are **carnivores** and occupy the third and higher trophic levels. The decomposers, primarily bacteria and fungi, exist on **detritus,** the waste products and dead remains of organisms, in a complex food web known as a **microbial loop.** Whatever their specialized feeding role may be, all heterotrophs metabolize the organic compounds synthesized originally by primary producers to gain available energy.

Organic compounds produced by autotrophs become the vehicle for the transport of usable energy to the other inhabitants of the ecosystem. A distinction must be made between the flow of essential nutrients and the flow of energy in an ecosystem. The movement of nutrient compounds and dissolved gases is cyclic in nature, going from autotrophs to consumers to decomposing bacteria and fungi back to the autotrophs (figure 2.7). Because there is limited input from outside ecosystems of most of these materials, the materials pass from one ecosystem component to another in cycles known as **biogeochemical cycles.** These cycles link living communities of organisms with nonliving reservoirs of important nutrients (figure 2.8).

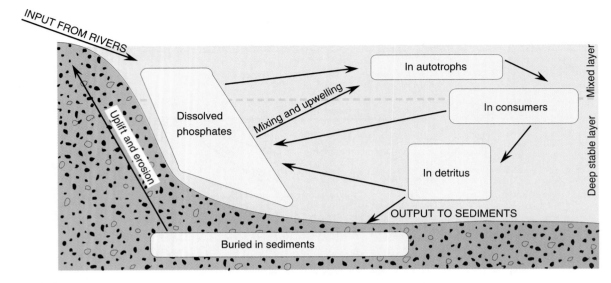

Figure 2.8

Biogeochemical cycle of phosphorus, showing its major marine reservoirs. Little phosphorus is available to autotrophs in the upper mixed reservoirs; much more is dissolved in deep, stable waters. Similar cycles exist for biologically important chemicals such as nitrogen and carbon.

In contrast to the cyclic flow of materials, the flow of energy in ecosystems is unidirectional, from the sun through the autotrophs to the consumers and decomposers. Living organisms, like most energy-consuming systems, are not highly efficient in their use of energy. Less than 1% of the solar energy available at the sea surface is absorbed by autotrophs. Furthermore, a portion of the energy captured in the photosynthetic process is used for cellular maintenance, growth, and reproduction. Thus, only a small fraction of the energy produced by photosynthesis is available to the consumers.

A similar decrease in available energy occurs between the herbivores and carnivores (figure 2.9). Laboratory and field studies of marine organisms place the efficiency of energy transfer from one trophic level to the next at between 6% and 20%. In other words, only 6% to 20% of the energy available to any trophic level is usually passed on to the next level. A widely accepted average efficiency is 10%; however, recent studies of some benthic communities and fish populations provide examples of energy efficiencies significantly higher.

The paths that nutrients and energy follow through the living portion of ecosystems are referred to as **food webs.** They may be grazing food webs, food webs commencing with autotrophs and progressing through a succession of grazers and predators; or they may be parallel detritus food webs, food webs based on waste materials and dead bodies from the grazing food webs. With only a few near-shore exceptions, the first trophic level of marine food webs is occupied by widely dispersed microscopic phytoplankton. The microscopic character of most of the marine primary producers imposes a size restriction on many of the occupants of higher marine trophic levels. Because very few animals are adapted to feed on organisms much smaller than themselves, marine herbivores are usually quite small. Large marine animals are carnivores and usually occupy higher

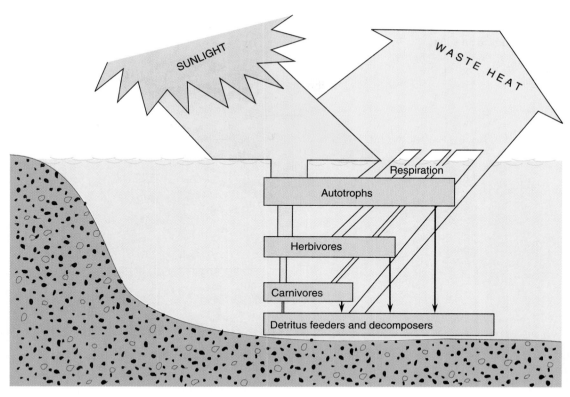

Figure 2.9

Energy flow in a marine ecosystem. Sunlight first captured by autotrophs is eventually degraded by cellular respiration and lost as waste heat.

levels in the food web. In contrast, the plants of the terrestrial ecosystem are generally quite large. As a result, most large land animals are herbivores. Food web can be simplified and arranged in a linear fashion to illustrate the decrease in available energy and material from lower to higher trophic levels. Figure 2.10 illustrates such a food pyramid, proceeding from phytoplankton to herring at the third trophic level.

Marine communities seldom have straight-line food chains. Therefore, *food web* is an accurate metaphor for the complex feeding relationships of marine organisms. Figure 2.11 outlines the major trophic relationships of the members of a typical marine community. The herring, like many of the other organisms of this food web, is an opportunistic feeder and does not specialize on only one type of food organism. Because of the complex feeding relationships of the herring, it is very difficult to place the herring in a particular trophic level. The adult herring occupies the third level when feeding on *Calanus* copepods, the fourth level when feeding on sand eels, and either the fourth or fifth trophic level when feeding on the amphipod *Themisto.* Even the complex of feeding relationships outlined in figure 2.11 is an oversimplification, for it ignores other marine animals that compete with the herring for the same food sources. Such a confusion of interrelated feeding patterns often becomes quite complex; therefore, we often simplify the food web concept for a more easily understood, though admittedly incomplete, explanation of the

Figure 2.10

Food pyramid that leads to an adult herring.

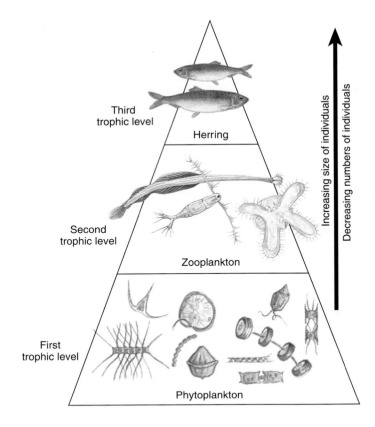

Third
trophic level

Herring

Second
trophic level

Zooplankton

First
trophic level

Phytoplankton

Increasing size of individuals

Decreasing numbers of individuals

trophic relationships of marine organisms. Figure 2.12 outlines some of the more common energy and nutrient pathways of a pelagic marine ecosystem.

Other marine organisms obtain their food by establishing highly specialized symbiotic relationships. The term **symbiosis** denotes an intimate and prolonged relationship between two (or more) species in which at least one species obtains some benefit from the relationship. Most commonly, the benefit is food. Symbiotic relationships can be reduced to a spectrum of interactions (figure 2.13), ranging from **mutualism,** which provides an obvious benefit to the **symbiont** as well as to its **host,** to **parasitism,** which benefits the symbiont at the expense of the host. A parasite lives on or in the host and benefits from the host. Parasites do not usually kill their hosts (those that do might be considered imperfect parasites or very slow predators), but they make their presence felt by reducing the host's food reserves, resistance to disease, and general vigor. The infected host then is more likely to become a casualty of infection, starvation, or predation, but not of the parasite directly. Intermediate between mutualism and parasitism is a broad category of **commensal** interactions, in which the symbiont benefits but there is an insignificant, or at least poorly known, effect on its host. Later chapters examine some representative commensal and mutualistic symbiotic relationships.

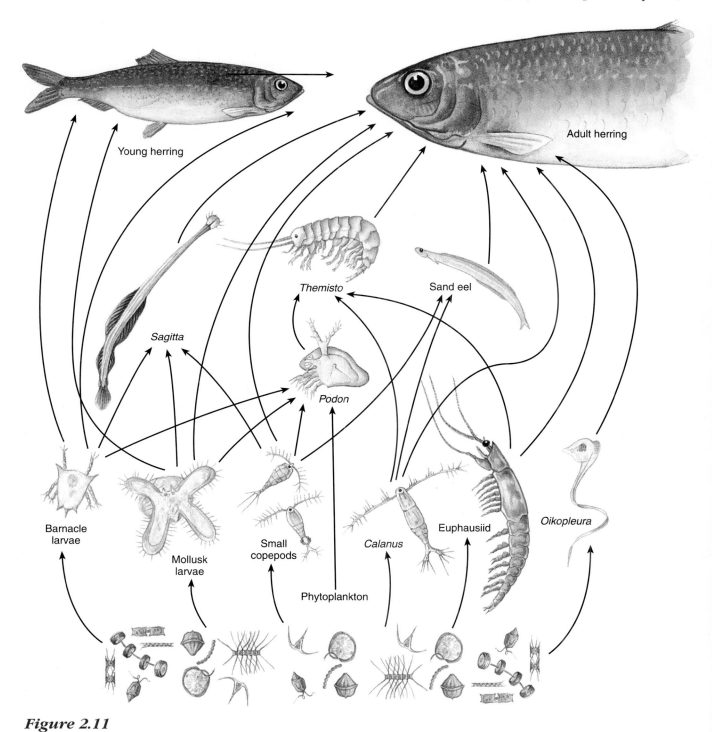

Young herring

Adult herring

Themisto

Sand eel

Sagitta

Podon

Barnacle
larvae

Mollusk
larvae

Small
copepods

Calanus

Euphausiid

Oikopleura

Phytoplankton

Figure 2.11

A marine food web, illustrating the major trophic relationships that lead to an adult herring.
Adapted from Hardy 1924.

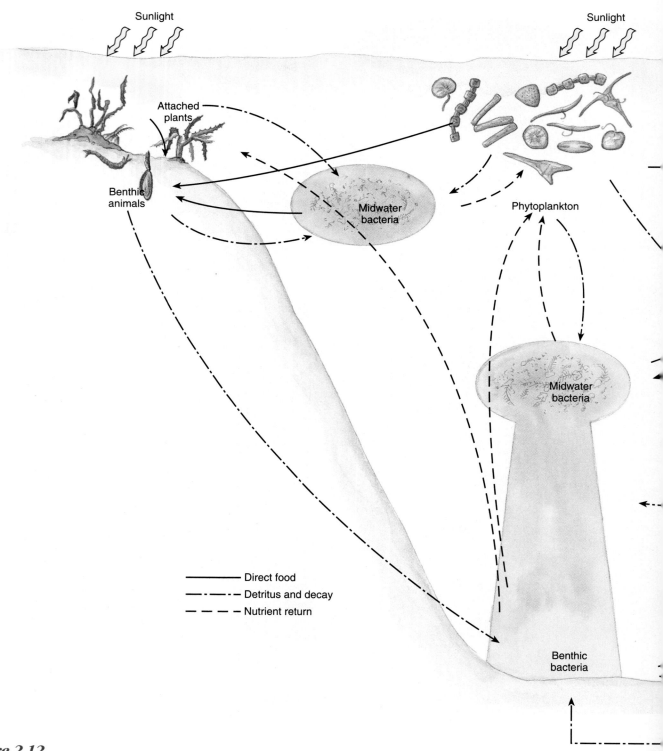

Sunlight

Sunlight

Attached
plants

Benthic
animals

Midwater
bacteria

Phytoplankton

Midwater
bacteria

——— Direct food
—·—·— Detritus and decay
— — — Nutrient return

Benthic
bacteria

ure 2.12

najor biotic components of a marine ecosystem with their interconnecting paths of energy and nutrient exchange.
ed from Russell-Hunter 1970.

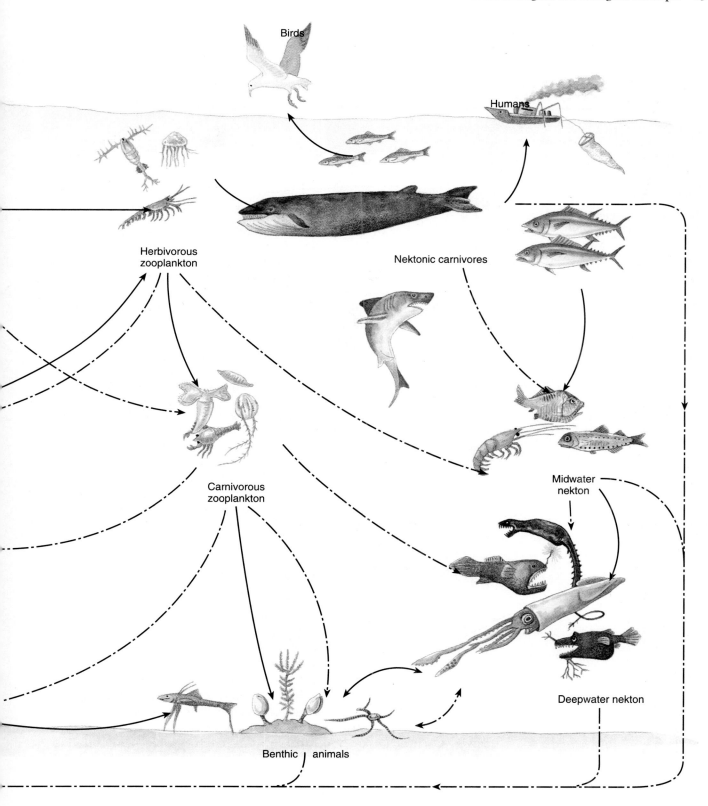

Birds

Humans

Herbivorous
zooplankton

Nektonic carnivores

Carnivorous
zooplankton

Midwater
nekton

Deepwater nekton

Benthic   animals

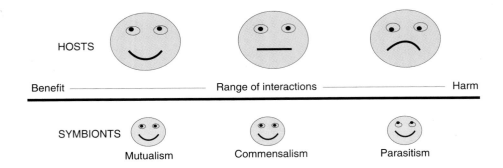

Figure 2.13

The range of common symbiotic interactions between a host and its symbiont.

Spatial Distribution

Natural systems are sometimes difficult to consider in their entirety. To cope with this complexity, we often divide these systems into smaller, more convenient units and then categorize the units and relate them to the whole system on the basis of certain characteristics. The classification of the marine environment (figure 1.37) is a good example. To be of value, any classification scheme must present the information in a generally accepted manner. Doing so requires an orderly framework to classify the available information so that it becomes more meaningful or useful. Whatever forms they assume, all classification schemes have one fundamental purpose: to provide an artificial, but accepted, means of treating information from complex natural systems in a useful and informative fashion.

A simple way to classify marine organisms is according to where they live (figure 2.14). The **benthos** includes the organisms living on the sea bottom **(epifauna)** or in the sediment **(infauna).** This definition is often extended to include those fishes and other swimming animals that are closely associated with the ocean bottom. Benthic photosynthesizers are restricted to the intertidal areas and shallow margins of the oceans. Below the photic zone, they disappear, and animals, bacteria, and fungi survive on organic material drifting down from above.

The large, actively swimming marine animals belong to the **nekton.** This group includes many fishes, marine mammals, and a few types of invertebrates such as squid and s ome crustaceans.

Plankton (derived from the Greek term *planktos,* which means "to wander") are defined by their movements and their small size. Carried about by water currents, they have little or no ability to swim horizontally, although some have remarkable abilities to swim vertically. Plankton are usually small, even microscopic organisms; however, some jellyfish have tentacles over 15 m long and a bell 2 m in diameter. Autotrophic members of the plankton are termed **phytoplankton.** They are nearly all microscopic, either a single cell or loose aggregates of a few cells, and are restricted to the sunlit, or photic, zones of the marine environment. The **zooplankton** are the heterotrophic plankton. Zooplankton range in size and complexity from microscopic single-celled organisms to large multicellular animals. The zooplankton are distributed throughout the pelagic division of the marine environment.

Sometimes, distinctions between these major groups are not clear-cut. Many fishes, for example, hatch from eggs as zooplankton and

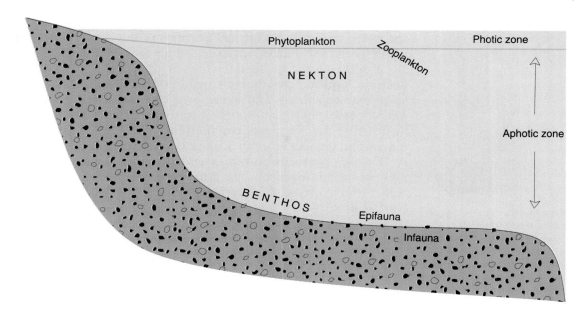

Figure 2.14

A spatial classification of marine organisms. Compare this figure with the classification scheme for the marine environment, figure 1.37.

then gradually develop into nektonic animals as their size increases and their swimming ability improves. Even so, this system is useful for referring to major groups of marine organisms living under similar environmental conditions.

The General Nature of Marine Life

Although modern marine organisms share many basic structural and behavioral characteristics with their terrestrial relatives, marine life is unique in several important ways. Marine organisms exist within a dense, circulating, interconnected seawater medium. The movement of seawater mixes and transports organisms, their food, and their waste products so that few of these organisms are isolated from the effects of other organisms. Populations of even the smallest unicellular planktonic organisms can become widely distributed by moving currents and water masses.

The biology of marine organisms is, to a large extent, the biology of the very small. It is the phytoplankton that initially establish much of the structural character of marine life. Even in very productive areas of the open ocean, the concentration of phytoplankton is thousands of times more dilute than a healthy cornfield. The dispersed nature, extremely small size, and rapid reproductive rates of the phytoplankton limit the size and abundance of other life in the sea. Most of the heterotrophs are congregated near the photic zone and its supply of food. At greater depths, the density of marine populations tends to decrease as the food supply diminishes. Below the photic zone, most marine life is dependent on the rain of detritus from above. Unlike the land, there are few plant-dominated communities in the sea. (A few notable exceptions are depicted in chapters 4 and 9.) Instead, the sea is occupied by coral reefs, mussel beds, and other communities characterized by their dominant animal members.

Many of the substances produced by marine primary producers are not consumed directly by herbivores but are dissolved into seawater. These substances, including lipids and amino acids, are eventually absorbed by suspended bacteria at all depths. The bacteria, in turn, become food for consumers capable of harvesting them. Thus, many small marine animals and many large ones are directly dependent on microscopic phytoplankton or even smaller bacteria for their nutrition. These are **suspension feeders** that employ numerous techniques and devices to extract minute food particles from seawater.

The sea provides buoyancy and structural support to many strikingly beautiful organisms. But if these organisms are removed from the water, they collapse into formless masses. It is only with the supportive aid of seawater that these organisms can continue to exist and function. Seawater also supports some extremely large animals. Deep-sea squids over 15 m long have been observed, and squids 20 or even 30 m in length are not improbable. Some blue whales approached weights of 200 tons before their populations were diminished by whaling. But these animals are exceptional and stand out in sharp contrast to the generally diminutive nature of life in the sea.

Summary Points

- All life forms are constructed on either a procaryotic or eucaryotic cellular plan. Within either of these cell types, living organisms process energy through cellular respiration for reproduction and growth, response capabilities, and regulatory processes.
- Marine organisms exhibit adaptations to numerous aspects of their physical environment, including salinity and temperature fluctuations. These adaptations occur on both ecological and evolutionary time scales and enable organisms to succeed in their reproductive efforts.
- The functional structure of marine communities is largely shaped by the trophic relationships that have evolved between community members. These relationships occur as parts of grazing food webs or detritus food webs or because of specialized symbiotic associations.
- The relatively dense fluid environment of marine organisms promotes the existence of microscopically small and dispersed phytoplankton. They, in turn, influence the general nature of all other forms of marine life. These features stand in sharp contrast to patterns of existence in terrestrial communities.

Review Questions

1. Compare the internal structure of a typical eucaryotic cell with that of a procaryotic cell.
2. Describe the fundamental differences between the flow of nutrients and the flow of energy in marine ecosystems. How is most of the energy lost from these ecosystems?
3. List and discuss the general conditions that cause some marine food webs leading to large animals to be much longer (have more trophic levels) than terrestrial food webs leading to animals of a similar size.

1. Describe conditions in which a small population might live where asexual reproduction would be advantageous over sexual reproduction.
2. What single property of water has limited the number of successful homeotherms in marine habitats? Why?
3. Freshwater crayfish and marine lobsters are closely related, yet each is incapable of surviving in the other's habitat. List and describe two likely osmoregulatory reasons that would account for this.

Challenge Questions

Books

Allen, T. F. H., and T. B. Starr. 1988. *Hierarchy: Perspectives for ecological complexity.* Chicago: University of Chicago Press.

Mader, S. S. 1992. *Biology.* Dubuque, IA: Wm. C. Brown Publishers.

Articles

Caron, D. A. 1992. An introduction to biological oceanography. *Oceanus* 35(3):10–17.

Groves, D. I., J. S. R. Dunlop, and R. Buick. 1981. An early habitat of life. *Scientific American* (October) 245:64–73.

Guttman, B. S. 1976. Is "levels of organization" a useful biological concept? *Bioscience* 26:112–13.

Lewin, R. A. 1982. Symbiosis and parasitism: Definitions and evaluations. *Bioscience* 32:254.

Pomeroy, L. R. 1992. The microbial food web. *Oceanus* 35(3):28–35.

Suggestions for Further Reading

Biology of Phytoplankton, Protozooplankton, and Bacteria
http://www.gov.au/boards/phytoplankton_board/2.html
© Australian Antarctic Division (Commonwealth of Australia),
http://www.antdiv.gov.au/
Contents: Studying Phytoplankton

MBREF: A Reference Source for Marine Biology Student Research
http://129.49.19.42/marinebio/mbref.html
Marine Biology Web: Department of Ecology and Evolution, State University of New York–Stony Brook
http://life.bio.sunysb.edu/marinebio/mbweb.html
An educational resource for marine biology students with reference lists organized by subject. Other links included to marine stations, tide information, and much more.

UCMP Topical Index
http://www.ucmp.berkeley.edu/help/topic.html
The University of California Museum of Paleontology and the University of California Regents
http://www.ucmp.berkeley.edu/index.html Marine Biology

Links to Other Web sites
http://www.lib.noaa.gov/
Click on this link for other NOAA and other oceanographic sites.

Library and Information Services
http://www.lib.noaa.gov/
Click on this link for Web pages of the NOAA Central Library, Oceanographic Data Center (NODC)

Internet Addresses

A California kelp forest.   Photo by T. Phillipp.

Marine Primary Producers

Chapter 3

Phytoplankton   73

Chapter 4

Marine Plants   96

Chapter 5

*Primary Production
in the Sea   114*

Mangrove trees occur in a predictable sequence along subtropical and tropical coasts. For example, in the Caribbean Sea, red mangroves (Rhizophora mangle) are intertidal, black mangroves (Avicennia germinans) occur just landward of the reds and are rarely in direct contact with seawater, white mangroves (Laguncularia racemosa) exist in a band behind the blacks, and gray mangroves, or buttonwoods (Conocarpus erectus), grow high and dry behind the whites. More than 50 years ago, it was suggested that this repeating sequence of mangrove species along Caribbean shorelines was due to an ecological succession (i.e., the progression from initial colonization of an area by organisms to a stable, climax community) that was initiated by the arrival of red mangrove seedlings on the shoreline. In this "just so" story, the many prop and aerial roots characteristic of red mangroves would slow water currents and cause a deposition of sediments in the area. The composition and chemistry of the sediment would in turn be altered by the decomposition of leaves shed by the red mangroves (3 ton per acre per year in some areas). Hence, the presence and biological activities of the red mangroves would alter the area such that it was no longer suitable for them, and they would be replaced by black mangroves, a species that does well in drier, more anoxic sediments than red mangroves. The red mangroves would "march" seaward in search of the intertidal zone. As the years passed, the red mangroves would continue to reclaim land from the sea, leading the way for the colonization of drier and drier land by black, white, and gray mangroves. This classical hypothesis is not consistent with recent data. First, the predicted patterns of sediment accumulation are not always seen. Sometimes mangroves occur on shorelines that are eroding, such as in South Africa. Second, the zonation patterns of some mangrove areas have existed for thousands of years without change. Third, studies of mangrove biogeography have shown that changes in vegetation occur because of changes in physical characteristics of the environment. Therefore, succession due to land building is no longer a widely accepted hypothesis. Today, geomorphic processes, physiological tolerance of inundation by seawater, interspecific competition, dispersal dynamics, soil physicochemistry, and predation of propagules are considered to be important factors in mangrove distribution. Hence, the distribution of mangrove species along subtropical and tropical shorelines seems to be the result of zonation not of succession. These plants will move seaward in depositional environments and landward in erosional areas, and they will remain unchanged along stable coasts.

Phytoplankton

A s indicated in the preceding chapter, much of the special nature of life in the sea is due to its smallness, a consequence of the small cell sizes of the most abundant marine primary producers, the phytoplankton. Marine phytoplankton include numerous members of two of the five kingdoms of life, so this is where we will begin our survey of the diversity of marine organisms.

The Five Kingdoms and Taxonomic Classification

The classification system used in this text, slightly modified from the system proposed by Margulis in 1978, groups living organisms into five kingdoms (each shaded in figure 3.2). In this system, the procaryotic organisms, Bacteria and Cyanobacteria, are placed in the kingdom Monera. These organisms consist of small cells of simple structural organization, lacking nuclei, multiple chromosomes, chloroplasts, and mitochondria. Bacteria are important decomposers in all marine habitats. All Cyanobacteria are photosynthetic, and many are important in phytoplankton communities.

The eucaryotic organisms that are single-celled or consist of simple aggregations of similar cells are placed in the kingdom Protista. Most nonparasitic protists are aquatic or live in soil. This kingdom includes the rest of the major phytoplankton (discussed in this chapter), a few terrestrial funguslike groups, as well as several nonphotosynthetic phyla sometimes referred to as protozoans. The marine protozoan groups are described in chapter 6.

Fungi form the third kingdom of living organisms. Fungi are complex organisms whose bodies are composed of **hyphae,** which are fine, threadlike tubes containing hundreds of haploid nuclei. The hyphae have cell walls made of chitin (similar to the material of arthropod exoskeletons, p. 168) and often lack cross-walls, so true cells of fungi are difficult to define. Diffuse masses of hyphae form a mycelium, the nonreproductive part of a fungus. Fungi are much more abundant and diverse on land, where their wind-blown spores effectively disperse them over long distances. Only a few hundred species of mostly inconspicuous fungi live in the sea. However, along with bacteria, marine fungi function as decomposers in most benthic environments, and a few, with their photosynthetic algal symbionts, exist as lichens on intertidal rocks (p. 217).

The last two kingdoms, Plantae and Animalia, include more familiar groups of organisms. Both consist of multicellular and, therefore, usually larger organisms. Plants are photosynthetic, nonmotile organisms with cellulose cell walls and life cycles that include alternating gametophyte and sporophyte generations. Most members of the kingdom Plantae live on land, yet several thrive in marine communities. Marine plants and their life cycles are described in chapter 4. Animals lack cell walls and have some muscle-contracting and nerve-conducting capabilities. These attributes provide flexibility and mobility, two of the hallmarks for evolutionary success of animals. Chapters 6 and 7 include a survey of the more common marine animals.

Biologists estimate that between 10 and 30 million different types, or **species,** of organisms exist on the earth today. Of these, only about 1.5 million have been identified and formally described, and the vast majority of those are biologically unknown. Although the great majority of species that have been described are land-dwelling insects, the diversity of life in the sea is immense. Because of the evolutionary processes that started in the sea and have operated for the past 3 to 4 billion years of earth history, each of these species exhibits some genetic relationship to all other species.

Sometimes, evolutionary relationships between organisms are obvious; for example, porpoises and dolphins. At other times, however, such relationships are more obscure (see Research in Progress: Molecular Systematics, p. 186). The **taxonomic method of classification** deals with this vast and often confusing array of diversity by reflecting these evolutionary, or **phylogenetic,** relationships of organisms. Taxonomic classification categorizes organisms into natural units. It traces the lines of evolution that have led to the diverse life forms of the present, and it identifies and describes similarities among existing groups of organisms.

The process of taxonomic classification consists of three basic steps. First, closely related groups of individual organisms must be recognized and described. Next, these groups, called **taxa** (singular, taxon), are assigned Latin (or latinized Greek) names according to procedures established by international convention. Finally, the described and labeled groups are fitted into a system of larger, more inclusive taxa.

The fundamental unit of taxonomic classification is the species. A species is a group of closely related individuals that are similar in appearance and that can and normally do interbreed and produce fertile offspring. The free exchange of genetic information between individuals of such groups connects each individual to a common gene pool and steers them along a common evolutionary path, with whole populations adapting to environmental influences over long periods of time.

This widely accepted definition of a species, however, poses special problems for the classification of marine organisms. Because of the environmental extremes occupied by many marine organisms, they are quite often difficult, or even impossible, to study alive, and little is known of their reproductive habits. In such cases, another somewhat circuitous definition is used: A species is a group of closely related individuals classified as a species by a competent taxonomist on the basis of anatomy, physiology, and other characteristics. Whichever definition is used, the species must be regarded as a functional biological unit that can be identified and studied.

Assigning names to species or larger groups of organisms is a process more regimented than merely recognizing and describing the species. Common names are often used in localized areas, but the

(a)

(b)

Figure 3.1

(a) Common dolphin, *Delphinus.*
(b) Dolphinfish, *Coryphaena,* also
known as a dorado or mahi-mahi.

lack of standardization in the use of common names detracts from
their widespread usefulness and acceptance. To some people, the
name *dolphin* refers to an air-breathing porpoiselike marine mammal
(figure 3.1a). To others, a dolphin is a tasty game fish (figure 3.1b).
These common-name drawbacks are eliminated when species and
other taxonomic groups are assigned scientific names that are accepted by international agreement as standard group names.

Following the scheme first introduced by the Swedish botanist
Linnaeus over two centuries ago, taxonomic names of organisms consist of two terms. The first is the genus name followed by the species
name. The genus name is always capitalized; the species name is not.
Both are either italicized or underlined. Each taxonomic name is
unique and represents only one species. Thus, there can be no confusion that *Delphinus delphis,* the common dolphin, refers only to a
mammalian species and not to the dolphinfish, *Coryphaena hippurus.*

The naming of a species does not complete the taxonomic classification process. The species is only part of a larger classification
scheme that consists of a hierarchy of taxonomic categories:

Kingdom
   Phylum (Division for photosynthetic groups)
      Class
         Order
            Family
               Genus
                  Species

Each category is constructed so that it encompasses one or more categories from the next lower level. All categories above the species level
are artificial; they were contrived for the convenient pigeonholing of

Research in Progress

Oceanography from Space

Observations of ocean conditions at sea have been the basis for amassing the extensive store of information that underlies our present understanding of oceanic structures and their functions. Typically, this information has been collected from ships or unmanned instrument buoys. These approaches were (and still are) appropriate for studying such conditions as water temperature, waves and currents, salinity, and the change in marine organisms from the sea surface downward; however, they were less effective in providing detailed views of how the same features varied through time or were distributed across oceanic distances. So, until the middle of this century, longtime series of data collections were combined and averaged to provide a "typical" view of any of these features at large oceanic scales. It was not possible to obtain a synoptic, or instant snapshot, view of how any major component of the oceans was behaving at any particular time.

In 1959, the first of the TIROS series of weather satellites was placed in orbit. A remote eye-in-the-sky view of weather patterns over whole oceans has become standard (and expected) fare for television weather reports. The TIROS satellites have since been replaced with two newer series, NIMBUS and NOAA. Their capability to detect and track every hurricane on either side of North America has proved invaluable.

In the 1970s, the ERTS and LAND-SAT series of satellites were launched to assess land-based resources. TRANSIT is a system of five satellites that provides global position-fixing capabilities anywhere on the globe. These position fixes are available every hour or two and are accurate within 40 m. Although none of these satellites were specifically designed for ocean observations, they have contributed to our understanding of the complex links between ocean conditions and weather, have given us our first detailed views of large-scale coastal features, and have provided oceanographers with a new and valuable ability to pinpoint the location of an experiment or a sample site at sea.

Since TIROS I, more than 40 nondefense satellites capable of studying the oceans have been placed in orbit by the United States. Only one of these, SEASAT, was specifically designed for ocean research. It lasted three months before failing, but in that short time, it provided an extensive global view of the variation in altitude of the sea surface. The radar beams directed from SEASAT could measure the distance between the satellite and the sea surface to within 5 to 7 cm (2 to 3 in). Useful information about the structure and composition of the seafloor was obtained by mapping the ocean surface from space. For example, the Gulf Stream varies about 100 cm in height across its width. Over seafloor trenches, the sea surface is as much as 60 m closer to the center of the earth, and a seamount causes the sea surface to bulge out about 5 m. SEASAT maps these variations in sea surface topography, and the maps are used to construct a composite view of how the ocean floor and surface current patterns vary over time periods of a few weeks.

With the shutdown of SEASAT, oceanography from space has continued with instruments placed aboard satellites dedicated to other remote-sensing tasks. The radar altimeter measurements initiated on SEASAT are being continued on the GEOS satellites. The coastal zone color scanner (CZCS) on NIMBUS 7 monitors chlorophyll concentrations in surface waters (figure 5.2 was obtained this way) and maps the large-scale distribution and abundance of marine phytoplankton, with a capability to repeat each observation every six days. Several satellites have radar, microwave, or infrared sensors that can measure sea-surface temperature, rain rate, wind speed, sea-surface roughness, and distribution of sea ice.

similar groups of organisms. These groups are not completely arbitrary, however. Each group reflects the evolutionary relationships known or assumed to exist between its component taxa on the basis of its anatomy, embryology, and biochemistry. It is because much of the evolutionary history of life is not known in detail that classification based on so little information is referred to as "artificial."

Ideally, each genus is composed of a group of very closely related, but genetically isolated, species. Families include related genera that

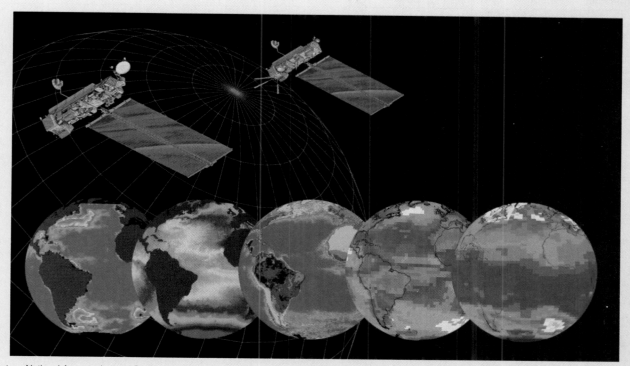

Courtesy National Aeronautics and Space Administration

The use of satellite-based remote-sensing devices has opened exciting possibilities for acquiring previously unattainable synoptic views of large expanses of the world ocean and has ushered in a new era of global oceanography. Presently, satellite remote sensing is the only method that can provide global-scale information about variations of sea-surface temperature, sea-surface altitude, and phytoplankton productivity. More ocean-oriented remote-sensing satellite missions, ranging from sea-surface temperature mapping to tracking of individual whales, are planned for the 1990s as part of NASA's Mission to Planet Earth.

For more information on this topic, see the following:
Stewart, R. H. 1985. Methods of satellite oceanography. *University of California Press.*
Williams, R. S., Jr., and J. G. Ferrigno, eds. 1994. Satellite image atlas of the world. *U.S. Geological Survey, Professional Paper 1386-B.*

have many features in common. The cat family is a familiar example (see Research in Progress: Molecular Systematics, p. 186). Orders include related families based on generalized characteristics. Classes, phyla (singular, phylum), and kingdoms are increasingly inclusive categories based on even more general features. The term *division* is used in place of phylum for plants, photosynthetic protists, and fungi.

Table 3.1 summarizes the taxonomy of several organisms mentioned in the first three chapters of this book. Dolphins and blue

Table 3.1
Taxonomic Classification of Some Marine Organisms

			Taxonomic Category				
Organism	Kingdom	Phylum/Division	Class	Order	Family	Genus	Species
Blue whale	Animalia	Chordata	Mammalia	Cetacea	Balaenopteridae	*Balaenoptera*	*B. musculus*
Dolphin	Animalia	Chordata	Mammalia	Cetacea	Delphinidae	*Delphinus*	*D. delphis*
Dolphinfish	Animalia	Chordata	Osteichthyes	Teleostei	Coryphaenidae	*Coryphaena*	*C. hippurus*
Copepod	Animalia	Arthropoda	Crustacea	Calanoida	Calanidae	*Calanus*	*C. finmarchicus*
Mangrove	Plantae	Anthophyta	Dicotyledones	Laminales	Avicenniaceae	*Avicennia*	*A. germinans*
Tintinnid	Protista	Ciliophora	Ciliata	Spirotricha	Tintinnidae	*Halteria*	*H. grandinella*

Figure 3.2

One scheme to illustrate the evolutionary relationships between kingdoms and among the divisions of autotrophic marine organisms. Each division is represented by a labeled box, and its relative evolutionary distance to other divisions is indicated by the branched lines. The kingdoms are shaded to indicate autotrophs (green) or heterotrophs (tan).

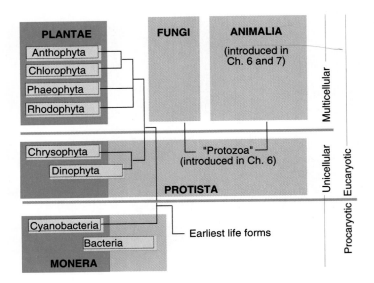

whales are more closely related to each other than to the other organisms listed in table 3.1, so they are placed in the same order, Cetacea, that includes other whales but excludes all other species of organisms. Copepods do not resemble whales or dolphins, yet their evolutionary connections are closer to both dolphins and blue whales (organisms in the same kingdom) than they are to mangroves in the kingdom Plantae. In this way, the taxonomic system of classification serves as a framework to support our understanding of the evolutionary relationships that exist between groups of organisms.

In figure 3.2, the major phyla and divisions of marine organisms are arranged to illustrate the presumed evolutionary relationships of each group. Only phyla or divisions with several free-living nonparasitic marine species are included. These phyla and divisions are then grouped into the five kingdoms described earlier and outlined in figure 3.2.

Phytoplankton Groups

Within the sunlit surface layer of the sea, marine primary producers thrive, ranging from extremely small cyanobacteria to tree-sized kelp plants. The remainder of this chapter describes the microscopic drifting primary producers, the phytoplankton.

Table 3.2
Major Divisions of Marine Phytoplankton and Their General Characteristics

Division (Common Name)	Approx. No. of Living Species	Percent of Species Marine	General Size and Structure	Photosynthetic Pigments	Storage Products	Habit
Cyanobacteria (blue-green algae)	200	~75	Unicellular, procaryotic, nonflagellated, microscopic	Chlorophyll a Carotenes Phycobilins	Starch	Mostly benthic
Chrysophyta (golden-brown algae) (coccolithophores) (silicoflagellates) (diatoms)	650 200 ? 6,000–10,000	~20 96 Most 30–50	Unicellular, often flagellated, microscopic	Chlorophyll a, c Xanthophylls Carotenes	Chrysolaminarin Oils	Planktonic and benthic
Dinophyta (dinophytes)	1,100+	93	Unicellular or colonial, flagellated, microscopic	Chlorophyll a, c Xanthophylls Carotenes	Starch Fats Oils	Planktonic

Adapted from Seagel et al. 1980, Dawson 1981, and Kaufman et al. 1989.

Almost all marine phytoplankton belong to three divisions in the kingdoms Monera and Protista (table 3.2). Consequently, they are all single-celled microscopic organisms. They are found dispersed throughout the photic zone of the oceans and account for the major share of primary productivity in the marine environment. Figure 3.3 shows the four size categories into which phytoplankton cells are grouped. Only in the last few years has it been possible to collect representative samples of the exceptionally small **picoplankton** and **ultraplankton**. As our knowledge of these very small phytoplankton groups improves, our understanding of their contribution to marine food webs is increasing. Presently, it is thought that the most important primary producers in all marine environments, but especially in oceanic waters, are **nanoplankton**-sized or smaller.

Cyanobacteria

Marine cyanobacteria have been the object of much recent study. Their small cell size (most are less than 5 µm) makes them very difficult to collect and study. Their cell structure (figure 3.4) is typical of procaryotes, with only a few of the complex membrane-bound organelles so obvious in larger eucaryotic cells (see figure 2.2). Unlike photosynthesis in bacteria, photosynthesis in cyanobacteria is similar to that in eucaryotic autotrophs. It is based on chlorophyll *a* and results in the production of oxygen.

Cyanobacteria are not newcomers to marine environments. Modern cyanobacteria have descended from some of the earliest forms of life on the earth. Fossil stromatolites made by cyanobacteria over 3 billion years ago are remarkably similar to modern ones found at the edges of tropical lagoons in Australia (figure 3.5) and the Bahamas.

Marine cyanobacteria are especially abundant in intertidal and estuarine areas, with a lesser role in oceanic waters. Some species of

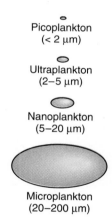

Picoplankton (< 2 µm)

Ultraplankton (2–5 µm)

Nanoplankton (5–20 µm)

Microplankton (20–200 µm)

Figure 3.3

Relative sizes of phytoplankton groups. All are enlarged 1,000×. At the same magnification, a human hair is as thick as this page is wide.

Figure 3.4

A transmission electron micrograph of a marine cyanobacterium, *Synechoccus.*

Reproduced by permission from J. B. Waterbury, *Canadian Bulletin of Fisheries and Aquatic Sciences* 214:71–120, 1986, ed. T. Platt and W. K.W. Li.

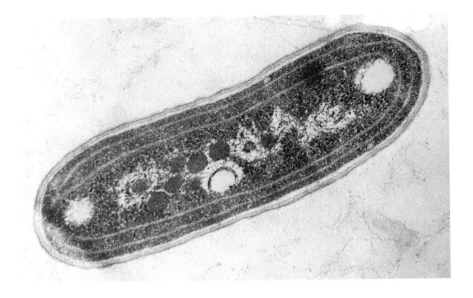

Figure 3.5

Stromatolites, resembling mushrooms 1 m high, grow on the shallow sandy bottom of Shark Bay, Australia.
Photo by D. Doubilet.

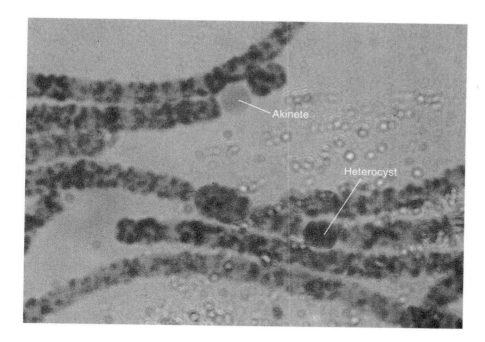

Figure 3.6

Micrograph of the cyanobacterium *Anabaena*, showing spores (akinetes) and nitrogen-fixing heterocysts among vegetative cells.

cyanobacteria produce dense blooms in warm-water regions. The red phycobilin pigment of *Oscillatoria* is responsible for the color and name of the Red Sea.

Benthic cyanobacteria can be found almost everywhere light and water are available. These organisms are individually microscopic and usually inconspicuous, but they may aggregate to produce macroscopic colonies. One abundant coral reef form, *Lyngbya,* develops long strands or hollow tubes of cells nearly a meter in length. Reproduction of cyanobacteria is usually accomplished by cell fission. Occasionally a growing colony will fragment to disperse the cells. More complex modes of reproduction, involving motile or resistant stages, are also known.

On temperate seashores, some species of cyanobacteria (figure 3.6) appear as tarlike patches encrusting rocks in the intertidal or splash zones. Other species can be found in abundance on mudflats of coastal marshes, estuaries, and in association with tropical coral reefs. One well-studied cyanobacterium, *Microcoleus,* is a major component of complexly laminated microbial mats that construct modern stromatolites (figure 3.5).

Several species of cyanobacteria are capable of utilizing atmospheric nitrogen (N_2) to satisfy their metabolic nitrogen needs. The process of **nitrogen fixation** is poorly understood, but it is known to be limited to cyanobacteria and several other types of bacteria. Field studies have shown that nitrogen-fixing cyanobacteria are fairly common in nearshore regions and are often associated with large marine plants, but they are rare in oceanic waters. Apparently, even though growth of most oceanic photosynthesizers is nitrogen limited, few planktonic N_2-fixing cyanobacteria have evolved to take advantage of the situation.

Cyanobacteria exhibit a strong tendency to form symbiotic associations with other organisms. Various examples of symbiosis with animals are common. Two genera can even be found inhabiting marine planktonic diatoms, such as *Rhizosolenia.* Many cyanobacteria live as **epiphytes,** attached to larger plants. Some epiphytic cyanobacteria, for example those inhabiting turtle grass beds along the Gulf Coast of the United States, are also nitrogen fixers. As such, they play an important role in the fertility and productivity of seagrass beds.

Chrysophyta

The division Chrysophyta consists of two classes: the Chrysophyceae and the Bacillariophyceae. The marine members of this division are single celled. Like all other eucaryotic autotrophs, their primary photosynthetic pigment is chlorophyll *a.* In addition, chrysophytes have accessory chlorophyll *c* and golden or yellow-brown xanthophyll pigments that are also characteristic of brown algae. Most have mineralized cell walls or internal skeletons of silica or calcium carbonate. Some species possess flagella for motility but, like other planktonic organisms, can do very little to counter horizontal transport by water currents.

Although most species of Chrysophyceae are found in fresh water, two groups, the coccolithophores and silicoflagellates, are relatively abundant in some marine areas. Most marine coccolithophores and silicoflagellates are nannoplanktonic. Only in recent years has the use of membrane filters, fine collection screens, and the wider application of scanning electron microscopic (SEM) techniques provided us with a better look at the very small nannoplankton.

Coccolithophores are unicellular, with numerous small calcareous plates, or **coccoliths,** embedded in their cell walls (figure 3.7). Coccolith remains from seafloor sediments have been studied, but it was not until 1898 that the photosynthetic cells producing them were observed. It has been suggested that coccoliths reflect much of the ambient light in clear tropical waters, permitting these organisms to thrive in areas of very high light intensity. Coccolithophores are found in all warm and temperate seas and may account for a substantial portion of the total primary productivity. In the Sargasso Sea, for instance, a single species, *Emiliania huxleyi* (figure 3.7a), seems to be responsible for most of the photosynthesis. However, the photosynthetic role of coccolithophores in global marine primary production is not yet well defined.

Figure 3.7

Scanning electron micrographs of three coccolithophore cells, each showing clearly their dense coverings of coccoliths. All magnified approximately 3,000×. *(a) Emiliania huxleyi, (b) Umbilicosphaera* sp., *(c) Anthosphaera* sp.
Courtesy F. Reid, Scripps Institution of Oceanography.

(a)

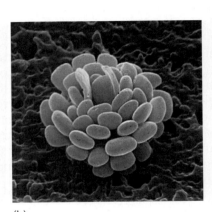
(b)

(c)

The silicoflagellates, like the coccolithophores, were first recognized and identified from fossil skeletons in marine sediments. Silicoflagellates have internal, and often ornate, silica skeletons. They have one or two flagella and many small chloroplasts (figure 3.8). The significance of silicoflagellates as marine primary producers has not been evaluated, but their contribution is thought to be small. Reproduction in coccolithophores and silicoflagellates is mostly by cellular fission (see figure 2.3).

The most obvious and often the most abundant members of the phytoplankton are the diatoms (class Bacillariophyceae). Although diatoms are unicellular, they may occur in chains or other loose aggregates of cells. Cell sizes range from less than 15 μm in length to 1 mm (1,000 μm) in length. Most diatoms are between 50 and 500 μm in size and are typically much larger than coccolithophores or silicoflagellates (figure 3.9). Diatoms have a cell wall, or **frustule,** composed of pectin with large amounts (up to 95%) of silica. The frustule consists of two closely fitting halves, an **epitheca** and a larger **hypotheca,** which fits tightly inside the epitheca (figure 3.10). Planktonic diatoms usually have many small chloroplasts scattered throughout the cytoplasm, but in low light intensities, the chloroplasts may aggregate near the cell ends.

Diatoms exist in an immense variety of forms derived from two basic cell shapes. The frustules of most planktonic species appear radially symmetrical from an end view. Circular, triangular, and modified square shapes are common. These are known as centric diatoms. Other diatoms, especially benthic forms, display varying types of bilateral symmetry and are termed pennate diatoms. Only pennate diatoms are capable of locomotion. The mechanism for locomotion is thought to involve a wavelike motion on the cytoplasmic surface that extends through a groove (the **raphe**) in the frustule. This flowing motion is accomplished only when the diatom is in contact with another surface. Diatoms capable of locomotion are generally restricted to shallow-water sediments or to the surfaces of larger plants and animals.

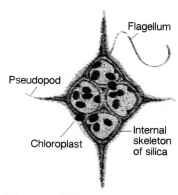

Figure 3.8

Dictyocha, a slightly, smaller common marine silicoflagellate.

Figure 3.9

This scanning electron micrograph illustrates the size difference between a typical centric diatom, *Thalassiosira,* and four coccolithophore cells, *Crenalithus.* Both the diatom and the coccolithophore cells adhere to a pad of coccoliths on the underside of the diatom cell. The association between these two species is thought to be symbiotic, but its precise nature is still unresolved.
Courtesy F. Reid, Scripps Institution of Oceanography.

Figure 3.10

Another scanning electron micrograph view of *Thalassiosira,* a coastal diatom, clearly showing the epitheca, hypotheca, and a connecting girdle of cell wall material.

Courtesy G. Fryxell.

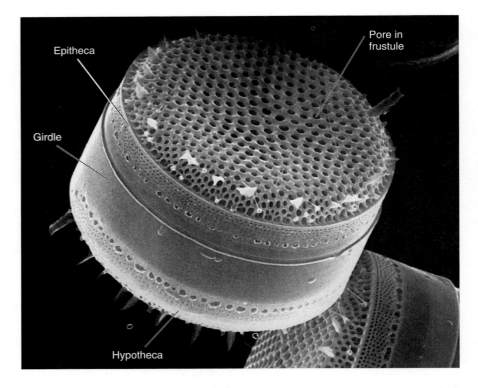

Figure 3.11

Scanning electron micrograph of a centric diatom, *Asteromphalus heptacles. (a)* The entire cell (4,700×). *(b)* A highly magnified portion of a similar cell, showing the character of perforations through the frustule.

Courtesy E. Venrick, Scripps Institution of Oceanography.

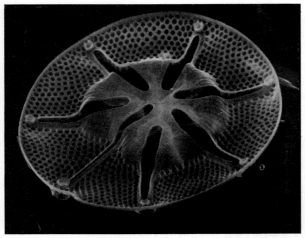

(a)

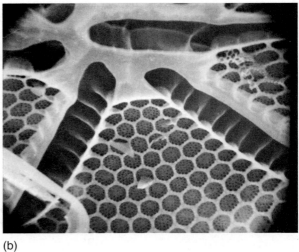

(b)

The silica frustules of diatoms exhibit large sculptured pits arranged irregularly or in striking geometric patterns (figure 3.11a). Each large pit penetrates a structural unit of the cell wall, usually hexagonal, called the **areolus** (figure 3.11b). The large outer pit connects with fine inner pores to facilitate water, nutrient, and waste exchange between the diatom's cytoplasm and the external environment.

Diatoms and most other protists reproduce asexually by simple cell division. An individual parent cell divides in half to produce two daughter cells (figure 3.12). This method of reproduction can yield a large number of diatoms in a short period of time. When conditions

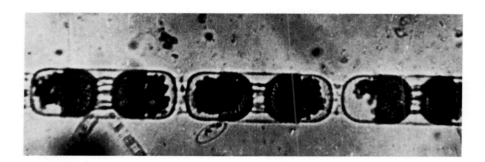

Figure 3.12

Cells in a chain of *Stephanopyxis* just after synchronized division was completed. The darker half of each cell is the newly formed hypotheca, still connected by a girdle of silica.

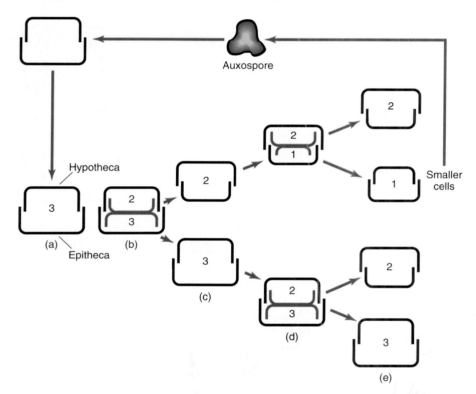

Figure 3.13

Diatom cell division and subsequent size reduction. Numerals represent distinct cell sizes.

for growth are favorable, a single diatom requires less than three weeks to produce 1 million daughter cells. Populations of diatoms and other rapidly dividing unicellular plants thus have an exceptional capacity to respond rapidly to improved growth conditions. Unfortunately, for them, their reproductive output usually is limited by predation or availability of light or nutrients.

Restrictions of size and shape imposed upon diatoms by the rigid frustule create a peculiar cellular reproduction problem (figure 3.13). During diatom cell division, two new frustule halves are formed inside the original frustule (figure 3.13b). One is the same size as the hypotheca of the parent cell (a) and is destined to become the hypotheca of the larger daughter cell (c). The other newly formed frustule half becomes the new hypotheca for the other daughter cell. Each daughter cell receives its epitheca from the original frustule of the parent cell. The daughter cells grow (figure 3.13c) and repeat the process (d, e). This method of cell division efficiently recycles the

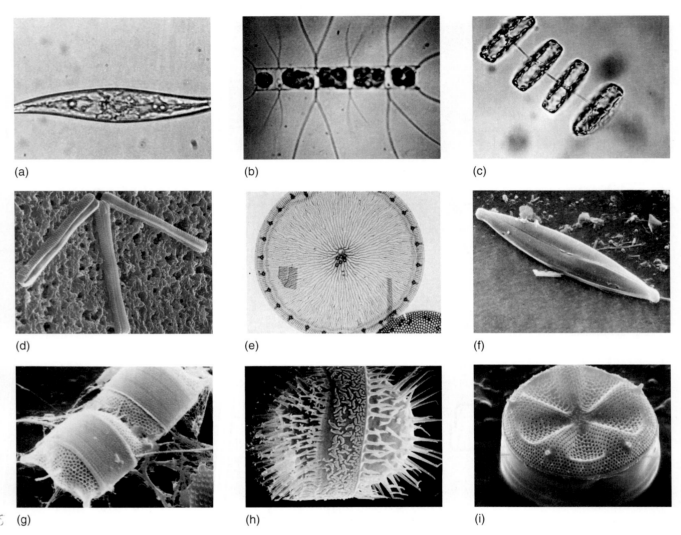

(a) (b) (c)

(d) (e) (f)

(g) (h) (i)

Figure 3.14

Light *(a–c)* and scanning electron *(d–i)* micrographs of several common types of temperate-water planktonic diatoms: *(a) Gyrosigma, (b) Chaetoceros, (c) Thalassiosira, (d) Thalassiothrix, (e) Coscinodiscus, (f) Nitzchia, (g) Biddulphia, (h) Chaetoceros,* resting spore; *(i) Actinoptychus.*
Courtesy F. Reid and K. Lang, Scripps Institution of Oceanography.

old frustules. However, a slight decrease in the average cell size results with each successive cell division. Reduction of cell size may continue for many months, eventually reaching a minimum of about 25% of the original cell size. This size reduction is not observed in all natural diatom populations, suggesting that continual readjustment of cell diameter occurs in some species. Diatoms also reproduce sexually.

When the minimal cell size is reached, the small diatom sheds its enclosing frustule, and the naked cell, known as an **auxospore,** flows out. The auxospore enlarges to the original cell size, forms a new frustule, and begins dividing again to repeat the entire sequence.

The variety of planktonic diatom species existing in temperate waters is impressive. Figure 3.14 illustrates a few of the more common types.

Benthic diatoms closely resemble their planktonic counterparts in cell structure, modes of reproduction, and other general characteristics. These organisms can be found on almost any solid substrate in shallow

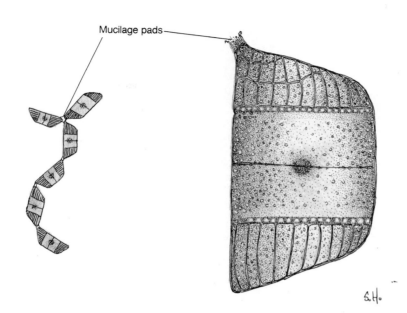

Mucilage pads

Figure 3.15

A benthic diatom, *Isthmia,* forming long, complex chains of cells. A close-up of a single cell is shown to the right.

seawater: on mud surfaces, rocks, larger marine plants, man-made structures, and the hard shells of marine animals. One type, *Cocconeis,* even makes a home on the undersides of blue whales. Many of the benthic diatoms secrete a mucilage pad that connects adjacent cells into complex chains and branching forms (figure 3.15). In this manner, macroscopic diatom colonies a few centimeters long are formed.

When compared with that of planktonic diatoms, the geographical distribution of benthic diatoms is severely restricted because of their need for light and for solid substrates. Still, benthic diatoms make a significant contribution to the total amount of primary production in estuaries, bays, and other localized shallow-water areas. Some species of these diatoms are also key components in the ecological succession of organisms that culminates in a rich growth of organisms on docks, boats, and other man-made structures (refer to figure 8.26). Several studies have indicated that marine bacteria are usually the first organisms to settle and grow on new or freshly denuded underwater structures. Development of a diatom film a few cells thick quickly follows and is succeeded by more complex populations of larger algae and invertebrate animals.

Dinophyta

The division Dinophyta includes a few species that are not photosynthetic; instead, they obtain energy from organic compounds dissolved in seawater or by ingesting particulate bits of food. However, most marine dinophytes are photosynthetic, and their share of the total marine plant production is significant; in warmer seas, it often surpasses that of diatoms.

Dinophytes (figure 3.16) are typically unicellular, with a large nucleus, two flagella, and several small chloroplasts containing photosynthetic pigments similar to those of diatoms. One broad, ribbon-like flagellum encircles the cell in a transverse groove. The other, a

Figure 3.16

This light micrograph of a dinophyte, *Oxytoxum,* illustrates its major cellular features.

Courtesy F. Reid, Scripps Institution of Oceanography.

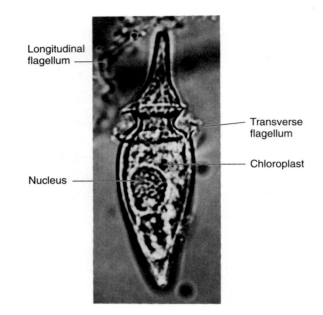

Longitudinal flagellum

Transverse flagellum

Chloroplast

Nucleus

longitudinal flagellum, projects forward and pulls the cell, providing forward motion. Cell sizes range from 25 to 1,000 µm. In armored forms, the cell wall consists of articulating cellulose plates arranged irregularly over the cell surface. The plates may be perforated by many pores. Spines, wings, horns, or other ornamentations also may decorate the cell wall. Figure 3.17 illustrates a few common marine dinophytes.

Dinophytes reproduce asexually by longitudinal cell division. Each new daughter cell retains part of the old cell wall and quickly rebuilds the missing part after cell division. Intermittent sexual reproduction has been reported in a few species; it is rapid and usually occurs in the dark, making it difficult to observe in natural conditions. The rate of reproduction is extremely rapid and approaches that of diatoms. Under optimal growth conditions, dense concentrations of dinophytes are produced quickly. Cell concentrations in these **blooms** are often so dense (1 million cells per liter) that they color the water red, brown, or green.

At night, dense blooms of luminescent forms (such as *Noctiluca* or *Ceratium*) become visible as a faint glow when disturbed by a ship's bow, a swimmer, or a wave breaking onshore. This luminescent glow is often highlighted by pinpoint flashes of larger crustaceans or ctenophores. This biological production of light, or **bioluminescence,** occurs in several species of dinophytes, some marine bacteria, and all major phyla of marine animals. Light is produced when luciferin, a relatively simple organic compound, is oxidized in the presence of the enzyme luciferase. The glowing reaction is a very efficient process, producing light but almost no heat. In some species of *Gonyaulax,* light production follows a daily rhythm, with maximal light output occurring just after midnight.

Along the East and Gulf coasts of the United States, dinophytes (*Ptychodiscus*) produce toxins that in bloom conditions are known as **red tides.** Red tides can cause high mortality in fishes and other

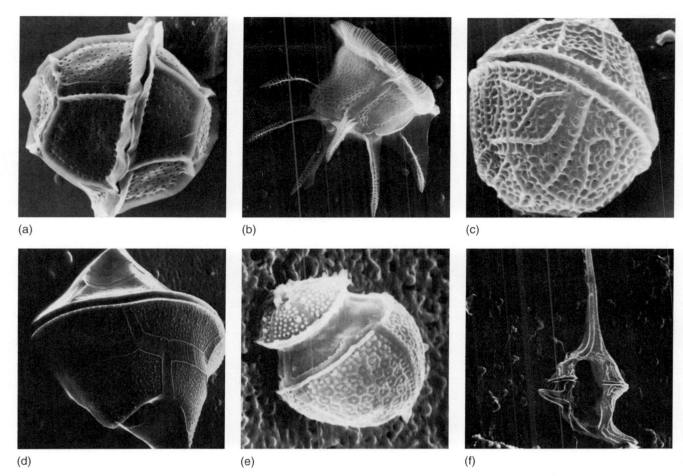

(a) (b) (c)

(d) (e) (f)

marine vertebrates. These dinoflagellate toxins either interfere with nerve functions, resulting in paralysis, or they irritate lung tissues of air-breathing vertebrates, including humans. Widespread mortality of coastal fish and mammal populations sometimes occurs after particularly intense red tides, fouling beaches and near-shore waters with their decomposing bodies.

Other species of dinophytes produce toxins that accumulate in animals (particularly shellfish) and render their flesh toxic. For instance, people who eat butter clams *(Saxodomus)* during the summer occasionally experience paralytic shellfish poisoning from **saxitoxin.** However, this toxin is actually produced by the dinophyte *Alexandrium,* which is ingested and concentrated by *Saxodomus.* Other toxic conditions caused by dinophytes include ciguatera fish poisoning (from *Gambierdiscus*) and neurotoxic shellfish poisoning (from *Ptychodiscus*).

A small group of specialized dinophytes (zooxanthellae) form a mutualism with a wide range of animals, including hermatypic corals, giant clams, anemones, and some flatworms. These symbiotic zooxanthellae account for substantial primary production in warm-water marine communities, and their role in reef-forming coral communities is described in chapter 10.

Figure 3.17

Scanning electron micrographs of some common marine dinophytes. All are 50–100 μm. *(a) Heteroaulacus, (b) Ceratocoup, (c) Alexandrium, (d) Protoperidinium, (e) Oxytoxum, (f) Ceratium.*
Courtesy F. Reid, E. Venrick, and Scripps Institution of Oceanography.

Table 3.3
Size Ranges of the Major Groups of Marine Phytoplankton

| | Cyanobacteria | Chrysophyta | | | Dinophyta | Chlorophyta |
		Diatoms	Silicoflagellates	Coccolithophores		
Picoplankton	+	+	+	+		+
Ultraplankton	+	+	+	+		+
Nannoplankton		+	+	+	+	+
Microplankton		+			++	

Adapted from Platt and Li 1986.

Other Phytoplankton

With sampling and microscopic techniques always improving, small phytoplankton from several other taxonomic groups are being recognized as important contributors to the trophic systems of many marine communities. Members of two additional classes of Chrysophyta and three unicellular classes of the division Chlorophyta are sometimes found in filtered samples of coastal seawater. Chlorophytes are much more common in fresh water, and that is where most of the research on this group is concentrated. Because most of the identified marine unicellular chlorophytes have been obtained from estuaries and coastal waters, freshwater origins for many of the species found in seawater samples is likely.

Table 3.3 summarizes the size distribution of the major groups of marine phytoplankton. Because many groups are extremely small, it is difficult to collect them and evaluate their relative roles in the total marine economy.

Special Adaptations for a Planktonic Existence

The evolutionary success of all phytoplankton hinges on their ability to obtain sufficient nutrients and light energy from the marine environment. Phytoplankton cells must be widely dispersed in their seawater medium to increase their utilization of dissolved nutrients, yet they must remain in the relatively restricted photic zone to absorb sufficient sunlight. These opposing conditions for successful planktonic existence have established some fundamental characteristics to which all phytoplankton and, indirectly, all other marine life have become adapted.

Size

One of the most characteristic features of all phytoplankton is their size. Almost without exception, they are microscopic, which suggests that a strong selective advantage accompanies smallness in phytoplankton. Why? In contrast to land plants, phytoplankton are constantly bathed in a medium that not only provides nutrients and water but also carries away waste products. Exchange of these materials in a fluid medium is accomplished by diffusion in either direction across the cell membrane.

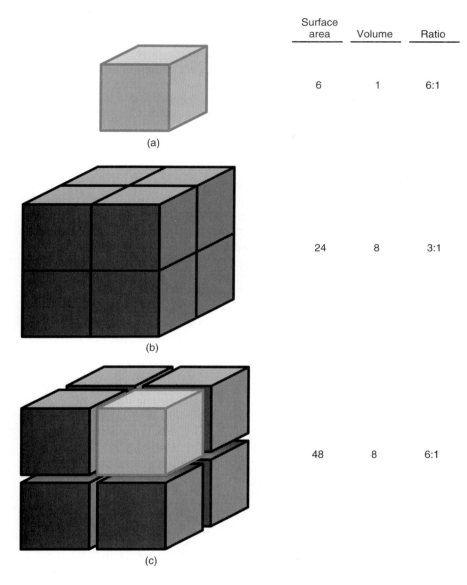

	Surface area	Volume	Ratio
(a)	6	1	6:1
(b)	24	8	3:1
(c)	48	8	6:1

Figure 3.18

With increasing size *(a, b)*, the ratio of surface area to volume decreases unless the larger structure *(c)* remains subdivided so that the interior surfaces are exposed.

The quantity of materials required by the cell depends on factors such as the rate of photosynthesis and growth. But if these and other variables are held constant, the basic material requirements of the cell are nearly proportional to the size or, more precisely, to the volume of the cell. However, the ability of the cell to satisfy its material requirements is not a function of the volume but of the extent of cell surface across which the materials can diffuse. Thus, the ratio of cell surface area to cell volume becomes quite important. Those cells with higher surface-area-to-volume ratios achieve an advantage in the competition to enhance diffusive exchange between their internal and external fluid environments (figure 3.18).

Reduction of cell size is an effective and widespread means of achieving high surface-area-to-volume ratios, but there are other means. Many phytoplankton cells have evolved complex shapes that

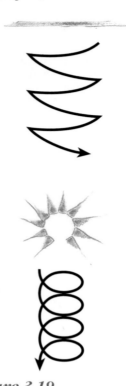

Figure 3.19

Sinking patterns of the elongate diatom *Rhizosolenia* (top) and the spiral chain-forming diatom *Asterionella* (bottom).

increase the surface area while adding little or nothing to the volume. Cell shapes resembling ribbons, leaves, or long bars and cells with bristles or spines are all common mechanisms to increase the amount of surface area relative to volume and thus increase diffusive exchange and frictional resistance to sinking. Cell vacuoles filled with seawater are common in diatoms. These cells are large, but the actual volume of protoplasm requiring sustenance is only a fraction of the total volume of the cell.

Sinking

Phytoplankton, with their heavy cell walls, are generally a bit denser than seawater and tend to sink away from surface waters and sunlight. The problem for phytoplankton is not to float, for floating would create intolerable crowding at the sea surface. Instead, phytoplankton need to slow their sinking rates so that a small fraction of any reproducing cell line has a few members carried upward by turbulent mixing even as most continue their slow downward slide through the photic zone.

Phytoplankton exhibit many adaptations that slow the sinking rate and prolong their trip through the photic zone. One of the most effective adaptations is to increase the frictional resistance to their passage through water by increasing the surface-area-to-volume ratio. Reduced cell sizes increase the ratio as does the production of horns, wings, or other cellular projections.

Other cells reduce their sinking rates with complex cell or chain shapes that trace zigzag or long spiral paths down through the water column. *Rhizosolenia* and *Asterionella,* shown in figure 3.19, demonstrate these adaptations. *Asterionella* forms long, curved chains of cells that spiral slowly through the water. *Eucampia, Chaetoceros,* and many other diatoms form similar twisted chains of cells. The asymmetrically pointed ends of *Rhizosolenia* create a "falling-leaf" pattern that prolongs its stay in the photic zone.

Adaptations for reducing the sinking rate are not limited to structural variations. Mechanisms that reduce the average cell density by "lightening the load" are also evident in some phytoplankton. Planktonic diatoms generally produce thinner and lighter frustules than do benthic diatoms. *Ditylum,* for example, is capable of excluding higher-density ions (calcium, magnesium, and sulphate) from its cell fluid and replacing them with less-dense ions. In addition, the production and storage of low-density fats and oils also help slow the rate of sinking.

Oscillatoria and some other planktonic cyanobacteria have evolved relatively sophisticated internal gas-filled vesicles to provide flotation. The walls of these vesicles are constructed of small protein units that can withstand outside water pressures experienced anywhere within the photic zone.

Adjustments to Unfavorable Environmental Conditions

Plankton have little or no capability of large-scale horizontal propulsion and must depend on the ocean's surface currents for dispersal. All of the adaptive features discussed above that extend the residence time of plankton in the horizontally moving surface currents also serve to increase their geographical distribution.

For protection, long spines and horns make phytoplankton relatively undesirable to herbivorous grazers. There is some evidence to suggest that copepods, for instance, prefer nonspiny diatoms to spiny ones. Slimy, gelatinous masses that sometimes surround clumps or chains of diatom cells also discourage some grazers. Spines, cell chaining, and cell elongation all may be economical methods of increasing apparent cell size to reduce mortality.

The optimal growth period for phytoplankton in nonupwelling temperate and polar seas is restricted by reduced sunlight in winter and limited nutrient supplies in summer. Faced with the prospect of weeks or months with reduced photosynthesis, phytoplankton in these regions have limited options. Some move, some switch to other energy sources, and some simply persist until conditions improve. The first strategy does not generally apply to diatoms, but motility, limited as it is, is extremely important to flagellated cells. A swim of merely one or two cell lengths is often sufficient to place the cell away from its excreted wastes and into an improved nutrient supply. Toxins of dinophytes also serve to discourage predation by herbivores and sometimes inadvertently improve their own nutrient supply by causing extensive fish kills and thus accelerating the renewal of crucial nutrients.

Strictly photosynthetic organisms must rely on stored lipids or carbohydrates for their short-term energy needs. When that source is depleted, some phytoplankton still have alternatives. Some species can improve their ability to harvest light by producing more chloroplasts that contain photosynthetic enzymes and pigments or by moving those chloroplasts closer to cell edges. Other species can absorb dilute but energy-rich dissolved organic material from surrounding seawater to tide them over. When these strategies have been exhausted, many diatoms produce dormant **cysts,** capsules that have reduced metabolic activity and increased resistance to environmental extremes (figure 3.20). It is likely that many near-shore species of dinophytes also produce dormant stages during periods of unfavorable growth conditions. With the return of improved growing conditions, these dormant cells germinate and commence photosynthesis and growth. At this point, the growing phytoplankton populations come under the regulatory influence of complex physical and biological factors considered in chapter 5.

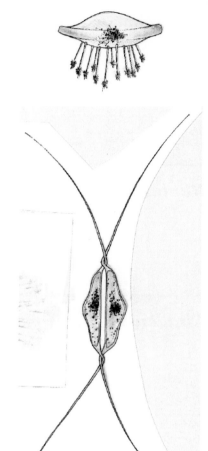

Figure 3.20

Inactive resistant stages of two species of *Chaetoceros.* (See figure 3.14b for the active growth form.)

Summary Points

- Living marine organisms include several hundred thousand species. To cope with this diversity, the taxonomic system of classifying and organizing groups of living organisms is used as a framework to begin our introduction to marine organisms.
- Marine phytoplankton consist mostly of species from three divisions, the Cyanobacteria, Chrysophyta, and Dinophyta, distributed in two kingdoms. Each phytoplankton group exhibits a unique combination of cell shapes, cell wall materials, photosynthetic pigments, and modes of locomotion.

- Cyanobacteria, coccolithophores, and silicoflagellates account for most of the smaller, less well-known picoplankton and ultraplankton; diatoms and dinophytes constitute the larger and more familiar nannoplankton and microplankton groups.
- Phytoplankton are microscopic and unicellular, and they generally exist throughout much of the photic zone. Most reproduce rapidly by asexual cell division. Their small size and complex cell shapes promote high surface-area-to-volume ratios that accelerate exchange of nutrients and waste materials as well as increase the cells' frictional resistance to sinking.

Review Questions

1. Diatoms exhibit many adaptations that function to slow their sinking rate. Why is it critical for diatoms to decrease their sinking rates?
2. What adaptations do marine phytoplankton possess that permit them to decrease their sinking rate?
3. What adaptations do dinophytes possess that permit them to succeed in warm waters where diatoms are less common?

Challenge Questions

1. Discuss some of the advantages and disadvantages inherent in grouping living organisms into the classification schemes outlined in this chapter.
2. Why do thriving populations of phytoplankton tend to be dispersed throughout the photic zone rather than crowded together at the sea surface?
3. Summarize the selective advantages for extremely small size in some phytoplankton groups. Why don't these advantages hold true for most seaweeds?

Suggestions for Further Reading

Books
Boney, A. D. 1975 *Phytoplankton.* Baltimore: University Park Press.

Humm, H. J., and S. R. Wicks. 1980. *Introduction and guide to the marine blue-green algae.* New York: Wiley.

Peterson, M. N. A., ed. 1993. *Diversity of Oceanic life: an evaluative review.* New York: Center for Strategic and International Studies.

Round, F. E., R. M. Crawford, and D. G. Mann. 1990. *The diatoms.* New York: Cambridge University Press.

Taylor, F. J. R., ed. 1987. *The biology of dinoflagellates.* London, England: Blackwell.

Articles
Fryxell, G. A. 1983. New evolutionary patterns in diatoms. *Bioscience* 33:92–98.

Hargraves, P. E., and F. W. French. 1983. Diatom resting spores: Significance and strategies. In: *Survival strategies of the algae,* ed. G. A. Fryxell, 49–68. New York: Cambridge University Press.

Krogmann, D. W. 1981. Cyanobacteria (blue-green algae)—Their evolution and relation to other photosynthetic organisms. *Bioscience* 31:121–24.

Margulis, L., D. Chase, and R. Guerrero. 1986. Microbial communities. *Bioscience* 36:160–70.

Marshall, H. G. 1976. Phytoplankton density along the eastern coast of U.S.A. *Marine Biology* 38:81–89.

Pickett-Heaps, J. 1976. Cell division in eukaryotic algae. *Bioscience* 26:445–50.

Platt, T., and W. K. W. Li, eds. 1986. Photosynthetic picoplankton. *Canadian Bulletin of Fisheries and Aquatic Sciences* 214:583.

Smayda, T. J. 1970. The suspension and sinking of phytoplankton in the sea. *Oceanography and Marine Biology Annual Review* 8:353–414.

Smith, W. O., Jr., and D. M. Nelson. 1986. Importance of ice edge phytoplankton production in the southern ocean. *Bioscience* 36:251–57.

Steidinger, K. A., and K. Haddad. 1981. Biologic and hydrographic aspects of red tides. *Bioscience* 31:814–19.

Phytoplankton
http://www.pmel.noaa.gov/bering/pages/env_phyt.html
The FOCI Project Office, NOAA/Pacific Marine Environment
Laboratory. Bering Sea and North Pacific Ocean.
http://www.pmel.noaa.gov/bering/ "Phytoplankton are free-
floating flora which convert inorganic compounds into complex
organic compounds. This process of primary productivity supports
the pelagic food chain" (Brittanica Online 1997).

Provasoli-Guillard National Center for Culture of Marine
Phytoplankton
http://ccmp.bigelow.org/
Bigelow Laboratory for Ocean Sciences, West Boothbay Harbor, ME
04575. Living stock culture collections for marine phytoplankton.
Includes searchable database and ordering and culturing
information.

Phytoplankton Pigment Concentration from U.S. Coast Guard
http://satori.gso.uri.edu/satlab/
Includes General Description of Plankton, Phytoplankton,
Photosynthesis and Respiration, Productivity, Seasonal
Phytoplankton Cycle, Upwelling and Primary Production, and much
more.

Light and Nutrients: Where Phytoplankton Live in the Water column
and Why
**http://www-ocean.tamu.edu/Quarterdeck/QD5.2/pariente-
light.html**
First published in *Quarterdeck* magazine, published by the
Department of Oceanography, Texas A&M University. "Sunlight and
nutrient concentrations determine where phytoplankton can
survive."

Phytoplankton
http://www.antdiv.gov.au/aad/p&p/is/v4/plankton.html
© Australian Antarctic Division (Commonwealth of Australia)
http://www.antdiv.gov.au/

Ocean Color Viewed from Space
http://www.athena.ivv.nasa.gov/curric/oceans/ocolor/
Provides links to: Satellite data, Sea level field analysis, Moored ocean
buoy data, Drifting ocean buoy data, Climatologies, and more.

Algae World
http://ghs.ssd.k12.wa.us/depts/ms/algae/index.html
Full, general discussion of seaweeds.

CHAPTER

4

Marine Plants

Benthic marine plants are probably more familiar to seashore observers than are most phytoplankton. They are more conspicuous because they are macroscopic, multicellular organisms often large enough to pick up and examine. As do phytoplankton, these organisms need sunlight for photosynthesis and are confined to the photic zone. But the additional need for a hard substrate on which to attach limits the distribution of benthic plants to that narrow fringe around the periphery of the oceans where the sea bottom is above the bottom of the photic zone (the inner shelf of figure 1.37). Some benthic plants inhabit intertidal areas and must confront the many tide-induced stresses that affect their animal neighbors. (These will be discussed in chapter 8.) Their restricted near-shore distribution limits the global importance of benthic plants as primary producers in the marine environment. But within the near-shore communities in which they live, they play a major role as first trophic level organisms.

The abundant plant groups so familiar on land—ferns, mosses, and seed plants—are poorly represented or totally absent from the sea. Instead, most marine plants belong to two divisions, Phaeophyta and Rhodophyta, that are almost completely limited to the sea. Two other divisions, Chlorophyta and Anthophyta, are found most commonly in fresh water and on land; yet they are important members of some shallow coastal marine communities. The characteristics of these divisions are summarized in table 4.1.

The Seaweeds

By far, the majority of large, conspicuous forms of attached marine plants are seaweeds. The term *seaweed* is often used to indicate a group of large marine plants attached to the bottom in relatively shallow coastal waters. The term is used here in a more restrictive sense, referring only to macroscopic members of the plant divisions Chlorophyta (green algae), Phaeophyta (brown algae), and Rhodophyta (red algae) (table 4.1).

Seaweeds are abundant on hard substrates in intertidal zones and commonly extend to depths of 30 to 40 m. In clear tropical seas, some species of red algae thrive at depths as great as 200 m, and one species has been reported as deep as 268 m in the Bahamas. Many seaweeds tolerate or even require extreme surf action on exposed rocky intertidal outcrops, where they are securely fixed to the solid substrate. Where they are abundant, seaweeds can greatly influence local environmental conditions for other types of shallow-water marine life by providing food, protection from waves, shade, and sometimes a substrate on which to attach.

Table 4.1
Major Divisions of Marine Plants and Their General Characteristics

Division (Common Name)	Approx. No of Living Species	Percent of Species Marine	General Size and Structure	Photosynthetic Pigments	Storage Products	Habit
Phaeophyta (brown algae)	1,500	99.7	Multicellular, macroscopic	Chlorophyll *a, c* Xanthophylls Carotenes	Laminarin and others	Mostly benthic
Rhodophyta (red algae)	4,000	98	Unicellular and multicellular, mostly macroscopic	Chlorophyll *a* Carotenes Phycobilins	Starch and others	Benthic
Chlorophyta (green algae)	7,000	13	Unicellular and multicellular, microscopic to macroscopic	Chlorophyll *a, b* Carotenes	Starch	Mostly benthic
Anthophyta (flowering plants)	250,000	0.018	Multicellular, macroscopic	Chlorophyll *a, b* Carotenes	Starch	Benthic

Adapted from Seagel et al 1980, Dawson 1981, and Kaufman et al. 1989.

Photosynthetic Pigments

Each seaweed division is characterized by specific combinations of photosynthetic pigments that are reflected in their color (figure 4.1) and in the common name of each division (refer to table 4.1). The bright, grass-green color of green algae is due to the predominance of chlorophylls over accessory pigments. Green algae vary in structure from simple filaments (figure 4.2a) to flat sheets and diverse complex branching forms (figure 4.2b). They are usually less than half a meter long, but one species of *Codium* from the Gulf of California occasionally grows to 8 m in length. When compared with brown and red algae, the Chlorophyta have fewer marine species. But in some locations, their limited diversity is compensated with dense populations of individuals from one or two species.

The photosynthetic pigments of the Phaeophyta can sometimes be seen as a greenish hue. But more often the green of the chlorophyll is partially masked by the golden **xanthophyll** pigments, especially fucoxanthin, characteristic of this division. This blend of green and brown pigments usually results in a drab, olive-green color. Many of the larger and more familiar algae of temperate seas belong to this division. A number of species are quite large and are sometimes collectively referred to as **kelp** (figure 4.3). In temperate and high latitudes, these species usually dominate the marine benthic vegetation. Numerous smaller, less obvious brown algae are also common in temperate and cold waters as well as in tropical areas.

Red algae, with red and blue **phycobilin** pigments as well as chlorophyll, exhibit a wide range of colors. Some are bright green, and others are sometimes confused with brown algae. However, most red algae living below low tide range in color from soft pinks

Figure 4.1

(*a*) Calcareous red algae, *Jania,* in a small tide pool. (*b*) Intertidal rocks covered with three green alga *Ulva.* (*c*) A brown alga, *Fucus,* in the middle intertidal.

(*a*) and (*b*) Courtesy G. Dudley.

to various shades of purple or red. Red algae are as diverse in structure and habitat as they are in coloration, and they seldom exceed a meter in length.

Structural Features of Seaweeds

Seaweeds are not as complex as the flowering plants. Seaweeds lack roots, flowers, seeds, and true leaves. Yet, within these structural limitations, seaweeds exhibit an unbridled diversity of shapes, sizes, and structural complexity. Microscopic filaments of green and brown algae can be found growing side by side with encrusting forms of red algae and flat sheetlike members of all three divisions. Many of the larger members of all three seaweed divisions develop into mature plants with similar general forms, each consisting of a **blade,** a **stipe,** and a **holdfast.**

The Blade

The flattened, usually broad, leaflike structures of seaweeds are known as blades. Seaweed blades often exhibit a complex level of branching

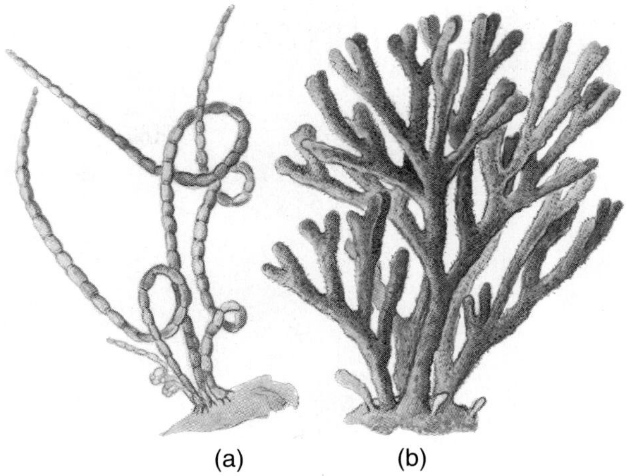

Figure 4.2
Two marine forms of green algae:
(a) Chaetomorpha, (b) Codium.

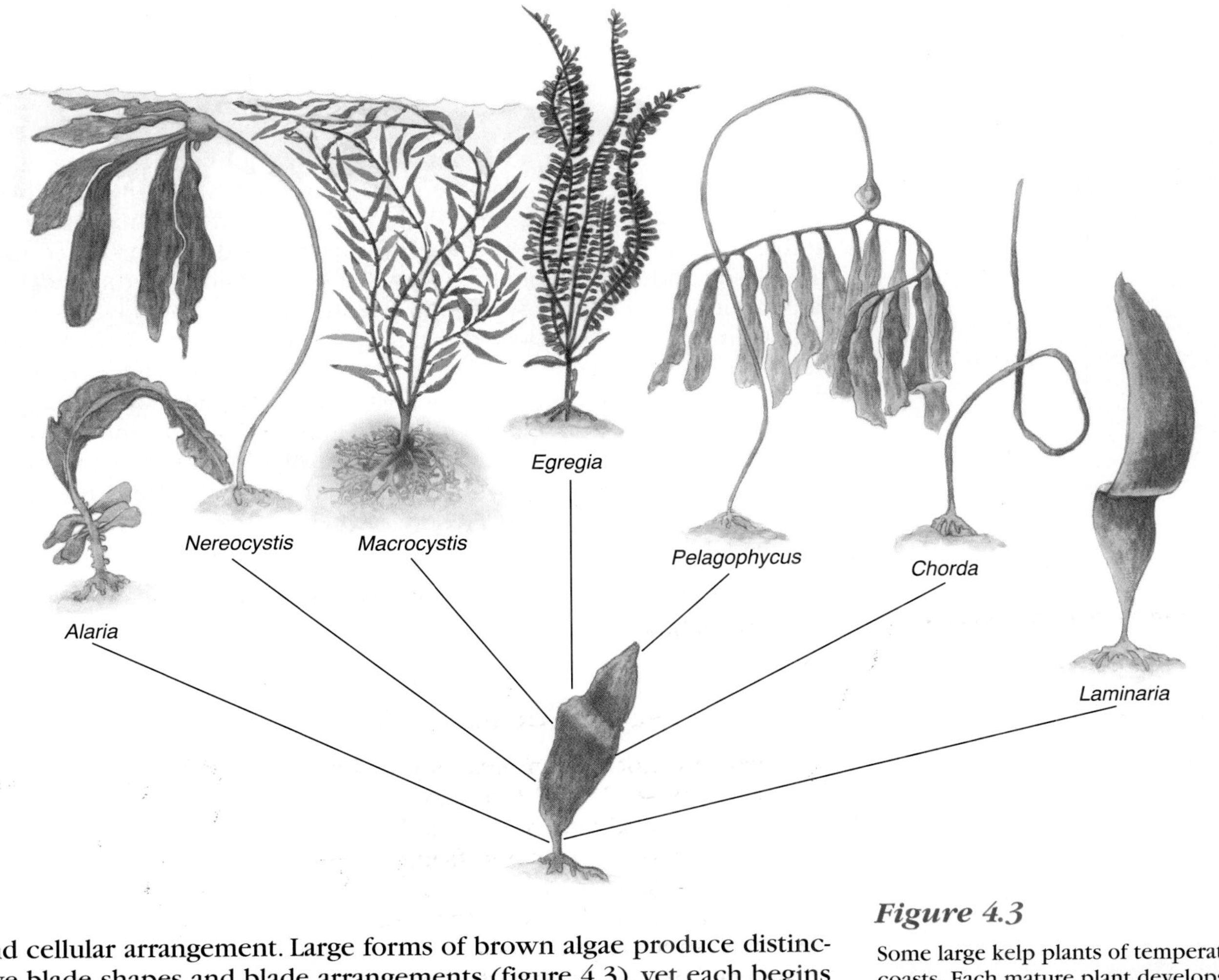

(a)   (b)

Alaria

Nereocystis   Macrocystis

Egregia

Pelagophycus

Chorda

Laminaria

Figure 4.3

Some large kelp plants of temperate coasts. Each mature plant develops from a young plant with a single flat blade.

and cellular arrangement. Large forms of brown algae produce distinctive blade shapes and blade arrangements (figure 4.3), yet each begins as a young plant with a single, unbranched, flat blade nearly identical to other young kelp plants.

The blades house photosynthetically active cells, but photosynthesis often occurs in the stipes and holdfasts as well. In cross section,

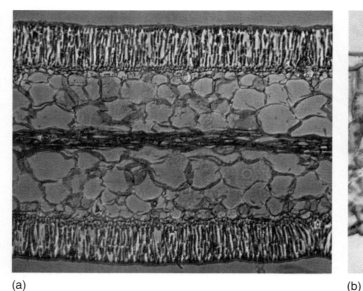

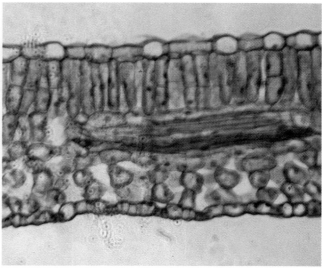

(a)   (b)

Figure 4.4

Cross sections of (*a*) a blade of a
typical marine alga, *Nereocystis,* and
(*b*) a flowering plant leaf. Note the
contrasting symmetry patterns.

Figure 4.5

A portion of the floating brown alga
Sargassum, with numerous small
pneumatocysts.
Courtesy D. Nelson.

seaweed blades (figure 4.4a) are structurally unlike the leaves of terrestrial plants (figure 4.4b). The cells nearer the surface of the blade are capable of absorbing more light and are photosynthetically more active than those cells near the center of the blade. "Veins" of conductive tissue and distinctions between the upper and lower surfaces are lacking in the blades of seaweeds. Because the flexible blades usually droop in the water or float erect, there is no defined upper or lower surface. The two sides of the seaweed blade are usually exposed equally to sunlight, nutrients, and water and are therefore equally capable of carrying out photosynthesis. Unlike seaweeds, the flowering plants (including sea grasses) exhibit an obvious asymmetry of leaf structure, with a dense concentration of photosynthetically active cells crowded near the upper surface (figure 4.4b). Below the upper surface is a spongy layer of cells separated by large spaces to enhance the exchange of carbon dioxide, which is often 100 times *less* concentrated in air than in seawater.

Pneumatocysts

Several large kelp species have gas-filled floats, or **pneumatocysts,** to lift the blades toward the sunlight at the surface. Pneumatocysts are filled with the gases most abundant in air: N_2, O_2, and CO_2. The pneumatocyst gases of some kelp plants also contain a few percent of carbon monoxide, CO. The CO is thought to be a by-product of metabolism. Again, there is a large diversity in size and structure. The largest pneumatocysts belong to *Pelagophycus,* the elkhorn kelp (figure 4.3). Each elkhorn kelp plant is equipped with a single pneumatocyst, sometimes as large as a basketball, to support six to eight immense drooping blades, each of which may be 1 to 2 m wide and 7 to 10 m long.

Sargassum produces numerous small pneumatocysts (figure 4.5). A few species of *Sargassum* lead a pelagic life afloat in the middle of the North Atlantic Ocean (the "Sargasso Sea"). In the Sargasso Sea,

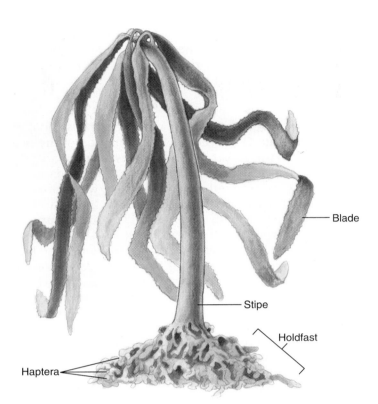

Figure 4.6

The northern sea palm *Postelsia* (Phaeophyta) is equipped with a relatively large stipe and a massive holdfast.

Blade

Stipe

Holdfast

Haptera

Sargassum produces large patches of floating plants that are the basis of a complex floating *Sargassum* community of crabs, fishes, shrimp, and other animals uniquely adapted to living among the *Sargassum.* Large masses of this plant community sometimes float ashore on the U. S. East and Gulf coasts, creating odor problems for beachgoers as the dying plants decompose. In the Sea of Japan, other species of attached intertidal *Sargassum* break off and also become free-floating for extended periods of time.

The Stipe

A flexible, stemlike stipe serves as a shock absorber between the wave-tossed upper parts of seaweeds and the securely anchored holdfast at the bottom. An excellent example is *Postelsia,* the sea palm (figure 4.6), which grows attached to rocks only in the most exposed, surf-swept portions of the intertidal zone. Its hollow, resilient stipe is remarkably well suited for yielding to the waves without breaking.

The blades of some seaweeds blend into the holdfast without forming a distinct stipe. In others, the stipe is very conspicuous and, occasionally, extremely long. The single long stipes of *Nereocystis, Chorda,* and *Pelagophycus* (figure 4.3) provide a kind of slack-line anchoring system and commonly exceed 30 m in length. The complex multiple stipes of *Macrocystis* (figure 4.3) are often even longer.

Special cells within the stipes of *Macrocystis* and a limited number of other brown and red algal species form conductive tissues strikingly similar in form to those present in stems of terrestrial plants. Radioactive tracer studies have shown that these cells transport the

Figure 4.7

Penicillus (Chlorophyta), with fine hairlike haptera for anchoring in loose sediments.

products of photosynthesis from the blades to other parts of the plant. In smaller seaweeds with photosynthetic stipes and holdfasts, the necessity for rapid, efficient transport through the stipe is minimal, and such internal transport is lacking.

The Holdfast

Holdfasts of the larger seaweeds often superficially resemble root systems of terrestrial plants. However, the basic function of the holdfast is to attach the plant to the substrate. The holdfast seldom absorbs nutrients for the plant as do true roots. Holdfasts are well adapted for getting a grip on the substrate and resisting violent wave shock and the steady tug of tidal currents and wave surges. The holdfast of *Postelsia* (figure 4.6), composed of many short, sturdy, rootlike **haptera,** illustrates one of several types found on solid rock.

Other holdfasts are better suited for loose substrates. The holdfast of *Macrocystis* has a large, diffuse mass of haptera to penetrate muddy or sandy bottoms and stabilize a mass of sediment for anchorage. The holdfast of *Penicillus* does the same thing on a much smaller scale, with many fine filaments embedded in sand or mud bottoms (figure 4.7).

A variety of small red algae are epiphytes and demonstrate special adaptations for attaching themselves to other marine plants. Figure 4.8 illustrates two common red algal epiphytes attached to a strand of surf grass. Using other marine plants as substrates for attachment is a common habit of many smaller forms of red algae.

Reproduction and Growth

Reproduction in seaweeds, as well as in most other plants, can be either sexual, involving the fusion of sperm and eggs, or asexual, relying on vegetative growth of new individuals (see figure 2.3). Some seaweeds reproduce both ways, but a few are limited to vegetative

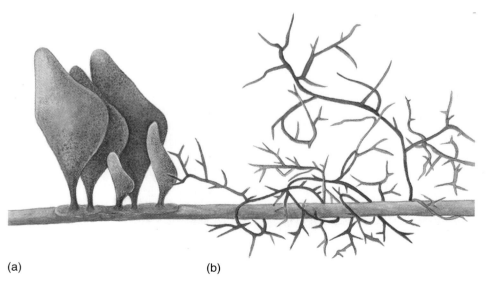

(a)   (b)

Figure 4.8

Two red algal epiphytes attached to a strand of surf grass, *Phyllospadix.* (*a*) *Smithora.* (*b*) *Chondria.*

reproduction only. The pelagic species of *Sargassum,* for instance, maintain their populations by an irregular vegetative growth followed by fragmentation. The dispersed fragments of *Sargassum* are capable of continued growth and regeneration for decades. Sexual reproduction is lacking in the pelagic species of *Sargassum* but not in the attached benthic forms of the same genus.

Much of the structural variety observed in seaweeds is derived from complex patterns of sexual reproduction, patterns that define the life cycles of seaweeds. For our purposes, these complex life cycles can be simplified to three fundamental patterns. The sexual reproduction examples of the first two types described here are not meant to cover the entire spectrum of seaweed life cycles but are used to develop a basic pattern that underlies the complexity and variation involved in sexual reproduction of seaweeds.

In the life cycle of most of the larger seaweeds, an alternation of **sporophyte** and **gametophyte** generations occurs. The green alga *Ulva* represents one of the simplest patterns of alternating generations (figure 4.9). This basic life cycle is, with minor modifications, common to plants but not to animals. The cells of the macroscopic sporophyte are diploid; that is, each cell contains two of each type of chromosome characteristic of that species. Some cells of the *Ulva* sporophyte undergo meiosis to produce single-celled, flagellated **spores.** As a result of meiosis, these spores contain only one chromosome of each pair present in the diploid sporophyte and are said to be haploid.

The spores are capable of limited swimming and then settle to the bottom. There they immediately germinate by a series of mitotic cell

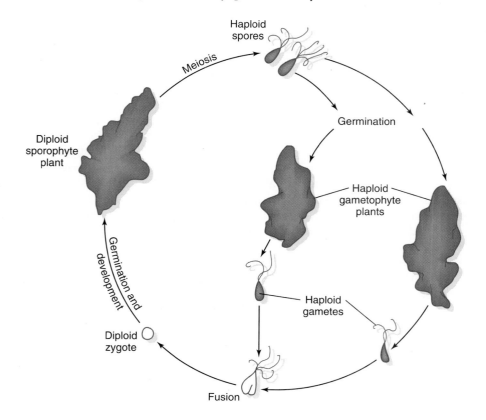

Figure 4.9

The life cycle of the green alga *Ulva,* alternating between diploid sporophyte and haploid gametophyte generations. Adapted from Dawson 1981.

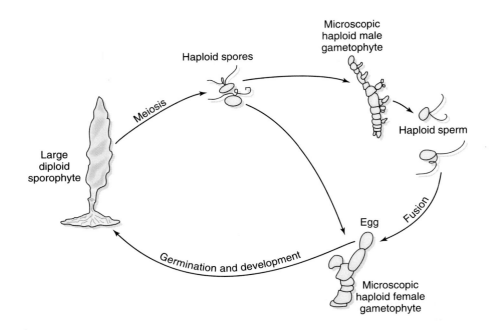

divisions to produce a large multicellular, still haploid, gametophyte. Cells of the gametophyte in turn produce flagellated, haploid gametes that are released into the water. When two gametes from different gametophytes meet, they fuse to produce a diploid, single-celled zygote. By repeated mitotic divisions, the zygote germinates and completes the cycle by producing a large sporophyte once again. In *Ulva,* the sporophyte and gametophyte generations are identical in appearance. The only structural difference between the two forms is the number of chromosomes in each cell; the diploid sporophyte has double the chromosomal complement of the haploid gametophyte cells.

The life cycles of numerous other seaweeds are characterized by a suppression of either the gametophyte or the sporophyte stage. In the green alga *Codium* and the brown alga *Fucus* the multicellular haploid generation is completely absent. The only haploid stages are the gametes. In other large brown algae, the gametophyte stage is reduced. The life cycle of *Laminaria* is similar to that of most other large kelp plants and serves as an excellent generalized example of seaweeds with a massive sporophyte that alternates with a diminutive gametophyte (figure 4.10). Special cells (called **sporangia**) on the blades of the diploid sporophyte undergo meiosis to produce several flagellated, microscopic spores. These haploid spores swim to the bottom and quickly attach themselves. They soon germinate into very small gametophytes composed of several cells. The female gametophyte produces large, nonflagellated eggs. The egg cells are fertilized in place on the female gametophyte by flagellated male gametes, the sperm cells produced by the male gametophyte. After fusion of the gametes, the resulting zygote germinates to form another large sporophyte.

The spores of green algae are characterized by four flagella; each gamete has two flagella that are equal in length and project from one end of the cell. The flagellated reproductive cells of brown algae

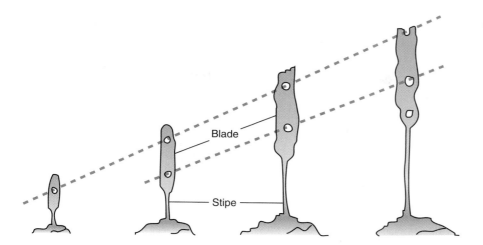

Figure 4.11

Generalized growth pattern of a kelp. Punched holes and blue lines indicate the pattern of blade elongation.
Adapted from Mann 1972.

Blade

Stipe

always have two flagella of unequal lengths, and they insert on the sides of the cells rather than at the ends.

Red algae lack flagellated reproductive cells and are dependent on water currents to transport the male gametes to the female reproductive cells. The most common life cycle of red algae has three distinct generations, somewhat reminiscent of the reproductive cycle outlined for *Ulva* (figure 4.9). A diploid sporophyte produces haploid spores that germinate into haploid gametophytes. Instead of producing a new sporophyte, however, the gametes from the gametophytes fuse and develop into a third phase unique to red algae, the **carposporophyte.** The carposporophyte then produces **carpospores** that develop into sporophytes, and the cycle is completed.

The development of a large multicellular seaweed from a single microscopic cell is essentially a process of repeated mitotic cell divisions. Subsequent growth and differentiation of these cells produce a complex plant with many types of cells, each specialized for particular functions. Once the plant is developed, additional cell division and growth occur to replace tissue lost to animal grazing or wave erosion. However, such cell division is commonly restricted to a few specific sites within the plant that contain **meristematic tissue** with a capacity for further cell division. These meristems frequently occur at the upper growing tip of the plant. In kelp plants and some other seaweeds, additional meristems situated in the upper and lower portions of the stipe provide for elongation of the stipe and blades. The meristematic activity of a cell layer at or near the surface of some kelp stipes provides lateral growth to increase the thickness of the stipe. The stipes of a few perennial species of kelp, including *Pterygophora* and *Laminaria,* retain evidence of this secondary lateral growth as concentric rings that resemble the annual growth rings of trees.

In the spring during periods of rapid growth, the rate of stipe elongation in *Nereocystis, Pelagophycus,* and *Macrocystis* often exceeds 30 cm per day. Growth rate estimates of some kelp plants on the east coast of Canada indicate that the kelp blades resemble moving belts of plant tissue (figure 4.11), growing at the base and eroding at the tips. At any one time, the plant itself (the standing crop) represents as little as 10% of the total plant material produced during a year.

Figure 4.12

Three common sea grasses from different marine climatic regions: (*a*) turtle grass, *Thalassia;* (*b*) eel grass, *Zostera;* (*c*) surf grass, *Phyllospadix.*

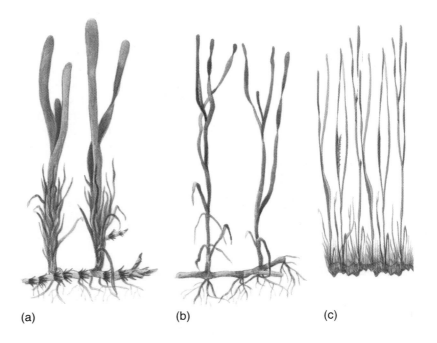

(a) (b) (c)

Division Anthophyta

Marine flowering plants are abundant in localized areas along some seashores and in backwater bays and sloughs. Sea grasses are exposed to air only during very low tides, whereas salt marsh plants and mangroves are emergent and seldom completely inundated by seawater. These plants represent a secondary adaptation to the marine environment by a few species of a predominantly terrestrial plant group—the flowering plants (division Anthophyta). In contrast to the seaweeds, flowering plants have leaves, stems, and roots, with water- and nutrient-conducting structures running through all three.

Twelve common genera, including about 50 species of sea grasses, are dispersed around coastal waters of the world. Half of seagrass species are restricted to the tropics and subtropics and are seldom found deeper than 10 m. Four common genera in the United States are *Thalassia, Zostera, Halodule,* and *Phyllospadix. Thalassia,* or turtle grass (figure 4.12a), is common in quiet waters along most of the Gulf Coast from Florida to Texas. *Zostera marina,* or eel grass (figure 4.12b and figure 9.12), is widely distributed along both the Atlantic and Pacific coasts of North America. *Zostera* normally inhabits relatively quiet, shallow waters but is occasionally found as deep as 50 m. Surf grass, *Phyllospadix* (figure 4.12c), is found on both sides of the North Pacific and inhabits the lower intertidal and shallow subtidal rocks that are subjected to considerable wave and surge action. *Halodule* prefers sandy areas with lower salinity.

Most sea grasses produce horizontal stems, or **rhizomes,** that attach the plants in soft sediments or to rocks (figure 4.12). From the buried rhizomes, many erect leaves develop to form thick green masses of vegetation. These plants are a staple food for near-shore marine animals and migratory birds. Densely matted rhizomes and roots also accumulate nutrients and organic debris within sediments to further alter the living conditions of the area.

Many sea grasses reproduce vegetatively by sprouting additional vertical leaves from the lengthening horizontal rhizome or from seeds produced in incomplete flowers. The purpose of most showy flowers on land plants is to attract insects or birds so that **pollen** grains are transferred from one flower to another and cross-fertilization occurs. Pollen grains contain the plant's sperm cells, but submerged sea grasses cannot utilize animals or the wind to transport their pollen. In all sea grasses, pollination occurs underwater, with water currents responsible for pollen transport.

Some sea grasses, including *Zostera,* produce threadlike pollen grains about 3 mm long (about 500 times longer than their cargo, the microscopic chromosome-carrying sperm cells). After release, these elongated pollen grains become ensnared on the **stigma,** the pollen-receptive structure of the female flower, and fertilization occurs if there is a species match between the pollen grain and the stigma. The problem of achieving a species match is handled by turtle grass in a manner somewhat reminiscent of broadcast spawners (p. 211). The small round pollen grains of turtle grass are released in a thread of sticky slime. When the slime lands on the appropriate stigma (also covered with a surface film of slime), the two slime layers combine to produce a firm bond between the pollen grain and the stigma, and fertilization follows. This two-component adhesive acts like epoxy glue to produce a strong bond after the separate components are mixed. It also provides a mechanism for selecting between compatible and foreign types of pollen grains. Only on contact with pollen of the same species will the stigma-pollen bond be formed. Foreign pollen grains do not adhere and are washed away, possibly to try again on another plant.

The seeds of each type of sea grass are well suited to their environment. Eel grass seeds drop into the mud and take root near the parent plant. In contrast, the fruits of *Thalassia* may float for long distances before releasing their seeds in the surf. The fruits surrounding individual seeds of *Phyllospadix* are equipped with bristly projections. When shed into the surf, these bristles snag small branches of algae, and the seeds germinate in place.

Several other species of flowering plants often exist partially submerged on bottom muds of quiet coastal salt marshes. These plants are usually situated so that their roots are periodically, but not constantly, exposed to tidal flooding. They are terrestrial plants that have evolved various degrees of tolerance to excess salts from sea spray and seawater. Some even have special structural adaptations for their semimarine existence. The cord grass, *Spartina* (figure 9.9a), for example, actively excretes excess salt through special two-celled salt glands on its leaves. Even so, several species of *Spartina* have higher experimental growth and survival rates in fresh water than in seawater. This difference strongly suggests that the salt marsh does not provide optimal growth conditions for *Spartina,* even though the salt marsh is its natural habitat. Competition with other land and freshwater plants may have forced *Spartina* and other salt-tolerant species into the restricted areas of the salt marshes.

Figure 4.13

Dense mangal thicket lining a tidal
channel.

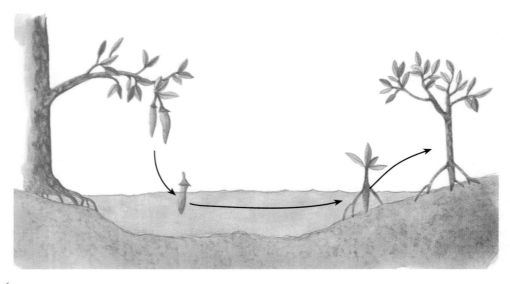

Figure 4.14

Germination cycle of a mangrove
seedling.

Several species of shrubby to treelike plants, the mangroves, create
dense thickets of tidal woodlands known as **mangals** (figure 4.13).
Members of these mangal communities are supported on their muddy
substrate by numerous prop roots that grow down from branches
above the water.

Mangroves are seed plants, and their pattern of development illustrates well a series of adaptations needed to exist on muddy tropical
shores (figure 4.14). Red mangroves *(Rhizophora)* produce seeds that
germinate while still hanging from the branches of the parent tree. As

the seedlings develop and grow longer, their bottom ends become heavier. When the seedlings eventually drop from the parent plant into the surrounding water, they float upright at the water's surface, are dispersed by winds or tides, and finally settle in shallow water on muddy shoreline. There, the seedlings promptly develop small roots to anchor themselves and continue to mature. The resulting tangle of roots traps additional sediments and increases the structural complexity of mangal communities. Birds, insects, snails, and other terrestrial animals occupy the upper leafy canopy of the mangroves, and a variety of fishes, crustaceans, and mollusks live on or among the root complex growing down into the mud.

Geographical Distribution

The interplay of a multitude of physical, chemical, and biological variables influences and controls the distribution of marine plants on a local scale. For instance, on an exposed rock in the lower intertidal zone on the Oregon coast, *Postelsia* may thrive, but 10 m away, the conditions of light, temperature, nutrients, tides, surf action, and substrate may be such that *Postelsia* cannot survive. Yet, on an oceanwide scale, only a few factors seem to control the presence or absence of major groups of seaweeds. Significant among these are water and air temperature, tidal amplitude, and the quality and quantity of light.

With these factors in mind, we can make a few generalizations concerning the geographical distribution of benthic plants. In marked contrast to the impoverished seaweed flora of the Red Sea, the tropical western coast of Africa, and the western side of Central America, seaweeds thrive in profusion along the coasts of southern Australia and South Africa, on both sides of the North Pacific, and in the Mediterranean Sea. The West Coast of the United States is somewhat richer in seaweed diversity than is the East Coast. From Cape Cod northward, the East Coast is populated with subarctic seaweeds. South of Cape Cod, the effects of the warm Gulf Stream become more and more evident, until a completely tropical flora is encountered in southern Florida.

Red algae are not rare in cold-water regions but are more abundant and conspicuous in the tropics and subtropics. Calcareous forms of red algae (and some brown and green as well) are characterized by extensive deposits of calcium carbonate ($CaCO_3$) within their cell walls (figure 4.1a). The use of calcium carbonate as a skeletal component by warm-water marine algae is apparently related to the decreased solubility of $CaCO_3$ in water at higher temperatures. In the tropics, plants expend less energy to extract $CaCO_3$ from the water, and here coralline red algae contribute to the formation and maintenance of coral reefs. Encrusting coralline algae grow over coral rubble, cementing and binding it into a mass that can resist the pounding of heavy surf. Some Indian Ocean "coral" reefs completely lack coral animals and are constructed and maintained entirely by coralline algae. The few calcareous forms of green algae that exist are also limited to warm water and play a large role in the production of $CaCO_3$ on some coral reefs.

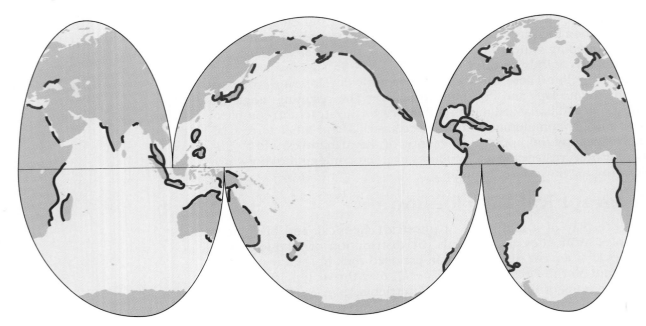

Figure 4.15

Distribution of salt marshes (green) and mangals (blue).

Kelp are temperate to cold-water species, with few tropical representatives. Extensive kelp beds are seldom found nearer to the equator than 30° N or 30° S latitude. The large kelps are especially abundant in the North Pacific. Cord grass and other grassy salt marsh plants generally are also found outside warm tropical or subtropical regions. In tropical and subtropical protected mud-bottom habitats, salt marshes give way to extensive mangal thickets. Most marine flowering plants are limited to soft bottoms in tidelands and shallow coastal waters and are discussed further in chapter 9.

Plant-Dominated Marine Communities

Some of the larger forms of benthic marine plants flourish in such profusion that they dominate the general biological character of their communities. Such community domination, a common feature of land plants, is exceptional in the sea. Away from the near-shore habitats occupied by benthic plants, the microscopic phytoplankton prevail as the major primary producers, and it is their consumers that define the visual character of their pelagic communities. In the near-shore fringe, however, mangals, salt marshes, sea grasses, and kelp beds thrive where the appropriate bottom conditions, light, and nutrients exist.

Mangals and salt marshes are emergent plant communities; most of the plant growth occurs above the sea surface. Both occupy protected muddy habitats. Mangals form warm-climate plant communities and seldom exist beyond 30° N and 30° S latitudes (figure 4.15). Because the leafy portions of these plants are above the water level, few marine animals graze directly on mangrove plants. Instead, leaves falling from these plants into the quiet waters surrounding their roots provide an important energy source for the detritus-based food webs of these communities.

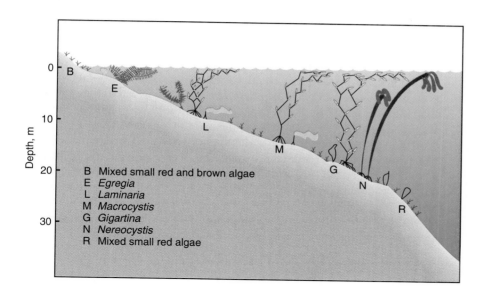

0

10

Depth, m

20

B Mixed small red and brown algae
E *Egregia*
L *Laminaria*
M *Macrocystis*
G *Gigartina*
N *Nereocystis*
R Mixed small red algae

30

B

E

L

M

G

N

R

Figure 4.16

Simplified cross section of a typical California kelp bed illustrating some of the complex, jungle-like, vertical structures created by the kelp bed components.

In cooler climates, mangals are replaced by a variety of salt marsh plants that also contribute heavily to detritus production in their protected environments as well as in nearby bays and estuaries. Some feature extensive stands containing several species of emergent grasses, particularly the various species of *Spartina*. At slightly higher elevations, these grasses give way to succulents (*Salicornia* and *Suaeda*), a variety of reeds and rushes, and the brush and smaller trees of the local woodland. These lush pastures are extremely productive and harbor a unique assemblage of organisms, including commercially important shellfish and finfishes. Yet as large urban centers develop near them, they have become popular sites for waste dumping, recreation, dredging and filling, and other detrimental uses. The degradation of salt marshes is a serious and worldwide problem that becomes more severe as human populations expand and place more pressure on these fragile habitats.

Well-developed sea grass and kelp beds seldom extend above the low tide line; they are submergent plant communities. Both sea grass and kelp beds abound with herbivores that graze directly on these plants and in turn become prey for higher trophic levels. The cool-water kelp plants form extensive layered forests of mixed species in both the Atlantic and Pacific oceans. The blades of the larger *Macrocystis, Laminaria,* or *Nereocystis* form the upper canopy and the basic structure of these plant communities. Shorter members of other brown algal and red algal species provide secondary understory layers and create a complex, three-dimensional habitat with a large variety of available niches (figure 4.16). The maximal depth of these kelp beds, usually 20 to 30 m, is limited by the light available for the young growing sporophyte. The larger kelp plants, with their broad blades streaming at the sea surface, create substantial drag against currents and swells and are susceptible to storm damage by waves and surge. Cast on the shore, these decaying plants are a major food source for beach scavengers.

Summary Points

- Benthic marine plants are subject to quite different environmental limitations than are phytoplankton. Most are attached to the sea bottom or to other organisms, and several achieve large sizes. The dominant groups are green algae (Chlorophyta), brown algae (Phaeophyta), red algae (Rhodophyta), and flowering plants (Anthophyta). Benthic plants are limited to the very narrow periphery of the sea, and therefore they produce less plant material on a global scale than do phytoplankton.
- The seaweeds (red, green, and brown algae) are usually characterized by a holdfast, stipe, and one or more blades. The few species of flowering plants that tolerate complete submergence in seawater are mostly sea grasses living on shallow soft sediments. In cooler climates, salt marsh plants and brown algae tend to dominate, with red algae and mangals increasing in importance in subtropical and tropical coastal areas.

Review Questions

1. A life cycle consisting of alternating gametophyte and sporophyte generations is characteristic of almost all plants. How do the basic features of that life cycle differ among the different divisions of seaweeds?
2. Name the three separate life stages in the life cycle of *Macrocystis* that are haploid, and list the life stage that produces each one.

Challenge Questions

1. What characteristics of green algae (Chlorophyta) support the hypothesis that they are ancestral to flowering plants (Anthophyta)?
2. How do local assemblages of kelp plants, sea grasses, and mangals influence and alter the physical characteristics of the shoreline on which they live?

Suggestions for Further Reading

Books
Abbott, I. A. 1978. *How to know the seaweeds*. Dubuque, IA: Wm. C. Brown Publishers.
Chapman, A. R. O. 1979. *Biology of seaweeds*. Baltimore: University Park Press.
Cole, K. M., and R. G. Sheath, eds. 1990. *Biology of the red algae*. New York: Cambridge University Press.
Dawes, C. J. 1981. *Marine botany*. New York: John Wiley & Sons.
Dring, M. J. 1982. *The biology of marine plants*. Baltimore: University Park Press.
Teal, J., and M. Teal. 1975. *The Sargasso Sea*. Boston: Little, Brown.
Tomlinson, P. B. 1986. *The Botany of mangroves*. New York: Cambridge University Press.

Articles
Duffy, J. E., and M. E. Hay. 1990. Seaweed adaptations to herbivory. *Bioscience* 40:368–75.

Estes, J. A., and J. F. Palmisano. 1974. Sea otters: Their role in structuring nearshore communities. *Science* 185:1058–60.
Foster, M. S. 1975. Algal succession in a *Macrocystis pyrifera* forest. *Marine Biology* 32:313–29.
Mann, K. H., and P. A. Breen. 1972. The relation between lobster abundance, sea urchins, and kelp beds. *Journal of the Fisheries Research Board of Canada* 29:603–9.
Margulis, L., D. Chase, and R. Guerrero. 1986. Microbial communities. *Bioscience* 36:160–70.
Pettitt, J., S. Ducker, and B. Knox. 1981. Submarine pollination. *Scientific American* (Mar.) 244:134–43.
Phillips, R. C. 1978. Sea grasses and the coastal environment. *Oceanus* 21:30–40.
Santelices, B. 1990. Patterns of reproduction, dispersal and recruitment in seaweeds. *Oceanography and Marine Biology* (Annual Review) 28:177–276.

Monterey Bay: Kelp Wrack
http://www.goodenuf.com/hembry/david/kelp_wrack.html
A discussion by David Hembry of the rotting kelp washed up on shore
and why it is good. Intelligent and well researched. Includes some
excellent photos.

Plants and Animals of Sea Grasses
http://www.pcug.org.au/~dfry/amanda/sgorgnsm.htm
Discusses the type of plants and animals found in sea grasses.
The Role of Seagrass Meadows in the Marine Environment
http://www.tip.net.au/~dfry/amanda/sgimport.htm
Visitors' Guide to Marine Ecosystems of Northern New South Wales by
Amanda Fry at **http://www.tip.net.au/~dfry/amanda/** Alternate
site: **http://www.pcug.org.au/~dfry/amanda/**

Introduction to the Anthophyta
http://ghs.ssd.k12.wa.us/depts/ms/algae/index.html
This web site, Algae World, discusses flowering plants.

Functional Marine Biology
http://life.bio.sunysb.edu/marinebio/subjectareas.html
Reproduction and Larval Ecology, Marine Environments, Speciation,
Biodiversity and Community Processes, Applied Marine Biology.
Links citing books, papers, and research in the topical area. The
Marine Biology Web at the Department of Ecology and Evolution,
State University of New York–Stony Brook,
http://life.bio.sunysb.edu/marinebio/mbweb.html

Identification of Florida's Sea Grasses
http://www.dep.state.fl.us/psm/webpages/grass~1.htm
Contains "Key to Marine Species" with sketches of various plants'
stems and leaves.

Internet
Addresses

5

Primary Production in the Sea

The two major categories of autotrophs in the sea, the attached benthic plants and the pelagic phytoplankton, differ in more than their physical appearance. These differences reflect adaptations to the diverse physical and chemical terrains of the benthic and pelagic divisions of the marine environment. The narrow sunlit benthic fringe of the ocean is home to a variety of large, relatively long-lived, attached plants. Yet these plants account for only about 5 to 10% of the total amount of plant material produced in the ocean each year.

From our shore-based perspective, benthic plants gain immediate attention because of their high **standing crop** (the amount of plant material alive at any one time), but this is a poor indicator of the amount of total primary production. The majority of marine primary production is accomplished by the small, dispersed pelagic phytoplankton. On an oceanic scale, the larger near-shore plants are minor players in the process of marine photosynthesis.

Primary production, a term interchangeable with autotrophy, is the biological process of creating high-energy organic material from carbon dioxide, water, and other nutrients. The organic material synthesized by the primary producers ultimately becomes transformed and is transferred to other trophic levels of the ecosystem.

Consider this example: A neatly trimmed lawn contains an easily measured amount of living plant material, its standing crop. If the lawn is maintained throughout a summer, it will be periodically mowed to maintain the same height, or, in other words, the same standing crop. During that summer, the lawn clippings will total much more than the standing crop, but the lawn clippings are not part of the lawn. They represent the primary production that occurred during the summer. In an analogous sense, it is this rapidly consumed production by phytoplankton, with typically very low standing crops, that fuels the metabolic processes of most of the consumers living in the sea.

The total amount of organic material produced in the sea by photosynthesis represents the **gross primary production** of the marine ecosystem. Gross primary production is occasionally difficult to measure in nature; nonetheless, it is useful as a base of reference for understanding the production potential of the marine ecosystem. A portion of the organic material produced by photosynthesis is utilized in cellular respiration to sustain the life processes of the photosynthesizers. Any excess production is used for growth and reproduction, and is referred to as the **net primary production.** Net marine primary production represents the amount of organic material available to support the consumers, nonphotosynthetic protists, and decomposers of the sea. Different types of living material contain varying proportions of water, minerals, and energy-rich components. To minimize the problems of comparing different types of primary producers, we commonly report standing crops in grams of organic carbon (gC). This unit represents approximately 10% of the live, or wet, weight of the standing crop. To compensate for differences in water content of different types of primary producers, we usually report production rates for phytoplankton in units of grams of organic carbon fixed by photosynthesis under a square meter of sea surface per day or per year ($gC/m^2/day$ or $gC/m^2/year$).

The following discussion of some global aspects of marine primary production will emphasize production by phytoplankton, but the general concepts are also valid for the larger forms of attached plants.

Measurements of Primary Production

Rates of primary production in the sea vary widely in time and space, and animals exploiting the autotrophs must adapt to the patterns. These production rates, and the ecological factors that affect them, have become clearer with the development of techniques for measuring primary production in the sea. Theoretically, the net photosynthetic rate of a phytoplankton population can be estimated by measuring the rate of change of some chemical component of the photosynthetic reaction (p. 58) such as the rate of O_2 production or CO_2 consumption by the phytoplankton.

For the first half of this century, the light and dark bottle technique was used to study primary production in marine phytoplankton. With this method, measured changes in O_2 consumption and production are used to compute phytoplankton respiration and photosynthetic rates. Figure 5.1 describes how the technique works in its simplest form.

To begin, representative samples of phytoplankton and their surrounding water are collected at specific preselected depths, and subsamples are placed in paired transparent (light) and opaque (dark)

Figure 5.1

The results of a hypothetical light and dark bottle experiment. Water samples from 10-m-depth increments are replaced at original depths in paired light and dark bottles. After a period of time, the bottles are retrieved and changes in O_2 are determined. Dark bottle (DB) values indicate O_2 decreases at each depth due to plant respiration without photosynthesis. Light bottle (LB) values represent O_2 changes from photosynthesis and respiration (net primary production) in the light. The difference between the two values (LB – DB) is the gross primary production. With these values, net and gross primary production curves can be drawn to represent the variation in photosynthesis attributable to depth.

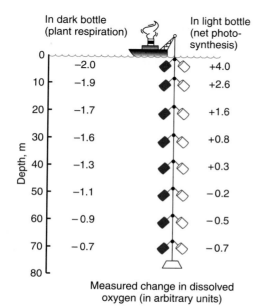

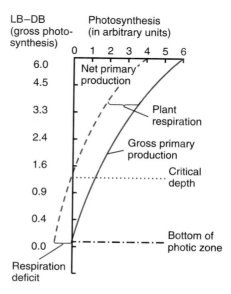

containers. Next, the containers are replaced at the depth from which each sample was obtained. After a few hours, the paired samples are retrieved, and the O_2 content of the water in each bottle is measured. Decreases in dissolved O_2 are used to estimate cellular respiration in the absence of light. In the light bottle, changes in O_2 concentration represent total (or gross) photosynthetic activity less the products of photosynthesis used in respiration. This is net photosynthesis, the excess of phytoplankton production over its respiration.

The precision of the light and dark bottle technique hinges on the assumption that respiratory O_2 consumption in the dark bottle is the same as that in the light bottle. Although this is an accurate assumption at optimal or below-optimal light conditions, it is usually not true for phytoplankton in high light intensities. Some O_2 may be consumed by zooplankton or bacteria included in the light or dark bottles. Some of the problems inherent in productivity estimates based on O_2 changes in light and dark bottles can be avoided by determining rates of CO_2 uptake instead. In the early 1950s, Steemann Nielsen first employed a procedure that used radioactive carbon (C^{14}) as a tracer in photosynthesis. Improved C^{14} measurements are now the preferred method for marine primary production studies.

The C^{14} procedure is similar to the O_2 production technique, but it is more sensitive when productivity is very low. Paired light and dark bottles are used. Each bottle is injected with a known quantity of bicarbonate containing the labeled C^{14}. After a period of incubation at the proper depth, the samples are recovered. The phytoplankton of each sample are collected on membrane filters and dried. The amount of radioactive carbon assimilated by the phytoplankton in the bottles is measured with a radioactivity counting device. Net primary production is then computed using an appropriate conversion factor. The resulting error in estimating primary productivity is about ±30%.

In photosynthetic plants, chlorophyll *a* is necessary to bring about photosynthetic reactions. One might assume, then, that the gross primary production in a volume of seawater is proportional to the amount of chlorophyll *a* contained in the living phytoplankton of the water sample. Such a relationship between gross primary production and chlorophyll *a* concentrations has been established, but only in an approximate fashion. Chlorophyll *a* concentration is a better indicator of the standing crop of phytoplankton. Standing crop sizes at any given moment are governed by a balance between crop increases (cell growth and division) and crop decreases (sinking and grazing). Most of the gross primary production of a healthy, actively growing phytoplankton population is not used in respiration but, instead, contributes to the existing standing crop. In contrast, old populations or healthy cells in poor growing conditions use a large portion of their gross production in respiration, and net production declines. If a net loss occurs, the population will eventually disappear.

The standing crop of a healthy phytoplankton population measured on successive days may demonstrate little or no increase, suggesting that no net production occurred from one day to the next. A more likely explanation is that significant net production did occur, but it replaced the portion of the crop lost to grazers and to sinking. Thus, the relationship between standing crop and productivity depends to a large degree on the **turnover rate** of newly created plant material.

The turnover rate of phytoplankton populations is extremely rapid. Many species of large phytoplankton divide once each day, and several of the smaller species divide even faster. The coccolithophore shown in figure 3.7a, for instance, undergoes almost two divisions per day. Its population is completely replaced, or turned over, twice each day, so comparatively little plant material exists in the water at any one time. Even higher turnover rates are expected for the smaller picoplankton and ultraplankton and in benthic algae.

For decades, O_2 or C^{14}-based measurements of primary productivity and of standing crop have been made from ships at widely spaced sampling stations. Environmental changes that occurred as the ship steamed from one station to the next could not be measured, nor were the details between stations examined. It simply had to be assumed that the data collected at the sample stations could be averaged over the areas between stations and between sampling periods. The complexity and richness of small- to moderate-scale spatial variations in phytoplankton abundance were missed, as were the day-to-day variations occurring at any sampling station.

In contrast to ship-based sampling, satellite sampling can provide a general, and instantaneous, overview of a large portion of ocean (figure 5.2). Satellites cannot directly measure marine primary

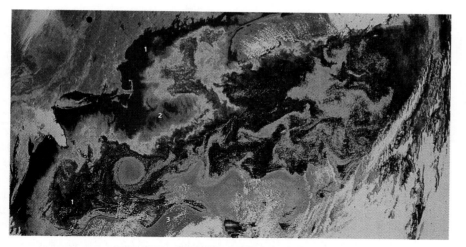

Figure 5.2

Composite satellite views of the North Atlantic Ocean. (Top) Phytoplankton concentrations, ranging from low (dark blue) to high (red). (Bottom) Sea surface temperature of the same area ranging from warm (red) to cold (dark blue).
Courtesy National Aeronautics and Space Administration

Figure 5.3

Patterns of population growth with
and without limiting resources.

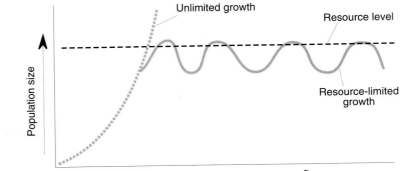

productivity. Instead, ocean surface color is measured with a coastal zone color scanner (CZCS) aboard the NIMBUS 7 weather satellite (see Research in Progress: Oceanography from Space, p. 76). The ocean color measurements are then used, with calibration from ship-based measurements, to estimate phytoplankton standing crops and growth rates and to extrapolate shipboard productivity measurements to large oceanic areas. Ultimately, it should be possible to use CZCS data alone to estimate primary production or growth rates.

Remote sensing of ocean color by satellite is the first technique to measure marine primary productivity on a global scale with enough resolution to permit analyses of phytoplankton changes over time scales of weeks or years. Since the CZCS was lofted into orbit in 1978, satellite imagery has revolutionized our view of primary productivity patterns in the ocean. As is apparent in figure 5.2, distribution patterns of phytoplankton are complex and show some similarities with sea-surface temperature distributions. Patches and eddies of phytoplankton are common. In upwelling areas, plumes of phytoplankton-rich water (known as squirts) extend as much as 200 km offshore. Before satellite observations, these dominant features of marine phytoplankton distribution were completely unknown.

Factors That Affect Primary Production

The continued synthesis of organic material by marine phytoplankton depends on a set of interacting biotic (biological) and abiotic conditions. If nutrients, sunlight, space, and other conditions necessary for growth are unlimited, phytoplankton population sizes increase in an exponential fashion (figure 5.3).

In nature, phytoplankton populations do not continue to grow unchecked as the unlimited growth curve in figure 5.3 suggests. Rather, their sizes are controlled by their tolerance limits to certain environmental factors (including predators) or by the availability of substances for which there is a minimum need. Any condition that exceeds the limits of tolerance or does not satisfy the basic material needs of an organism establishes a check on further population growth and is said to be a **limiting factor.** Phytoplankton populations limited by one or a combination of these factors are forced to deviate from the exponential growth curve shown in figure 5.3. Important limiting factors for phytoplankton are light, nutrient availability, and herbivore grazing. In the ocean, each major group of phytoplankton responds differently to combinations of these factors. In general, diatoms and silicoflagellates thrive in lower light intensities and colder water than do dinophytes and coccolithophores. Consequently, conditions that promote the growth of either group tend to exclude the other. These factors will be examined, first alone, then in concert, in an attempt to convey the complex dynamic interactions that exist between these living communities and their immediate surroundings.

Light

The requirement for light imposes a fundamental limit on the distribution of all marine photosynthetic organisms. To live, these organisms must remain in the photic zone. The depth of the photic zone is determined by a variety of conditions, including the atmospheric absorption of light, the angle between the sun and the sea surface, and water transparency (discussed in chapter 1).

Water transparency causes light in seawater to diminish from the sea surface downward. At some depth, the light intensity is so faint that no photosynthesis will occur. This depth defines the bottom of the photic zone and varies from a few meters deep in coastal waters to more than 200 m in clear tropical seas. At a depth somewhat above the bottom of the photic zone, the rate of photosynthesis is balanced by photorespiration. This depth of no net primary production is the **critical depth** (figure 5.1). The critical depth is approximately equivalent to the depth at which the available light is reduced to 1% of its surface intensity (see figure 1.17). In clear tropical waters, the critical depth often extends below 100 m throughout the year. In higher latitudes, it may reach 30 to 50 m in midsummer, but it nearly disappears during the winter. (These are average critical depths for mixed phytoplankton assemblages composed of many different species; each species has its own peculiar critical depth.)

In moderate and low light intensities, photosynthesis by phytoplankton exhibits a direct relationship to light intensity (figure 5.4). At higher light intensities, photosynthesis ceases to follow the light-intensity curve; it may stabilize or even decrease nearer the sea surface because of **photoinhibition** by strong light. Between the light-limited and light-inhibited portions of the photosynthetic curve *b* in figure 5.4 is a zone of **saturation light intensity.** At this point, photosynthesis no longer increases in proportion to increasing light intensities. The photosynthetic machinery of phytoplankton cells is saturated with light, and higher light intensities nearer the sea surface fail to elicit proportionate increases in photosynthesis.

Phytoplankton from different environments exhibit some degree of photosynthetic adjustment to varying light intensities. Therefore, the saturation light intensity for any phytoplankton population changes with changing sets of environmental conditions. Variations in saturation light intensities are also found among major phytoplankton groups. Dinophytes seem to be better adapted than diatoms to intense light (figure 5.5). As a result, their relative contribution to the total marine primary production is much greater in tropical and subtropical regions.

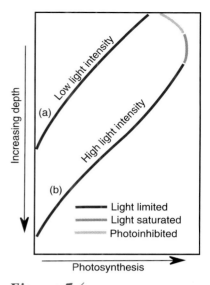

Figure 5.4

Relationship between photosynthesis and depth at low and high light intensities.

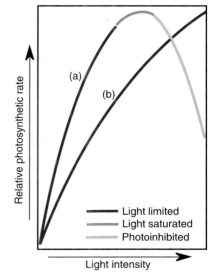

Figure 5.5

A comparison of photosynthetic responses of two phytoplankton species to varying light intensities. The diatom *Planktonella* (curve *a*) shows marked photoinhibition at high light intensities; the photosynthetic rate of *Dinophysis,* a dinophyte (curve *b*), continues to increase even at high light intensities. Adapted from Qasim et al., 1972.

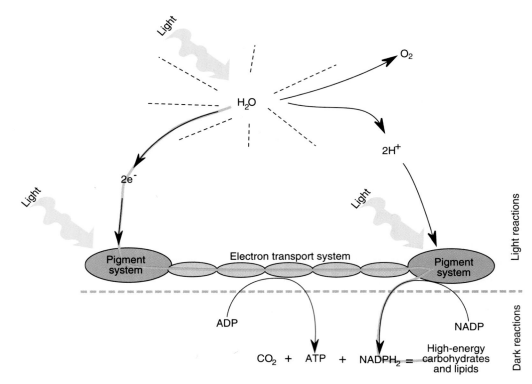

Figure 5.6

Diagrammatic representation of the photosynthetic mechanism of eucaryotic plants. The colored line traces the path of electrons initially activated by light energy.

Photosynthetic Pigments

The photosynthetic apparatus of all marine primary producers except cyanobacteria is located in the chloroplasts of actively photosynthesizing cells. It is in the chloroplasts (or the whole cells of bacteria and cyanobacteria) that the pigment systems are located, containing chlorophyll and varying amounts of other photosynthetic pigments listed in tables 3.2 and 4.1. There, they absorb light energy and convert it to forms of chemical energy that can be used by the plant and by other organisms.

Both cyanobacteria and eucaryotic autotrophs employ an elaborate two-part photosynthetic process involving complex pigment systems and two distinct sets of chemical reactions. In the first set, the **light reaction** portion of photosynthesis (figure 5.6), photons of light are absorbed by chlorophyll molecules located in two separate pigment systems. The photons energize electrons and pump them through a series of other enzymes whose function is to manage some of that electron energy and transfer it to ATP and another high-energy carrier molecule, $NADPH_2$. As the term implies, light is needed to drive the light reaction; without light, the reaction ceases.

The pigment systems and enzymes involved in the light reaction are housed within flattened sacs, which are stacked to form numerous **grana** within each chloroplast (figure 5.7). The **stroma** surrounds the grana and contains the enzymes needed for the next step of photosynthesis, the **dark reaction.** Light energy is not necessary to maintain the dark reaction, but the high-energy ATP and $NADPH_2$ produced by the light reaction are. Energy from these substances is

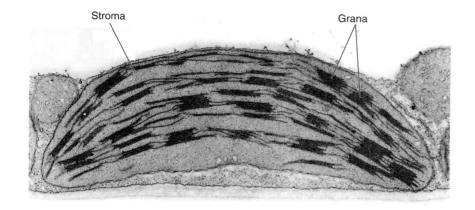

Stroma   Grana

Figure 5.7

A transmission electron micrograph
of a chloroplast with stacked sets of
parallel membranes (narrow dark
bands).
Courtesy Herbert W. Israel, Cornell
University.

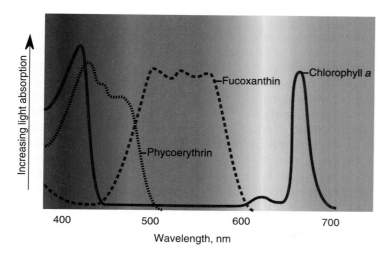

Increasing light absorption

Fucoxanthin

Chlorophyll *a*

Phycoerythrin

400   500   600   700

Wavelength, nm

Figure 5.8

Patterns of light absorption for three
photosynthetic pigments:
phycoerythrin (an accessory pigment
found in Rhodophyta and
Cyanobacteria), fucoxanthin (an
accessory pigment found in
Phaeophyta, Chrysophyta, and
Dinophyta), and chlorophyll *a*.
Adapted from Saffo, 1987.

used in the dark reaction to synthesize sugars and a variety of other
organic compounds needed by the cell.

Chlorophyll appears green for the same reason coastal seawater
appears green. Both absorb more of the available light energy from the
violet and red ends of the spectrum, leaving the green light to be reflected back or to penetrate more deeply (see chapter 1). Chlorophyll
serves as the basic energy-absorbing pigment for land plants. However, a few meters of seawater absorbs much of the red and violet portions of the visible spectrum before it reaches the chloroplasts of most
marine plants. Because chlorophyll best absorbs energy from red and
violet light, its effectiveness is reduced in seawater.

The evolutionary response of most marine plant groups has been to
supplement the light-absorbing ability of chlorophyll with **accessory
pigments** (figure 5.8). These pigments absorb light energy over a wide
range of wavelengths and then transfer the energy to chlorophyll for introduction into the light reaction. Accessory pigments absorb light from
spectral regions where chlorophyll cannot. Figure 5.8 illustrates the
complementary effect of chlorophyll *a* and accessory pigments such as
fucoxanthin, which is found in brown algae and diatoms. Fucoxanthin
absorbs light primarily from the blue-green region of the spectrum, the
region where chlorophyll absorbs least effectively. In combination,

chlorophyll and fucoxanthin are capable of absorbing energy from most of the visible light spectrum. Another group of accessory pigments, the phycobilins, are found in red algae and cyanobacteria. These pigments have absorption spectra much like that of fucoxanthin. These and other accessory pigments listed in tables 3.2 and 4.1 have enabled various groups of marine plants to adapt to the limited conditions of light availability in seawater.

The depth changes normally experienced by marine phytoplankton expose them to a wide variety of submarine light conditions. Their diverse pigments are capable of absorbing light energy at almost any depth within the photic zone. Patterns of vertical distribution of attached benthic plants are more complex. At first glance, it might appear that the green algae and sea grasses, with their preponderance of chlorophyll pigments, do not fare well at moderate depths because of their limited ability to absorb the deeper-penetrating green wavelengths. But plants can adapt to low or limited wavelength light conditions in other ways. For example, because some green algae have dense concentrations of chlorophyll that appear almost black, they are able to absorb light at essentially all visible wavelengths. In addition, most green plants have chlorophyll *b* as well as chlorophyll *a*. Chlorophyll *b* has a strong light-absorbing peak in the blue region of the visible spectrum and can collect a good fraction of the deep-penetrating blue light available in tropical waters. Still, red and brown algae, with their abundant xanthophyll and phycobilin pigments working in concert with chlorophyll, appear to have a slight competitive advantage in occupying the deeper portions of the photic zone in turbid coastal waters and function at no disadvantage in shallow waters or intertidal zones.

Nutrient Requirements

The nutrients required by all primary producers are a bit more complex than might be indicated by the general photosynthetic equation:

$$6CO_2 + 12H_2O \rightarrow C_6H_{12}O_6 + 6H_2O + 6O_2$$

Proper growth and maintenance of cells depend on the availability of more than just water and carbon dioxide because plants are composed of compounds that cannot be assembled from C, H, and O alone.

These nutrient requirements can be best understood by determining the basic composition of the cell itself. Chemical analysis of a hypothetical "average" marine primary producer might yield the results shown in figure 5.9.

In general, marine primary producers experience no difficulty in securing an adequate supply of water. Most are continuously and completely bathed by seawater, and few cells of any marine plant are seriously isolated from the external water environment. Some species living in the intertidal zone are subject to desiccation (dehydration) at low tide. These plants exhibit both resistance to and tolerance of water loss. Occasionally, a combination of low tides and dry winds from the land occur and seriously damage intertidal plants.

Coccolithophores and some seaweeds are equipped with cell walls or internal skeletons of calcium carbonate ($CaCO_3$). Carbon dioxide for carbonate formation and for photosynthesis exists in

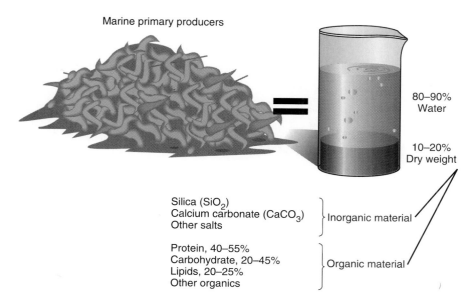

Marine primary producers

80–90% Water

10–20% Dry weight

Silica (SiO$_2$)
Calcium carbonate (CaCO$_3$)
Other salts } Inorganic material

Protein, 40–55%
Carbohydrate, 20–45%
Lipids, 20–25%
Other organics } Organic material

Figure 5.9

Chemical composition of typical marine autotrophs.

seawater as carbonic acid (H$_2$CO$_3$), bicarbonate (HCO$_3^-$), and carbonate (CO$_3^{-2}$). The abundance of these ions in seawater is influenced by photosynthesis, respiration, water depth, and pH balance (see p. 28). However, the concentration of total CO$_2$ present in seawater is not low enough to inhibit photosynthesis or the formation of CaCO$_3$. Calcium ions (Ca^{+2}) necessary for calcium carbonate formation are also very abundant in seawater at all depths (see table 1.3). Silica (SiO$_2$) is required by silicoflagellates and diatoms, and concentrations of dissolved silica occasionally become so depleted that the growth and reproduction of these phytoplankton groups are inhibited.

Organic matter is a widely used term collectively applied to those biologically synthesized compounds that contain C, H, usually O, lesser amounts of N (nitrogen), and P (phosphorus), and traces of vitamins and other elements necessary to maintain life. Proteins, carbohydrates, and lipids are the most abundant types of organic compounds in living systems. Each contains carbon, hydrogen, and oxygen in varying ratios. Figure 5.10 summarizes the generalized nutrient needs of photosynthetic cells.

How much of each of these elements do primary producers require? Elemental analyses of phytoplankton whole-cell cultures grown under various light conditions provide an average atomic ratio of approximately 110(C):230(H):75(O):16(N):1(P). Carbon, hydrogen, and oxygen are abundantly available from carbonate (CO$_3^{-2}$) or bicarbonate ions (HCO$_3^-$) and water (H$_2$O). Nitrogen is much less plentiful but is present in seawater as nitrate (NO$_3^{-2}$), with lesser amounts of nitrite (NO$_2^-$), and ammonium (NH$_4^+$). High concentrations of molecular nitrogen (N$_2$), which constitutes 78% of the earth's atmosphere, are also dissolved in seawater. However, most marine organisms are not metabolically equipped to utilize this latter source of N. They may, however, contribute a substantial portion of the total N used by the phytoplankton in nutrient-depleted seas. Phosphorus, present

Figure 5.10

A simplified photosynthetic cell, illustrating the chemical requirements and products of several components of the cell.

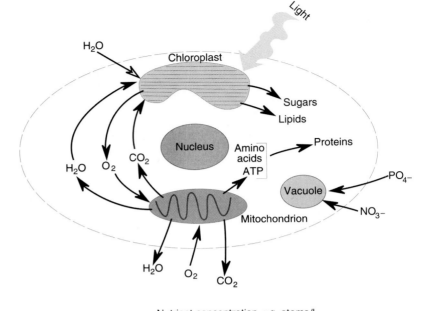

Figure 5.11

Distribution of dissolved silicate (dotted lines), nitrate (dashed lines), and phosphate (solid lines) from the surface to 3,000 m in the Atlantic (black) and Pacific (blue) oceans.

From H. U. Sverdrup, Martin W. Johnson, and Richard H. Fleming, *The Oceans: Their Physics, Chemistry, and General Biology,* © 1942, renewed 1970. By permission of Prentice Hall, Inc. Englewood Cliffs, New Jersey.

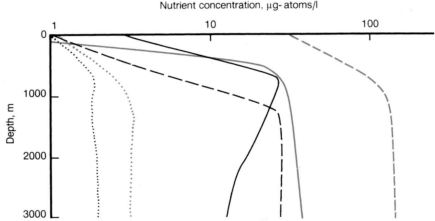

principally as phosphate (PO_4^{-3}), is less abundant in seawater than is nitrate, but the biological demands on phosphate are also less but just as critical (for example, in the synthesis of ATP and cell membranes). The ratio of usable N and P in seawater is similar to the ratio of 16N:1P found in living cells of marine primary producers.

Figure 5.11 shows the vertical distribution patterns of silicate, nitrate, and phosphate in seawater. These nutrients are usually in short supply in the photic zone during the growing season because of continual utilization by primary producers. In periods of rapid phytoplankton growth, needed quantities of one or more of these nutrients may not be available. In such circumstances, continued growth is limited by the rate of nutrient regeneration.

In addition to the nutrient elements just mentioned, marine autotrophs require several other elements in minute amounts. These **trace elements** include iron, manganese, cobalt, zinc, copper, and others. Depletion of iron in English Channel waters has been observed

Figure 5.12

A simplified marine nutrient cycle.

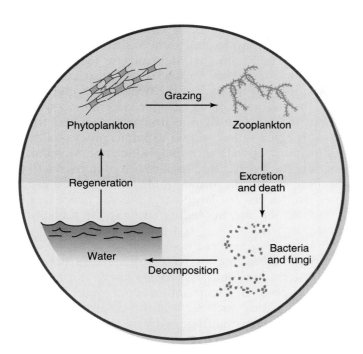

during spring diatom blooms, suggesting that iron availability may limit the size or composition of phytoplankton populations. But in most cases, it is not known whether natural concentrations of other trace elements dip to values low enough to inhibit growth of marine primary producers.

Vitamins too are crucial for the proper growth and reproduction of primary producers. Some species of diatoms, for example, require more vitamin B_{12} during auxospore formation than at other times. Some can synthesize their own vitamins; others must rely on free-living bacteria to provide this and other critical vitamins that they cannot synthesize for themselves.

Nutrient Regeneration

Most of the biomass produced by marine photosynthesis is eventually consumed by herbivores to be converted to more herbivore bodies or to be formed into fecal wastes. In either case, these compact particles quickly become colonized by bacteria and sink as "marine snow" to depths well below the photic zone. Other nutrients are excreted as nitrogen-containing urea and ammonia wastes and are taken up to the phytoplankton as quickly as they are excreted. Regeneration of the nutrients initially used to produce phytoplankton cells or marine plants is dependent on respiration by consumers and on decomposition of organic material by bacteria and fungi living in the water column and on the sea bottom. Bacterial action decomposes organic material and returns phosphates, nitrates, and other nutrients to seawater in inorganic form for reuse by the primary producers (figure 5.12). Bacteria also absorb dissolved organic compounds from seawater and convert them to living cells that become an additional food source for many benthic and small planktonic animals.

Figure 5.11 indicates that major concentrations of limiting nutrients reside below the photic zone, where they cannot be utilized by photosynthesizers. Their combined demands for light from the sea surface and nutrients from below impose severe restrictions on the rates of primary production. For much of the ocean, the sunlit photic zone is isolated from the nutrients of the deeper waters by a well-developed and permanent thermocline. Marine primary producers thrive only in those parts of the sea where active dynamic processes move colder nutrient-laden waters upward into the photic zone. Molecular diffusion does not account for a very substantial return of nutrients to the photic zone. Much more significant to the rapid and continued growth of marine primary producers are large-scale mixing processes, including small-scale turbulence and upwelling that rapidly transport nutrient-rich deep water upward.

Wind, waves, and tides create turbulence in near-surface waters and mix nutrients from deeper water upward. Turbulent mixing is most effective over continental shelves, where the shallow bottom prevents the escape of nutrients into deeper water. Tidal currents in the southern end of the North Sea and the eastern side of the English Channel, for example, are sufficient to mix the water almost completely from top to bottom. As a result, summer phytoplankton productivity there remains high as long as sunlight is sufficient to maintain photosynthesis.

In tropical and subtropical latitudes of most oceans, the strong year-round thermocline and associated pycnocline near the base of the photic zone act as a strong barrier to inhibit upward mixing of deep nutrient-rich waters (figure 5.13 top). Consequently, these regions have very low rates of primary production, comparable to terrestrial deserts.

Thermoclines also develop in temperate waters to restrict the return of deep-water nutrients, but only on a seasonal basis (figure 5.13 center). During winter, the surface water cools and sinks. The thermocline disappears (figure 5.13a) and deeper nutrient-rich water is mixed with the surface water. As solar radiation increases in the spring, the surface water warms and the thermocline is reestablished (b). A well-developed summer thermocline (c) resembles the permanent thermocline of tropical and subtropical waters and creates an effective barrier blocking nutrient return to the photic zone. With shorter days and cooler weather in autumn (d), the thermocline weakens and then disappears in winter. This process of **convective mixing** is a seasonal phenomenon in temperate regions, continuing from late fall to early spring. But in high latitudes, continuous heat loss from the sea to the atmosphere and low amounts of solar radiation produce year-round convective mixing (figure 5.13 bottom). Low light conditions rather than scarce nutrients usually limit the primary production in these polar regions.

Subsurface water is carried to the photic zone by several processes collectively termed upwelling. One type, coastal upwelling, is produced by winds blowing surface waters away from a coastline. The surface waters are replaced by deeper waters rising to the surface (figure 5.14). Near-shore currents, which veer away from the shoreline, produce the same result. Four major coastal upwelling areas occur in the California, Peru, Canary, and Benguela currents, and

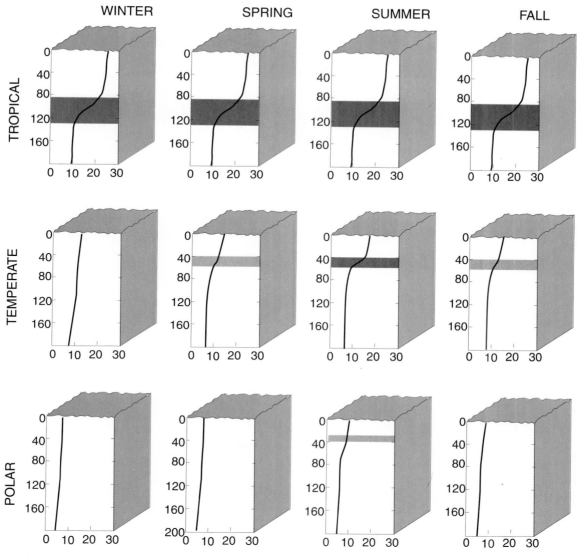

Figure 5.13

Seasonal growth and decline of thermoclines in tropical (top), temperate (center) and polar ocean waters (bottom). Depth, in m, is on vertical axes, temperature, in °C, on horizontal axes.

lesser ones occur along the coasts of Somalia and western Australia. All of these currents are on eastern sides of subtropical current gyres (figure 5.15) and flow toward the equator. Figure 5.16 illustrates the influence of upwelling on nutrient availability in the photic zone. Note that the nitrate concentrations at a depth of about 50 m are 5 to 10 times higher in the upwelling systems than at similar depths in adjacent nonupwelled water.

Another type of upwelling is more limited in extent and normally exists only in the central Pacific Ocean. The Pacific Equatorial Current flows westward, straddling the equator. The Coriolis effect causes a slight displacement to the right for the portion of the current in the Northern Hemisphere and to the left for the portion of the current in the Southern Hemisphere (see figure 1.31). The resultant divergence of water away from the equator creates an upwelling of deeper water to replace the water that has moved away.

Figure 5.14

Coastal upwelling in the Northern Hemisphere.

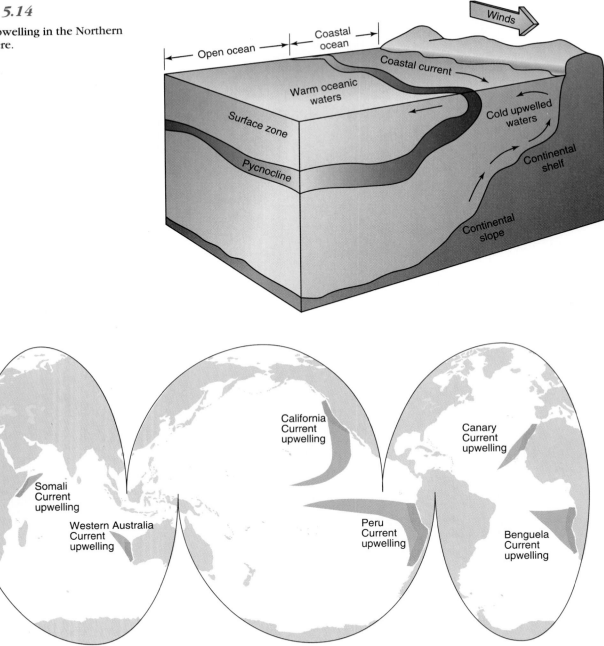

Figure 5.15

Principal regions of coastal upwelling (blue) and downcurrent areas of increased primary productivity (green).

Langmuir cells (named after Irving Langmuir, who first clarified their structure after he observed *Sargassum* in the North Atlantic floating in long rows parallel to the wind direction) are parallel pairs of small counter-rotating convection cells driven by surface winds. Langmuir convection cells set up alternating zones of divergence and convergence and sometimes sweep detritus and plankton to the lines of convergence between adjacent cells. This material is often evident

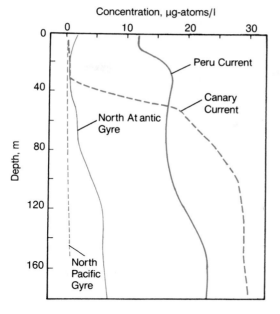

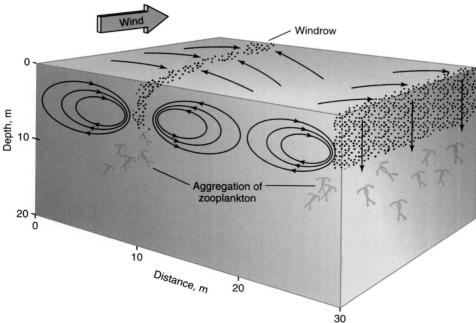

Figure 5.16

A comparison of the vertical distribution of nitrate in upwelling areas (heavy curves) and in adjacent nonupwelling central ocean regions (light curves).
Redrawn from Walsh 1984.

Figure 5.17

Circulation patterns of wind-driven Langmuir convection cells.

at the surface as long parallel "slicks," foam lines, or rows of floating debris (figure 5.17). Similar accumulations may also be created by internal waves and other factors. Langmuir cells extend only a few meters deep and are not important for nutrient upwelling from deep water. However, these convection cells may create nutrient traps under the convergences. Phytoplankton and particulate debris that accumulate under the convergences attract grazing zooplankton in concentrations often 100 times as dense as those in adjacent areas.

Figure 5.18

Changes in diatom population within the given time frame with *(a)* grazing rate equal to cell division rate, *(b)* grazing rate doubled, and *(c)* grazing rate increased 5 times. Adapted from Fleming 1939.

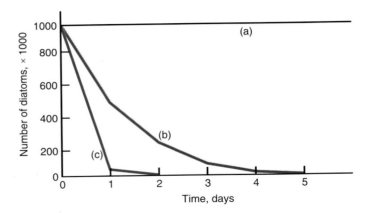

Grazing

The trophic interrelationships of marine phytoplankton and small herbivores (mostly zooplankton and small fishes) can be complex, and are described in more detail in chapter 12 (p. 332–335). Grazing may decrease the standing crop and sometimes the productivity of a phytoplankton population. The capacity of herbivorous zooplankton to quickly reduce a phytoplankton population is shown in figure 5.18. In an experimental setting, with initial conditions of a phytoplankton population having a density of 1 million cells per liter and reproducing once each day, a herbivore population density was adjusted to achieve a grazing rate that just held the phytoplankton population constant. When the zooplankton population density was doubled, the phytoplankton population was reduced to 27,000 cells in five days. At a density 5 times that necessary to hold the phytoplankton population constant, the zooplankton reduced the phytoplankton cell density to 24 cells per liter in three days, and the culture was essentially eliminated in five days.

Ideally, grazing rates should adjust to the magnitude of primary productivity to establish a balance between producer and consumer populations. Photosynthetic rates do limit the average size of the animal populations that primary producers support, yet short-term fluctuations of both phytoplankton and grazer populations often occur. The magnitude of these fluctuations tends to be moderated somewhat by stabilizing **feedback mechanisms** between all trophically related populations. An abundant food supply permits the grazers to reproduce and grow rapidly (figure 5.19). Eventually, however, they consume their prey more quickly than the prey can be replaced. Overgrazing reduces the phytoplankton population and its photosynthetic capacity, causing food shortages, starvation, and consequent reductions of the enlarged herbivore populations. When grazing intensity is reduced after herbivore populations crash, the phytoplankton population may recover, increase in size, and again set the stage with an abundant food source to cause a repeat of the entire cycle. Such oscillations of population size may extend through many trophic levels of the food web.

In most cases, a delay exists between population peaks of the consumed and the consumer. The length of the time lag depends largely on the consumer's reproductive response to an increasing food supply. In favorable conditions, asexually reproducing phytoplankton can

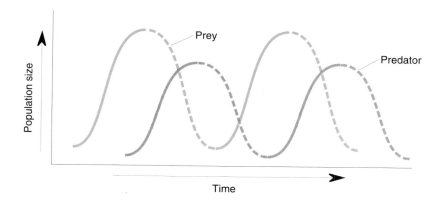

Figure 5.19

Generalized population changes of a prey species and its predator, oscillating between unlimited (solid lines) and limited (dashed lines) phases of population growth.

divide rapidly and can increase their population size more quickly than larger, sexually reproducing zooplankton. Phytoplankton, therefore, can achieve a greater population size before the zooplankton catch up and may experience larger population fluctuations than the more slowly breeding herbivores.

In addition to the large-scale geographical variations in phytoplankton density observed from satellites (figure 5.2), marine phytoplankton also exhibit much smaller scale localized patchiness. Dense patches of phytoplankton tend to alternate with concentrated patches of zooplankton. The inverse concentrations of phytoplankton and zooplankton densities stem in part from the effects of grazing and because of differences in their reproductive rates. A dynamic model of phytoplankton growth, grazing, and subsequent zooplankton migration that establishes and effectively maintains the alternating patchy distribution of marine phytoplankton and zooplankton is suggested in figure 5.20. This is but one of many models used to explain phytoplankton patchiness. Initially, a dense patch of phytoplankton provides favorable growth conditions for herbivores attracted from adjacent water into the phytoplankton patch (figure 5.20a). The grazing rate increases in the area of the patch and declines elsewhere. Production soon decreases in the original patch and increases in adjacent areas. Eventually, the original phytoplankton patch is eliminated by the dense concentration of grazers. The adjacent areas become the new phytoplankton patches (figure 5.20b) and attract herbivores from the recently overgrazed region, thus repeating the entire sequence.

Some species of zooplankton are attracted by particular phytoplankton species and repelled by others. External metabolites secreted by phytoplankton have been suggested as one cause for selective grazing. Detection of phytoplankton patches by the quality of light passing through them is another possibility. Some copepods, when exposed to predominantly red light, display a "red dance," with most of their movements oriented vertically. In light with a strong proportion of blue, the same copepods exhibit a "blue dance," with most of their movements oriented horizontally. The horizontal movements have been interpreted as hunting or searching. Horizontal motions eventually bring the copepods into or under a phytoplankton patch, where increased chlorophyll concentrations decrease the relative proportion of blue light. The copepods then shift to the vertical red dance, which maintains their position within the phytoplankton patch.

(a)

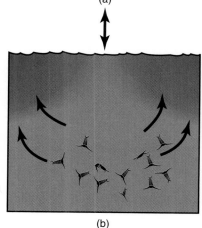

(b)

Figure 5.20

A dynamic model for establishing and maintaining patches of marine phytoplankton and zooplankton.
Adapted from Bainbridge 1957.

Figure 5.21

The seasonal variation of light intensity at the sea surface sets in motion a cascading series of changes in the photic zone. Eventually, these factors influence primary production, either directly (solid arrows) or through feedback links (dashed arrows).

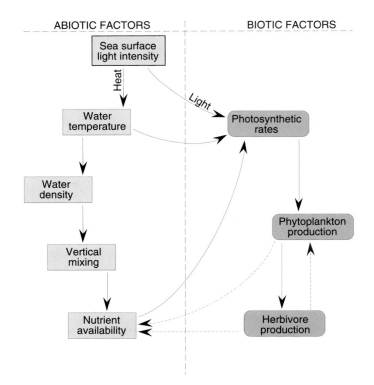

Seasonal Patterns of Marine Primary Production

The spatial patchiness of phytoplankton is related on a large scale to areas of nutrient abundance and on much smaller scales to the local influences of grazers, near-surface turbulence, and nutrient patches. Seasonal phytoplankton variations, or patchiness in time, occur in response to changes in light intensity, nutrient abundance, and grazing pressure. The underlying pulse for these time changes is the predictable seasonal variation in the intensity of sunlight reaching the sea surface. Figure 5.21 outlines the major links between factors involved in defining the actual pattern of phytoplankton production through time. With these in mind, one can develop the seasonal pattern of phytoplankton production for several marine production systems.

Temperate Seas

Figure 5.22 depicts a somewhat idealized graphical summary of major physical, chemical, and biological events in temperate areas well away from the effects of coastlines. A prominent feature in the production cycle of temperate seas is the spring diatom increase, or diatom bloom. Diatom blooms are the result of combined seasonal variations of water temperature, light and nutrient availability, and grazing intensity. In early spring, water temperature and available light increase, nutrients are abundant in near-surface waters, and

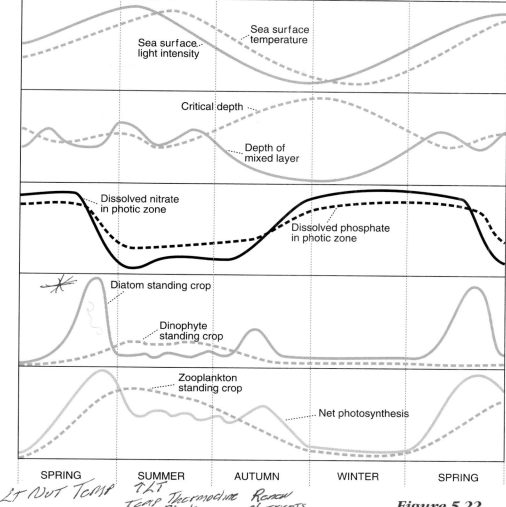

Handwritten notes below the chart:

LT NUT TEMP   *↑LT Temp Thermocline Renew Nut Block Nutrients*

SPRING   SUMMER   AUTUMN   WINTER   SPRING

Figure 5.22

Seasonal fluctuations in the major features of a temperate-water marine primary productivity system.

grazing pressure is diminished. Conditions are ideal for rapid and abundant growth of primary producers. If the bottom of the mixed layer extends below the critical depth (determined by light penetration), near-surface turbulence will distribute the phytoplankton cells randomly throughout the mixed layer, and primary production will remain low. Cells in the deeper portions of the mixed layer will receive insufficient light, and no net production will occur. The spring bloom will commence only after the thermocline thins the mixed layer to a level above the critical depth. In general, bloom conditions in the open ocean occur as a broad band of primary production sweeping poleward with the onset of spring. The standing crop of diatoms increases quickly to the largest of the year and begins to deplete nutrient concentrations. The grazers respond to the additional forage by increasing their numbers.

As spring warms into summer, sunlight becomes more plentiful, but the now strongly developed seasonal thermocline effectively blocks nutrient return from deeper water. Coupled with increased grazing, the diatom population peaks and then declines and remains low throughout the summer. With food more scarce, the summer zooplankton population also drops off. Unlike diatoms, dinophyte populations increase slowly during the spring, remain healthy throughout the summer, and decline in autumn because of diminished light intensity. This replacement of diatoms by dinophytes is a form of seasonal succession resulting from some basic ecological differences between the two principal groups of phytoplankton. Recall that diatoms lack flagella, cannot swim, are more readily inhibited in high-light intensities, perform better in low-light intensities (figure 5.5), and have a nutrient need for silicate. These features give diatoms a competitive advantage in less-well-lit, colder, denser, nutrient-rich waters and dinophytes an advantage in warmer, better-lit waters that may be deficient in silicate.

Cooler autumn air temperatures begin to break down the summer thermocline and allow convection to renew nutrients to the photic zone. The phytoplankton respond with another bloom, which, although not as remarkable as the spring bloom, is often sufficient to initiate another upswing of the zooplankton population. As winter approaches, the autumn bloom is cut short by decreasing light and reduced temperatures. As production goes down, resistant overwintering stages of both phytoplankton and zooplankton become more abundant. Convective mixing continues to recharge the nutrient load of the surface waters in readiness for a repeat of the entire performance the following spring. It is now estimated that, on average, about 120 gC/m^2/yr is produced in oceanic temperate and subpolar areas, with most of that total occurring during the spring diatom bloom.

Warm Seas

In temperate regions, the production characteristics of tropical and subtropical waters closely resemble those of continuous summer. Sunlight is available in abundance, yet production is low (about 40 gC/m^2/yr) owing to a strong permanent thermocline that blocks vertical mixing of nutrients. The low rate of nutrient return is partially compensated for by a year-round growing season and a deep photic zone. Even so, production and standing crops are low, and dinophytes are usually more abundant than diatoms. Coral reefs and regions of equatorial upwelling are more productive (up to 1,000 gC/m^2/yr for coral reefs), but both are very limited in geographical extent.

Coastal Upwelling

Coastal upwelling in temperate seas alters the generalized picture presented in figure 5.22 by replenishing nutrients during the summer, when they would otherwise be depleted. As long as light is sufficient and upwelling continues, high phytoplankton production occurs and is reflected in abundant local animal populations. In some areas, the duration and intensity of coastal upwelling fluctuate with variations in atmospheric circulation. Along the Washington and Oregon coasts, the variability of spring and summer wind patterns produces sporadic upwelling interspersed with short periods of no upwelling and lower primary productivity. In the Peru Current, upwelling is massive and is interrupted only by El Niño conditions (see Research in Progress: El Niño, p. 136). In the absence of major disturbances such as El Niño events, coastal upwelling zones have average productivity rates of about 300 gC/m^2/yr.

Polar Seas

Sea-surface temperatures in polar regions are always low. The thermocline, if one exists at all, is poorly established and its associated pycnocline is not an effective barrier to nutrient return from deeper waters. Light, or more correctly the lack of it, is the major limiting factor for plant growth in polar seas. Sufficient light to sustain high phytoplankton growth rates lasts for only a few months during the summer. Even so, photosynthesis can continue around the clock during those few months to quickly produce huge phytoplankton populations. As the light intensity and day length decline, the short summer diatom bloom declines rapidly. Winter conditions closely resemble those of temperate regions except that in polar seas the conditions endure much longer. There, the complete cycle of production consists of a single short period of phytoplankton growth, equivalent to a typical spring bloom immediately followed by an autumn bloom and decline that alternates with an extended winter of reduced net production.

In both the Arctic Ocean and around the Antarctic continent, the formation and melting of sea ice play a central role in shaping patterns of primary productivity. As ice melts in the spring, the low-salinity meltwater forms a low-density layer near the sea surface. This increases vertical stability, which encourages phytoplankton to grow near the sunlit surface. The melting ice also releases temporarily frozen phytoplankton cells into the water to initiate the bloom. As the sea ice continues to melt in early summer, the zone of high phytoplankton productivity follows, creating a very productive, seasonal, migrating ice edge community of diatoms, krill, birds, seals, fishes, and whales. Animals that do exploit this production system must be prepared to endure long winter months of little primary production. The annual average productivity rates for polar seas are low (about 25 gC/m^2/yr) because so much of the year passes in darkness with almost no phytoplankton growth.

Around the Antarctic continent, however, upwelling of deep, nutrient-rich water supports very high summertime primary production rates and annual average rates of about 100 gC/m^2/yr. In these regions, water that sinks in the Northern Hemisphere and flows south at mid-depth surfaces between 50° and 60° S latitude (see figure 1.35), bringing with it a thousand-year accumulation of dissolved nutrients. The extraordinary fertility of the Antarctic seas stands in sharp contrast to the barrenness of the adjacent continent. Consequently, almost all Antarctic life, whether terrestrial or marine, depends on marine food webs supported by this massive upwelling.

Research in Progress

El Niño

The cold and productive waters of the Peru Current flow in marked contrast to the dry and barren land of the adjacent coast. Each summer around Christmastime, a warm current from the north (called El Niño for the child Jesus) flows south, bringing rain and warmer temperatures. Occasionally though, this intrusion of warm tropical water penetrates much farther south, stays longer, and brings heavy rains to the coastal desert. These strong El Niño events bring an explosion of plant growth to the coastal land deserts, yet this same southerly flow of warm, less-dense water blocks upwelling of nutrient-rich waters, and coastal marine populations decline. During severe El Niño years, some fish and dependent seabird populations disappear altogether.

It was not until a few decades ago that oceanographers became aware that the seemingly isolated El Niño events of coastal Peru were linked by atmospheric conditions to dramatic weather changes all over the globe. In 1982–1983, we experienced a severe El Niño, with a complete cessation of upwelling in the Peru Current, heavy flooding along the California coast, drought in Australia, and record-breaking cold in Europe.

It now seems clear that both the causes and effects of El Niño extend throughout the tropical Pacific Ocean as a fluctuation of atmospheric pressures, wind fields, surface ocean currents, and rainfall now called the Southern Oscillation. The term *El Niño* is now used to describe one-half of the Southern Oscillation, with weak trade winds, warm-water temperatures in the eastern tropical Pacific, and small differences in surface air pressures across the tropical Pacific Ocean. The other phase of the Southern Oscillation, La Niña, is characterized by oceanographic and atmospheric conditions opposite those of El Niño. La Niña is generally described as the more benign or normal phase of the Southern Oscillation, but it is thought to have been a contributor to the severe drought conditions of the central North American plains states in 1988. El Niño and La Niña are simply two contrasting expressions of the same global event, the Southern Oscillation. When the Pacific Coast was experiencing an intense El Niño in 1982–1983, La Niña conditions prevailed over the Atlantic Ocean. Another, possibly more intense, El Niño-Southern Oscillation (ENSO) event

developed in 1997 and was continuing as this was being written.

The detailed mechanics of the Southern Oscillation are far from understood, but we do know that it has an irregular time scale of a few years and that it is driven by large-scale changes in ocean surface winds and water temperatures of the tropical Pacific Ocean. The atmosphere over the tropical Pacific works like a convection engine to transport enormous loads of heat from the warm sea surface to the relative cold of the upper atmosphere 15 km above the sea surface. These zones of atmospheric convection normally shift back and forth across the equator with the sun each year. In El Niño years, the area of warm water drifts eastward along the equator, and the zone of convection follows. The trade winds relax, causing the Equatorial Currents to slow and the Equatorial Countercurrents to increase their flow of warm surface water eastward. The eastern tropical Pacific experiences warmer sea-surface temperatures (right figure), blocked upwelling, and increased rainfall. Eventually, the area of warm tropical water contracts westward, and El Niño conditions are

Global Marine Primary Production

High latitudes, shallow regions, and zones of upwelling generally support large populations of marine primary producers, but most of this production is accomplished during the warm summer months, when light is not a growth-limiting factor. Open-ocean regions, especially in the tropics and subtropics, where a strong thermocline and pycnocline are permanent features, and in polar seas, where light is limited through much of the year, have low (<50 gC/m^2/yr) rates of primary production.

replaced by the typical La Niña features of cooler eastern tropical Pacific surface temperatures (left figure), low rainfall, and well-developed coastal upwelling along Peru and northern Chile.

For more information on this topic, see the following:
Jacobs, G. A., et al. 1994. *Decade-scale trans-Pacific propagation and warming effects of an El Niño anomaly.* Nature 370:360–63.

Philander, S. G. 1990. El Niño, La Niña, and the Southern Oscillation. *San Diego: Academic Press.*

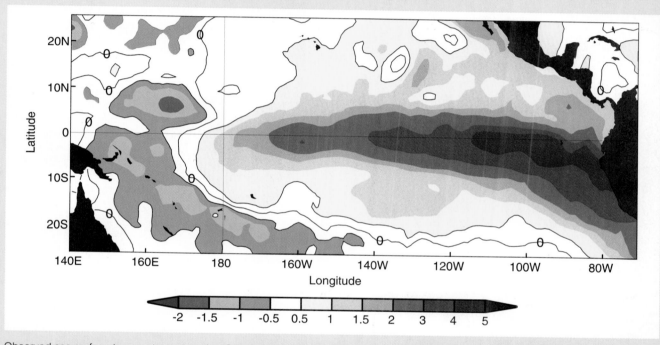

Observed sea surface temperature anomaly, °C in the Equatorial Pacific Ocean (based on 7-day average in mid-September, 1997)

Table 5.1 lists and compares several marine production zones (see p. 140). Figure 5.23 presents a more general picture of marine primary production. Published values included in syntheses such as this vary somewhat, depending on the sets of assumptions used to make productivity estimates for poorly studied regions of the globe, but most published values are within 20% of each other. In the past few decades, the estimates of marine primary production have slowly crept upward, and further revisions are expected. Seventy-two percent of the total primary production occurs in the open ocean, spread thinly over 92% of the ocean's area. The more productive regions are

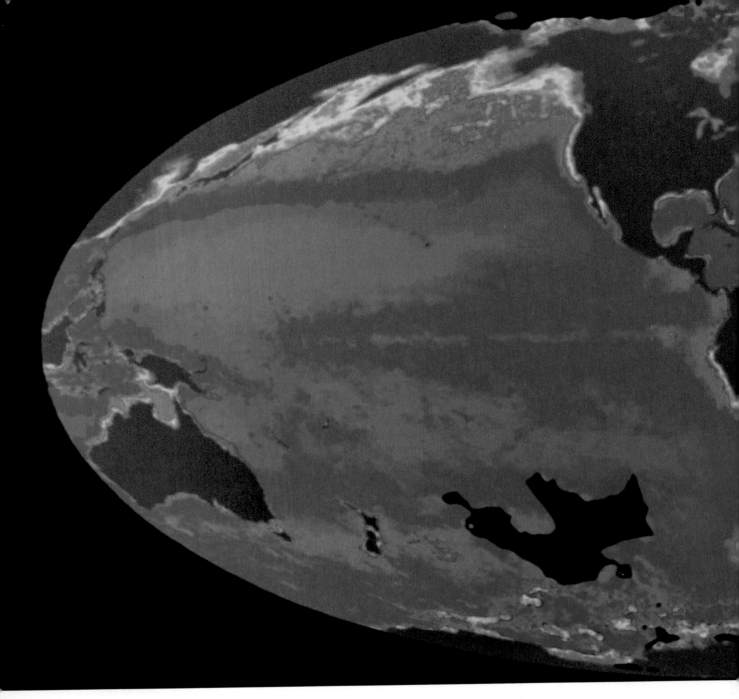

Figure 5.23

The geographical variation of marine primary production, compiled from more than three years of observations by the satellite-borne coastal zone color scanner. Primary production is low (less than 50 gC/m²/yr) in the central gyres (magenta to deep blue), moderate (50–100 gC/m²/yr) in the light blue to green areas, and high (greater than 100 gC/m²/yr) in coastal zones and upwelling areas (yellow, orange, and red).

Photo by Gene Felderman NASA/GSFC Space Data Computing Div., Greenbelt, MD.

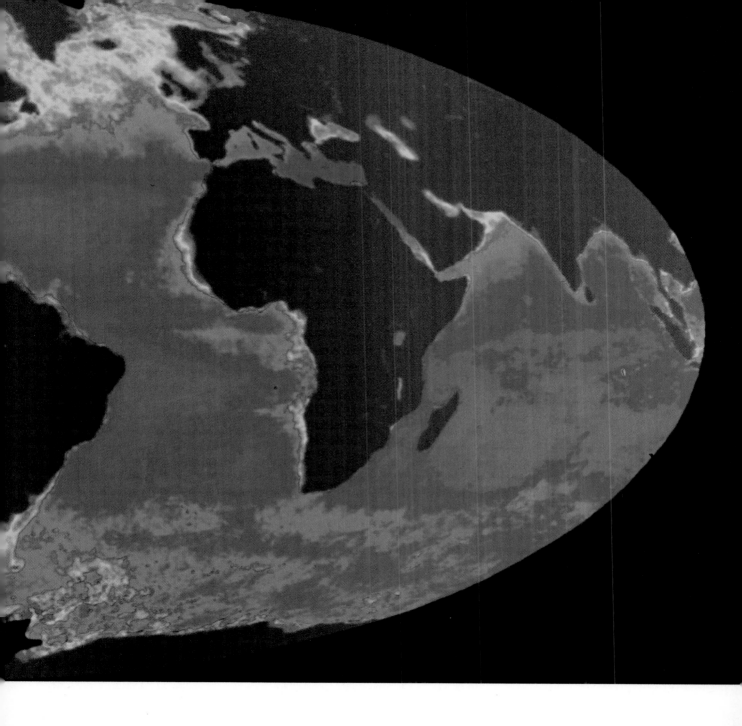

Table 5.1
Rates of Net Primary Production for Several Ocean Regions

Region	Area ($\times 10^6$ km)	% of Ocean	Average gC/m^2/yr	Total net 10^9 tonnes, of C
Open ocean				
Tropics and subtropics	190	51	40	7.6
Temperate and subpolar (including Antarctic upwelling)	100	27	120	12.0
Polar	52	14	25	1.3
Continental shelf				
Nonupwelling	26.6	7.2	200	5.3
Coastal upwelling	0.4	0.1	300	0.1
Estuaries and salt marshes	1.8	0.05	800	1.4
Coral reefs	0.1	—	1,000	0.1
Seagrass beds	0.02	—	1,000	0.02
				27.8

Data from Falkowski 1980, Walsh 1984, and Smith and Nelson 1986.

very limited in geographical extent. Collectively, estuaries, coastal upwelling regions, and coral reefs produce only about 1 billion of the 28 billion tonnes of carbon produced each year.

The productivity numbers of table 5.1 indicate, as previously noted, that nearly 28 billion tonnes of carbon, or 250–300 billion tonnes of photosynthetically produced material, are generated each year in the world ocean. For comparison, the entire human population of the earth requires 5 billion tonnes of food annually to sustain itself. And many members of that population are hungrily eyeing the bounty of the sea. But for several reasons to be discussed in chapter 15, this abundant profusion of plant material will probably never be utilized on a scale sufficient to alleviate the serious nutritional problems already rampant in much of our population. Instead, this vast amount of organic material will continue to fuel the metabolic machinery of the animal members of marine trophic organizations.

Summary Points

- Rates of photosynthesis by phytoplankton and marine plants (gross primary production) and rates of net primary production are measured with light and dark bottle techniques using radioactive C^{14} tracers and are supplemented with oceanic-scale satellite monitoring of phytoplankton abundance. With these techniques, the effect of several environmental factors on the rates of primary production in the sea can be studied. Significant among these factors are light intensity and quality, nutrient availability, and grazing pressure.
- Several groups of marine autotrophs complement the light-absorbing capabilities of chlorophyll with various accessory pigments that give them their characteristic coloration.
- The light and nutrient requirements of primary producers are usually spatially separated in the sea; light is available at the sea surface, yet

great stores of dissolved nutrients are concentrated in waters below the photic zone. Consequently, primary production rates in the sea are substantially less than those on land. Only in areas where upwelling and vertical mixing return the deep-water nutrients to the photic zone do marine primary producers thrive.

- The cumulative effects of cyclic grazing pressures and varying light intensity and nutrient availability create a complex pattern of seasonally and geographically variable primary production. Taken as a single production system, the marine environment generates 250–300 billion tonnes of photosynthetically produced material each year, about one-half of what is produced on land.

Review Questions

1. Name three ocean current regions characterized by coastal upwelling.
2. For each of the following processes of returning nutrients to the photic zone, list the general geographical regions where the named process exerts the major influence on the rate of nutrient return: diffusion, coastal upwelling, convective mixing.
3. List and describe the physical and chemical factors that initiate the spring and fall diatom blooms in temperate ocean waters.
4. Describe the absorption spectra of seawater, chlorophyll *a*, and an accessory pigment; explain why large attached marine plants have concentrations of accessory pigments and relate this to their general depth distribution.

Challenge Questions

1. In areas of upwelling, the rate of marine primary productivity may often be comparable to good farmland, although the phytoplankton population sizes may be very low. Using phytoplankton standing crop and turnover rates, explain why this is true.
2. What advantage does the radioactive carbon method for measuring primary productivity have over the oxygen production method used earlier? What advantages do satellite-borne sea-surface color scanners have over the C^{14} method?
3. Describe two different factors that permit chlorophyll-laden green algae to thrive at greater depths in clear tropical water than they do in temperate coastal water.
4. How would you account for the relatively small contribution that seaweeds are thought to make to the total marine plant production system?

Suggestions for Further Reading

Books

Bougis, P. 1976. *Marine plankton ecology.* New York: Elsevier.

Falkowski, P. G., ed. 1980. *Primary productivity in the sea.* New York: Plenum Press.

Lembi, C. A., and J. R. Waaland, eds. 1988. *Algae and human affairs.* Cambridge University Press.

Rowan, K. S. 1989. *Photosynthetic pigments of algae.* New York: Cambridge University Press.

Steele, J. H. 1974. *The structure of marine ecosystems.* Cambridge: Harvard University Press.

Steeman Nielsen, E. 1975. *Marine photosynthesis.* New York: Elsevier.

Articles

Baker, J. D., and W. S. Wilson. 1986. Spaceborne observations in support of earth science. *Oceanus* 29(4):76–85.

Brown, O. B., et al. 1985. Phytoplankton blooming off the U.S. east coast: A satellite description. *Science* 229:163–67.

Chisholm, S. W. 1992. What limits phytoplankton growth? *Oceanus* 35(3):36–46.

Correll, D. L. 1978. Estuarine productivity. *Bioscience* 28:646–50.

King, R. J., and W. Schramm, 1976. Photosynthetic rates of benthic marine algae in relation to light intensity and seasonal variations. *Marine Biology* 37:215–22.

Landry, M. R. 1976. The structure of marine ecosystems: An alternative. *Marine Biology* 35:1–7.

Malone, T. C. 1971. The relative importance of nannoplankton and netplankton as primary producers in tropical oceanic and neritic phytoplankton communities. *Limnology and Oceanography* 16:633–39.

Mann, K. H. 1973. Seaweeds: Their productivity and strategy for growth. *Science* 182:975–81.

Mcpeak, R. H., and D. A. Glantz. 1984. Harvesting California's kelp forests. *Oceanus* 27(1):19–26.

Nicol, S., and I. Allison. 1997. The frozen skin of the southern ocean. *American Scientist* 85:426–47.

Perry, M. J. 1986. Assessing marine primary productivity from space. *Bioscience* 36:461–67.

Philander, G. 1989. El Niño and La Niña. *American Scientist* 77:451–59.

Richardson, L. L. 1996. Remote sensing of algal bloom dynamics. *Bioscience* 46:492–501.

Saffo, M. B. 1987. New light on seaweeds. *Bioscience* 37:654–64.

Smith, W. O., Jr., and D. M. Nelson. 1986. Importance of ice edge phytoplankton production in the southern ocean. *Bioscience* 36:251–57.

Walsh, J. J. 1984. The role of ocean biota in accelerated ecological cycles: A temporal view. *Bioscience* 34:499–507.

Internet Addresses

Phytoplankton
http://www.pmel.noaa.gov/bering/pages/env_phyt.html
Bering Sea and North Pacific Ocean
http://www.pmel.noaa.gov/bering/
The FOCI Project Office, NOAA/Pacific Marine Environment Laboratory.

Phytoplankton Pigment Concentration from U.S. Coast Guard
http://satori.gso.uri.edu/satlab/
Includes General Description of Plankton, Phytoplankton, Photosynthesis and Respiration, Productivity, Seasonal Phytoplankton Cycle, Upwelling and Primary Production, and much more.
Excellent site, with links to many more relevant topics.

Morris. B. D., 1992:"The Biological Importance of Physical Oceanographic Processes: Spatial and Temporal Scales."
http://www.maths.unsw.edu.au/~brad/essay/essay.html
The marine ecology is made up of two important areas: biological processes that occur in the sea and physical processes of the world's oceans. This essay discusses the interrelations of the two.

Ocean Biodiversity from the State Darwin Museum
http://www.darwin.museum/ru/Bio/oceane.htm
State Darwin Museum **http://www.darwin.museum.ru.** Discussion of the history and the present of ocean biodiversity.

NOAA El Niño Theme Page
http://www.pmel.noaa.gov/toga-tao/el-nino/institutions.html
National Oceanic and Atmospheric Administration (NOAA). Provides links to: Satellite data, Sea level field analysis, Moored ocean buoy data, Drifting ocean buoy data, Climatologies, and more.

Ocean Climate Laboratory
http://www.lib.noaa.gov/
Click on link for the Laboratory. National Oceanographic Data
  Center (NODC) National Oceanic and Atmospheric
  Administration (NOAA)

Ecosystem Modeling Group
http://www-ocean.tamu.edu/~ecomodel/Welcome.html
Adrian Burd, Department of Oceanography Texas A&M University.
  Study of the interactions between physical, biological, and
  chemical processes in a variety of marine environments by
  constructing mathematical models of the systems being studied.
  Contains links to Marine particles, Larval transport, Macrophytes,
  and Ecosystem models.

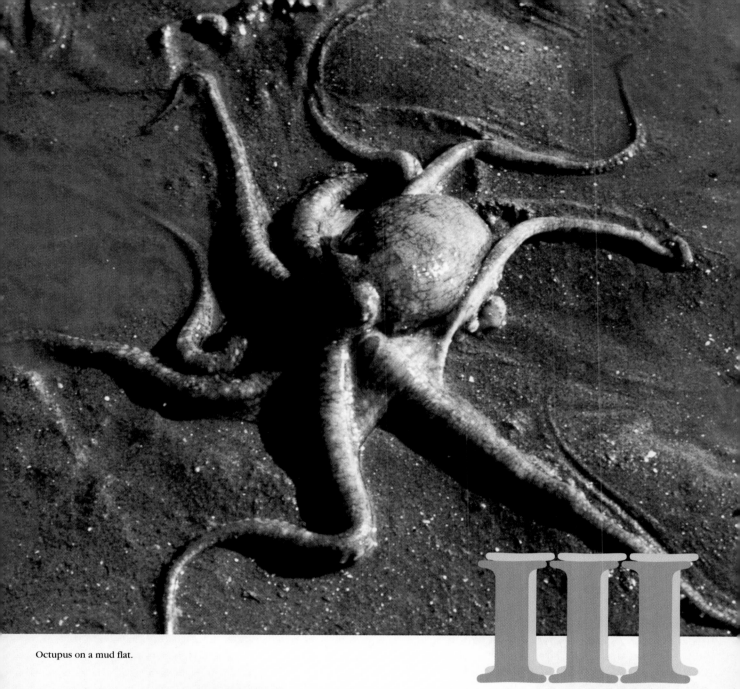

Octupus on a mud flat.

Diversity of Marine Animals

Teleost fishes are considered to be one of the most successful groups of living marine vertebrates because of their abundance, diversity, and ability to survive and reproduce in most marine environments. In contrast, only seven living species of sea turtles exist, and all of them are threatened or endangered throughout some or all of their range. Comparison of the numbers and diversity of these two groups raises an interesting question. Since there are so many more teleosts than sea turtles, are the former better adapted to life in the ocean than the latter? One answer is Yes; species diversity and numerical abundance are commonly used to compare the relative ecological and evolutionary success of different groups of organisms (species that are not adapted do not exist). From an evolutionary perspective, however, a better answer would be Not necessarily. Groups represented by many species cannot always be assumed to be better adapted ecologically than those with fewer species. The seven species of sea turtles, for example, are considered by many to be the most highly adapted of all marine vertebrates. To understand this second, more equivocal answer, it is necessary to appreciate the differences and evolutionary consequences of the terms adaptedness and adaptiveness. Highly adapted species are "experts" within the range of environmental conditions experienced by their ancestors, and their descendants exhibit little genetic and phenotypic variation in environments that change gradually over many generations. Natural selection reduces the number of young with phenotypes dissimilar to those of their parents, and adaptedness gradually increases. Sea turtles, which may be too highly adapted to cope with the rapid environmental changes presently occurring as a result of human activities, may be at risk for extinction. Many species of teleost fishes are also highly adapted to certain marine habitats and they face the same hazards as sea turtles, but the bony fishes, as a group, retain the adaptive potential to meet the challenges of new and changing marine environments created by humans. The cost of adaptiveness is the production of young that do not survive in the environments of their parents. The potential benefits may outweigh this cost in the long run, however. An ecological catastrophe is said to occur when species cannot adapt quickly enough to accommodate environmental changes. Are sea turtles or teleosts better adapted to life in the sea? The best answer may be that it will depend to a great extent on us.

Protozoans and Invertebrates

I n this chapter, the remainder of the kingdom Protista, the nonphotosynthetic protozoans, is introduced, followed by the invertebrate phyla of the kingdom Animalia. Only the more abundant and obvious phyla are described (groups that are primarily or wholly parasitic are not discussed). Of the many phyla introduced in this chapter, only a few—in particular, protozoans, cnidarians, nematodes, mollusks, annelids, echinoderms, and arthropods—clearly dominate the composition of and monopolize the energy flow in most marine communities. These phyla are presented in an order that generally corresponds to a movement up the right, or heterotrophic side, of the phylogenetic tree in figure 3.2 and expanded for more detail in figure 6.1. This sequence represents several significant trends in animal evolution. Increased complexity and specialization of structures are evident, especially in the systems involved in oxygen exchange, excretion, feeding and digestion, circulation, and reproduction. In the more complex phyla, there is a greater dependence on sexual reproduction and less dependence on asexual budding or fragmentation. Improved sensory systems and increasingly complex brains able to integrate sensory information have led to expanding patterns of behavioral responses. Table 6.1 lists these phyla and some of their major taxonomic subgroups.[1]

Animal Beginnings—The Protozoans

The term *protozoa* encompasses a variety of nonphotosynthetic, microscopic members of the kingdom Protista. (The photosynthetic protists were described in chapter 3.) They are included in this chapter for convenience and because many biologists casually consider them "single-celled animals." Protozoans consist of a single cell or loose aggregates of a few cells, and they share some other features of animals described later in this chapter, including ingesting food for nutrition and an absence of cell walls.

Seven protozoan phyla are usually described in modern classification systems. Of these, four are mostly or completely parasitic; the other three include marine nonparasitic species thriving in both benthic and planktonic communities. Asexual reproduction by cell division is common. Sexual reproduction, when it does occur, is often quite complex, with the process of meiosis separated from that of nuclear fusion. These three free-living protozoan phyla are commonly distinguished by their different methods of locomotion. When moving in water, these small cells encounter very high viscous forces between

---

[1]For more detailed taxonomic information regarding local species in your coastal area, see Appendix C for a list of several field-oriented identification guides.

Table 6.1
A Partial (and Brief) Taxonomy of the Marine Animal and Nonphotosynthetic Protist Groups Included in Figure 6.1*

Kingdom: Protista
   Phylum: Sarcomastigophora (5,000, all habitats)—unicellular animals; locomotion with flagella or pseudopodia
   Phylum: Ciliophora (5,000, all habitats)—unicellular animals; locomotion with numerous cilia
   Phylum: Labyrinthomorpha—(5–10, all habitats) small colonial encrusting networks of cells
Kingdom: Animalia
   Phylum: Porifera (10,000, mostly marine)—simple multicellular animals found attached to solid substrates in benthic habitats; reproduction is both asexual and sexual and results in free-swimming larval stages
   Phylum: Placozoa (1, marine)—small asymmetrical plate of cells
   Phylum: Cnidaria (9,000, mostly marine)—radially symmetrical animals with mouth, tentacles, nematocysts, and simple sensory organs and nervous system; common in both benthic and pelagic habitats; reproduction is both sexual and asexual (by budding or fission)
     Class: Hydrozoa—solitary or colonial, with both polypoid and medusoid forms
     Class: Scyphozoa—free-swimming medusoid forms (most jellyfishes)
     Class: Anthozoa—attached benthic polypoid forms (corals and anemones)
   Phylum: Ctenophora (90, marine)—biradially symmetrical, pelagic swimming animals with rows of cilia (ctenes)
   Phylum: Platyhelminthes (12,700, all habitats)—free-living and parasitic flatworms
     Class: Turbellaria—small free-living flatworms with incomplete digestive tracts and ciliated undersides; found in benthic habitats
   Phylum: Nemertina (650, mostly marine)—most are small, inconspicuous wormlike benthic animals with complete digestive tracts
   Phylum: Gastrotricha (175, mostly marine)—microscopic, with elongated bodies; in benthic habitats
   Phylum: Kinorhyncha (64, marine)—elongated, less than 1 mm in length; in benthic habitats
   Phylum: Gnathostomulida (80, marine)—small benthic worms
   Phylum: Priapulida (8, marine)—small, benthic worms
   Phylum: Nematoda (10,000, all habitats)—parasitic and free-living roundworms a few millimeters in length; mostly benthic
   Phylum: Entoprocta (60, mostly marine)—nearly microscopic benthic animals that form colonial encrustations on hard substrates
   Phylum: Ectoprocta (4,000, marine and fresh water)—superficially resemble Entoprocta
   Phylum: Phoronida (70, marine)—tube-dwelling benthic worms
   Phylum: Brachiopoda (260, marine)—benthic animals; bodies covered with hinged shell
   Phylum: Mollusca (65,000, mostly marine)—unsegmented body usually covered with external shell of 1, 2, or 8 pieces
     Class: Aplacophora—rare benthic mollusks without shells
     Class: Monoplacophora—rare, benthic
     Class: Amphineura—shallow-water benthic animals known as chitons; 8-piece shell
     Class: Gastropoda—mostly benthic; shell usually absent or of 1 piece; includes slugs, snails, and limpets
     Class: Scaphopoda—benthic; shell of 1 piece and elongated; known as tusk shells
     Class: Bivalvia—benthic; shell of 2 pieces; clams, oysters, and other bivalves
     Class: Cephalopoda—benthic and pelagic; shell usually absent, foot modified as tentacles with suckers; octopuses and squids
   Phylum: Sipuncula (250, marine)—benthic worms a few centimeters long; known as peanut worms
   Phylum: Echiurida (60, marine)—benthic; cylindrical worms
   Phylum: Pogonophora (80, marine)—deep-water benthic; tube-dwelling worms; to several centimeters in length
   Phylum: Hemichordata (80, marine)—elongated benthic worms; acorn worms
   Phylum: Chaetognatha (50, marine)—pelagic, active predators; a few millimeters in length; known as arrowworms
   Phylum: Annelida (8,700, marine, fresh water, and terrestrial)—segmented worms, to several centimeters in length
     Class: Polychaeta—mostly benthic, free-living
     Class: Hirudinea—leeches; some parasitic
   Phylum: Arthropoda (920,000, all habitats)—segmented animals with bodies covered by exoskeleton of chitin; most a few centimeters or less in length; several classes not found in marine habitats

water molecules and, regardless of their mode of locomotion, do not swim so much as crawl through their watery environment.

Sarcomastigophora

A large and widespread phylum, the sarcomastigophorans use either whiplike flagella or extensions of their cellular protoplasm, pseudopodia (figure 6.2), or both, for locomotion. The foraminiferans

Class: Merostomata—horseshoe crabs; benthic near-shore animals
Class: Pycnogonida—sea spiders; benthic animals with 4 pairs of elongated legs
Class: Crustacea—mostly marine; with 2 pairs of antennae; numerous pelagic and benthic species
    Subclass: Branchiopoda—brine shrimps
    Subclass: Ostracoda—seed shrimps; pelagic animals usually less than 1 cm
    Subclass: Copepoda—abundant animals in pelagic and benthic habitats; microscopic to about 1 cm
    Subclass: Cirripedia—barnacles; larger benthic, attached animals
    Subclass: Malacostraca
        Order: Mysidacea—mysids; benthic and pelagic, size to a few centimeters
        Order: Cumacea—burrow in mud and sand; size to a few centimeters
        Order: Isopoda—benthic; body flattened dorsoventrally; size to a few centimeters
        Order: Amphipoda—benthic and pelagic; body laterally flattened; size to a few centimeters
        Order: Stomatopoda—mantis shrimps; benthic; size to 30 cm
        Order: Euphausiacea—krill; pelagic; size to several centimeters
        Order: Decapoda—crabs, shrimps, and lobsters; mostly benthic; several centimeters to 1 m in size
Phylum: Echinodermata (5,300, marine)—5-sided radial symmetry; most benthic
    Class: Echinoidea—sea urchins, sand dollars
    Class: Asteroidea—sea stars
    Class: Ophiuroidea—brittle stars
    Class: Crinoidea—feather stars, sea lilies
    Class: Holothuroidea—sea cucumbers
    Class: Concentricycloidea
Phylum: Chordata (39,000, all habitats)
  Subphylum: Urochordata
    Class: Ascidiacea—sea squirts; benthic; solitary or colonial
    Class: Larvacea—pelagic; less than 1 cm
    Class: Thaliacea—salps; pelagic; gelatinous
  Subphylum: Cephalochordata—slender, laterally compressed; benthic
  Subphylum: Vertebrata—fishes and tetrapods
    Class: Agnatha—lampreys and hagfishes
    Class: Chondrichthyes—sharks, skates, and rays
    Class: Osteichthyes—bony fishes; includes about 30 orders with marine species
    Class: Amphibia—frogs, toads, and salamanders
    Class: Reptilia—marine turtles, iguanas, crocodiles, and sea snakes
        Order: Testudinata—turtles
        Order: Squamata—iguanas and snakes
        Order: Crocodilia—caymens and crocodiles
    Class: Aves—marine birds
        Order: Sphenisciformes—penguins
        Order: Procellariiformes—albatrosses, petrels, fulmars, shearwaters
        Order: Pelecaniformes—pelicans, cormorants, gannets, boobies
        Order: Charadriiformes—gulls, sandpipers, puffins
    Class: Mammalia
        Order: Carnivora—sea lions, seals, walruses, sea otters
        Order: Cetacea—whales
        Order: Sirenia—manatees and dugongs

*The numbers in parentheses refer to the approximate numbers of described species in that group.
Data mostly from Villee et al. 1984 and Hickman et al. 1984.

and radiolarians are members of this phylum, as are a large variety of amoeboid and flagellated forms. Sometimes, the photosynthetic dinophytes (considered with phytoplankton in chapter 3) are also included in this phylum.

About one-half of all named protozoans are foraminiferans. Foraminiferans are shelled amoebae that are mostly marine. They are common in the plankton, but most are benthic or live attached to plants and other animals. Most foraminiferans are microscopic, although

Figure 6.1

One scheme to illustrate the evolutionary relationships between eucaryotic kingdoms and among the phyla of protozoans and animals introduced in this chapter. Each phylum is represented by a labeled box, and its relative evolutionary distance to other phyla is indicated by the branched lines. The kingdoms are shaded to indicate autotrophs (green) or heterotrophs (tan).

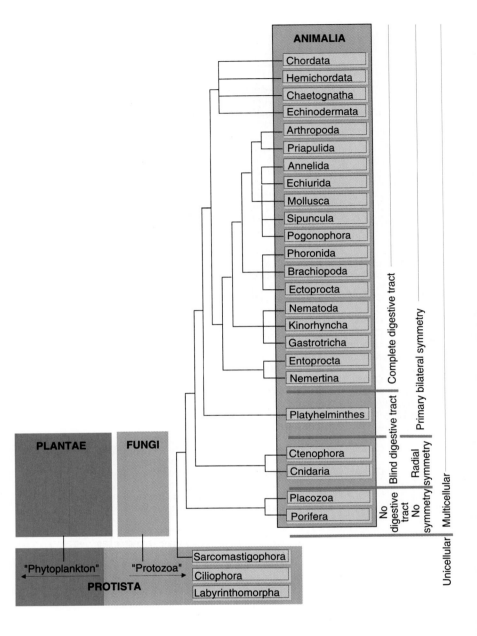

individuals of a few species are several millimeters in size. They have internal chambered shells, or tests, usually composed of either $CaCO_3$ or cemented sand grains. Penetrating this shell, or test, are numerous pseudopodia (figure 6.2a). The pseudopodia are used for locomotion, for attachment, and for collecting food. Some planktonic foraminiferans, such as *Globigerina* (figure 6.2b), are so widespread and abundant that their tests blanket large portions of the seafloor. After thousands of years of accumulation, this globigerina ooze may form deposits tens of meters thick. The famous chalk cliffs of Dover, England, are composed

(a)

(b)

Pseudopodia

Figure 6.2

A planktonic foraminiferan, *Globigerina. (a)* Drawing of an intact animal. *(b) Photograph of* Globigerina test.

(a) From Brady 1884; *(b) Courtesy Deep Sea Drilling Project.*

mainly of foraminiferan tests that accumulated on the seafloor and were subsequently lifted above sea level.

Radiolarians are entirely marine, and most members are planktonic. They are similar in size to planktonic foraminiferans. An internal skeleton of silica (SiO_2) forms the beautiful symmetry often associated with radiolarians (figure 6.3).

Ciliophora

Members of the phylum Ciliophora are known as ciliates, as they possess **cilia** as their chief means of locomotion. Structurally, a cilium is much like a short flagellum; however, cilia are typically much more numerous than flagella, and they move in a coordinated manner (figure 6.4). Cilia move water parallel to the cell surface (flagella move water perpendicular to the surface of the cell). Tintinnids are probably the most abundant of the marine ciliophores. The tintinnid cell is partially enclosed in a vase-shaped lorica made of cemented particles or of a material secreted by the cell (figure 6.5). Ciliated tentacles at one end of the cell are used for feeding. A large variety of other ciliates exist in the sea. Most are parasitic on or in other marine animals.

Figure 6.3

Scanning electron micrograph of the silicate skeletons of two planktonic radiolarians. *(a) Euphysetta elegans,* 230×. *(b) Elatomma pinetum,* 110×.
Courtesy Kuzo Takahashi, Scripps Institution of Oceanography.

Figure 6.4

The pattern of ciliary movement, appearing as waves of alternating recovery and power strokes sweeping over the cell surface.

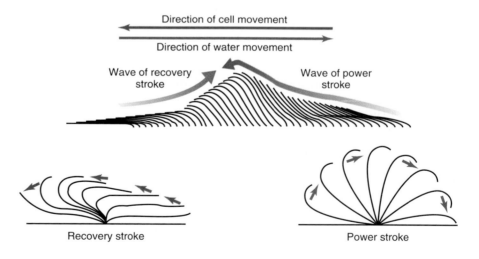

Figure 6.5

A marine tintinnid with a crown of cilia at one end.

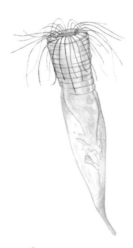

Labyrinthomorpha

The small phylum Labyrinthomorpha includes a group of free-living organisms characterized by a network of slime through which colonies of small (about 10 μm long), spindle-shaped cells live and move. Most species are marine, forming their colonial networks on the surfaces of sea grasses and benthic algae. Their mechanism of gliding through their slime networks is not understood and is accomplished without pseudopodia, cilia, or flagella. Reproduction is both sexual and asexual in this phylum.

Defining Animals

The boundary separating the kingdoms Protista and Animalia are vague (see the criteria listed on p. 74), and some colonial protozoans seem on the verge of crossing that boundary with their impression of multicellularity. Yet these protozoans lack two of the hallmark features of animals: contractile muscles and signal conducting neurons. In addition, animals usually reproduce sexually, producing a diploid zygote that develops by successive cleavages to a hollow ball of cells, the **blastula.**

The first two animal phyla described, the Porifera and Placozoa, might be considered a sideline to the major trends of animal evolution. They are animals, yet they lack the coordinated patterns of cell functions seen in other animal phyla, with individual cells functioning independently rather than as parts of more specialized tissues or organs.

Porifera

The Porifera is one of the few animal phyla with a widely accepted common name: the sponges. The sponges are among the simplest multicellular animals. Each sponge consists of several types of loosely aggregated cells organized into a recognizable organism. Despite their structural simplicity, sponges share several advantages with other multicellular animals. Unlike protistans, cells within each individual sponge can divide repeatedly to permit larger size and longer life span of the organism. In addition, specialization of cells, although limited in sponges, can promote more efficient handling of food, protection, and other diverse chores of survival (figure 6.6).

The name Porifera stems from the numerous pores, holes, and channels that perforate the bodies of sponges. Water is circulated through these openings into an internal cavity, the spongocoel, where food and oxygen are extracted by flagellated choanocytes lining the spongocoel. The water then exits through a large excurrent pore, the osculum.

Sponges are mostly marine and are usually found attached to hard substrates such as rocks, pilings, or animal shells. Sometimes they are radially symmetrical, but more commonly they conform to the shape of their substrate or to the sculpting influence of waves and tides. Some sponges are supported internally by a network of flexible **spongin** fibers. (The commercial bath sponge is actually the spongin skeleton with all living material removed.) Other sponges have skeletons composed of hard mineralized **spicules.** The spicules are either calcareous ($CaCO_3$) or siliceous (SiO_2). The spicule skeleton of the deep-water glass sponge *Euplectella* is one of the most complex and beautiful of all sponges (figure 6.7).

Figure 6.6

A group of marine finger sponges and several of the specialized cell types that make up the sponge wall.

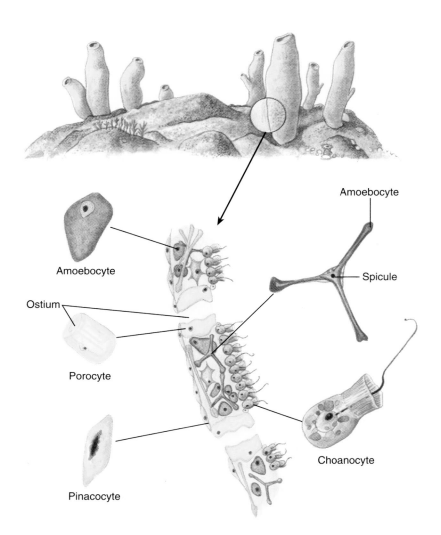

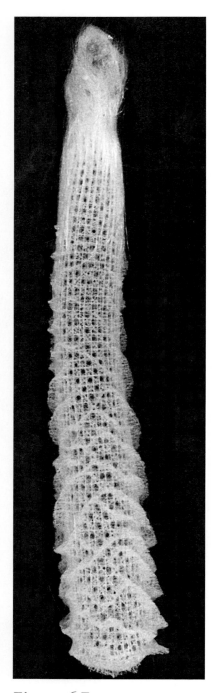

Figure 6.7

Silicate skeleton of a glass sponge, *Euplectella.*

Placozoa

The phylum Placozoa is represented by only one known species, *Trichoplax adhaerens,* and until 1971 it was misidentified as a larva from another phylum. Each animal is 2–3 mm long and consists of a few thousand cells shaped into a flattened plate. Outside aquaria, it is found throughout the Pacific Ocean free-swimming for at least part of its life. As it can move in any direction, it exhibits no symmetry. Both surfaces are covered with cilia for swimming. Digestion of food is accomplished by secreting enzymes externally then absorbing the digested molecules. Like sponges, this animal exhibits limited specialization and organization of cells and can regenerate a complete animal from a single cell. Sexual reproduction is unknown in this species.

Radial Symmetry

Members of the phyla Cnidaria and Ctenophora exhibit radially symmetrical body plans. The circular shape of radially symmetrical animals

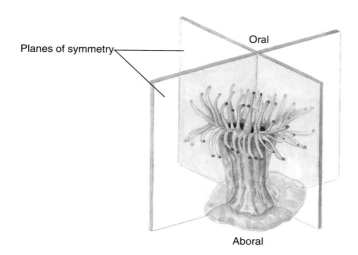

Planes of symmetry

Oral

Aboral

Figure 6.8

Planes of symmetry in a radially
symmetrical animal.

provides several different planes of symmetry to divide the animal
into mirror-image halves (figure 6.8). The mouth is located at the center of the body on the **oral** side; the opposite side is the **aboral** side.
Radially symmetrical animals possess a relatively simple diffuse nerve
net that lacks a central brain to process sensory information or to or-
ganize complex responses.

Cnidaria

The phylum Cnidaria includes a large, diverse group of relatively simple, yet versatile, marine animals, such as jellyfishes, sea anemones,
corals, and hydroids. The inner and outer body walls of all cnidarians
are separated by a gelatinous layer called the **mesoglea.** A centrally
located mouth leads to a baglike digestive tract, the **gastrovascular
cavity.** The mouth is surrounded with tentacles capable of capturing and ingesting a wide variety of marine animals. The tentacles and,
to a lesser extent, other parts of the body, are armed with batteries of
microscopic structures, the **nematocysts.** Nematocysts are produced in special cells, the **cnidoblasts,** and are a characteristic of
this phylum. They are discharged when stimulated by contact with
other organisms. Some nematocysts are adhesive and stick to the
prey, others become entangled in the prey's bristles or spines, and
still others (figure 6.9) pierce the prey and inject a paralyzing toxin.

Cnidarians exist as free-swimming **medusae** or as sessile benthic
polyps. The two forms have essentially the same body organization.
The oral end of the medusa, bearing the mouth and tentacles, is oriented downward. The mesoglea of most medusae is well developed
and is jellylike in consistency, thus earning them the descriptive, if inappropriate, name of jellyfish. In the polyp, the mouth and tentacles
are directed upward. Many species of cnidarians alternate between a
swimming medusoid generation and an attached benthic polypoid
generation. In a generalized cnidarian life cycle (figure 6.10), polyps
can produce medusae or additional polyps by budding. The medusae
in turn produce eggs and sperm that, after fertilization, develop into
the polyps of the subsequent generation.

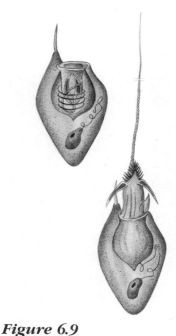

Figure 6.9

(Top) Undischarged nematocyst.
(Bottom) Discharged penetrant
nematocyst.

Figure 6.10

Generalized cnidarian life cycle.

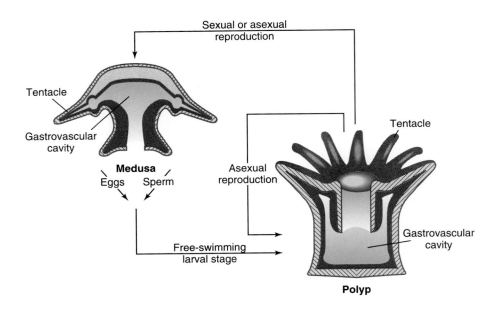

The phylum Cnidaria consists of three classes, each characterized by its own variation of the basic cnidarian life cycle shown in figure 6.10. The Hydrozoa includes colonial hydroids and siphonophores, such as the Portuguese man-of-war, *Physalia.* Hydrozoans usually have well-developed medusoid and polypoid generations. Various individuals of the polyp colony are specialized for particular functions, such as feeding, reproduction, and defense.

In the class Scyphozoa, the polyp stage is reduced or completely absent. This class includes most of the larger and better-known medusoid jellyfishes (figure 6.11). In the third class, the Anthozoa, the polyp form dominates and the medusoid generation is absent. Many anthozoans, such as corals and sea fans, are colonial, but some anemones exist as large solitary individuals (figure 6.12). Unlike most cnidarians, the corals and some other anthozoans (and a few hydrozoans) produce external, often massive, deposits of $CaCO_3$.

Ctenophora

The phylum Ctenophora consists of about 90 species. All are marine and most are planktonic, usually preying on small zooplankton. Most individuals are smaller than a few centimeters in size, but one tropical genus *(Beröe)* may be found up to 20 cm in length.

Ctenophores are closely related to cnidarians. Ctenophores have radial body symmetry, a gelatinous medusa-like body, and, in some, colloblast cells that superficially resemble cnidarian nematocysts but are sticky rather than barbed. In fact, one ctenophore species does possess true nematocysts.

Members of this phylum have external longitudinal bands of cilia, called **ctenes** (figure 6.13), that provide propulsion. Tentacles armed with colloblasts capture food.

Figure 6.11

A large jellyfish, *Pelagia.*
Photo by T. Phillipp.

Figure 6.12

Sea anemones, with nematocysts showing as white beadlike structures on tentacles.
Photo by T. Phillipp.

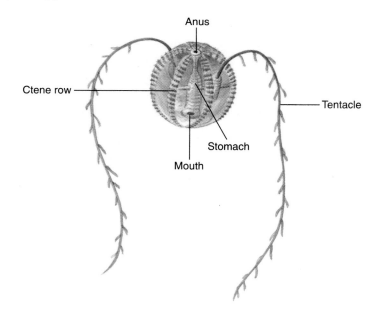

Figure 6.13

A ctenophore, *Pleurobranchia*, with tentacles and four radial rows of visible ctenes.

Bilateral Symmetry

With one exception (the echinoderms), the remainder of the animal phyla exhibit **bilateral body symmetry.** Bilateral symmetry refers to a basic animal body plan in which only one plane of symmetry exists to create two mirror-image halves (figure 6.14). Such animals exhibit definite head (anterior) and rear (posterior) ends, right and left sides,

157

Figure 6.14

Plane of symmetry in a bilaterally
symmetrical animal.

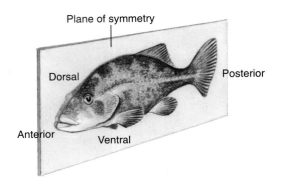

and a top (dorsal) and bottom (ventral) surface. These animals possess
sophisticated sensory systems capable of one-way conduction of
nerve impulses, a more complex mass of nerve cells necessary to
process the widening scope of sensory information. Accompanying
this evolutionary trend toward an anterior brain has been the development in the head region of specialized sensory receptors for vision,
smell or taste, and hearing.

The simplest groups of bilaterally symmetrical animals are composed of small, often overlooked inhabitants of soft mud and sand.
Seven phyla have elongated, mostly wormlike body shapes that easily
slip through mud or between sand grains. (The reasons for the apparent success of "worms" in soft-bottom benthic habitats are discussed
on p. 165.)

Platyhelminthes

Most Platyhelminthes, or flatworms, are parasitic (this group includes
flukes and tapeworms). Only members of the class Turbellaria are free-
living. Turbellaria are primarily aquatic, and the great majority are marine. There are a few planktonic species of flatworms, but most dwell
in sand or mud or on hard substrates.

Marine turbellarians are usually less than 10 cm long, thin, leaf-
shaped, and sometimes quite colorful (figure 6.15a). Cilia, best developed on the flatworm's underside, cover its outer surface. These cilia
provide a gliding type of locomotion for moving over the bottom. The
mouth is usually centrally located on the underside and leads to a
baglike digestive tract. Turbellarians are carnivorous, preying on other
small invertebrates.

Nemertina

The nemertines are benthic animals, known as ribbon worms, that are
closely related to the flatworms but have a more elaborate body structure (figure 6.15c). They have a simple circulatory system, a complex
nervous system, and a complete digestive tract. Individuals of one
species are over 2 m long, but most are much smaller. These shallow-
water animals are equipped with a remarkable proboscis for defense
and food gathering. The proboscis can be everted rapidly from the

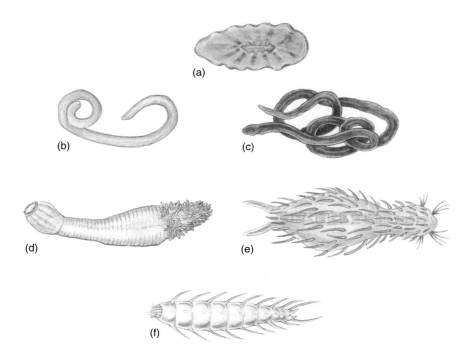

Figure 6.15
Some simple wormlike animal phyla:
(a) flatworm, *(b)* nematode,
(c) nemertine, *(d)* priapulid,
(e) gastrotrich, *(f)* kinorhynch.

anterior part of the body to ensnare prey. The proboscis of some nemertine worms has a piercing stylet to stab prey and inject a toxin.

Gnathostomulida

Gnathostomulids, or jaw worms, are a small group of marine worms also closely related to turbellarian flatworms. Like turbellarians, jaw worms have a mouth but no anus. Individuals seldom exceed 1 mm in length. They live in bottom deposits, where they scrape bacteria and algal film off sediment grains with their jaws and associated basal plate. Of the 80 or so species, most are hermaphroditic, and free-swimming larvae are absent.

Gastrotricha

Gastrotrichs include a large variety of marine species, but most are so small (usually less than 1 mm) that they go unnoticed by most observers. They are cylindrical and elongated, with a mouth, feeding structures, and sensory organs at the anterior end (figure 6.15e). Marine gastrotrichs inhabit sand and mud deposits in shallow water and feed on detritus, diatoms, or other small animals.

Kinorhyncha

Kinorhynchs are exclusively marine and resemble gastrotrichs. They also are cylindrical and elongated, but are covered with a cuticle that is segmented (figure 6.15f). Marine kinorhynchs occur in sand and mud deposits, sometimes in densities of more than 1 million individuals in each square meter of seafloor, where they feed exclusively on bacteria.

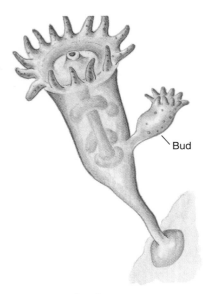

Bud

Figure 6.16

A solitary entroproct, *Loxosomella,* with its crown of feeding tentacles.

Priapulida

More wormlike in appearance are the priapulid worms (figure 6.15d). Less than ten species occur in the phylum Priapulida. They live buried in intertidal sediments of polar and subpolar waters and seldom exceed 10 cm in size. Priapulid worms are detritus feeders or are predatory, feeding on soft-bodied invertebrates they capture with their eversible proboscis. Priapulids are found in hypersaline ponds and anoxic muds, as well as more moderate benthic habitats.

Nematoda

Nematode worms (figure 6.15b) are among the most common and widespread multicellular animals. Some are parasitic, but many are free-living. Most marine nematodes live in the bottom sediments and are found at virtually all water depths. In fact, nematodes are probably the most abundant multicellular animals in the marine benthic environment. Locomotion is not well developed; nematodes depend on quick bending movements of their small bodies. Cylindrical in cross section and greatly elongated, nematodes seldom exceed a few centimeters in length.

Entoprocta

Entoprocts are benthic, living on rocks, shells, sponges, and seaweeds. Most are colonial, secreting thin calcareous encrustations over rocks, seaweeds, and the hard shells of some other animals (figure 6.16). Superficially, they resemble small colonial hydroids. Their external appearance is also quite similar to that of members of another phylum, the Ectoprocta.

Entoprocts and ectoprocts are still sometimes combined in an outdated phylum, the Bryozoa. But studies of internal structures have shown that these two groups are only distantly related, and two separate phyla are warranted. Individuals of both groups have U-shaped digestive tracts and a crown of tentacles projecting from the upper surface. The mouth and anus of entoprocts open within the ring of tentacles (hence the name Entoprocta—inner anus).

The Lophophore Bearers

The **lophophore** is a crown of ciliated feeding tentacles found in three structurally dissimilar phyla of marine animals: the Ectoprocta, Phorondia, and Brachiopoda. These and the other phyla to follow are characterized by a true internal body cavity, or **coelom.** This cavity originates during embryonic development and causes the digestive tract to separate from the body wall. This separation allows the coelomic fluids to move and aid circulation of oxygen, wastes, and nutrients. With a coelom, the digestive tract has become specialized, and its efficiency improved. Body wall muscles function independently of the digestive tract and have a greater range of specialized actions. The coelom was thus a major step in the evolutionary development of more complex animal phyla.

Ectoprocta

The ectoprocts are a major animal phylum, with 4,000 freshwater and marine species. They are primarily members of shallow-water benthic

Figure 6.17

A magnified view of a branched
colony of ectoprocts with extended
feathery lophophores (arrow).

communities, occupying the same general habitats as entoprocts. The
ectoproct mouth is located within the tentacles, but the anus is not.
Like entoprocts, ectoprocts are colonial and form encrusting or
branching masses of small individuals (usually less than 1 mm in size;
figure 6.17).

Ectoprocts and entoprocts provide excellent examples of how the
evolutionary pathways of quite different animal groups converge to
similar adaptive forms and habits when exposed to similar environmental stresses. Such convergent evolution is common in several taxonomic groups, and additional examples will be encountered elsewhere in this book.

Phoronida

The phylum Phoronida consists of about 70 species of elongated, burrowing animals. All are marine and live in tubes in shallow water. During feeding, the lophophore projects out of the tube, but it can be
rapidly retracted for protection. The phoronids seldom exceed 20 cm
in length and have no appendages except for the lophophore.

Brachiopoda

The brachiopods, or lamp shells, were very successful in the past,
with more than 30,000 extinct species described. Fewer than 300 still
survive. *Lingula,* for example, has an unbroken fossil history that extends back over the past one-half billion years of the earth's history!
All brachiopods are benthic and live attached to the sea bottom by a
muscular stalklike pedicle (figure 6.18). The outer calcareous shell superficially resembles that of a bivalve mollusk, but the symmetry of
the shells is quite different. Bivalve shells are positioned to the left
and right of the soft internal organs. In contrast, brachiopod shells are
not symmetrical and are located on the dorsal and ventral sides of the
soft organs.

Figure 6.18

Brachiopods (*Neothyris;* at arrow)
among scattered shell fragments at a
depth of 40 m.

Courtesy J. Richardson, Victoria Museum,
Melbourne.

Living brachiopods occupy a wide variety of seafloor niches, from shallow-water rocky cliffs to deep muddy bottoms. As in the phoronids and ectoprocts, the ciliated lophophore gathers minute suspended material for nutrition from seawater.

The Mollusks

Mollusca

The mollusks are among the most abundant and easily observable groups of marine animals because they have adapted to all the major marine habitats. It is difficult to characterize such a large and diverse group as the phylum Mollusca, but some common traits are observable. Mollusks are unsegmented animals. Most mollusks have a hard external shell surrounding the soft body and use a large muscular foot for locomotion, anchorage, and securing food. Most mollusks have an array of specialized sense organs in the anterior region of their body near the brain. (This pattern of body organization, known as **cephalization,** is most apparent in squids and octopuses.)

This phylum is composed of seven classes. Representatives of five of these classes are quite common and are shown in figure 6.19. In four of these classes, the early planktonic larval form is the **trochophore** larva (figure 6.20). This type of larva is also found in annelid worms. As the trochophore grows, a ciliated tissue, the velum, develops, and the larva is then known as a **veliger.** The velum is used to collect food and to swim.

Chitons belong to the class Amphineura. They are characterized by eight calcareous plates embedded in their dorsal surfaces. These animals, found in rocky intertidal areas, use their large muscular foot to

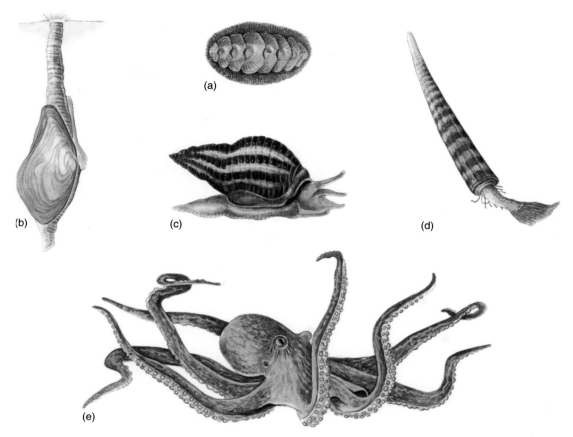

Figure 6.19

Representatives of the common classes of mollusks: *(a)* Amphineura, *(b)* Bivalvia, *(c)* Gastropoda, *(d)* Scaphopoda, *(e)* Cephalopoda.

cling to protected depressions in rocks. Chitons feed by grazing algae from rocks with a rasping tonguelike organ, the **radula** (figure 6.21). Radulas are also found in three other classes of mollusks: the gastropods, scaphopods, and cephalopods.

The class Gastropoda includes snails, slugs (marine, freshwater, and terrestrial), limpets, abalones, and nudibranchs. Although one-piece shells are characteristic of this class, several types of gastropods lack shells. Most gastropods are benthic; only a few without shells, or with very light ones, successfully assume a pelagic lifestyle. Like chitons, many gastropods graze on algae; others feed on detritus and organic-rich sediments. Numerous gastropods are also successful predators of other slow-moving animal species.

The 200 species of tusk shells, class Scaphopoda, are found buried in sediments in a wide range of water depths. As the common name implies, the shells of these animals are elongated and tapered, somewhat like an elephant's tusk but open at both ends. The head and foot project from the opening at the larger end of the shell. Microscopic organisms from the sediment and water are captured by adhesive tentacle-like structures.

The Bivalvia, which includes mussels, clams, oysters, and scallops, have hinged two-piece, or bivalve, shells. As adults, most are slow-moving benthic animals. But some, such as mussels and oysters,

Figure 6.20

Oyster life cycle. About two weeks are needed to develop from egg to spat.

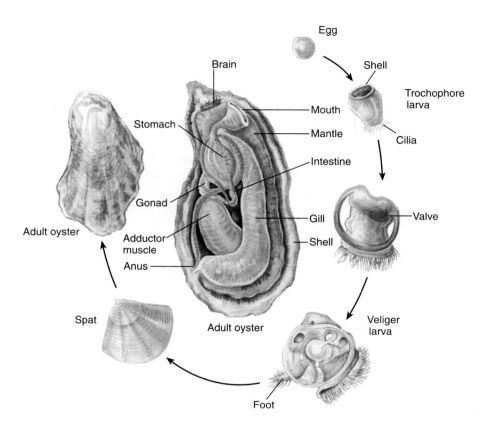

Figure 6.21

Scanning electron micrograph view of the radula of a gastropod mollusk.
Courtesy Dr. Carole S. Hickman.

are cemented to hard substrates. This class has an extensive depth range, from intertidal areas to below 5,000 m. All bivalves lack radulas and are unique among mollusks in having large and sometimes elaborate gills. These gills are covered with cilia (figure 6.22) that circulate water for gas exchange while sorting extremely small food particles entrapped on a thin mucous film secreted over the gill surfaces. Consequently, bivalves are specialized to feed on suspended bacteria, phytoplankton, and microscopic detrital particles found in sediment deposits.

Molluscan evolution has reached its zenith in the class Cephalopoda, which includes the squids, octopuses, cuttlefish, and nautiluses. Members of this class are specialized carnivorous predators with sucker-lined tentacles, well-developed sense organs and large brains, and reduction or loss of the external shell typical of most other mollusks. The eyes of squids and octopuses, for example, are remarkably similar to our own, with a retina, cornea, iris, and lens focusing system. A unique propulsion system, using high-speed jets of water, provides swimming speeds greater than those of any other marine invertebrate. Octopuses also use their eight arms to gracefully crawl over the seafloor. Cephalopods are larger than most other invertebrates. Octopuses sometimes have arm spans exceeding a few meters. The giant squid, *Architeuthis,* which may reach 20 m in length and weigh over 1 ton, is by far the largest living invertebrate species.

(a)

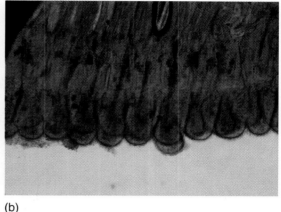
(b)

Figure 6.22

(a) Photo of the intact gill of the mussel *Mytilus. (b)* Micrograph of the extensive gill cilia.

More Wormlike Phyla

The wormlike body structure of the phyla previously discussed was an extremely successful evolutionary adaptation. We will examine the reasons for this success before introducing the remaining phyla of marine "worms."

Most wormlike marine invertebrates live in soft mud or sand deposits. Their elongated body forms permit effective burrowing movements despite a lack of rigid internal skeletons to support the muscles of locomotion. Muscles in the body wall work against the enclosed fluid contents of the body to allow burrowing actions and other body movements. The fluids cannot escape and are essentially incompressible. As such, they provide a **hydrostatic skeleton** for the muscles of the body wall. In the more effective burrowing worms, these muscles are arranged in two sets; circular muscle bands around the body and longitudinal muscles extending the length of the body. Like all other muscles, these muscles can work only by contracting. Thus, when the circular muscles of a worm's body contract, its diameter decreases, squeezing its hydrostatic skeleton and forcing the body to elongate. If the rear of the body is anchored, contracting the circular muscles results in a forward movement of the anterior end. The circular muscles resume their precontraction state by relaxing, allowing the longitudinal muscles to shorten the body and make it fatter. These two types of muscles continue to work in opposition to each other to provide an effective sediment-burrowing motion for a large variety of marine worms.

Sipuncula

Sipuncula, another phylum of wormlike creatures, are found throughout the world ocean. The 250 species of sipunculids, or peanut

Figure 6.23

Sipunculid worms, *Sipunculus.*
Courtesy R. Brusca.

worms, are entirely marine. Most are found in the intertidal zone, but their distribution extends to abyssal depths. Peanut worms are benthic. They live in burrows, crevices, or other protected niches and are often in competition with other wormlike animals. Sipunculids range from 2 mm to over 50 cm in size and have a cylindrical body that is capped by a ring of ciliated tentacles surrounding the anterior mouth (figure 6.23).

Echiurida

The Echiurida are a small phylum of benthic marine worms that resemble peanut worms in size and general shape. Echiurids are common intertidally, but they are occasionally found at depths exceeding 6,000 m. Most echiurids live in burrows in the mud. One remarkable feature of echiurids is their extensible proboscis, a feeding and sensory organ that projects from their anterior. The proboscis of some species is longer than the remainder of the body and is quite effective for gathering food by "mopping" the sediment while the worm remains in the protected confines of its burrow.

Urechis, an echiurid of the California coast known as the fat innkeeper, has a very short proboscis. The proboscis secretes a mucous net from the animal to the wall of its U-shaped burrow (figure 6.24). Usually, the burrow is also inhabited by small crabs, shrimps, or other casual guests. Water is pumped through the burrow by repeated waves of contractions along the worm's body wall. As water passes through the mucous net, extremely fine particles are trapped. When the net is clogged with food, the worm consumes it and constructs another.

Pogonophora

Pogonophorans are almost exclusively deep-water, tube-dwelling marine worms. (Eighty percent of the known species live below 200 m.)

Sediment surface

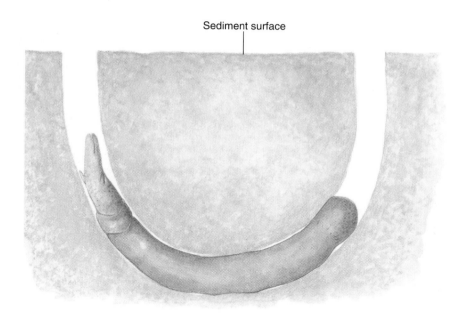

Figure 6.24

The fat innkeeper, *Urechis,* shown within its burrow.

This obscure phylum was not even described until 1900. Since then, about 70 species have been described. Adults are generally only a few centimeters in length. Pogonophorans are noted for their complete lack of an internal digestive tract and are thought to absorb dissolved organic material or to use symbiotic bacteria to provide their energy needs. The remarkably large tube worms recently discovered in deep-sea hot spring communities (figure 11.13) were initially assigned to the phylum Pogonophora. However, ongoing studies suggest that they may be sufficiently distinct to warrant the erection of their own separate phylum.

Hemichordata

Hemichordates (acorn worms) are a small group of benthic marine worms closely related to the pogonophorans. They have an anterior proboscis and a soft, flaccid body that is up to 50 cm long. These worms are generally found in shallow water and live in protected areas under rocks or in tubes or burrows.

Chaetognatha

In contrast to the general benthic habitat of most marine wormlike animals, chaetognaths, or arrowworms, are torpedo-shaped planktonic carnivores (figure 6.25). Although they seldom exceed 3 cm in length, they are voracious predators of other zooplankton. Arrowworms swim with rapid darting motions and capture prey with the bristles that surround their mouth. Only about 50 species of arrowworms exist, but they are frequently very abundant in the zooplankton. Certain species of arrowworms apparently respond to and associate with subtle chemical or physical characteristics of seawater. As such, they serve as useful biological indicators of particular oceanic water types.

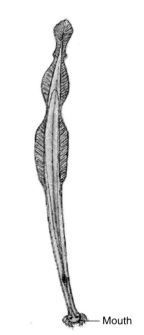

— Mouth

Figure 6.25

Sagitta, a chaetognath.

Figure 6.26

The filtering structures of a tube-dwelling polychaete worm.
Photo by T. Phillipp.

Segmented Animals

Annelida

The annelids are usually represented by the familiar terrestrial earth-worm. However, this phylum also contains a diverse and successful group of marine forms with more than 5,000 species, the class Poly-chaeta. Polychaete worms, like other annelids, are segmented, as are members of the next two phyla. The body cavity and internal organs contained within polychaete worms are subdivided into a linear series of structural units called **metameres.** The result of segmentation is a sequential compartmentalization of the worm's hydrostatic skeleton and surrounding muscles. This permits a greater degree of localized changes in body shape and a more controlled and efficient form of locomotion.

Some polychaetes ingest sediment to obtain nourishment, others are carnivorous, and many use a complex tentacle system (figure 6.26) to filter microscopic bits of food from the water. Suspension-feeding polychaetes often occupy partially buried tubes and are common in intertidal areas; however, they are also found in deeper water. One polychaete, *Tomopteris,* is planktonic throughout its life cycle.

Arthropoda

Like annelids, arthropods are segmented linearly. In addition to the advantages of segmentation, arthropods possess a distinctive hard **exoskeleton** that consists of a complex organic substance, **chitin.** This rigid outer skeleton serves not only as an impermeable barrier against fluid loss and bacterial infection but also as a complex lever system for muscle attachment. Its structure resists deformations caused by contracting muscles, allowing faster responses and greater control of movements. Flexing of the body and appendages is limited

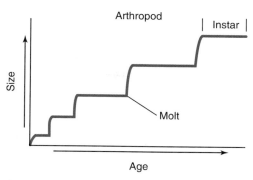
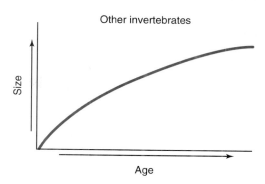

Arthropod · | Instar |

Other invertebrates

Size

Size

Molt

Age

Age

Figure 6.27

Patterns of arthropod and nonarthropod growth. In contrast to the smooth curve of other animals, arthropods rapidly increase their body size in steps following each molt of the exoskeleton.

Figure 6.28

Two mating horseshoe crabs, *Limulus.*

Courtesy A. Kuzirian, Marine Biological Laboratory.

to thin membranous joints located between the rigid exoskeletal plates. In addition, the exoskeleton cancels the shape-changing advantages of segmentation and hydrostatic skeletons found in annelid worms. The exoskeleton also restricts continuous growth. Periodically, arthropods shed their old exoskeleton, and it is replaced by a new, larger one as the animal quickly expands to fill it (figure 6.27).

Members of the phylum Arthropoda account for over 75% of all living animal species identified so far. Most belong to the class Insecta, a group abundant on land and in freshwater habitats. Only a few species of insects, including pelagic water striders (figure 1.12), several types of sand and kelp flies, and water beetles have evolved to thrive in seawater environments. However, three other classes of this phylum, the Crustacea, Merostomata, and Pycnogonida, are primarily or completely marine in distribution.

Two classes of arthropods are completely marine, but their diversity is very restricted. The first class, Merostomata, has an extensive fossil history that includes extinct water scorpions 3 m long. Modern representatives include the horseshoe crab, *Limulus,* an inhabitant of the Atlantic and Gulf coasts of North America (figure 6.28). The sea spiders of the class Pycnogonida are long-legged bottom dwellers with very reduced bodies (figure 6.29). Small pycnogonids only a few

Figure 6.29

A deep-sea pycnogonid with a leg span of about 30 cm.

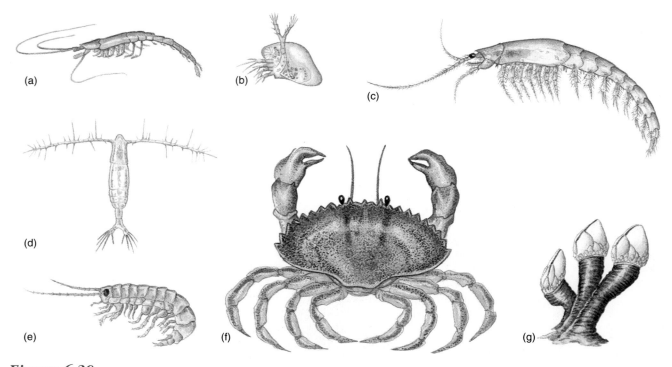

Figure 6.30

A variety of marine crustaceans:
(a) mysid, *(b)* cladoceran,
(c) euphausiid, *(d)* copepod,
(e) amphipod, *(f)* crab, *(g)* barnacle.

millimeters in size are quite common intertidally. They can be collected from hydroid or ectoproct colonies or from the blades of intertidal algae. Deep-sea pycnogonids are often much larger and may have leg spans of 60 cm.

The third class, the Crustacea, is an extremely abundant and successful group of marine invertebrates. Obvious and well-known crustaceans include shrimps, crabs (figure 6.30), lobsters, and barnacles (figure 6.30). These large, mostly benthic, crustaceans are not representative of the entire class, however. Most marine crustaceans are very small and are major components of the zooplankton.

Crustaceans are arthropods with two pairs of **antennae.** Few other useful generalizations can be made concerning this class. Its members exhibit a tremendous diversity in body structure (figure 6.30) and mode of feeding. The range of habitats also varies greatly, from burrowing ghost shrimp to planktonic copepods and parasitic barnacles. Representatives of several of the crustacean subgroups listed in table 6.1 are included in figure 6.30. Two of these groups figure significantly in so many marine trophic associations that they merit special attention.

The copepods (subclass Copepoda; figures 2.10, 2.11, and 6.30) are small crustaceans, seldom larger than 1 cm. Despite their size, their efficient filter-feeding mechanisms (described on p. 333) and overwhelming numbers in pelagic communities dictate that much of the energy available from the first trophic level is channeled through copepods. They, in turn, are consumed by predators as diverse as minute fish larvae and huge right whales.

Euphausiids (figures 2.10, 2.11, and 6.30) are somewhat larger than copepods, but they fill similar trophic roles in pelagic communities. These crustaceans have a global distribution (see figure 12.7). The largest species of this group, *Euphausia superba,* grows to 6 to 7 cm. Found in cold waters around Antarctica, this species aggregates in enormous dense shoals sometimes tens of kilometers long. They are a favorite prey of fish, whales, seals, and penguins and are fast becoming the world's largest single-species commercial fishery (chapter 15 examines this topic).

Radial Symmetry Revisited

Echinodermata

The echinoderms, an exclusively marine phylum, are widely distributed throughout the sea. They are common intertidally and are also abundant at great depths. Almost all forms are benthic as adults. Most are characterized by a calcareous skeleton, external spines or knobs, and a five-sided, or pentamerous, radial body symmetry (figure 6.31). Because echinoderms develop from bilaterally symmetrical larval stages, radial body symmetry is a secondary condition in this phylum. This and other aspects of their evolutionary history separate them on the phylogenetic tree of animal groups (figure 6.1) from those phyla characterized by primary radial body symmetry. A unique internal water-vascular system hydraulically operates the numerous tube feet. The tube feet extend through the skeleton to the outside and serve as respiratory, excretory, sensory, and locomotor organs.

Six classes of echinoderms currently exist. Representatives of each are shown in figure 6.32. The Echinoidea are spiny herbivores or sediment ingesters variously known as sea urchins, heart urchins, and sand dollars. The Asteroidea, or sea stars, are usually five-armed, but the number of arms may vary. Six-, ten- and twenty-one-armed sea stars are known. Most sea stars are carnivorous, but a few use cilia and mucus to collect fine food particles. Feather stars and sea lilies (class Crinoidea), usually attached to the sea bottom with the mouth oriented upward, trap plankton and detritus with their arms and the mucous secretions. Sea cucumbers of the class Holothuroidea are sausage-shaped and have their mouth located at one end of their body. The body wall is muscular, with reduced skeletal plates and spines. A few sea cucumbers feed on plankton, but most ingest sediment. The brittle stars (class Ophiuroidea) are smaller than most other echinoderms but are very common animals in soft muds, rocky bottoms, and coral reefs. A recently described class, the Concentricycloidea, has been created for a few species of echinoderms known as sea daisies (figure 6.32). These animals are essentially flattened, armless sea stars and have been collected only at depths greater than 1,000 m.

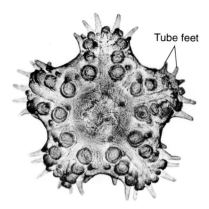
Tube feet

Figure 6.31

A very young sea star, with numerous tube feet projecting from its body.
Courtesy Carolina Biological Supply Company.

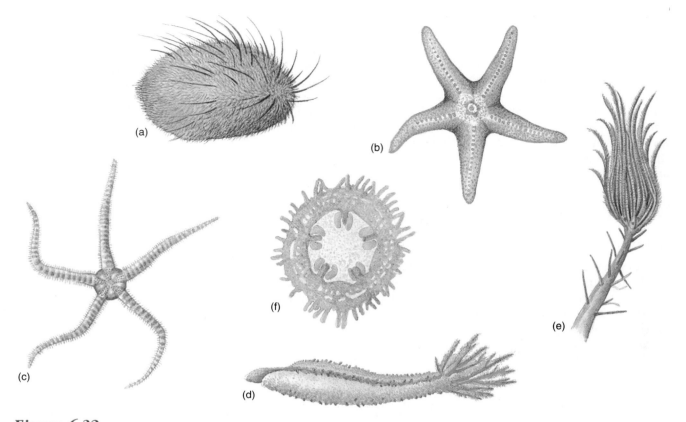

Figure 6.32

Representatives of the six living echinoderm classes: *(a)* Echinoidea, *(b)* Asteroidea, *(c)* Ophiuroidea, *(d)* Holothuroidea, *(e)* Crinoidea, *(f)* Concentricycloidea.

The Invertebrate Chordates

Chordata

Like mollusks and arthropods, the phylum Chordata represents a pinnacle in animal evolution. Chordates exhibit a remarkable variety of body forms, from small gelatinous zooplankton to large fishes and whales. Yet at some stage in their development, all members of this phylum possess a supportive **notochord** (made of a cartilage-like material), a hollow dorsal nerve cord, and pharyngeal arches.

A unique feature of one subphylum, the Vertebrata, is the vertebral column that replaces the notochord as it extends through the main axis of the body for support. Members of this subphylum include fishes and tetrapods and are described in the next chapter. The remaining two subphyla, the Urochordata and Cephalochordata, are relatively small and are completely marine in distribution.

The subphylum Urochordata includes animals such as benthic filter-feeding sea squirts (figure 6.33) and the planktonic gelatinous salps. Another subphylum of the chordates, Cephalochordata, includes the lancelet *Branchistoma*. These animals are small and tadpole-shaped and live partially buried tailfirst in near-shore sediments.

Figure 6.33
Nearly transparent sea squirts, each with a small incurrent and a large excurrent opening for circulating water through its body cavity.
Photo by T. Phillipp.

Summary Points

- Protozoan protists and invertebrate animals are well represented in the marine environment. All major and most minor phyla have at least some marine species, and several phyla are found only in the sea.
- Each phylum and its major subgroups are briefly described to provide the means for acquiring a working familiarity with marine protozoan and animal groups.
- The phyla are introduced in an order generally corresponding to an ascent of the phylogenetic tree (see figures 3.2 and 6.1). Multicellularity, bilateral body symmetry, the evolution of a coelom, cephalization, and increased species diversity appear as major trends along the evolutionary paths of the animal kingdom.
- Protozoans, cnidarians, mollusks, and arthropods are especially abundant in marine communities.

Review Questions

1. Many common marine animals have wormlike body forms. What survival advantages might this body shape create for mud or sand dwellers?
2. List and discuss the advantages and disadvantages of the rigid arthropod exoskeleton in comparison with the fluid hydrostatic skeleton of annelid worms.
3. List the genus names of two common local intertidal animals that exhibit radial body symmetry.

Challenge Questions

1. What survival advantages and disadvantages might an animal such as a sea anemone with radial body symmetry have over an animal with bilateral symmetry?
2. Why do you think many critical sense organs are concentrated in the head region of "higher" animals rather than in other parts of their bodies?
3. Why are protists placed in the same kingdom (in a five-kingdom classification system) as photosynthetic diatoms and dinophytes?

Suggestions for Further Reading

Books

Margulis, L., and K. V. Schwartz. 1987. *Five kingdoms: An illustrated guide to the phyla of life on Earth.* San Francisco: W. H. Freeman.

National Research Council. 1981. *Marine invertebrates: Laboratory animal management.* Washington, D.C.: National Academy Press.

Niesen, T. M. 1982. *The marine biology coloring book.* New York: Harper and Row.

Pechenik, J. A. 1991. *Biology of the invertebrates.* Dubuque, IA: Wm. C. Brown Publishers.

Warner, G. F. 1977. *The biology of crabs.* New York: Van Nostrand Reinhold.

Wells, M. J. 1978. *Octopus: Physiology and behavior of an advanced invertebrate.* New York: John Wiley.

Willmer, P. 1990. *Invertebrate relations.* Cambridge: Cambridge University Press.

Articles

Atema, J. 1980. Senses in the sea: An introduction. *Oceanus* 23(2):2–4. Also see other articles in this issue describing sensory capabilities.

Buzas, M. A., and S. J. Culver. 1991. Species diversity and dispersal of benthic foraminifera. *Bioscience* 41:483–89.

Erwin, D., et al. 1997. The origin of animal body plans. *American Scientist* 85:126–137.

Hadley, N. F. 1986. The arthropod cuticle. *Scientific American* (July) 244:104–12.

Levinton, J. S. 1992. The big bang of animal evolution. *Scientific American* (September) 252:84–91.

Roper, C. F. E., and K. J. Boss. 1982. The giant squid. *Scientific American* (April) 246:96–105.

Internet Addresses

Micscape Article: Protozoa—The Stentor
http://www.microscopy-uk.org.uk/mag/articles/stentor.html
Microscopy UK, Micscape and contributors. Explore the miniature world at Microcopy UK
http://www.microscopy-uk.org.uk/ Describes where you can find the Stentor and how different types of illumination under the microscope can be used.

ETI Biodiversity Center
http://www.eti.uva.nl/network/cnidaria/default.shtml
Classification (Phyla: Marine Organisms)
http://www.eti.uva.nl/database/urmo/clssfctn/default.shtml.
"The recognized phyla in the Register are as listed below (for the time being multicellular animals only). Phyla that have been (partly) incorporated in the database part of the Register are highlighted." Institute of Systematics and Population Biology.

Introduction to the Cnidaria
http://www.ucmp.berkeley.edu/cnidaria/cnidaria.html
The Phylum Cnidaria
http://www.utm.edu/~rirwin/cnidaria2.htm
Brief description and sketches.

The Lophophore
http://www.ucmp.berkeley.edu/glossary/gloss7/lophophore. html
Descriptions and photos of Brachiopoda, the Bryozoa, and the Phoronida.

Conchology 101 by Dr. Gary Rosenberg
http://coa.acnatasci.org/conchnet/c-101a.html
Conchology 101 by Dr. Gary Rosenberg is part of the Conchologist's Information Network (Conch-Net) maintained by the Conchologists of America and hosted by the Academy of Natural Sciences of Philadelphia. What is a mollusk?, Molluscan classes, Establishing relationships, Origins of diversity, Species, Classification of the Mollusca. Contains comprehensive summaries.

Marine Vertebrates

Because we described so many (three protozoan and 25 invertebrate) phyla in the last chapter, including the invertebrate chordates, it may seem a distortion to devote an entire chapter to the remaining chordate subphylum, the Vertebrata. Yet the species and ecological diversity of marine vertebrates rivals that of all other animal phyla but mollusks, arthropods, and perhaps the annelid polychaetes. Vertebrates occupy all major marine habitats and are especially dominant in the pelagic realm of the sea. As a group, adult vertebrates express an enormous range of body sizes, from 3-cm-long deep-sea lightfishes to whales and sharks over 15 m long. Consequently, vertebrates exploit food sources of almost all sizes found in the sea and are important in almost all marine food webs.

Vertebrate Features

All chordates with a vertebral column are members of the phylum Chordata, subphylum Vertebrata. All chordates have the hallmark structural features of the phylum at some time during their development. A hollow nerve cord extends along the dorsal midline of the body, and below this structure lies a cartilaginous notochord. A postanal tail extends well beyond the posterior opening of the complete digestive tract. Pharyngeal pouches develop as openings on each side of the pharynx. In addition to these basic chordate features, vertebrates also have specialized sense organs and a brain consisting of a concentration of nerves positioned at the enlarged anterior end of the nerve cord. Skeletal muscles are segmented into **myomeres,** which permit controlled and efficient movements. During development, the notochord becomes segmented into a linear series of articulated skeletal units, the **vertebrae.** These structural features are exhibited by the generalized cyclostome (figure 7.1).

The details of the origin and early evolution of vertebrates have been blurred by the passage of nearly one-half billion years. Most evolutionists agree that early vertebrates evolved from filter-feeding chordate ancestors that had characteristics resembling those of larval sea squirts (subphylum Urochordata) and the larval stages of modern lancelets such as *Branchistoma* (subphylum Cephalochordata) and of lampreys (primitive jawless fishes). The earliest vertebrates, the now-extinct jawless ostracoderms, appeared in the fossil record about 450 million years ago. Whether they first evolved in freshwater streams or in seawater remains unclear, because the fossils of early vertebrates have been found in both ancient marine and freshwater habitats of nearly the same age. Further fueling the debate is the fact that the body fluids of most marine invertebrates are nearly isotonic to seawater, whereas those of most groups of vertebrates

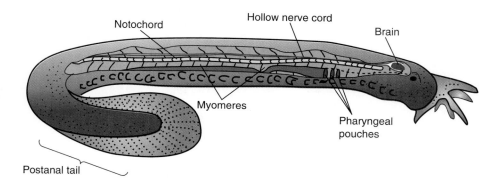

Figure 7.1
Generalized structure of a chordate.

have ionic concentrations less than half that found in seawater. The reduced ion concentrations in the body fluids of vertebrates living in low-salinity estuarine or freshwater conditions have the advantage of lessening the osmotic gradient and of reducing the metabolic energy required for osmoregulation. For proper nerve and muscle function, sodium ion concentration of vertebrate body fluids needs to be maintained at 30–50% its concentration in seawater. However, the hagfish (another jawless fish) is believed to be the most primitive surviving vertebrate and, therefore, most likely to be representative of the primitive vertebrate osmoregulatory condition. Unlike other vertebrates, hagfishes have body fluids isotonic to seawater, supporting arguments for a marine origin for early vertebrates.

One reasonable scenario proposes that early vertebrates moved between marine and freshwater habitats, possibly for spawning in fresh water as do salmon, sea lampreys, and other **anadromous** fishes that migrate in from the sea when adult. Such migrations would place the young stages in nutrient-rich environments, where they could feed and grow with perhaps fewer large, active predators than would be encountered in the sea. In these brackish and freshwater environments, the evolution of reduced-ion body fluids would be expected to conserve substantial amounts of energy that could be used for rapid growth. Only after gaining some size would the fishes move into marine habitats to compete for larger prey items. This is precisely what several species of lampreys do now. From this scenario for early vertebrates, it is suggested that each major group has evolved its own solution to the problem of maintaining osmoregulatory balance. These adaptations are described for each group in the next few pages.

Marine Fishes

Marine fishes include three of the seven classes of vertebrates (figure 7.2). Collectively, fishes are difficult to characterize precisely; typically they live and grow in water, swim with fins, and use gills for oxygen and carbon dioxide exchange. The classes exhibit definite differences in body structure, in specialization of sense organs, in solutions to osmoregulatory changes, and, ultimately, in species diversity.

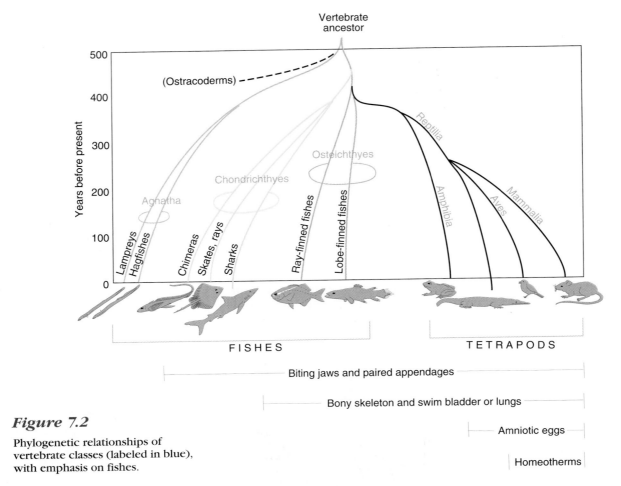

Figure 7.2

Phylogenetic relationships of
vertebrate classes (labeled in blue),
with emphasis on fishes.

Agnatha—The Jawless Fishes

Agnathans are vertebrates that are often characterized in negatives;
they lack the paired fins, biting jaws, and skin scales so noticeable in
most other fishes. Only two orders of these primitive fishes exist
(table 7.1), the hagfishes and the lampreys. Hagfishes are entirely marine benthic scavengers that secrete a thick, viscous slime from skin
gland to deter potential predators. Hagfishes periodically tie their flexible bodies into sliding knots to clean themselves of excessive slime
buildup (figure 7.3). This knot-tying behavior also provides stability
and leverage while tearing apart pieces of food.

As indicated in the previous section, most lampreys are anadromous (a few remain in fresh water). In both the freshwater and marine
portions of their life cycles, adults of most species of lampreys para-
sitize various species of bony fishes (figure 7.4). Parasitic lampreys use
numerous hook-shaped teeth within their oral disc to attach to their
hosts. Then they rasp through the skin with a toothed tongue. Anticlot-
ting salivary enzymes assist the continued flow of blood from the host
in a manner reminiscent of vampire bats. The parasitic life-style of lampreys is likely a specialized form of feeding not representative of the
way early vertebrates obtained their food. Neither hagfishes nor lampreys play major roles in present-day marine communities.

Table 7.1
Classes and Orders of Marine Vertebrates

Class	Order	Common Examples	No. of Marine Families
Agnatha			2
	Myxiniformes	hagfishes	1
	Petromyzontiformes	lampreys	1
Chondrichthyes			50
	Hexanchiformes	frill sharks	2
	Squaliformes	dogfish sharks	4
	Heterodontiformes	horn sharks	1
	Orectolobiformes	whale sharks	5
	Lamniformes	thresher and mackerel sharks	7
	Carcharhiniformes	swell, tiger, and whitetip sharks	8
	Pristiophoriformes	sawsharks	1
	Pristiformes	sawfishes	1
	Squatiniformes	angel sharks	1
	Rhinobatiformes	guitarfishes, thornbacks	3
	Rajiformes	skates	4
	Torpediniformes	electric rays	4
	Myliobatiformes	stingrays, eaglerays	8
	Chimaeriformes	chimaeras	1
Osteichthyes			~209
	Elopiformes	tarpons, bonefishes	3
	Anguilliformes	moray, conger, snake, and snipe eels, gulpers	20
	Notacanthiformes	deep-sea eels	3
	Salmoniformes	salmon, smelts, hatchetfishes, dragonfishes	15
	Clupeiformes	sardines, anchovies, herring, menhaden	5
	Myctophiformes	laternfishes, lizardfishes, barracudinas	17
	Ophidiiformes	cusk-eels, eelpouts	5
	Gadiformes	cods, hakes, pollock, moras	12
	Batrachoidiformes	toadfishes, midshipmen	1
	Gobiesociformes	clingfishes	1
	Lophiiformes	anglers, frogfishes	16
	Atheriniformes	silversides, sauries, halfbeaks, flying fishes	16
	Lampridiformes	oarfishes, ribbonfishes, opahs	6
	Beryciformes	deep-sea bigscales, whalefishes	7
	Gasterosteiformes	sticklebacks, sea horses, pipefishes	10
	Scorpaeniformes	rockfishes, scorpionfishes, sculpins, poachers	20
	Perciformes	basses, groupers, remoras, jacks, dolphinfishes, pomfrets, snappers, croakers, surfperches, barracudas, wrasses, gobies, tunas, billfishes	~60
	Pleuronectiformes	flatfishes	6
	Tetraodontiformes	triggerfishes, boxfishes, puffers, molas	9
Reptilia			7
	Squamata	sea snakes, iguana	3
	Crocodilia	estuarine crocodiles, caymen	2
	Testudines	sea turtles	2
Aves			27
	Podicipediformes	grebes	1
	Sphenisciformes	penguins	1
	Procellariiformes	albatrosses, petrels, shearwaters, fulmars	4
	Pelecaniformes	tropic birds, cormorants, boobies, gannets, pelicans, frigate birds	5
	Anseriformes	ducks, geese	1
	Ciconiiformes	herons	2
	Falconiformes	osprey, eagles	2
	Gruiformes	rails, coots	1
	Charadriiformes	stilts, avocets, plovers, sandpipers, skuas, turnstones, phalaropes, jaegers, gulls, terns, skimmers, auks, murres, puffins	10
Mammalia			17
	Carnivora	seals, sea lions, walruses, sea otters, polar bears	5
	Cetacea	whales, dolphins, porpoises	10
	Sirenia	manatees, dugongs	2

Examples and approximate number of marine families are listed as indications of group diversity.

Figure 7.3

Slime removal behavior of a hagfish.

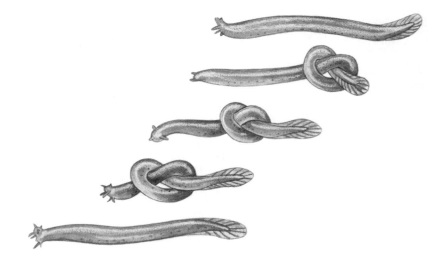

Figure 7.4

Two sea lampreys, *Petromyzon*, attached to a host fish.
Courtesy National Biological Survey, Great Lakes Science Center.

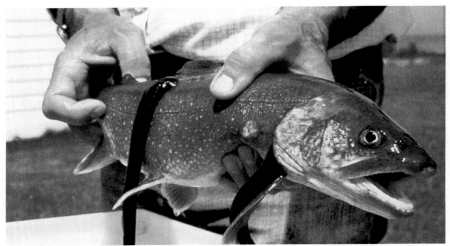

Chondrichthyes—Sharks, Rays, and Chimaeras

The 14 orders (table 7.1) included in the class of vertebrates called Chondrichthyes exhibit three basic body plans: streamlined sharks, dorsoventrally flattened rays, and the unusual chimaeras (figure 7.2). The evolution of paired fins and biting jaws armed with teeth (features found in all groups of marine fishes except agnathans) provide sharks, rays, and their allies with the structures needed for more effective predation, better maneuverability, and faster swimming speeds.

Almost all species of Chondrichthyes are marine. As in most other vertebrate groups, the concentration of salts in their body fluids are much less (about 50%) than that of seawater. To achieve osmotic equilibrium with seawater, marine Chondrichthyes accumulate high concentrations of urea and trimethylamine oxide (TMAO) in their body fluids. Urea is a toxic nitrogen-containing waste product synthesized by the liver, and TMAO is produced to protect proteins from its potentially damaging effects. The urea, together with TMAO and salt (NaCl), provides a total internal ion concentration equal to that of seawater

Figure 7.5

A coelacanth from the western Indian Ocean. The fish is about 1 m long.
Courtesy Scripps Institution of Oceanography (UCSD).

outside the body. Ammonia, a breakdown product of urea, is responsible for the sharp smell associated with dead sharks.

Members of this class are often referred to as the cartilaginous fishes for, although their skeletons may be very strong, rigid, and highly mineralized, the vascular, calcium-rich, stronger tissue known as bone is absent in these vertebrates. As a group, species of Chondrichthyes tend to be larger than members of the other two classes. Adult body lengths range from less than 20 cm for some of the deep-sea sharks to more than 10 m for whale sharks and basking sharks. With the exception of some whales, these are the largest living animals.

Most sharks are fast-moving, pelagic predators, whereas most rays are lie-and-wait ambush specialists. Both sharks and rays have sophisticated sensory systems and complex central nervous systems to process and integrate sensory information about their environment. These animals are discussed further in later chapters, especially shark swimming (chapter 13) and sensory capabilities (chapter 14).

Osteichthyes—The Bony Fishes

One of the two subclasses of Osteichthyes illustrated in figure 7.2, the ray-finned fishes (Actinopterygii), reach their peak diversity in marine habitats, with more than 13,000 species. In contrast, the lobe-finned fishes (Sarcopterygii) have but one living marine member, the coelacanth of the Indian Ocean (all other living Sarcopterygii are freshwater lungfishes). The coelacanth (figure 7.5) shares some common characteristics with sharks and rays, including urea accumulation for osmoregulation. It obtains buoyancy from low-density fats in its single vestigial lung. The coelacanth first appeared in the fossil record some 400 million years ago and was assumed to have become extinct about 60 million years ago. Then, in 1938, a fresh (but dead) specimen was collected off the island of Madagascar and examined. Other individuals of this rare, but definitely not extinct, species have since been collected from the Comoro Archipelago, where they live in water depths greater than 100 m. Both the coelacanth and the freshwater lungfishes have been proposed as possible ancestors of the land-dwelling tetrapod vertebrates (figure 7.2).

The ray-finned fishes are divided into three major groups, of which the infraclass Teleostei is the largest. Of 39 orders of teleost fishes, 26 orders encompassing hundreds of families and thousands of species occupy marine habitats (table 7.1). Thus, a majority of the bony fishes that live in the sea are teleosts. Marine and brackish-water representatives of the other two actinopterygian infraclasses are represented only by freshwater sturgeons (Chondrostei) and gars (Holostei). Regardless of where they live, all teleost fishes maintain body fluids hypoosmotic to seawater (refer to figure 2.5 for osmoregulatory processes) and have skeletons of bone. Most teleosts also have a gas-filled hydrostatic, or buoyancy, organ, the **swim bladder.** There are some exceptions, however. Some deep-sea teleost species employ other means to regulate their density relative to that of seawater. Buoyant swim bladders free the pectoral fins from the need to provide lift, as they must in sharks, which must swim to maintain or change their depth (more on that topic in chapter 13). These two features, swim bladders and bony skeletons, complement the membranous fins supported by rays for fine control of swimming movements. Freed of some of the structural constraints seen in the remarkably uniform body plan of cartilaginous fishes, teleosts have adapted to enormously varied aquatic habitats and exhibit a diversity of species unmatched in the phylum Chordata.

Most of the species of living teleost fishes (about 58% of the total number) live in marine habitats and dominate the flow of energy in their communities. Several of the groups listed in table 7.1 are familiar to most of us; others are not commonly seen except by professional biologists. A few representative species of seven of the common marine orders of teleost fishes are illustrated in figure 7.6. These and other teleost fishes will be referred to throughout the remainder of this text.

Marine Tetrapods

Sharing the pelagic realm with the numerous fishes are three of the four living classes of air-breathing tetrapods (figure 7.2): the reptiles, the birds, and the mammals. With the exception of a crab-eating frog that lives in the mangrove swamps of Southeast Asia, amphibians are strictly nonmarine. Tetrapods are four-limbed, air-breathing vertebrates with a terrestrial predecessor somewhere in their distant evolutionary past. Within each class, various groups have adapted to a marine existence independently of each other. Each of these groups depends on the sea for food and spends a good portion of time in the sea. Despite the obvious specializations of each of the three classes, these groups share several important adaptations. They all use lungs to breathe air in an environment where air is available only at the sea surface. Many successfully prey on other active animals even though they cannot use their sense of smell underwater and have only limited vision. Like teleost fishes, tetrapods have body fluids that are hypoosmotic to seawater; they, too, lose water to the environment by osmosis, and, with the exception of some marine reptiles and birds whose salt-secreting glands enable them to drink seawater, marine tetrapods must satisfy all metabolic water needs from their food. Also, two of these tetrapod classes, the birds and the mammals, are homeothermic in an environment perpetually colder than their bodies. Despite these limitations imposed by their terrestrial ancestry, several groups of tetrapods have invaded the sea and have done so very successfully.

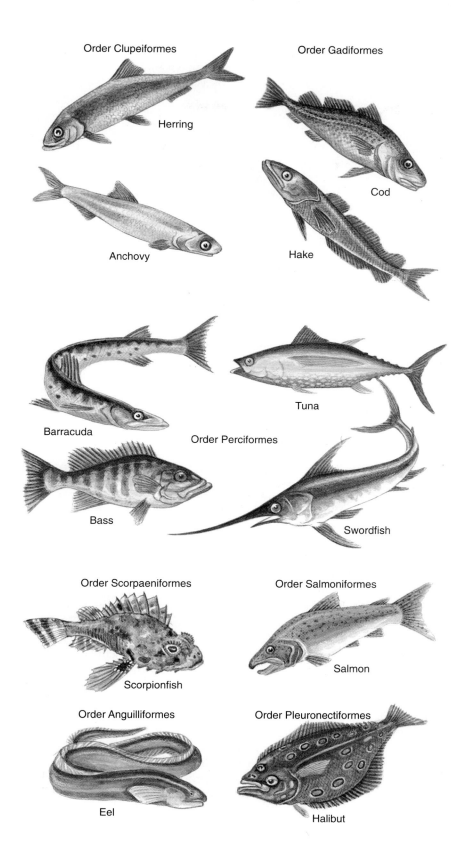

Order Clupeiformes

Herring

Anchovy

Order Gadiformes

Cod

Hake

Barracuda

Order Perciformes

Tuna

Bass

Swordfish

Order Scorpaeniformes

Scorpionfish

Order Salmoniformes

Salmon

Order Anguilliformes

Eel

Order Pleuronectiformes

Halibut

Figure 7.6

Some representative body types of seven common orders of marine fishes listed in table 7.1.

(a)

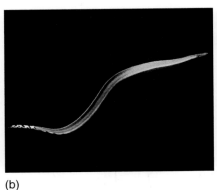

(b)

(c)

Figure 7.7

Some marine reptiles: *(a)* marine iguana, *Amblyrhynchus,* of the Galápagos Islands; *(b)* the sea snake; *(c)* green sea turtle, *Chelonia*.
Courtesy Corel Professional Photos *(a)* J. Graham, Scripps Institution of Oceanography *(b)* and T. Phillipp *(c)*.

Figure 7.8

A Lesser Albatross, with prominent nasal openings for salt excretion. Photo by J. Harvey.

Marine Reptiles

A few species of snakes, turtles, and iguana from three reptilian orders (table 7.1) are succeeding quite well in the sea, and some caymens, alligators, and crocodiles occupy coastal estuarine habitats. The single species of marine iguana (figure 7.7a) is restricted to rocky haulouts on the shores of the Galápagos Islands. There, they graze on seaweeds exposed at low tide or dive in shallow near-shore water for deeper marine plants. Four families and more than 100 species of snakes, most of them venomous, live in the ocean or in brackish-water estuaries. The 61 species of true sea snakes (figure 7.7b) are related to cobras and live in the warmer latitudes of the Indian and Pacific oceans. The yellow-bellied sea snake, *Pelamis platuris,* is the world's most abundant reptile and, like many of its relatives, is highly toxic. Most sea snakes are **ovoviviparous;** that is, they retain their eggs internally until they hatch, then give birth to live young. All seven species of sea turtles are primarily tropical and subtropical in distribution, although leatherback turtles have been tracked with radio transmitters into the cool southern waters of the Peru Current. Most species of sea turtles eat a variety of invertebrates, but the green turtle (figure 7.7c) feeds exclusively on sea grasses in shallow waters. All sea turtles come ashore only to lay eggs in nests dug in sandy beaches above the high tide line.

Marine reptiles and birds have evolved a double-barreled solution to deal with the shortage of fresh water and with the extra salt loads associated with feeding at sea: special **nasal glands** and kidneys. Both reptiles and birds have complex salt-excreting glands (figure 7.8), one above each eye, that concentrate NaCl to twice its concentration in seawater. This concentrated salty solution drips down or is blown out of the nasal passages. The kidneys of both birds and reptiles convert toxic nitrogen wastes from body metabolism to **uric acid** rather than to urea as the kidneys of mammals and many fishes do. Uric acid is a nearly nontoxic substance that requires only about 2 g of water for each gram of uric acid excreted. The white uric acid paste is mixed with feces for elimination, so urine production (and its associated water loss) as we know it in mammals does not occur in either birds or reptiles.

The ability to produce uric acid is related to another major evolutionary advancement of birds and reptiles, the shelled egg, or **amniotic egg** (figure 7.9). Although most sea snakes are ovoviviparous, most reptiles and all birds lay large, shelled eggs that require some period of

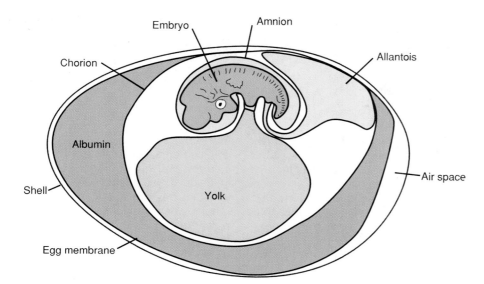

Chorion · Embryo · Amnion · Allantois · Albumin · Shell · Yolk · Air space · Egg membrane

Figure 7.9

Cross section of a typical amniotic egg, showing the major internal membranes. The amnion, for which this type of egg is named, surrounds and protects the developing embryo.

Figure 7.10

Marine turtle hatchling emerging from its egg.
Photo by Scott A. Eckert.

incubation before hatching. During this period, the enclosed developing embryo floats in its own water-filled sac, the amnion, while its nitrogen wastes are stored as concentrated uric acid in the allantois because these wastes cannot be eliminated across the protective outer eggshell. Gas exchange occurs freely during development across a third membrane, the chorion. When the egg hatches, the shell, inner membranes, and accumulated uric acid are abandoned (figure 7.10).

Fertilization of shelled eggs must occur before the protective shell is in place, so all birds and reptiles fertilize their eggs internally. Yet shelled eggs, uric acid excretion, and even internal fertilization were early adaptations for life out of water. For birds, sea turtles, and iguanas that have returned to the sea, laying and incubating their eggs ashore are still the norms.

Research in Progress

Molecular Systematics

Before we can assign an organism to its proper taxonomic category, we must interpret what we know of its evolutionary history. This history must be reconstructed from whatever clues are available; typically, the most reliable and widely used clues are general anatomical features (especially shell, bones, or teeth), fossils, and embryonic or larval development patterns. For each organism, it must be established whether a similar structure or developmental state seen in different taxonomic groups indicates a common evolutionary origin (and a common ancestor) or whether that feature has evolved independently in several groups. Let's begin with a familiar example. Most people would probably correctly guess that dogs and foxes (placed in the family Canidae) are more closely related than either is to cats (family Felidae). But it is less obvious that bears (family Ursidae) are more closely related to dogs than to cats.

A relatively new technique, **molecular systematics**, has been developed to assist in unraveling the complex knot of ancestral information derived from anatomical features, fossils, and developmental studies.

Differences in the biochemical structure of proteins, for example, should be at least as meaningful as differences in bone or tooth structures. Recent improvements in the techniques for determining the sequences of amino acids in proteins have enabled researchers to detect and compare differences in proteins of related groups of organisms. Cytochrome *c*, collagen, albumin, hemoglobin, and other blood proteins are frequently used. Cytochrome *c* consists of a chain of 100 amino acids. Twenty-seven are common to all the species studied so far. The amount of difference seen in the other 73 amino acids of cytochrome *c* taken from different groups of organisms is a clue to their degree of relatedness. Because the rate of change of a protein like cytochrome *c* is thought to be relatively constant through time, the amount of structural difference in a protein found in different groups of organisms can also serve as a crude clock to estimate the amount of time that has passed since those groups diverged from a common ancestor.

For example, elephants and the now extinct mastodons and mammoths (order Proboscidea) appear to have shared a relatively recent common ancestor. How long ago that common ancestor existed is best answered with molecular systematic methods. These methods help to clarify relationships between animal groups where similarities are not apparent, as between elephants, manatees, and dugongs (order Sirenia).

Similar approaches can be used with other proteins and with short segments of DNA. These techniques have revealed that a few species of mammals, such as cheetahs and northern elephant seals, possess very little of the genetic diversity thought to be critical to their long-term ability to adapt to changing environmental conditions.

Molecular systematic methods are also being used to justify reclassifying larger taxonomic groups. Recent comparisons of DNA sequences of pinnipeds (seals, sea lions, and walruses) with those of cats and dogs have yielded strong evidence in support for a new suborder of mammals, the Pinnipedia, within the order Carnivora. Molecular approaches to taxonomic questions are one of the latest developments in our understanding of evolution.

Marine Birds 8600 3%

Birds (class Aves) are tetrapods with feathers and front appendages adapted for flight. Several of the basic avian adaptations for flight in a three-dimensional aerial environment, including streamlined and insulated bodies, have been adapted to exploit marine habitats as well. There is a greater diversity of birds (table 7.1) than of either reptiles or mammals, and their role in marine communities is substantial. The term *marine* has a variety of meanings when applied to birds. A few birds, such as penguins (figure 7.11a), spend most of their lives at sea, going ashore only to breed and raise their young. At the other extreme are some ducks, geese, and coots that are common on inland ponds and

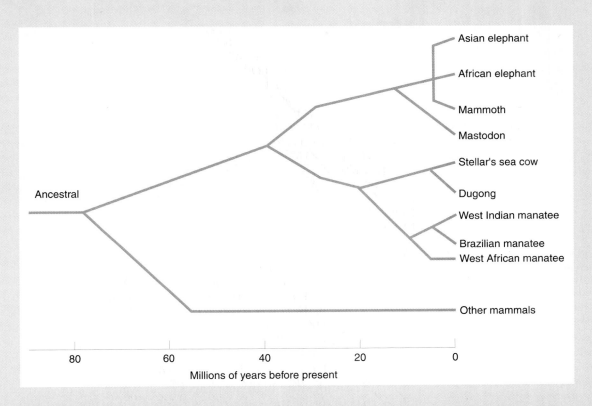

Ancestral

Asian elephant
African elephant
Mammoth
Mastodon
Stellar's sea cow
Dugong
West Indian manatee
Brazilian manatee
West African manatee
Other mammals

80 60 40 20 0

Millions of years before present

For more information on this topic, see the following: Davis, J. I. 1996. Phylogenetics, molecular evolution, and species concepts. Bioscience 46:502-11.

Lowenstein, J. M. 1985. Molecular approaches to the identification of species. American Scientist 73:541-47. Martin, A. P., G. J. P. Naylor, and S. R. Palumbi. 1992.

Rates of mitochondrial DNA evolution in sharks are slow compared with mammals. Nature 357:153-54

lakes, but some species do move into coastal waters to feed. Herons, stilts, sandpipers, turnstones, and other shorebirds (figure 7.11b) venture into shallow coastal waters only to feed on benthic animals. Others, including albatrosses, petrels, gannets, pelicans, gulls, terns, and murres, are more pelagic. These birds forage extensively at sea and often rest on the sea surface rather than returning to land to roost.

Pelagic birds, from penguins to pelicans, prey extensively on animals living in neritic waters. Their diving styles, patterns of pursuit, and even bill shapes differ greatly (figure 7.12), depending on their prey preferences. Shorebirds also show a great deal of variation in bill shape, again depending on their food preferences (figure 7.13).

(a)

(b)

Figure 7.11

(a) Emperor Penguins on an Antarctic ice ridge. (b) A mixed flock of shorebirds, including Sandwich Terns, Forster's Terns, and a Laughing Gull.

(a) Courtesy Sea World, Inc., San Diego;
(b) Courtesy Sneed B. Collard, Tyndall Air Force Base, Florida.

Birds and mammals are the only homeotherms in the animal kingdom. Although some large fishes, such as tunas, and the leatherback sea turtle can maintain body temperatures somewhat higher than the waters in which they swim, they lack the integrated physiological adaptations of true homeotherms. While at sea, birds and mammals live in direct contact with seawater much colder than their body temperatures. Most live in high-latitude, food-rich waters where water temperatures always hover near the freezing point. But even in more temperate latitudes, the high heat capacity of water (about 25 times as high as that of air of the same temperature) is a major heat sink and makes serious inroads into the heat budgets of these homeotherms.

Marine birds and mammals exhibit several adaptations that reduce their body heat losses to tolerable levels. The streamlined bodies of marine birds and mammals adapt them to move through air and water with minimal resistance and may also reduce the extent of body surface in contact with seawater and the amount of heat transferred to the water. The major muscles of propulsion (which generate considerable heat) are located within the animal's trunk rather than on the exposed parts of the flippers, wings or feet. A surface layer of dense feathers, fur, or blubber also streamlines and insulates the bodies of many species. Body heat losses are further limited by restricting the flow of warm blood from the core of the body to the cooler skin. Vascular countercurrent heat exchangers found in feet, flippers, and, in cetaceans, flukes (figure 7.14) also conserve heat. Arteries penetrating these appendages are surrounded by several veins carrying blood in the opposite direction. Heat from the warm blood of the central artery is absorbed by the cooler blood in the surrounding veins and carried back to the warm core of the body before much of it can be lost through the skin of the appendages. Collectively, these structures allow the relatively large-bodied birds and mammals to exploit cold, productive marine communities.

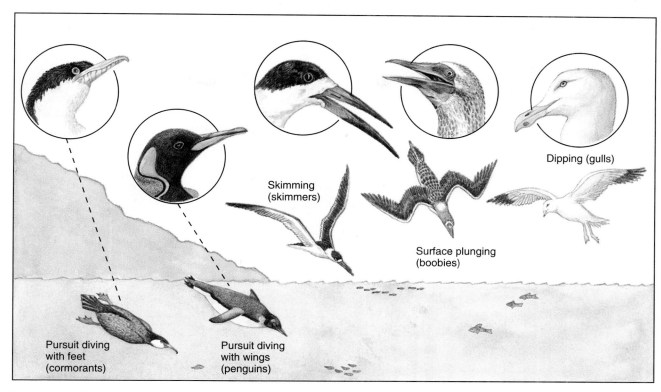

Figure 7.12

Bill shapes and pursuit patterns of birds that feed at sea.

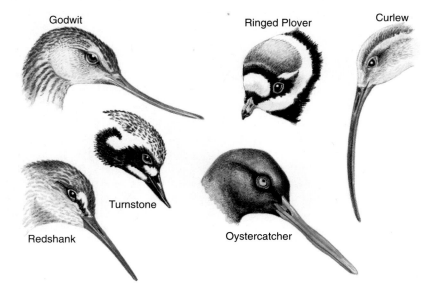

Figure 7.13

Bill shapes of some common wading birds.

Marine Mammals

Of all the tetrapod classes, only the mammals (class Mammalia, subclass Eutheria) are characterized by **viviparity,** the internal nourishment and development of a fetus (not an egg, as in some cartilaginous fishes and sea snakes). Mammals also have body hair, milk-secreting mammary glands, specialized teeth, and an external opening for the reproductive

Figure 7.14

A cross section of a small artery from the tail fluke of a bottle-nosed dolphin. The muscular artery in the center is surrounded by several thin-walled veins carrying blood in the opposite direction.

From Sam H. Ridgway, *Mammals of the Sea, Biology and Medicine*, 1972.

Courtesy Charles C. Thomas, Publisher, Springfield, Illinois.

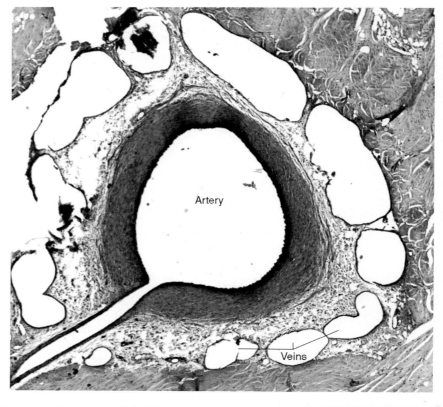

Figure 7.15

A sea otter, *Enhydra*.

© Pat and Tom Leeson/Photo Researchers, Inc.

tract separate from that of the digestive system. Thus, eutherian mammals lack the cloaca that is characteristic of all nonmammalian vertebrates.

The three orders of marine mammals listed in table 7.1 have experienced varying degrees of adaptation in their evolutionary transition from life on land to life in the sea. Seals, sea lions, walruses, sea otters, and polar bears (order Carnivora) are quite agile in the sea, yet all except the otter must leave the ocean to give birth. Sea otters (figure 7.15) have retained a strong resemblance to their fish-eating relatives of freshwater lakes and streams. Sea otters prefer to eat benthic invertebrates

(a)

(b)

Figure 7.16

Two types of pinnipeds hauled out at low tide: *(a)* harbor seals, *Phoca;* and *(b)* Steller sea lions, *Eumatopia.* *(b)* Courtesy B. Mate, Oregon State University.

they find along the shallow edges of the North Pacific. At some point in their evolutionary past, sea otters entered a tool-using stone age of their own. Using rocks carried to the surface with their food, they float on their backs and crack open the hard shells of sea urchins, crabs, abalones, and mussels to get at the soft insides.

Recent studies comparing molecular structures of blood and eye lens proteins and of DNA from seals, walruses, and sea lions indicate that they all share a common evolutionary ancestry and are placed together in the suborder Pinnipedia. Pinnipeds evolved from terrestrial carnivores and, in the sea, have maintained their predaceous habits. Only one species of walrus survives today in shallow Arctic waters, where they feed on benthic mollusks. Seals and sea lions (figure 7.16) are not as easily distinguished from each other as they are from the walrus. In the open water, sea lions swim using a slow, underwater "flying" motion of their front flippers. Seals propel themselves underwater with side-to-side movements of their rear flippers.

Manatees, dugongs, and sea cows (order Sirenia, figure 7.17) are large, ungainly creatures with paddle-like tails (manatees) or horizontal flukes (dugongs and sea cows) and no pelvic limbs. They are

Figure 7.17

Manatee cow and calf, *Trichechus.*
Courtesy Sirenia Project, U.S. Fish and
Wildlife Service.

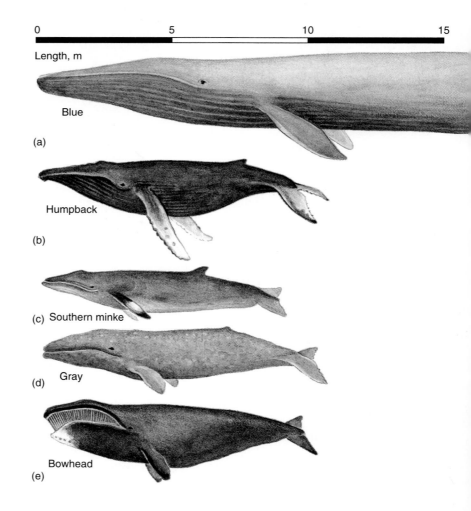

docile, herbivorous animals and are now completely restricted to shallow tropical and subtropical coastal waters, where they can secure an abundance of macroscopic marine and freshwater vegetation. They inhabit coastal regions along both sides of Africa, across southern Asia and the Indo-Pacific, and across the western Atlantic from South America to Florida and the Gulf of Mexico. At one time, the Steller's sea cow occupied parts of the Bering Sea and the Aleutian Islands. It took hunters and whalers less than 30 years, from the time the explorer Bering discovered these slow, quiet animals in 1741, to exterminate the species.

The evolution of cetaceans from terrestrial ancestors has culminated in a remarkable assemblage of structural, physiological, and behavioral adaptations to a totally marine existence. In contrast to typical mammals, cetaceans lack pelvic appendages and body hair, breathe through a single or a pair of dorsal blowholes, are streamlined, and propel themselves with broad, horizontal tail flukes. Body lengths vary greatly, from dolphins less than 2 m long to blue whales exceeding 30 m in length and weighing over 100 tons (figure 7.18). Convincing evidence of the tetrapod ancestry of whales can be seen during embryonic development.

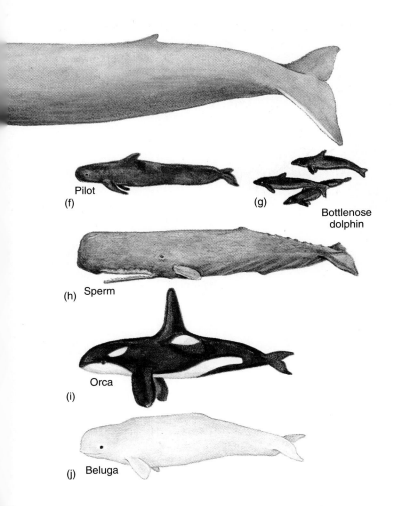

(f) Pilot

(g)

Bottlenose dolphin

(h) Sperm

(i) Orca

(j) Beluga

Figure 7.18

A few species of cetaceans, showing the immense range of body sizes at maturity. Baleen whales are on the left, toothed whales on the right: *(a)* blue whale, *Balaenoptera;* *(b)* humpback whale, *Megaptera;* *(c)* southern minke whale, *Balaenoptera; (d)* gray whale, *Eschrichtius; (e)* bowhead whale, *Balaena; (f)* pilot whale, *Globicephala; (g)* bottle-nosed dolphin, *Tursiops; (h)* sperm whale, *Physeter; (h)* orca, *Orcinus;* *(j)* beluga whale, *Delphinapterus.*

Figure 7.19

A 70-day embryo of a gray whale. Note the definite rear limb buds (arrow).

From Dale Rice and Al Wolman 1971. *The Life History and Ecology of the Gray Whale (Eschrichtius robustus)*. Spec. Publ. No. 3, American Society of Mammalogists, Courtesy D. Rice.

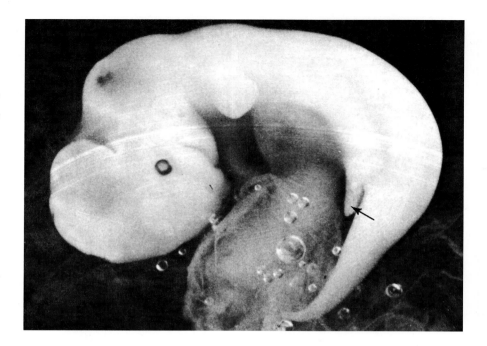

Figure 7.20

An adult gray whale in its winter breeding lagoon. The baleen plates attached to the upper jaw of its open mouth are evident.

Courtesy B. Reitherman.

Limb buds develop (figure 7.19) and then disappear before birth, leaving only a vestigial internal remnant of pelvic appendages.

The modern whales (order Cetacea) are of two distinct types. The filter-feeding baleen whales (suborder Mysticeti) lack teeth and, in their place, have rows of comblike **baleen** that project from the outer edges of their upper jaws (figure 7.20). All other living whales

Figure 7.21

Pilot whales, *Globicephala*, stranded on a New England beach.
Courtesy New England Aquarium.

(including porpoises and dolphins) are toothed whales (suborder Odontoceti). They are generally smaller than mysticetes and are well equipped with numerous teeth to catch fish, squid, and other slippery morsels of food. The smaller odontocetes, especially, are very social and are thought by many researchers to be highly intelligent animals (see Research in Progress: Cetacean Intelligence, p. 406). Ongoing studies evaluating cetacean intelligence and communication capabilities remain highly visible aspects of marine mammal research. These studies are necessarily biased toward smaller species that are easily maintained in captivity. They are complemented by information derived from examination of singly or mass-stranded animals (figure 7.21), and from research programs conducted in the animal's natural habitat.

Summary Points

- All living vertebrates share several structural features, including bilateral body symmetry, a postanal tail, an anterior brain, specialized sense organs, and a centrally located vertebral skeleton for support.
- There are more species of fishes than all other vertebrate groups combined, and a majority of these are marine. The members of class Agnatha exhibit a mixture of primitive and specialized characteristics. Agnathans lack paired fins and biting jaws. As adults, the living agnathans—lampreys and hagfishes—are scavengers, and some of the lampreys are parasites.
- The class Chondrichthyes includes sharks, rays, and chimaeras. Bony fishes (class Osteichthyes) account for all other living fishes except agnathans.
- Three classes of air-breathing marine tetrapods— reptiles, birds, and mammals—live in the sea. Sea snakes, turtles, the marine iguana, marine crocodiles, and large numbers of birds and mammals forage in the sea. Reptiles and birds share several adaptations, including nasal salt glands and uric acid excretion, for life in a salty environment.
- Marine mammals include two abundant and widespread groups, the cetaceans (whales, porpoises, and dolphins) and the pinnipeds (seals,

195

sea lions, and walruses) and the less common sea otters, manatees, and dugongs. Each has evolved from a terrestrial mammalian ancestor. Cetaceans are very streamlined, breathe through one or two dorsal blowholes, and lack hair and rear legs or flippers. Even so, they are mammals, breathing air, giving birth to live young, and maintaining elevated body temperatures.

- Large size and reduced body-surface-to-volume ratios of diving birds and marine mammals are obvious adaptations to reduce heat loss. Less apparent, but also significant, are insulating layers of feathers, blubber, or fur and countercurrent heat exchangers in blood vessels of appendages.

Review Questions

1. Arrange the following classes of the phylum Chordata in order of increasing evolutionary advancement or complexity: Osteichthyes, Chondrichthyes, Mammalia, Reptilia, Agnatha.
2. Compare the osmoregulatory adaptations of marine fishes, reptiles, and mammals.
3. List and describe the major evolutionary structural advances exhibited in the three classes of fishes. How do these morphological adaptations influence the size and swimming ability of members of each class?

Challenge Questions

1. What types of marine birds in your locality are also common in ponds and lakes? In fields? Where do they roost at night?
2. Discuss the major structural and physiological adaptations that have occurred in present-day killer whales that were not present in their terrestrial tetrapod ancestors.
3. Describe the adaptive significance of nasal salt glands and uric acid excretion for reptiles and birds feeding at sea.
4. List two specific structural features that distinguish each of the following marine mammal groups from the others: baleen whales, toothed whales, sea lions, seals, manatees.

Suggestions for Further Reading

Books

Alderton, D. 1988. *Turtles and tortoises of the world.* New York: Facts on File, Inc.

Bond, C. E. 1979. *Biology of fishes.* Philadelphia: W. B. Saunders Co.

Bonner, W. N. 1982. *Seals and man: A study of interactions.* Seattle: University of Washington Press.

Compagno, L. J. V. 1988. *Sharks of the order Carcharhinoformes.* Princeton, NJ: Princeton University Press.

Croxall, J. P. 1987. *Seabirds: Feeding ecology and role in marine ecosystems.* New York: Cambridge University Press.

Dunson, W. A., ed. 1975. *The biology of sea snakes.* Baltimore: University Park Press.

Ernst, C. H., and R. W. Barbour. 1989. *Turtles of the world.* Washington, D.C.: Smithsonian Institution Press.

Gaskin, D. E. 1982. *The ecology of whales and dolphins.* Portsmouth, NH: Heinemann.

Gilbert, P. W., ed. 1963. *Sharks and survival.* Boston: D. C. Heath and Company.

Herman, L. M., ed. 1980. *Cetacean behavior: Mechanisms and functions.* New York: John Wiley & Sons.

Lofgren, L. 1984. *Ocean birds.* London: Croom Helm.

Love, J. A. 1990. *The sea otter.* London: Whittet Books.

Marshall, N. B. 1966. *The life of fishes.* New York: World Publishing Company.

Martin, A. R. 1990. *Whales and dolphins.* London: Salamander Books.

Reynolds, J. E., and D. K. Odell. 1991. *Manatees and dugongs.* New York: Facts on File.

Riedman, M. 1990. *The pinnipeds: Seals, sea lions and walruses.* Berkeley: University of California Press.

Stirling, I., and D. Guravich. 1988. *Polar bears.* Ann Arbor: University of Michigan Press.

VanBlaricom, G. R., and J. A. Estes. 1988. *The community ecology of sea otters.* New York: Springer-Verlag.

Waller, G., ed. 1996. *Sea life.* Washington, D.C.: Smithsonian Institution Press.

Wrootton, R. J. 1990. *Ecology of teleost fishes.* New York: Chapman and Hall.

Articles

Berta, A. 1994. What is a whale? *Science* 263:180–82.

Geraci, J. R. 1978. The enigma of marine mammal strandings. *Oceanus* 21 (Spring):38–47.

Irving, L., and J. Krog. 1954. Body temperatures of Arctic and subarctic birds and mammals. *Journal of Applied Physiology* 6(11):667–80.

Owens, D.W. 1980. The comparative reproductive physiology of sea turtles. *American Zoologist* 20:549–63.

Romer, A. S. 1967. Major steps in vertebrate evolution. *Science* 158:1629–37.

Schmidt-Nielson, K. 1959. Salt glands, *Scientific American* 300:109–16.

Steele, J. H. 1991. Marine functional diversity. *Bioscience* 41:470–74.

Würsig, B. 1988. The behavior of baleen whales. *Scientific American* 258(4):102–7.

Internet Addresses

Chordata: More on Morphology
http://www.ucmp.berkeley.edu/chodata/chordatamm.html
Descriptions of Chordates.

Tetrapods: More on Morphology
http://www.ucmp.berkeley.edu/vertebrates/tetrapods/ tetramm.html
Descriptions of Tetrapods. The University of California Museum of Paleontology at Berkeley and the UC Regents.

All about Regeneration
http://darwin.bio.uci.edu/~mrjc/regen.html
States that the tetrapod limb is the model system best studied for regeneration. Much about regeneration on this site with links to other sites. Dr. Susan V. Bryant, Professor of Biological Sciences, University of California, Irvine, e-mail: svbryant@uci.edu. Welcome to the Laboratory of Regeneration Studies
http://darwin.bio.uci.edu/~mrjc/index.html

Marine Mammal Science
http://pegasus.cc.ucf.edu/~smm/strat.htm
"The field of marine mammal science has a growing appeal. Yet, many students do not clearly understand what the field involves. This brochure addresses questions commonly asked by people seeking a career in marine mammal science in the United States and provides suggestions on how to plan education and work experience." Courtesy of The Society for Marine Mammalogy, Dr. Daniel K. Odell e-mail: odell@pegasus.cc.ucf.edu. Marine Mammal Science
http://pegasus.cc.ucf.edu/~ssm/strat.htm

Anadromous Fishes
http://bonita.mbnms.nos.noaa.gov/sitechar/fish.html
Overview of Anadromous Fishes of the Monterey Bay National Marine Sanctuary, and Anadromous Salmonids of the MBNMS: Coho salmon, Chinook salmon, Steelhead. Monterey Bay National Marine Sanctuary. Pages on this site cover materials that will help students who are learning Marine Biology. Monterey Bay National Marine Sanctuary Site Characterization
http://bonita.mbnms.nos.noaa.gov/sitechar/index.html

Florida Marine Research Institute
http://www.fmri.usf.edu
Click on "Current Issues" for Manatee Mortality Statistics, Manatee
Rescue Statistics, Marine Turtle Stranding Statistics, Red Tide
Status—S.W. Coast, and Red Tide Status—East Coast, and follow links
for more information. © Florida Marine Research Institute.

Manatees
http://www.seaworld.org/manatee/manatees.html
Sea Turtles
http://www.seaworld.org/Sea_Turtle/seaturtle.html
Animal Resources
http://www.seaworld.org/infobook.html © Busch Entertainment
Corporation. The Sea World/Busch Gardens Animal Information
Database **http://www.seaworld.org**

U.S. Fish and Wildlife Service
http://bluegoose.arw.r9.fws.gov/
Contains searchable database and other links with information on
National Wildlife Refuges and Wetland Management.

Cowry and urchin on a shallow rocky bottom. Photo by Tom Phillipp.

Benthic Communities

IV

The intertidal region of a rocky shoreline is a wonderfully complex community of marine organisms vying for existence in a very harsh environment. The key point of the above statement is that all residents of the rocky intertidal, such as barnacles, mussels, sea anemones, fishes, snails, chitons, and limpets, are marine species. Therefore, they all rely on seawater to keep them cool and hydrated, to provide them with oxygen and food, and to sustain and disperse their gametes and offspring. Unfortunately, because they live on intertidal rocks, they are periodically exposed to air for as long as 12 hours. During these times of low tide, they risk heat stress, desiccation, suffocation, and starvation while waiting for the sea to return. Moreover, during any 12-hour period, they may experience extreme variations in temperature, salinity, or concentration of oxygen. For example, consider a small dent in the rocky shoreline of a tropical coast that forms a tide pool when it is abandoned by the receding tide. This ephemeral, intertidal microhabitat is the home of all organisms that were stranded in it, and it will exist for 2–12 hours while the tide retreats, turn around, and returns. While this puddle is isolated from the rest of the sea, profound changes can occur. For example, suppose this tide pool was formed just before a torrential rainfall. The large volume of rainwater falling into the tide pool would result in rapid and large decreases in temperature and salinity. The species calling this tide pool home must be tolerant of such changes. But this arduous period of isolation from the sea is not over. What if the rain storm ended as quickly as it began? (This rapid onset and end are not uncommon in tropical latitudes.) Then the tide pool would bake in the tropical sunshine for several hours. Water temperature would increase greatly, evaporation would cause salinity to climb, and, because the solubility of a gas in water is inversely proportional to temperature, oxygen concentration within the tide pool would decrease. Again, the tide pool residents would need to withstand these stressful variations to survive until the next high tide. The cyclical nature of the intertidal results in fluctuating but predictable environmental conditions. Intertidal species respire, feed, and reproduce during favorable periods (i.e., while submerged during high tides) and reduce activity during low tides, returning to a specific refuge or retreating into a closed shell. Hence, it seems that this community is well adapted to the stresses of intertidal regions. Nevertheless, an obvious question concerning these intertidal species arises: Why do they live there at all? Why not crawl several meters down the rocks to a level that is always subtidal and seemingly much less stressful? The answer seems to be that the harshness of the intertidal region results in a decrease in both predation pressure and competition. Perhaps these benefits are enough to compensate for life in a presumably suboptimal habitat.

The Intertidal

O ver 90% of the animal species found in the ocean and nearly all of the larger marine plants live in close association with the sea bottom. These organisms form the benthos. Benthic primary producers exist only in the shallow, near-shore fringe where the sea bottom coincides with the photic zone, introduced in chapter 5. Benthic animals range from high intertidal zones to cold, perpetually dark trenches more than 10,000 m deep.

Benthic organisms occupy the interface between the sea bottom and the overlying water. The characteristics of the sea bottom and the overlying water, the exchange of substances between the sediments and the overlying water, and conditions established by the other members of their communities define the environmental demands benthic organisms must meet for survival.

This chapter will examine the general conditions of life on the seafloor, with emphasis placed on temperate climate intertidal shorelines. The following three chapters will focus on different benthic communities: estuaries, coral reefs, and subtidal seafloor communities.

Living Conditions on the Seafloor

Benthic animals that crawl about on the surface of the sea bottom or sit firmly attached to it are referred to as the epifauna. Epifauna are associated with rocky outcrops or the surface of firm sediment deposits. Other benthic animals, the infauna, find food or protection within the substrate forming the bottom. Figure 8.1 shows the relative abundance of epifauna and infauna at different marine climatic zones. Note that epifauna seem to be more sensitive than infauna to climatic differences.

Infaunal clams, worms, and crabs are macroscopic and are familiar to anyone who has spent a few moments digging in a sandy beach or mudflat. These **macrofauna** either swallow or displace the sediment particles around them as they move. Less obvious, but no less important, are **microfauna,** microscopic infauna less than 100 µm in size. Intermediate in size between the macrofauna and microfauna are the **meiofauna,** a very interesting and abundant group of animals. The meiofauna are also referred to as **interstitial animals,** because they occupy the spaces (the interstices) between sediment particles.

Seafloor Characteristics

The sea bottom supports the weight of many organisms considerably denser than seawater. Some animals excavate burrows or construct tubes of soft sediments. On hard bottoms, animals and plants secure a firm attachment so that they can resist the tug of waves and currents. Benthic organisms are adapted for life on or in particular bottom types, and the character of life there, to a large extent, is dependent on the properties of the bottom material, which varies from solid rock surfaces to very soft, loose deposits.

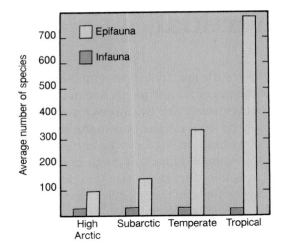

The sea bottom also accumulates plankton, waste material, and other detritus sinking from the sunlit waters above. In some regions, fallout of organic detritus from the photic zone is the only source of food for the inhabitants on the bottom. A variety of worms, mollusks, echinoderms, and crustaceans obtain their nourishment by ingesting accumulated detritus and digesting its organic material.

The composition of the sea bottom is determined by its constituent materials and, in shallow water, by the amount of energy available in the wind-driven waves and currents at the sea surface. Chapter 1 explained that the energy of the waves (and their ability to move particles) decreases from the sea surface downward (figure 1.26) and disappears at depths equal to one-half the wavelength. The coastline shape also strongly influences the amount of energy expended on the shore when waves break. Wavefronts approaching shore start to slow down just as they begin interacting with shallow reefs and bars. As a result, the wavefronts lag behind the rest of the wave as it approaches shore and changes shape to approximately match the curvature of an irregular coastline (figure 8.2). This process redistributes the energy in the breaking waves, concentrating wave energy on headlands while spreading out and diminishing the energy of waves entering bays, coves, and other coastal indentations. Taken together, the overall behavior of wind waves causes large amounts of energy to be expended on headlands in shallow waters (shallower than one-half the wavelength of the waves) and lesser amounts of energy to be expended in coastal indentations and in deeper waters. On exposed headlands, erosion is the principal result of breaking waves. These high-energy environments are continuously swept clean of fine sediment particles, detritus, and anything not securely attached to the bottom. This debris is washed offshore into deeper water or alongshore into bays or low-energy coastal features, where it settles to the bottom and adds to the growing deposits already there.

On an oceanic scale, rocky bottoms are scattered around the edges of the ocean basins where erosional processes dominate. Rocky outcrops also occur in association with deep-sea ridges, rises, and volcanoes, but the outcrops are comparatively small and geographically isolated.

Most of the sea bottom is covered with small sediment particles and other debris (e.g., planktonic skeletons) that have settled from the

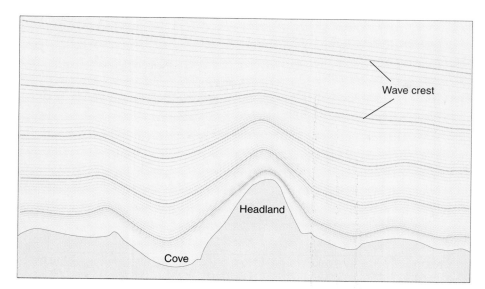

Figure 8.2

In coves and bays, refraction of advancing ocean waves spreads out the wave crests (and their energies) and concentrates them on headlands and other projecting coastal features.

Figure 8.3

Manganese nodules scattered on the surface of the seafloor in the Pacific Ocean.

Courtesy Scripps Institution of Oceanography.

surface. Marine sediments are derived from several sources. A few minerals precipitate from their dissolved state in seawater to produce irregular deposits on the seafloor. Manganese nodules (figure 8.3) are a well-known example of this type of deposit. Such deposits may eventually have some commercial importance as a source of minerals, but they also may have an important influence on the structure of some benthic communities.

Marine sediments near shore and on the continental shelves are largely the products of erosion on land and subsequent transport by rivers (and, to a lesser extent, winds) to the sea (figure 8.4). Once in the ocean, suspended sediment particles are carried and sorted by current and wave action according to their size and density. Large,

Figure 8.4

Particle size ranges for some common sources of marine sediments. Biogenic particles are shown in blue, terrigenous particles in tan.

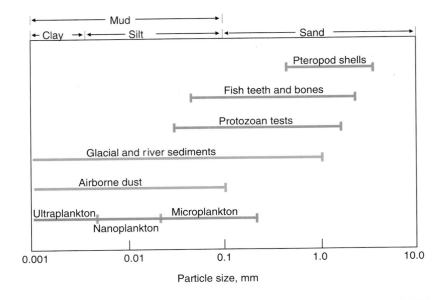

Figure 8.5

Sandstone erosion pits created by the rasping actions of small chitons.
Courtesy G. Dudley

dense sand grains quickly settle to the bottom near shore. Very fine clay particles are often carried several hundred kilometers out to sea before settling.

Animal-Sediment Relationships

Benthic animals mix and sort the sediments through their burrowing and feeding activities. Oxygen and water from the sediment surface circulate down into the sediment through their burrows. Sedimentary characteristics are further modified when particles are cemented together to form tubes and when sediments are compacted to form fecal pellets and castings. On rocky bottoms, the grazing activities of chitons, gastropods, and sea urchins aid the erosive processes of waves by scraping away rock particles as well as food (figure 8.5).

Figure 8.6

Two large pycnogonids dredged from the deep sea bottom.

There are even a few benthic animals, such as boring clams, that are especially adapted for boring into solid rock.

The distributional patterns of benthic plants and animals are strongly influenced by the firmness, texture, and stability of their substrate. These features govern the effectiveness of locomotion or, for nonmotile species, the persistence of their attachment to the bottom. Epifauna are most frequently associated with firm or solid bottom material. When found on softer muds, some display elongated, stiltlike legs or fins that extend into or along the sediment for better support and traction (figure 8.6).

The particle size and organic content of the bottom material limit the versatility and distribution of specialized feeding habits. Suspension feeders depend on small plankton or detritus for nutrition. Filtering devices (figure 8.7) or sticky mucous nets or sheets collect minute suspended food particles from the water. Suspension feeders generally require clean water to avoid clogging their filters with indigestible particles. Therefore, they are usually found on rocks or are associated with coarse sediments.

In deep-ocean basins, mudflats, and other soft-bottom areas, **deposit feeding** is common. Deposit feeders engulf sediments and process them through their digestive tracts. They extract nourishment from the organic material in the sediment in much the same manner as earthworms do. Deposit feeders indiscriminately ingest any available sediments; detritus feeders select organically rich substrates for consumption.

Several animal species are capable of extracting sufficient nourishment from sediments by conducting digestive processes outside their bodies. These animals absorb the products of digestion either through specialized organs or across the general body wall. Deep-sea

Figure 8.7

Barnacles, *Balanus,* attached to a snail shell, with their feathery filtering appendages extended.
Photo by T. Phillipp.

pogonophorans, small worms related to the large tube worms recently discovered in association with deep-sea hot springs (figure 11.15), depend on this type of **absorptive feeding,** as do numerous echinoderms. Sea stars are usually carnivorous, but a few species are quite opportunistic. When feeding on bottom sediments, the bat star extrudes its stomach outside its body and then digests and absorbs organic matter from the sediments. Omnipresent bacteria also depend on extracellular digestion. As they absorb nutrients and their population grows, they in turn become a significant source of particulate food for deposit feeders.

Benthic environments abound with **predators** and **scavengers** who feed on the residents of the bottom or on their remains. Most bottom predators and scavengers are permanent members of the benthos and are eventually eaten by other benthic consumers. Fishes, however, often make serious inroads into intertidal animal populations during high tides, and shorebirds replace them as predators at low tide (figure 8.8).

Regardless of the feeding habit employed by benthic animals, the ultimate source of food is the primary producers of the photic zone. Intertidal and shallow-water benthic plants provide direct sources of nutrition for the abundant herbivorous **algal grazers.** Some algal grazers nibble away bits of the larger seaweeds (figure 8.9). Most, however, rasp filmy growths of diatoms, cyanobacteria, and small encrusting plants from rocky substrates. Sea urchins use their five-toothed Aristotle's lantern to remove algal growths. Herbivorous gastropods and chitons accomplish similar results with their file-like radula.

Figure 8.8

A Godwit probing for food in the lower intertidal zone.

Figure 8.9

A large snail, *Norrisia,* grazing on a kelp stipe.
Photo by T. Phillipp.

Larval Dispersal

A sluggish benthic life-style does not limit sedentary bottom creatures to narrow geographical ranges. One-fifth of the common shallow-water animal species found at San Diego, California, for example, can also be found along the entire West Coast of the United States and in

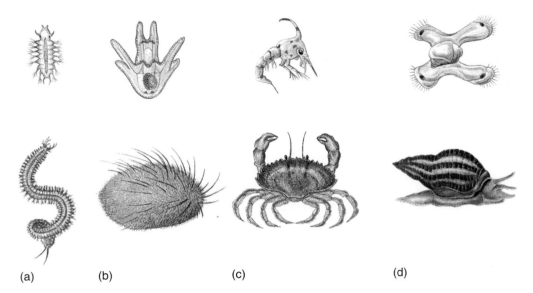

(a) (b) (c) (d)

Figure 8.10

Planktonic larval forms (above) and adult forms (below) of some common benthic animals: *(a)* polychaete larva, *(b)* sea urchin pluteus, *(c)* crab zoea, and *(d)* snail veliger.

British Columbia. Several other benthic species are even more widely dispersed, often in similar ecological conditions on opposite sides of the same ocean basin. *Mytilus edulis,* variously known as the bay mussel, blue mussel, or edible mussel, is common to temperate coasts of both sides of the Pacific and Atlantic oceans.

A few animals, including a small percentage of barnacles, often hitch rides on floating debris, on the hulls of ships, or in the ballast water of ships to travel transoceanic distances. An Australian barnacle, *Elminius modestus,* was apparently introduced to England by supply ships during World War II. It has since colonized the coasts of Ireland, France, Belgium, the Netherlands, Germany, and Denmark. In many sheltered reaches of these coastlines, *E. modestus* is competing with and replacing native barnacle populations.

A far more common adaptation for extending the geographical range of temperate- and warm-water benthic species involves the production of temporary planktonic larval stages that account for the majority of temporary plankton forms, the **meroplankton.** These small, feeble swimmers, bearing little resemblance to their parents (figure 8.10), drift with the ocean's surface currents for some time before they metamorphose and assume their benthic life-styles.

It has been estimated that about 75% of shallow-water benthic invertebrate species produce larvae that remain planktonic for two to four weeks. Over 5% of the species examined had planktonic larval stages exceeding three months, with a few as long as six months in duration (figure 8.11). Our understanding of ocean currents suggests that none but the most prolonged larval stages can make direct transoceanic trips before settling to the bottom. For each extra day the larvae remain in the plankton, they are exposed to additional threats of predation, increased pressures of finding food, and greater possibilities of being carried by the currents to areas where survival is unlikely. Yet the perils of a planktonic existence provide several advantages to offset the enormous mortality experienced by these larvae.

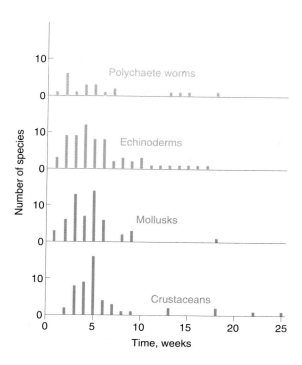

Figure 8.11

Typical duration of planktonic existence for four common groups of marine benthic invertebrates.
Adapted from Thorson, 1961.

Even in very slow ocean currents, drifting planktonic forms may spread far beyond the geographical limits of the adult population. Many are swept into unfavorable areas and perish. But any survivors may expand their parents' original range or settle into and mix with other populations and reduce their genetic isolation.

During their planktonic existence, many types of larvae react positively to sunlight and remain near the sea surface and their food supply, the phytoplankton. As their planktonic life draws to a close and they seek their permanent homes on the bottom, some larvae remain near the sea surface and ride into intertidal shorelines on waves and tides. Other larvae shun the light and swim near the bottom. Most enter a swimming-crawling phase and settle to the bottom, investigate it, and if it is not suitable, swim up to be carried elsewhere.

Just how larvae know when a suitable substrate is encountered is an important, but as yet unanswered, question. Chemical attractants, current speeds, types and textures of bottom material, and the effects of light are only partial answers to the question. Specific bottom types, such as sand or hard rock, do not attract larvae from a distance. Once the appropriate bottom type is encountered, larvae may be induced to remain and quickly metamorphose into a bottom-living stage. In contrast, chemical substances diffusing from established populations of some attached animals, including oysters and barnacles, attract larvae of their own species. This attraction may be beneficial for oyster and barnacle larvae, for the presence of adults in the settling site ensures that physical conditions are appropriate for survival. Also, the larvae's eventual reproductive success may be enhanced if they are in the vicinity of several other members of their species. For many other larvae, however, settling among their adults is catastrophic. Older, established individuals generally have relatively lower demands

Figure 8.12

Several major environmental factors that influence the selection of suitable bottom types by planktonic larvae.

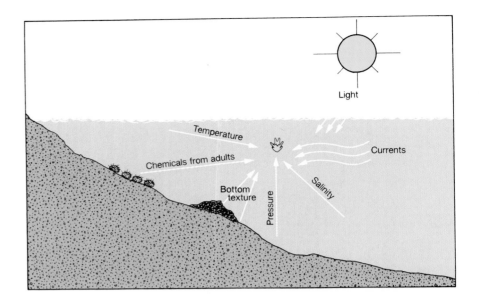

for food and oxygen, have more stored energy, and in general, can effectively compete for resources with the newly settled young. In times of shortages, the younger or smaller individuals are usually the first to suffer. Figure 8.12 illustrates the more obvious environmental features that may guide and influence larval settling.

Until they settle to the bottom, planktonic larvae are not in competition for food or space with the adults of their species. Even so, competition for food among plankton is often rigorous. About 10% of the species with long larval phases produce large yolky eggs that provide larvae with most or all of their nutritional supply. However, most species hatch from small eggs with little stored food and must begin feeding and competing with each other almost immediately.

Some benthic species, especially those in the tropics and the deep sea, spawn all year long; others have short and well-defined spawning seasons. In the latter group, the timing of reproduction or spawning is geared to produce young at times most advantageous to their survival. The spawning periods of many species of benthic animals are timed to place their larvae in the plankton community when phytoplankton is abundant and readily accessible. In shallow waters, temperature and day length provide two of the more obvious cues for timing reproduction. The gonads of spring and summer spawners develop in response to rising water temperatures and lengthening days. Oysters, for instance, refrain from spawning until a particular water temperature is reached.

Regardless of how many eggs are produced, the reproductive success or fitness of an individual requires that its **fecundity** (the production of eggs or offspring) exceed its offspring **mortality** (the rate at which individuals are lost). Any population whose mortality consistently exceeds its fecundity will shrink and eventually disappear. Adaptations that increase fecundity or reduce mortality improve the chances for successful reproduction.

The fecundity of some shallow-water benthic animals is truly amazing. Each female of many of these species produce more than 10 million eggs annually. A sea slug, *Aplysia,* weighing a few kilograms produced an estimated 478 million eggs during five months of laboratory observations. Such excessive reproduction enthusiasm would quickly place any shoreline knee deep in sea slugs if all of the spawning efforts of only a few adults survived.

Obviously, the egg and larval mortality of these species are extremely high. Of the millions of potential offspring produced, very few attain sexual maturity. The eggs of some benthic animals are fertilized internally before they are released into the water. Some species of snails, crabs, sea stars, and other invertebrates retain their larvae internally or in special brood pouches until the larvae are at reasonably advanced stages of development. For some of these invertebrates, brooding may be an adaptive consequence of small adult size. Smaller species of benthic invertebrates, with correspondingly smaller gonads, are less likely to produce sufficient planktonic larvae to equal their larger competitors. Consequently, they employ internal fertilization and larval incubation to reduce offspring mortality to some extent.

Many of the abundant and familiar seashore animals are **broadcast spawners;** they spew great quantities of eggs and sperm into the surrounding water, and fertilization occurs. These numerous eggs are necessarily small and hatch quickly into planktonic larval forms ready to experience the benefits and perils of a temporarily pelagic life-style. Each larva, then, is a relatively low-cost genetic insurance policy for spawning adults, there to ensure that the genes of the parents survive another generation.

Fertilization in these broadcast spawners is not as haphazard as it may seem. Chemical substances (referred to as **pheromones**) are present in the egg or sperm secretions of sea urchins, oysters, and many other marine invertebrates. When shed into seawater, these pheromones induce other nearby members of the same population to spawn, and they in turn stimulate still others to spawn until much of the population is spawning simultaneously (see chapter 10). As you might guess, spawning pheromones are quite specific. For instance, the sperm secretions of one sea urchin induce members of the same species to spawn, but the secretions have no effect on oysters or other species of sea urchins. A structural protein, contained in the heads of sea urchin sperm cells, binds only with other proteins on the surface coat of sea urchin eggs (figure 8.13). Together with pheromones, these substances regulate the timing of spawning, the specificity of sperm for eggs of the same species, and ultimately the overall prospects for successful fertilization by broadcast spawners.

It is distressing to note that, for many years, smashing sea urchins has been a popular and sporting method of removing these "pests" from the kelp beds on which they feed. Diving clubs frequently organized competitions, with prizes going to the club that killed the most urchins in a specific time. The impact of these eradication programs on sea urchin populations is questionable, for a smashed sea urchin (if it was ready to spawn) releases not only its own eggs or sperm but also pheromones that cause its neighbors to do likewise. Thus, a Sunday afternoon's efforts to eliminate a few thousand urchins may, in fact, trigger a mass spawning of billions of sea urchin eggs (enough to replace the animals destroyed, with more than a few left over). Worse

Figure 8.13

A scanning electron micrograph of sea urchin eggs covered with sperm cells.
Courtesy M. Tegner, Scripps Institution of Oceanography.

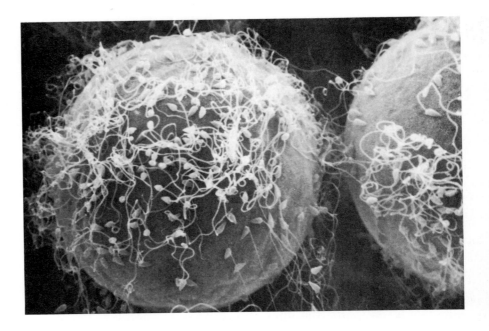

yet, such activities promote the attitude that humans are capable of interfering with and correctly altering the character of natural systems to fit their own needs without first understanding the basic features of those systems.

Intertidal Communities

The coastal strip where land meets the sea is home to some of the richest and best-studied marine communities found anywhere. Although this coastal strip is narrow, its influence is enhanced by the wealth of marine organisms present. Typically, the total biomass in a square meter at the low tide line is at least 10 times as high as that of a comparable area on the bottom at 200 m and is several thousand times higher than that found in most abyssal areas.

The periodic rise and fall of the tides have a dramatic effect on a portion of the coastal zone known as the intertidal, or **littoral,** zone. In the littoral zone, the sea, the land, and the air all play important roles in establishing the complex physical and chemical conditions to which all intertidal plants and animals must adapt.

Tidal fluctuations of sea level often expose intertidal plants and animals to severe environmental extremes, alternating between complete submergence in seawater and nearly dry terrestrial conditions. Local characteristics of the tides, including their vertical range and frequency, determine the amount of time intertidal plants and animals are out of water and exposed to air. Still, regardless of their locations, most intertidal regions have exposure curves that resemble figure 8.14.

For most intertidal plants and animals, low tide is a time of physiological stress. When the tide is out, exposed organisms are subjected to wide variations of atmospheric conditions. The air may dry and overheat their tissues in hot weather or freeze them during cold weather. Rainfall and freshwater runoff create osmotic problems as well. Predatory land

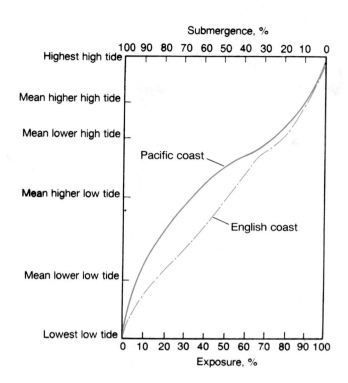

Submergence, %

Highest high tide

Mean higher high tide

Mean lower high tide

Pacific coast

Mean higher low tide

English coast

Mean lower low tide

Lowest low tide

Exposure, %

Figure 8.14

Exposure curves for the Pacific Coast of the United States and the Atlantic Coast of England.

Adapted from Ricketts and Calvin, 1968, and Lewis, 1964.

animals, such as birds, rats, and raccoons, also make their presence felt in the intertidal zone at low tide. Only at high tide are truly marine conditions restored to the intertidal zone. The returning waters moderate the temperature and salinity fluctuations brought on by the previous low tide. Needed food, nutrients, and dissolved oxygen are replenished, and accumulated wastes are washed away.

Accompanying the beneficial effects of seawater is the physical assault of waves and surf. The influence of wave shock on the distribution of intertidal plants and animals is apparent on all exposed coastlines of the world. Surf, storm waves, and surface ocean currents shift and sort sediments, transport suspended food, and disperse reproductive products. Much of the wave energy expended on the shore and the organisms living there eventually serves to shape and alter the essential character of the shoreline itself. Continually modified by the power of ocean waves, shorelines assume a variety of forms (figure 8.15). Rocky shorelines are constantly swept clean of finer sediments by heavy surf or strong currents. Waves on beaches remove fine silt and clay particles but leave sand grains behind. The finer materials are washed out to sea or are deposited in the quiet, protected waters of bays and lagoons.

The variety of tidal conditions, bottom types, and wave intensities along the shore create a boundless assortment of living conditions for coastal plants and animals. It is difficult to characterize the prevailing conditions on long stretches of shoreline without risking overgeneralization. The West Coast of North America, for instance, has many rugged rocky cliffs exposed to the full force of wave action. Yet interspersed between these cliffs and headlands are numerous sandy beaches and quiet mud-bottom bays and estuaries. On the East Coast, conditions vary from the spectacular rugged coastline of northern New England

Figure 8.15

An aerial photograph of a portion of the Oregon coast with protected coves, exposed headlands, sandy beaches, and offshore rocky reefs. Note the complex pattern of wave refraction around the offshore reefs.

Courtesy U.S. Geological Survey.

and the Canadian Maritime Provinces to extensive sandy beaches in the mid-Atlantic states. From Chesapeake Bay south to Florida, numerous coastal marshes are protected from extensive wave action by long, low barrier islands that parallel the mainland. Similar conditions with smaller tidal ranges exist along much of the Gulf Coast (see chapter 9).

The southern tip of Florida is the only shoreline on the continental United States to experience tropical conditions. Tropical shorelines are typically marked by large coral reefs (see chapter 10) or by extensive swampy woodlands of mangroves (see chapter 4). The coral reef system off the Florida Keys, fairly typical for the Caribbean, is unlike most other parts of the continental United States. Mangroves grow in profusion along Florida's southern coast and recently have become established along the extreme southern Texas coast. Included among these mangroves are several types of shrubby and treelike plants that grow together to form impenetrable thickets. Their branching prop roots trap sediments and detritus to extend existing shorelines or to form new low-lying islands.

Within a particular type of coastal environment, the interrelated influences of tidal exposure, bottom type, and intensity of wave shock produce an infinitely varied set of vertically arranged habitats. The vertical distribution of coastal plants and animals reflects the vertical changes in the shoreline's environmental conditions. Different species tend to occupy different levels or zones within the intertidal shoreline. Quite often, each zone is sharply demarcated from adjacent zones by the color, texture, and general appearance of the species living there. The result is a well-defined vertical series of horizontal life zones, sometimes extending for substantial horizontal distances along a coastline.

The vertical distribution of intertidal plants and animals is governed by a complex set of environmental conditions that vary along gradients above and below the sea surface. Temperature, wave shock, light intensity, and wetness are some of the more important physical factors that vary along such gradients. Breaking waves can impose very large forces on intertidal organisms (figure 8.16). These organisms in turn

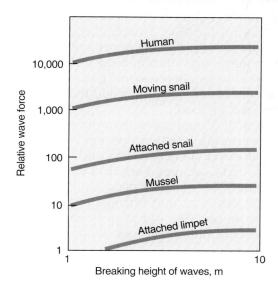

Figure 8.16

The relative force of breaking waves over a range of wave heights for several different intertidal animals.
Adapted from Denny, 1985.

demonstrate a remarkable variety of adaptations to deal with the forces associated with large waves. Several biological factors, including predation and competition for food and space, are superimposed on the physical gradients to further delineate the life zones of the shoreline. The combination of physical and biological factors frames and limits the range of an organism's existence. It, as well as the biological role the organism plays in its habitat, defines the organism's **niche.** The complex interplay of physical and biological conditions and the variety of shore life itself create an abundance of niches for intertidal organisms.

As the tides advance and recede over the intertidal portions of the shoreline, so too do the vertically graded changes in temperature, light intensity, degree of predation, and other environmental factors. Ocean tides are not the sole cause of vertical zonation within the intertidal shoreline. They do, however, modify and compress the pattern of zonation to make it more pronounced. (Vertically arranged sequences of plant and animal populations also can be found on mountainsides, in the sea beneath low tide, and in tideless lakes and ponds.)

For years, marine biologists have wrestled with the intricate problems of defining and identifying specific subzones within the intertidal shoreline, but the multiplicity of factors influencing intertidal zonation have seriously compounded the task. As a result, each of the major subzones frequently subdivided on the basis of locally abundant plant or animal populations. The makeup of intertidal populations may vary from place to place, causing the patterns of zonation to change; yet vertical zonation remains a visible unifying feature on all shorelines. It therefore seems a logical and appropriate basis on which to compare and describe the marine life of a few selected shores.

Rocky Shores

Chapter 2 emphasized that ecosystems have two complementary pathways for energy transfer: a grazing food web and a detritus food web. The grazing food web routes energy and nutrient material from the primary producers through the grazers and predators. Detritus eaters utilize the bits and pieces of dead and decaying matter available within the ecosystem. Within the coastal zone, neither rocky shores nor sandy beaches appear to be complete ecosystems by themselves. The trophic relations within rocky shore communities exhibit well-developed and complicated grazing food webs but very little in the way of detritus food webs. The erosional nature of rocky shores simply prohibits the accumulation of detritus and the existence of those animals dependent on it for food.

In rocky intertidal communities, patterns of distribution and abundance, as well as trophic relationships, are complex and sometimes change dramatically over short distances and from season to season. A shaded northern exposure may harbor several species absent from nearby sunny slopes. Tide pools contain an assemblage of plants and animals quite different from that surrounding well-drained platforms. The variety of life on one side of a boulder may differ markedly from life on the other side. And if you look under the boulder, still other species may be found.

With such a bewildering array of niches available on a small stretch of shoreline, it might seem improbable to find recurring themes of vertical zonation on widely separated shorelines. Yet similar patterns of zonation do exist on temperate rocky shores, whether in New England, Australia, British Columbia, or South Africa. Vertical zonation is such a compelling feature of life on rocky shores that considerable effort has been expended in devising schemes to identify and describe distinct intertidal subzones and their inhabitants. For those of you who desire a detailed account of life on intertidal rocky shores, several works with regional emphasis are listed at the end of this chapter.

The following discussion examines some of the more conspicuous intertidal plants and animals and the adaptations that permit them to remain conspicuous. Three rather ill-defined zones, the upper, middle, and lower intertidal zones, are used for reference. Because the boundaries separating these zones are artificial, they are frequently violated by their residents. You are likely to see species that dominate one zone scattered throughout other zones as well.

The Upper Intertidal

In the upper intertidal, living conditions are sometimes nearly as terrestrial as they are marine. Often only a vague demarcation separates land and marine vegetation. The area is wetted infrequently by extremely high tides and splash from breaking waves and is sparsely inhabited by marine organisms. Dark mats of the cyanobacterium *Calothrix* or the lichen *Verrucaria* frequently form a band or series of tarlike patches to mark the uppermost part of the rocky intertidal. Small tufts of *Ulothrix*, a filamentous green algae, may also extend into the highest parts of the intertidal. These plants are tolerant to large temperature changes and are adapted for resisting desiccation. The small, tangled filaments of *Calothrix* are embedded in a gelatinous mass to maintain their store of water and to reduce evaporation. Lichens, such as *Verrucaria*, are symbiotic associations of a fungus and a unicellular alga (figure 8.17). In the case of *Verrucaria*, the fungal part absorbs and holds several times its weight of water, water

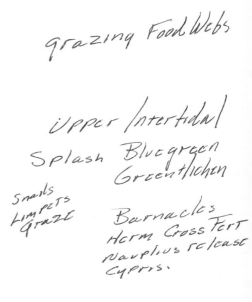

Handwritten margin notes: grazing Food Webs; Upper Intertidal; Splash Bluegreen Green/lichen; Snails Limpets Graze; Barnacles Herm Cross Fert Nauplius release Cypris.

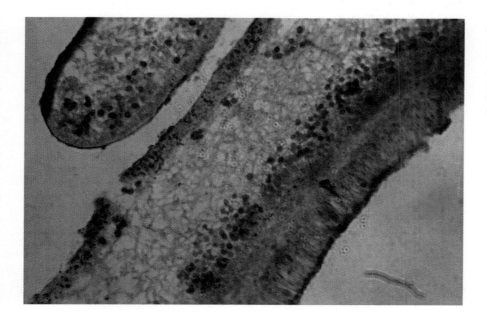

Figure 8.17

Magnified cross section of a lichen with algae cells (dark spots) embedded in fungal filaments (light strands). The cup-shaped feature is an ascocarp, a reproductive structure.

Figure 8.18

Two members of a sparsely populated rocky upper intertidal: the shore crab, *Pachygrapsus,* and the periwinkle snail, *Littorina.*

used by the fungus as well as by the photosynthetic algal cells that produce food for the entire lichen complex.

Only a few species of snails, limpets, and occasional crustaceans graze on the sparse and scattered vegetation of the upper intertidal (figure 8.18). Unlike its other close marine relatives, the small littorine snail *Littorina* is an air breather. *Littorina* use a highly vascularized mantle cavity in much the same manner as land snails do for gas exchange. Some species of littorine snails are so well adapted to an air-breathing existence that they drown if forced to remain underwater. Like littorines, limpets of the upper intertidal (especially *Acmaea*) are amazingly tolerant of temperature changes. Both littorines and limpets can seal the edges of their shell openings against rock surfaces to anchor themselves and to retain moisture. They are algal grazers and use their file-like radulae (see figure 6.21) to scrape the small algae and lichens from the rocks.

A conspicuous zone of small barnacles (figure 8.19) frequently appears just below the lichens and blue-green algae. Barnacles are filter feeders, but in the high intertidal, they are able to feed only when wetted by high spring tides a few hours each month. While submerged, their feathery feeding appendages extend from their volcano-shaped shell and sweep the water for minute plankton. Between high tides, a set of hinged calcareous plates blocks the entrance to the shell and seals in the remainder of the animal.

Most barnacles are hermaphroditic; they contain gonads of both sexes. Yet most generally refrain from fertilizing their own eggs. During mating, a long tubular penis is extended into a neighboring barnacle, and the sperm are transferred to fertilize the neighbor's eggs. The eggs develop and hatch within the barnacle's shell and are released as microscopic free-swimming planktonic organisms known as **nauplii** (figure 8.20a). After several molts of its exoskeleton, the nauplius develops into a **cypris larva** (figure 8.20b). The cypris eventually settles to the bottom, selects a permanent settling site, and then cements itself to the bottom with a secretion from its an-

Figure 8.19

Stunted acorn barnacles, *Chthamalus*, survive in the shallow depression of carved letters.

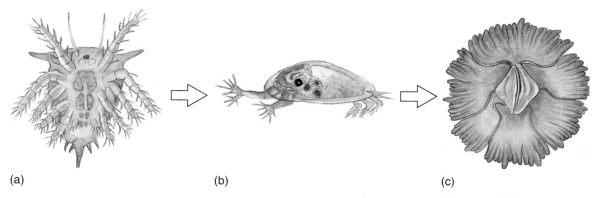

(a)   (b)   (c)

Figure 8.20

Planktonic and early benthic stages of the barnacle *Balanus:* *(a)* nauplius stage, *(b)* cypris stage, *(c)* early benthic stage.

tennae. Cypris larvae are attracted by the presence of other barnacles, thus ensuring settlement in areas suitable for barnacle survival and for obtaining future mates. Soon after settling, the cypris turns over, loses its larval appearance, and begins to surround itself with a wall of calcareous plates (figure 8.20c).

Connell's work on two species of intertidal barnacles provides a clear example of how the interplay of physical and biological factors influences the eventual vertical distribution of adult barnacles. In England, the larvae of one barnacle, *Chthamalus*, settle principally in the upper half of the intertidal, whereas the larvae of the other barnacle, *Balanus,* settle throughout the entire intertidal range (figure 8.21). Desiccation (dehydration) in the upper extremes of the intertidal rapidly eliminates a good number of the settled *Balanus* but has little effect on *Chthamalus* at the same levels. Below the level of significant desiccation effects, however, *Balanus* is clearly the better competitor for space, overgrowing and undercutting *Chthamalus* wherever the two species overlap. The resulting distribution of adult forms of both species provides a useful generalization applicable to other attached animals living in space-limited intertidal situations: The upper limit of species distribution is restricted by the species' ability to cope with

Figure 8.21

The limiting effects of desiccation and competition on the vertical distribution of two species of intertidal barnacles: *Chthamalus* (blue bars) and *Balanus* (green bars).
Adapted from Connell 1961.

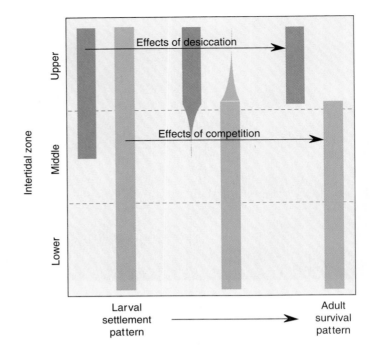

environmental stresses and other physical factors, such as temperature or desiccation, whereas a species' lower vertical range is limited by biological factors, especially competition with superior species.

The Middle Intertidal

The middle intertidal is occupied by greater numbers of individuals and species than is the upper intertidal zone. This zone, sufficiently inundated by tides and waves, provides an abundance of plant nutrients, oxygen and plankton food for filter-feeding animals. The lush growths of green, red, and brown algae also furnish a bountiful supply of locally produced food for grazers.

Occasional small, water-filled, tide pools protect hermit crabs, snails, nudibranchs, anemones, and a few small fish species from exposure and the physical assault of the surf. The upper range of anemones is usually determined by their tolerance to desiccation. Aggregate anemones of the Pacific Coast (*Anthopleura elegantissima*) are known to withstand internal temperatures as great as 13° C above the surrounding air temperature. Yet serious water loss will destroy them. These anemones combat extreme desiccation and temperature fluctuations by retracting their tentacles and attaching bits of light-colored stone and shell to themselves, presumably to reflect light and heat (figure 8.22).

The clumped mats characteristic of the aggregate anemone are the result of a peculiar mode of asexual reproduction. To divide, these anemones pull themselves apart by simultaneously creeping in opposite directions. Each half quickly regenerates its missing portion, producing two new individuals to replace the original. All the members of a clump resulting from this asexual fission are **clones,** or genetically identical individuals, with the same sex and color patterns. The clonal clumps of anemones are uniformly spaced and are separated

Middle Intertidal

Find ↑ Species
Than upper
Tides - Nutrients, O₂
    Plankton

Algae Lush
Tide Pool) Snails + Hermits
    Anemones
    Nudibranchs
    (Tolerant of Some
        Dessication)

] [acernf]

Mussels } Compete For
Barnacles } Solid Substrate
Fucus } + H₂O

Sea Stars \
Snails > EAT
Carn Fish / Barnacle

Sub mussel
    ↑ # Species

Figure 8.22

The aggregate sea anemone, *Anthopleura.* Individuals above the waterline have retracted their tentacles and covered themselves with light-reflecting shell fragments.

Figure 8.23

Close-up view of mussels, *Mytilus,* and acorn barnacles, *Balanus,* of the middle intertidal.

from adjacent clones by bare zones about the width of a single anemone. These anemones also have separate sexes and can reproduce sexually by releasing eggs and sperm into the water.

These bare zones between anemones result from the subtle warfare between dissimilar individuals from adjacent clones. They are armed with special tentacles, or **acrorhagi,** that can inflict serious damage to anemones of opposing clones but that have no effect on individuals of the same clump. In these border wars, anemone clumps rely on a mechanism of self-recognition so that the aggressive response is directed only to members of genetically dissimilar clones.

The dominant and conspicuous members of the middle intertidal zone (figure 8.23) are mussels *(Mytilus),* barnacles (usually *Balanus),* some chitons and limpets, and several species of brown algae (especially *Fucus* or *Pelvetia).* These animals securely anchor themselves to the substrate and generally present low, rounded profiles to minimize resistance to breaking waves (see figure 8.16). The plants are secured by strong holdfasts and usually have sturdy but flexible stipes to absorb much of the wave shock. *Fucus* and *Pelvetia* have thickened cell walls to resist water loss during low tide (figure 8.24).

In the densely populated middle intertidal zone, mussels, barnacles, brown algae, and other sessile creatures are limited by two commonly shared resources: the solid substrate on which they live and the water, which provides their dissolved nutrients and suspended food. Mussels, barnacles, and algae compete for these critical resources in different ways. In regions where physical factors permit each to survive, these three competing groups interact by dominating the available attachment space or by overgrowing their competitors and monopolizing the resources available from the water (figure 8.25).

Figure 8.24

Fucus, a brown alga, thrives on a Maine intertidal rock.

Figure 8.25

A bed of acorn barnacles (*Balanus*) severely deformed because of crowding. Because the barnacles were inhibited from expanding laterally by their neighbors, they grew upward instead.

Available space on the rock surfaces of the middle intertidal is crucial for survival, yet it is seldom fully utilized. A number of interacting biological and physical processes occasionally create patches of open space. Sea stars and predatory snails continually remove small areas of barnacles and mussels. Seasonal die-offs of algae, and even battering by heavy surf and drifting logs, also clear patches for future settlement and competition.

Chance events and their relationship to seasonal patterns of reproduction wield appreciable influence in settlement patterns. Most of the middle intertidal animals have free-swimming larval stages and are capable of settling almost anywhere within the intertidal zone. Algal spores and barnacle larvae simultaneously settling on bare rock eventually grow and compete for available space. Because of the space limitations that may exist on exposed coasts, the algae are usually squeezed out as the barnacles increase in diameter and dislodge or overgrow them. On some sheltered rocky coasts, though, recruitment of barnacle larvae is prevented by sweeping action of wave-tossed algal blades. Only in this manner can *Fucus* achieve and maintain spatial dominance over barnacles. If the barnacles are not removed before they are securely cemented into place, they escape the adverse effects of algal blades because of growth and may eventually force *Fucus* off the rocks.

Barnacles are not necessarily safe once they have outgrown or overgrown their algal competitor. They are consumed in prodigious numbers by sea stars, carnivorous snails, and certain fishes. Even herbivorous limpets have a detrimental, and sometimes severe, impact on barnacle populations. Young barnacles are eaten or dislodged by limpets as limpets bulldoze their way through their grazing activities. Limpets seem to have less effect on the small crack-inhabiting *Chthamalus* than on the larger more exposed *Balanus*. Thus, in the presence of limpet disturbance, *Chthamalus* gains a slight competitive advantage over the otherwise dominant *Balanus*.

The larval stages of mussels do not require bare rock exposures; they will settle on algae and barnacles and among aggregates of adult mussels. After settling, the young mussels crawl over the bottom, seeking improved conditions before they attach themselves to the substrate with several strong elastic **byssal threads.** Byssal threads are formed from a fluid secreted by an internal byssal gland. The fluid flows down a groove in the small tongue-shaped foot and onto the substrate. On contact with seawater, the fluid quickly toughens to form an attachment plate and a thread. Then the foot is moved slightly and additional plates and threads are formed.

If left undisturbed, mussels eventually overgrow barnacles and algae. Seldom, however, do rocky intertidal conditions remain undisturbed for long. Mussels are extensively preyed on by sea stars (such as *Pisaster* on the Pacific Coast and *Asterias* on the Atlantic Coast). These sea stars are quite sensitive to desiccation and are limited to sites that remain underwater. Consequently, their impact on mussel populations is much more severe in the lower portions of the mussels' intertidal range. Young mussels also are devoured by *Nucella* and other predatory snails. Some mussels survive those predatory onslaughts by numerically swamping an area with more individuals than

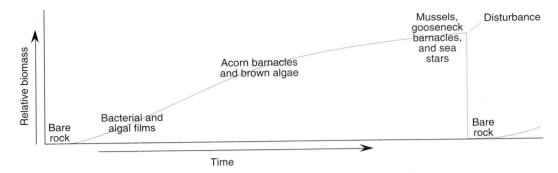

Figure 8.26

General pattern of succession through time on temperate rocky shores. Blue curve indicates relative biomass.

the local predators can consume. In time, the mussels may escape, too, through growth, becoming so large that snails are incapable of drilling through their shells to consume the soft flesh within.

As patches of mussels are cleared out by predators or broken off by waves, they are temporarily replaced by algae or barnacles, but gradually the mussels regain their ascendancy. In this way, diversity of species is maintained. It is ultimately the dynamic balance resulting from competition between these dominant organisms and the patterns of disturbance that affect their survival that shapes the biological character of the middle intertidal. These organisms, in turn, influence the distribution and abundance of other plant and animal species.

A cursory examination of mussel beds often reveals two other conspicuous animal species, filter-feeding gooseneck barnacles and predatory sea stars. But living on the mussel shells or among the thick masses of byssal threads underneath is an extremely complex community of more fragile and often unseen animals. This submussel habitat protects several common species of clams, worms, shrimps, crabs, hydroids, and many types of algae. Many of these species utilize mussel shells as available substrate for attachment. Others exist because they are unable to survive in the same area without the protection afforded by the canopy of mussels overhead.

This complex association of organisms is wholly dependent on the existence of thick masses of well-anchored mussels. This is the culmination, the **climax** stage, of a long succession of plant and animal populations that begins on bare rock and progresses through a sequence of population changes (figure 8.26). Eventually, this process of biological succession may reach the climax stage: in this case, the mussel bed–gooseneck barnacle–sea star community. When natural disturbances remove the mussels and disrupt the stability of the community, the mussels are quickly replaced by a predictable succession of nonmussel populations (figure 8.26). Thus, the structure of communities such as those dominated by mussels are seldom stable for long; rather, they achieve a state of dynamic equilibrium between the stabilizing effect of succession and the many disruptive factors that reduce that stability.

The Lower Intertidal

The biological character of lower intertidal rocky coasts differs markedly from the zones above it. It is difficult for some species and impossible for others to tolerate the exposed conditions found in the

Figure 8.27

Surf grass interspersed with brown algae in the lower intertidal.

[Handwritten notes in right margin:]
Lower Intertidal
↑ Species diversity
Species do not tolerate exposure to air.
Seaweed-
Br. Gr. Red
Anemone Metridium
Echinoids, nudibranchs
snails
seaspiders   Feed on
✳ All Echinoderms
Stars, B. star, urchin, etc.
Sens. to Sal. Change +
Dessication.

Competition
Predation
Other Bio Interactions

upper and middle intertidal zones. The few species that do are often present in vast numbers. In the lower intertidal, the emphasis changes to a community with a high diversity of species, often without the conspicuous dominant types so characteristic of the middle and upper intertidal.

The lower intertidal of rocky shores abounds with seaweeds. Brown, red, and even a few species of green algae of moderate size spread a protective canopy of wet blades over much of the zone. In other places, extensive beds of sea grasses achieve a similar effect (figure 8.27). Tufts of small filamentous brown and red algae carpet many of the rocks. Calcareous red algae become especially prolific at these levels. The pinkish hue of *Lithothamnion* encrusting rocks and lining the sides of tide pools is a common sight.

The animals of the lower intertidal include a few species from many animal phyla. It is here that the diversity, complexity, and sheer beauty of intertidal marine life abound. On the East Coast, a large white anemone, *Metridium*, occurs in tide pools and on exposed portions of the lower intertidal. *Anthopleura*, a beautiful green anemone (figure 8.28), occupies a similar habitat on the West Coast. Securely anchored by discs at their bases, anemones are predators of planktonic animals and small fishes. They capture prey by discharging many microscopic nematocysts from special cells in their tentacles. When touched with a finger, nematocysts of sea anemones produce a slight tingling, sticky sensation.

The batteries of nematocysts found on anemone tentacles effectively discourage the hostile intentions of most predators, but they do not guarantee complete immunity against predation. A few snails and sea spiders

Figure 8.28

The green anemone *Anthopleura.*
Photo by T. Phillipp.

penetrate the sides of anemones and feed on the unprotected tissues. Eolid nudibranchs also commonly graze on anemones and the closely related hydroids. These nudibranchs possess mechanisms, not yet completely understood, that block the discharge of the toxic nematocysts. During digestion, the undischarged nematocyst-containing cnidocytes are preserved and passed to special storage sacs in the rows of fingerlike **cerata** along the back of the nudibranch (figure 8.29). There the cnidocytes possibly serve as defensive mechanisms against predators.

The echinoderms are another familiar group of animals in the lower intertidal. Sea stars, sea urchins, brittle stars, and sea cucumbers are all quite sensitive to desiccation and salinity changes and are seldom seen in abundance above the lower intertidal zone. Although slow-moving, sea stars are voracious predators of mussels, barnacles, snails, an occasional anemone, and even other echinoderms.

Sea stars continuously leak substances that initiate alarm reactions in their prey; actual contact usually leads to even more vigorous escape movements. Prey species apparently recognize and identify their sea star predators by the substances they exude. They react violently to the touch or presence of sea stars that usually prey on them, but they seldom respond to those not encountered in their normal habitat. When approached by some species of sea stars, many normally sessile species execute remarkable escape responses. Scallops swim jerkily away, clams and cockles leap clear of the sea star, and sea urchins and limpets crawl away relatively rapidly. Some sea anemones detach themselves and somersault or roll aside when touched by certain sea stars (figure 8.30).

Figure 8.29

A nudibranch, *Hermissenda,* with long, fingerlike cerata projecting from its upper surface.
Photo by T. Phillipp.

Figure 8.30

A swimming anemone, *Actinostola,* evading a predatory sea star. The anemone has detached itself from the bottom and is somersaulting away.

Photo by C. Birkland, courtesy P. Dayton, Scripps Institution of Oceanography.

Figure 8.31

Vertical zonation patterns on a 3-m-high rock on the west coast of Vancouver Island, British Columbia.

Upper intertidal

Middle intertidal

Lower intertidal

In summary, rocky intertidal zones of temperate shores have a dynamic pattern of organization dominated by physical forces at the upper extreme that diminish in influence downward and are gradually replaced by competition, predation, and other biological interactions. Figure 8.31 shows the general pattern of vertical zonation of intertidal plants and animals on temperate rocky shores.

At the low tide line, the lower intertidal merges with the uppermost part of the inner shelf zone. Where rocky bottoms extend below the low tide line and are not covered by sediments, the transition from intertidal to subtidal is gradual. Many plant and animal species common to the lower intertidal also are abundant in neighboring shallow subtidal regions. However, rocky substrates eventually give way to soft sediments. In protected stretches of coastlines or below the sea surface, wave action is diminished and loose sediments and detritus begin to accumulate. Organisms of the rocky shore disappear and are replaced by those typical of sand or mud bottoms.

Sandy Beaches and Muddy Shores

Beaches and mudflats, the ecological complement to rocky shores, are the depositional features of the coastal zone and are best developed along sinking (or subsiding) coastlines. They are unstable and tend to shift and conform to conditions imposed by waves and currents. Large plants find the shifting nature of soft sediments difficult to cope with, and few exist there. The few plants that have managed to adapt support even fewer grazers. Detritus food webs dominate on these depositional shores. Bits of organic material washed off adjacent rocky shores and the surrounding land or drifted in from kelp beds farther offshore sustain the detritus feeders of sandy and muddy shores.

Beaches are made of whatever loose material is available. Quartz grains, black volcanic sand, or pulverized carbonate plant and animal skeletons are most common. Beaches occur where waves are sufficiently gentle to allow sand to accumulate but still strong enough to wash the finer silts and clays away. A good portion of the sand on many beaches is eroded away by large winter waves and deposited as underwater sandbars offshore. Smaller waves the following summer move the sand back on shore. Longshore currents also slowly move beach sands parallel to the shore. In response, populations of beach inhabitants may fluctuate widely from season to season and from one year to the next.

Mudflats are somewhat more stable than beach sands, but they too may be altered by seasonal variations in current patterns and wave activity. Intertidal mudflats are usually found in estuaries or quiet reaches of bays and lagoons. Only in these protected coastal environments can significant amounts of finer silt and clay particles settle out. Mudflats contain some sand, but the sand is mixed with varying amounts of finer silt and clay particles to produce mud. The terms *sand, silt,* and *clay* have widely accepted general meanings, but each also refers to a specific range of sediment particle sizes (see figure 8.4).

Several properties of marine sediments are established by the size and shape of sediment particles. The size of spaces between sediment particles (the interstitial spaces) decreases with finer sediments. Interstitial space size, in turn, regulates the porosity and permeability of sediments to water. In coarse sands, water flows freely between sand grains, recharging the supply of dissolved oxygen and flushing away wastes. Beaches, on the other hand, are usually steeper than mudflats, so they drain and dry out more quickly. In fine-grained muds, sediment particles are packed so tightly that little water can percolate through. With the exchange of interstitial water restricted, oxygen used by mud dwellers is not rapidly replenished, and their wastes are not quickly removed.

Fine-grained muds with small interstitial spaces are effective traps for particles of organic debris. Much of the accumulated organic material is found in a thin, brownish, oxygenated surface layer about 1 cm thick. Below this layer, the organic content of the muds usually decreases as animals and decomposing bacteria and fungi consume it. Respiration by the inhabitants of the muds also reduces the available dissolved oxygen supply of the interstitial waters. The lower limit of oxygen penetration in organic-rich sediments is usually apparent as a color change, from a light color in the oxygenated surface layer to a dark or even black color in the anaerobic zone below. The anaerobic conditions of deeper muds inhibit, but do not halt, decomposition of the organic material.

Bacteria and fungi are the major groups of marine organisms capable of utilizing the rich organic accumulations in the anaerobic portion of muddy sediments. Without oxygen, these anaerobic decomposers must use other available elements for their respiratory processes. Sulphate, an abundant ion in seawater, is commonly reduced to hydrogen sulfide (H_2S) in a process much the reverse of the chemosynthetic pathways used by deep-sea hot spring bacteria (p. 313). The H_2S is responsible for the memorable rotten-egg odor and black color so characteristic of anaerobic marine muds.

Not all benthic bacteria and fungi are anaerobic, however. Aerobic decomposers dominate the surface oxygenated layer of mud, but their numbers decline rapidly with depth. Beneath the oxygenated layer, the anaerobes are active down to 40 to 60 cm, where their numbers also dwindle rapidly. The overwhelming abundance of both aerobic and anaerobic decomposers is responsible for most of the chemical changes that occur in marine sediments. The results of these chemical reactions include the decomposition of organic material, consumption of dissolved oxygen near the bottom, and the recycling of critical plant nutrients back to the water (figure 8.32).

Figure 8.32

Nitrogen cycle of a soft bottom marine community. Several types of bacteria sequentially reduce and excrete nitrogen compounds for reuse by marine autotrophs.

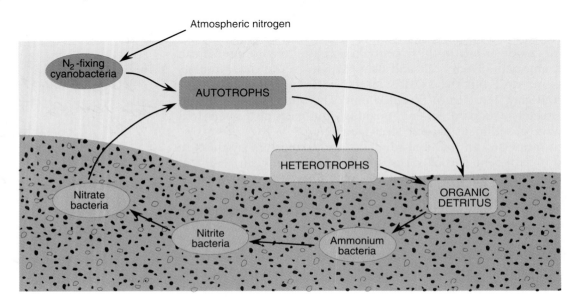

When compared with the teeming populations of the rocky intertidal, beaches appear to be quite desolate. Macroscopic algae and large, obvious epifauna are rare. Shifting, unstable sands are unsuitable platforms for surface anchorage, and nearly all the permanent residents of the beach dwell underground. Patterns of zonation are more difficult to demonstrate, yet, under the sand, there are distinguishable life zones comparable to those on rocky shores (figure 8.33).

The upper portions of sandy beaches along temperate coasts are occupied by a few species of amphipods, particularly *Talorchestia*. The common name of "beach hopper" reflects the unusual bounding mode of locomotion these small crustaceans use. Beach hoppers prefer to burrow a few centimeters into the sand during the day, but are most active at night. Occasionally, they make excursions down the beach face as the tide recedes.

In the upper parts of subtropical and tropical beaches, talitrid amphipods are replaced by ghost crabs. Like the talitrids, ghost crabs are nocturnal scavengers. They live in burrows and return infrequently to water to dampen their gills. Like the nearly terrestrial littorine snails of the upper rocky intertidal zone, ghost crabs are well adapted for long stays on the upper beach without contact with water.

The middle beach is frequently populated by a variety of other amphipods, lugworms *(Arenicola),* dense concentrations of isopod crustaceans, and the sand crab, *Emerita.* These amphipods demonstrate the fundamental feeding methods employed by the larger members of the beach community. *Arenicola* occupies a U-shaped burrow, with its

Figure 8.33

Sandy beach zonation along the East Coast of the United States. The species change rapidly from the portion permanently underwater at left to the dry part of the beach above high tide at right.

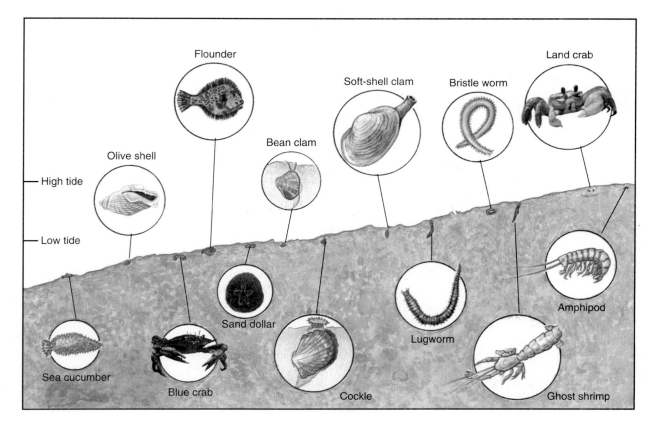

Figure 8.34

A few examples of the interstitial fauna of sandy beaches. Each is of a different phylum, yet all exhibit the small size and worm-shaped body characteristic of interstitial fauna: *(a)* a polychaete, *Psammodrilus,* *(b)* a copepod, *Cylindropsyllis, (c)* a gastrotrich, *Urodasys,* and (d) a hydra, *Halammohydra.*
Redrawn in part from Eltringham, 1972.

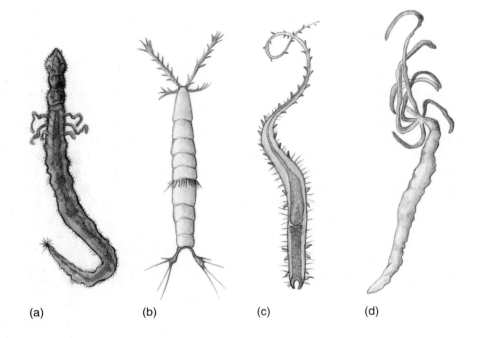

(a)   (b)   (c)   (d)

head usually buried just below a sand-filled surface depression. The burrows are more or less permanent, for waves stir up and move sediment and detritus to the head region, where they are consumed. Mounds of coiled castings indicate the location of the other end of this sediment ingester.

Small isopods are usually less than 1 cm long. They actively prey on smaller interstitial animals that inhabit the pore spaces between sand grains. Many animal phyla are represented in the interstitial fauna of beaches, and a few groups such as harpacticoid copepods and gastrotrichs are practically confined to the interstices of beach sands. Despite their divergent backgrounds, most interstitial animals exhibit the basic adaptations needed for life between sand grains. They are elongated, small (no more than a few millimeters), and move with a sliding motion between sand grains without displacing them. Examples of interstitial animals from different phyla are shown in figure 8.34. Some interstitial animals are carnivorous; others feed on detritus deposits and material in suspension. A specialized feeding habit, unique to interstitial animals, is sand licking. Individual sand grains are manipulated by the animals' mouthparts to remove minute bacterial growths and thin films of diatoms.

The sand crab illustrates a third feeding mode common to many beach macrofauna. When feeding, *Emerita* burrows tail first into the sand and faces down the beach (figure 8.35). Only its eyes and a pair of large feathery antennae protrude above the sand. When a wave breaks over the crab and begins to recede, the antennae are extended against the rush of water. Entrapped phytoplankton (and possibly even large bacteria) are swept into the filtering antennae, then moved to the mouth by other feeding appendages.

In the lower portion of intertidal beaches, the diversity of life increases. Polychaete worms, still other amphipods, and an assortment

Figure 8.35

The sand crab, *Emerita,* with feeding antennae extended.
Courtesy A. Wenner.

of clams and cockles appear. Many of these lower beach inhabitants, such as soft-shelled clams and cockles of the Atlantic Coast, represent the upper fringes of much larger subtidal populations. The small wedge-shaped bean clam, *Donax,* of the Atlantic and Gulf coasts (but not the Pacific Coast species) migrates up and down the beach with the tides, yet it is usually considered an inhabitant of the lower beach. *Donax* responds to the agitation of incoming waves of rising tides by emerging from the sand to feed. After the wave carries the clam up the beach, it digs in to await the next wave and another ride. During ebb tides, the behavior is reversed. *Donax* emerges only after a wave breaks and begins to wash back down the beach. So, with little energy expenditure of its own, this small clam capitalizes on the abundance of available wave energy to carry it up and down the beach face.

Donax is one of many sandy beach inhabitants to exhibit a rhythmic behavior that corresponds to the tidal cycle. Fiddler crabs quietly sit out submergence by high tides, then emerge from their burrows at low tide to feed or engage in social activities. When fiddler crabs are removed to the laboratory, their activity rhythms remain in concert with the changing tidal cycle for some time despite the absence of tidal cues.

Most, and possibly even all, organisms have an innate time sense, a "biological clock." Rhythmic cycles of body temperatures, activity levels, oxygen consumption, and a host of other physiological variations occur independently of changes in the external environment. The internal "mechanism" of the clock is not known, but its existence has been demonstrated in a wide variety of organisms ranging from single-celled diatoms to humans. The tide-related cycle of activity exhibited by the fiddler crab is known as a **circalunadian rhythm;** tidal cycles repeat every lunar day (24.8 hours). The coloration of fiddler crabs depends on **circadian rhythms** (based on a solar day of 24 hours). They are light-colored at night but darken during the day.

Clean sandy beaches commonly grade into intermediate muddy sand and eventually to the muddy shores of bays and estuaries (see chapter 9). The concepts of vertical zonation as they are applied to the sloping faces of beaches and rocky shores become inappropriate on these nearly level expanses of mudflats. The epifauna of mudflats are dominated by mobile species of gastropod mollusks, crustaceans, and polychaete worms. These organisms sometimes range over wide areas of the mudflat and demonstrate only blurred, weakly established patterns of lateral zonation.

Mud-dwelling organisms, however, do occupy vertically arranged zones in the sediment. A few centimeters below the mud surface, the interstitial water is generally devoid of available oxygen, and the infauna must obtain their oxygen from the water just above the mud or do without. The numerous openings of tubes and burrows on the surfaces of estuarine mudflats (figure 8.36) attest to an unseen wealth of animal life underneath. Bivalve mollusks extend tubular siphons through the anaerobic mud to the oxygenated water above. The depth to which these animals can seek protection in the mud is limited largely by the

Figure 8.36

Barren surface of a mudflat, with tubes, openings, burrows, and other evidence of abundant animal life beneath the surface.

lengths of their siphons and thus, indirectly, by their ages. Other infauna utilize the sticky consistency of fine-grained organically rich muds to construct permanent burrows with connections to the surface.

During high tides, the submerged portions of intertidal beaches and mudflats are visited by shore crabs, shrimps, fishes, and other transients from deeper water. Some come to forage for food; others find the protected waters ideal for spawning. Their forays are only temporary, however, for they leave with the ebbing tide and surrender the intertidal to shorebirds and coastal mammals. During low water, the long-billed curlews and whimbrels probe for the deeper infauna, while sandpipers and other short-billed shorebirds concentrate on the shallow infauna and small epifauna (see figure 7.13). Bats, rats, raccoons, and coyotes also patrol the shore at night during low tides.

When the tide is out, the infauna of muddy shores must also cope with an absence of available oxygen and with changing air temperatures. Some switch from aerobic to anaerobic respiration. In doing so, many encounter a dilemma in trying to match the relative inefficiency of anaerobic respiration as an energy-yielding process with the increased metabolic rates forced by increasing tissue temperatures. Larger infauna exist anaerobically only temporarily and revert to aerobic respiration as soon as they are covered by the tides. But for many of their smaller burrowing neighbors with no direct access to the oxygen-laden waters above, anaerobic respiration is a permanent feature of their infaunal existence in intertidal muds.

Summary Points

- The sea bottom is one of the most varied and, in places, rigorous habitats on the earth. The variety of benthic life forms attests to the diverse range of niches available on the bottom.
- The seafloor, the water just above the bottom, and other living plants and animals all shape the environmental conditions faced by the benthos. Feeding, locomotion, and reproduction are all strongly influenced by characteristics of the sea bottom.
- The periodic rise and fall of the tides along coastlines adds to the environmental complexity of the intertidal sea bottom. Intertidal organisms must deal with the effects of exposure to the atmosphere as well as to seawater and to conditions of the ocean bottom.
- This exposure, in concert with other physical and biological factors, produces graded sets of life zones. In the upper intertidal, desiccation and an intermittent food supply are major problems. The organisms of the more densely populated middle intertidal experience a much less restrictive physical environment but encounter increased interspecific competition. In the lower intertidal, the emphasis on a few well-adapted and dominant species is replaced by a much more diversified assemblage of plants and animals.
- Sandy beaches and muddy shores are depositional environments characterized by deposits of loose sediments and organic detritus. Patterns of vertical zonation do occur, but they are more apparent within the sediment than along its surface.

Vertical Zonation

Quite Apparent on Rocky Beach

Less apparent on Sandy (need To dig)

Review Questions

Deposit, Graze,
Absorb, Pred, Scav
Filter

1. List two specific advantages and two disadvantages of broadcast spawning by shallow-water benthic animals.
2. List the various feeding methods of benthic animals and relate those to the type of substrate the animals occupy.
3. What are the major factors that influence the vertical distribution of intertidal plant and animal species? How are these factors affected by tidal variations on a shoreline?
4. Discuss the ecological relationships between mussels, barnacles, *Fucus*, and sea stars on a temperate intertidal coast.
5. Why are benthic epifauna and attached plants seldom found on sandy beaches exposed to wave action?

Challenge Questions

1. Discuss the advantages and disadvantages of living on solid bottom substrates rather than on mud or sand.
2. Compare the species diversity of the middle and lower intertidal zones on rocky shores. On sandy beaches. What accounts for these differences?
3. List and discuss the ecological advantages of planktonic larval stages for benthic animals living in shallow water.

Suggestions for Further Reading

Books

Brafield, A. E. 1978. *Life in sandy shores.* Studies in biology No. 89. London: Edward Arnold.

Carson, R. L. 1979. *The edge of the sea.* Boston: Houghton Mifflin.

Eltringham, S. K. 1972. *Life in mud and sand.* New York: Crane, Russak & Co.

Little, C., and J. A. Kitching. 1996. *The biology of rocky shores.* New York: Oxford University Press.

Moore, P. G., and R. Seed. 1986. *The ecology of rocky coasts.* New York: Columbia University Press.

Newell, R. C. 1979. *Biology of intertidal animals.* Faversham, Kent, U.K.: Ecological Surveys Ltd.

Reise, K. 1985. *Tidal flat ecology.* New York: Springer-Verlag.

Ricketts, E. F., J. Calvin, and J. W. Hedgpeth. 1986. *Between Pacific tides,* 4th ed. Revised by D.W. Phillips. Stanford, CA: Stanford University Press.

Articles

Abele, L. G., and K. Walters. 1979. Marine benthic diversity: A critique and alternative explanation. *Journal of Biogeography* 6:115–26.

Armstrong, R. A., and R. McGehee. 1980. Competitive exclusion. *American Naturalist* 115:151–70.

Connell, J. H. 1961. The influence of interspecific competition and other factors on the distribution of the barnacle *Chthamalus stellatus. Ecology* 42:710–23.

Dayton, P. 1971. Competition, disturbance and community organization: The provision and subsequent utilization of space in a rocky intertidal community. *Ecological Monographs* 41:351–89.

Denny, M. W. 1985. Wave forces on intertidal organisms: A case study. *Limnology and Oceanography* 30(6):1171–87.

Feder, H. M. 1972. Escape responses in marine invertebrates. *Scientific American* (July)227:92–100.

Harger, J. R. E. 1972. Competitive coexistence among intertidal invertebrates. *American Scientist* 60:600–07.

Koehl, M. A. R. 1982. The interaction of moving water and sessile organisms. *Scientific American* (December) 247:124–34.

Lubchenco, J. 1978. Plant species diversity in a marine intertidal community: Importance of herbivore food preference and algal competitive abilities. *American Naturalist* 112:23–39.

Menge, B. A. 1975. Brood or broadcast? The adaptive significance of different reproductive strategies in the two intertidal sea stars *Leptasterias hexactis* and *Pisaster ochraceus. Marine Biology* 31:87–100.

Morse, A. N. C. 1991. How do planktonic larvae know where to settle? *American Scientist* 79:154–67.

Palmer, J. D. 1975. Biological clocks of the tidal zone. *Scientific American* (February)232:70–79.

Peterson, C. H. 1991. Intertidal zonation of marine invertebrates in sand and mud. *American Scientist* 79:236–48.

Schneider, D. C. 1978. Equalization of prey numbers by migratory shorebirds. *Nature* 271:353–54.

Whitlatch, R. B. 1981. Patterns of resource utilization and coexistence in marine intertidal deposit-feeding communities. *Journal of Marine Research* 38:743–65.

Internet Addresses

Rocky Intertidal Habitats
http://bonita.mbnms.nos.noaa.gov/sitechar/rocky.html
Biological Diversity, Distribution Patterns and Temporal Changes.
Sandy Beaches
http://bonita.mbnms.nos.noaa.gov/sitechar/sandy.html
Structure and Formation, Meiofauna and Macrofauna and Zonation of
    Beach Macrofauna. Monterey Bay National Marine Sanctuary.
    Monterey Bay National Marine Sanctuary Site Characterization
    http://bonita.mbnms.nos.noaa.gov/sitechar/index.html
ECOFLAT, The eco-metabolism of an estuarine tidal flat
http://www.nioo.knaw.nl/cemo/ecoflat/Ecoflat.html
Thorough in-depth study of the role of intertidal flats in the ecology of
    estuaries. "ECOFLAT is a research project funded by the European
    Commission in the framework of ENVIRONMENT & CLIMATE and
    ELOISE (European Land Ocean Interactions Studies) programmes
    (1996–1999)."
Sanctuary Marine Life
http://cinms.nos.noaa.gov/animals/animals.stm
Site provides information about intertidal life, subtidal invertebrates,
    fish, kelp, pinnipeds, cetaceans, and sea birds; also links to
    underwater slides (photos), marine mammal sightings and
    migration. Channel Islands National Marine Sanctuary
    http://cinms.nos.noaa.gov/
Underwater Push Pin Pictures
http://www.skio.peachnet.edu/noaa/pictures.html
Two dozen expandable photos of various life forms.
Invertebrates of the Reef
http://www.skio.peachnet.edu/noaa/inverts.html
Pictures of various marine invertebrates.
Vertebrates of the Reef
http://www.skio.peachnet.edu/noaa/verts.html
Pictures of various marine vertebrates.
Ecology Of Gray's Reef
http://www.skio.peachnet.edu/noaa/grhb/ecology.html
Full discussion of reef, with sketches. Excerpts from "Gray's Reef
    National Marine Sanctuary: An Educational Handbook"
    http://www.skio.peachnet.edu/noaa/grhb.html by
    contributing writers Jay R. Calkins, EdD and Carol A. Johnson.
    NOAA/Gray's Reef National Marine Sanctuary. Gray's Reef National
    Marine Sanctuary
    http://www.skio.peachnet.edu/noaa/grnms.html

Photo of sea cucumber
http://www.human.com/sos/pic7a.gif
Surface photo of kelp bed
http://www.human.com/sos/pic6a.gif
Pictures from The Monterey Bay
http://www.human.com/sos/piclist.html
Links to more marine photos.
Launch Pad
http://www.human.com/sos/launch.html
Links to other Monterey Bay related sites. Save Our Shores, Vicki
  Nichols, Executive Director. Save Our Shores
  http://www.human.com/sos/

Estuaries

Estuaries are semienclosed coastal embayments where freshwater rivers meet the sea. Here fresh water and seawater mix, creating unique and complex ecosystems. Such familiar places as the Chesapeake Bay, San Francisco Bay, Great South Bay, Tampa Bay, Puget Sound, and the Mississippi River delta are among the hundred or so bodies of water officially designated as estuaries in the United States. More than one-third of the people in the United States live within the river drainage areas of our estuaries. The confluence of river, sea, land, and, more than in any other ecosystem discussed in this book, humans, shapes and influences the character of these important transitional coastal habitats.

Estuaries are highly variable ecosystems that continually change because of local physical, geological, chemical, and biological factors. The transition from fresh water to saltwater occurs over an area that is unique to each estuary. The estuaries of large rivers such as the Columbia River may extend many miles inland, whereas the estuaries of small streams may be only a few hundred meters in extent. The fresh water and seawater of an estuary may be well mixed, partially mixed or highly stratified. The location of estuaries varies from areas such as steep-sided coastal fjords to shallow and flat bays and sloughs.

The size and shape of estuaries are influenced by the relative amounts of fresh water and seawater entering the estuary and by the geological history of the area. Geological movement of the earth's crust has elevated and lowered coastal areas, and the formation and melting of glaciers have alternately removed and returned water from the ocean basins. The resulting changes in sea level alter the size and shape of estuaries by altering the water depth and the extent of submerged coastal features.

Physical forces at work also influence the chemistry of an estuary. When fresh water draining from a coastal watershed mixes with the ocean tides, the fate of sediments and pollutants being carried downstream becomes complicated. Saline water pushes upstream during high tides and encounters suspended river sediments, causing much of the sediment load to be deposited near the head of the estuary. Because many pollutants are transported downstream in the river water or adsorbed onto sediment particles, unique estuarine conditions can also influence the fate and availability of pollutants to the inhabitants of the ecosystem.

In their natural states, estuaries are among the most biologically productive ecosystems on the earth. Their rates of primary productivity rival and often exceed those of coral reefs, rain forests, and even intensively cultivated corn fields. These special habitats are created by the combination of turbulent mixing, daily fluctuating tidal cycles, and the downstream flow of fresh water that usually changes seasonally in velocity and volume. When these forces meet in an estuary, they exert considerable and complicated effects on the system, creating diverse terrestrial and aquatic habitats atypical of either the river or the sea. More than two-thirds of the species of fish and shellfish harvested by both commercial and sports fishers depend on estuaries for feeding or as nursery areas. Estuaries also provide habitat for thousands of species of terrestrial and aquatic organisms, including many threatened, endangered, and rare species.

Estuaries were often ignored by both freshwater ecologists and oceanographers for many years, because the study of estuaries was not a vital part of either specialty. During that time, the overall role of estuaries in coastal primary productivity was seriously underestimated. Within the past several decades, however, attitudes about estuaries have changed considerably; estuaries and their surrounding wetlands are recognized as fragile environments that have been heavily used and disturbed. Estuaries have become an endangered type of natural habitat. Many estuaries that used to be rich sources of fish, game, and shellfish have become stagnant and unproductive as a result of unregulated or poorly regulated economic exploitation and pollution. Dredging navigation channels in estuarine ports, filling unique estuarine wetlands for development, disposing wastewaters from coastal communities, diverting rivers for irrigation purposes, and allowing pesticide-contaminated rainwater to run into coastal watersheds have changed the character of estuaries and threatened their ecological integrity. Restoration and enhancement efforts are under way to reverse some of the environmental degradation and biological devastation of many of the world's major estuaries. Unfortunately, the double-edged sword of runaway human population growth coupled with a geometric increase in per capita human consumption make these efforts a truly daunting challenge.

Types of Estuaries

The characteristics of individual estuaries depends largely on their recent geological history. Most estuaries owe some of their present configuration to ancient patterns of river or glacial erosion that occurred during the LGM (last glacial maximum), when sea level worldwide was about 150 m lower than at present. These scoured river or glacial channels assumed their present configuration when the great continental ice sheets melted and gradually flooded them. Some estuaries remain very sensitive to slight changes in sea level; increases of only a few meters could drown small estuaries, and comparable decreases could shrink others (figure 9.1) or even cause them to disappear.

Estuaries are found in some form along most coastlines of the world, but most are evident in wetter climates of temperate and tropical latitudes. In such areas, drainage of inland watersheds provides the necessary freshwater input at the head of the estuary to keep salinities below those of adjacent open-ocean waters. In North America, excellent examples of all major types of estuaries exist. Estuaries are classified by both their modes of formation and their patterns of water circulation. **Coastal plain estuaries,** also known as drowned river valleys (figure 9.2a), lie along the northern and central Atlantic

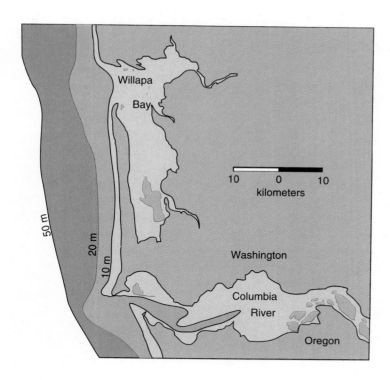

Figure 9.1

The Columbia River estuary and Willapa Bay, with shorelines (green), and at 10 m (light blue), 20 m (medium blue), and 50 m (dark blue) depth contours.

Courtesy Earth Resources Technology Program.

Coast, the Canadian Maritime region, and many areas of the West Coast of North America; among them are the Columbia, Chesapeake, and Delaware estuaries. These estuaries are broad, shallow embayments that formed from deeper V-shaped channels as the sea level rose and flooded river mouths after the last episode of continental glaciation. Sea level was lower because polar ice caps were larger then, and more of the world's water was trapped in the ice caps. Changes in climate after the LGM melted ice caps and caused sea level to rise to its current level.

The extent to which the sea invaded these coastal river valleys since the LGM is determined by the steepness and size of the valley, the rate of river discharge, and the range and force of the tides of the adjacent sea. This type of estuary continues to be gradually modified as wave erosion cuts away some existing shorelines and deposits of river sediment create new shorelines by building mudflats.

Bar-built estuaries are common along the southern Atlantic Coast and the Gulf of Mexico (figure 9.2b) in North America and along the coastal lowlands of northwestern Europe. These estuaries are formed as near-shore sand and mud are moved by coastal wave action to build an obstruction, or barrier island, in front of a coastal area fed by one or more coastal streams or rivers. Often these small coastal rivers and streams have little freshwater flow, so the estuary may be partially or completely blocked by sand deposited by ocean waves. During rainy seasons, however, the increased runoff often temporarily reopens the estuary mouth.

Not all estuaries have restricted mouths. Some estuaries have broad, poorly defined fan-shaped mouths called **deltas.** The Mississippi River delta and other similar delta estuaries are created as very

Figure 9.2

Satellite views of three types of
estuary: *(a)* Chesapeake Bay and
Delaware Bay, both coastal plain
estuaries on the U.S. East Coast.
(b) Matagorda and San Antonio Bays
on the Texas coast, both bar-built
estuaries. *(c)* deeply incised fjords
between Glacier Bay National Park
and Juneau, Alaska.

(a) Courtesy NASA; *(b)*and *(c)* courtesy M.
Miller, Earthsat.

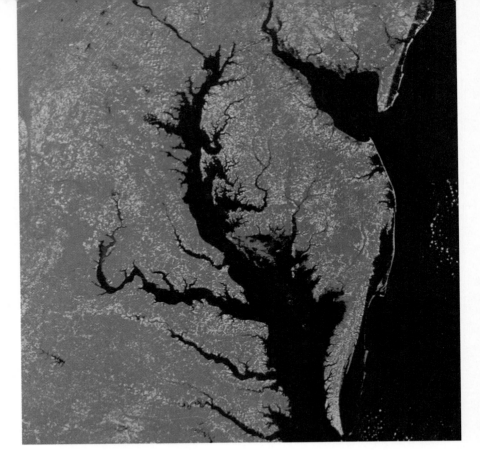

(a)

(b)

Figure 9.2 Continued

(c)

heavy loads of sediments eroded from the upstream watersheds are deposited at the river mouth. Different still are **tectonic estuaries** such as San Francisco Bay, created when the underlying land sank in response to crustal movements of the earth. As coastal depressions created by these movements sank below sea level, they filled with water from the sea and also became natural land-drainage channels, directing the flow of land runoff into the new estuary basin.

From the central West Coast northward, estuaries become more deeply incised into coastal landforms and gradually merge into the deep, glacially carved fjords characteristic of British Columbia, southeastern Alaska (figure 9.2c), Norway, and southern Chile. In cross section, fjords resemble the Black Sea (see figure 1.36). In general, fjords are deeper than other types of estuaries, and the deepest regions of fjords are in the upstream reaches. The shallow sills at their mouths partially block the inflow of seawater and lead to stagnant conditions near the bottoms of deeper fjords (see figure 1.36).

Estuarine Circulation

In addition to their structural differences, estuaries exhibit different patterns of freshwater and seawater mixing within the basin. The upstream-to-downstream variations in salinity, water temperature, turbidity, and current action are complex and change markedly during a tidal cycle and in response to seasonal changes in the volume of freshwater stream discharge. Salinity values typically increase from

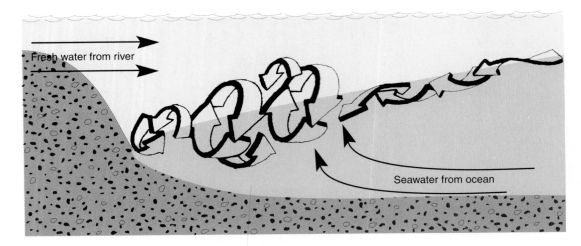

Fresh water from river

Seawater from ocean

Figure 9.3

The general pattern of fresh water and seawater mixing in an estuary.

the surface downward and downstream from the estuary head. As tides change sea level in a typical estuary, higher-density seawater moves in and out along the estuary bottom and is gradually mixed upward into the outflowing low-salinity surface water (figure 9.3). Because of this mixing, there is an inward flow of nutrient-rich water along the bottom of the estuary and a net outward flow at the surface. This upward mixing generates, on a localized scale, a process of estuarine upwelling that replenishes nutrients and promotes growth of estuarine primary producers.

The shape of an estuary's basin is a major factor in determining its mixing pattern. A triangular estuary with a wide, deep mouth allows seawater to move farther upstream. The currents may be strong and the water is well mixed, so salinity and water density are nearly the same from the surface downward at any location within the estuary (figure 9.4a). In narrow-mouthed estuaries, circulation is decreased, creating more-pronounced vertical salinity gradients. These narrow-mouthed estuaries are often highly stratified. They have a pronounced seawater wedge under the less-dense fresh water on the surface (figure 9.4b), with an abrupt salinity change where seawater and fresh water meet. Lines of equal salinity, or **isohalines,** can be plotted by analyzing water samples taken at fixed depths throughout a tidal cycle and connecting points with the same salinity values. The shape of the isohalines is useful in classifying types of estuaries and in understanding the distribution of estuarine organisms.

In addition to the more predictable effects of tides and river discharge, circulation in estuaries often changes rapidly and less predictably in response to short-term influences of heavy rainfall or changing winds. The Coriolis effect also exercises its influence on circulation patterns within estuaries (see figures 1.31 and 9.15) by forcing seawater farther upstream on the left sides (when facing seaward) of estuaries in the Northern Hemisphere and on the right sides of estuaries in the Southern Hemisphere.

The time necessary for water in an estuary to be moved out to sea is called the **flushing time.** Complete flushing, which replaces all of an estuary's water volume, may take from days to years depending on the

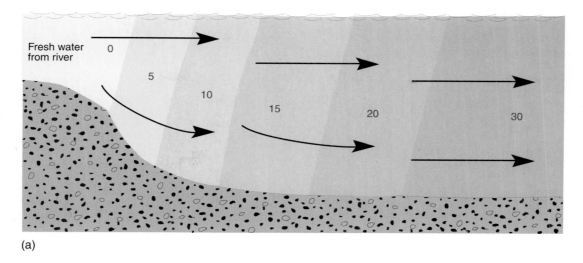

(a)

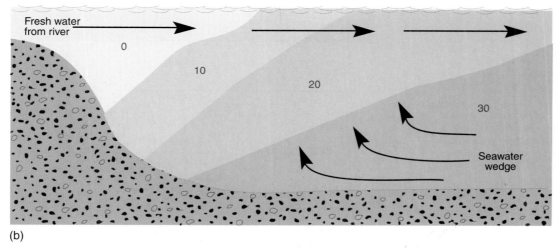

(b)

combination of tides, river flow, wind, and salinity gradients. The flushing time in an estuary strongly influences the transport of nutrients and is a crucial factor in determining the fates of pollutants in estuaries.

Figure 9.4

Cross sections of a well-mixed (*a*) and stratified (*b*) estuary, with the resulting salinity (‰) shown in color.

Salinity Adaptations

To survive in the variable environment of most estuaries, organisms, particularly sessile benthic ones, must be able to tolerate frequent changes in salinity and internal osmotic pressure. The abundance and distribution of estuarine organisms are continually influenced by changing salinity associated with short-term events—such as seasonal patterns of freshwater runoff, daily tidal cycles, and pulses of fresh water from heavy rains—and with long-term events—such as the advance and retreat of sea level as continental glaciers grow and recede through geological time.

A small fraction of animal species that live in estuaries, especially insect larvae, a few snails, and polychaete worms, have their closest relatives in fresh water. The great majority, however, are derived from

Figure 9.5

Relative contributions of freshwater,
brackish water, and marine species
to estuarine fauna.
Redrawn from Remane 1934.

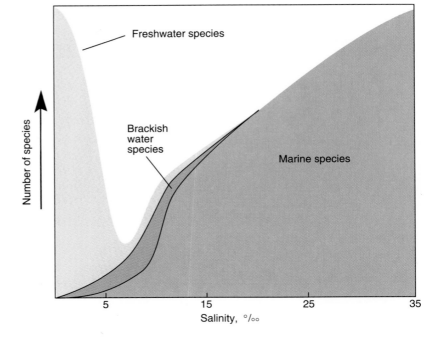

Figure 9.6

Comparison of salinity variations
through a typical tidal cycle of
interstitial water (tan) with that of
the overlying water (blue) in
Pocasset Estuary, Massachusetts.
Adapted from Mangelsdorf 1967.

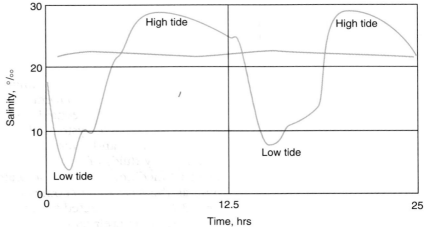

marine forms (figure 9.5) and include some of the same species found
on nearby beaches and nonestuarine mudflats. Some animal species
have poorly developed osmoregulatory capabilities and avoid osmotic
problems by not venturing too far into estuaries. Others employ adaptive strategies to overcome the osmotic problems of recurring exposure to low and variable salinities of estuarine waters. Some of these
adaptations are modifications of structural or physical systems already
imperative for survival on exposed intertidal shorelines. Oysters and
other bivalve mollusks, for instance, simply stop feeding and close
their shells when subjected to the osmotic stresses of low-salinity
water. Isolated within their shells, they switch to anaerobic respiration
and await high tide, when water higher in salinity and oxygen returns.
Other animal species retreat into mud burrows, where salinity fluctuations due to tidal cycles are usually much less severe (figure 9.6).

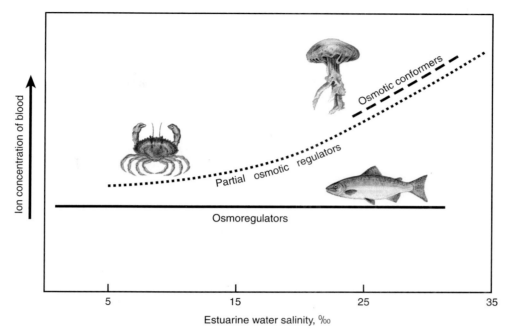

Figure 9.7

Variations in ion concentrations of
body fluids or blood with changing
external water salinities for osmotic
conformers, partial osmotic
regulators, and osmoregulators.

A few species of tunicates, anemones, and several other soft-bodied
estuarine epifauna are **osmotic conformers.** Osmotic conformers are
unable to control the osmotic flooding of their tissues when subjected
to low salinities, so their body fluids fluctuate and remain isotonic with
the water around them (figure 9.7).

The most successful and abundant groups of estuarine animals
have evolved mechanisms to stabilize the water and ion concentrations
of their body fluids despite external variations. These mechanisms are
as varied as the organisms themselves, yet all involve systems that acquire essential ions from the external medium and excrete excess
water as it diffuses into their bodies. The body fluids of estuarine crabs
remain nearly isotonic with their external medium in normal seawater
but become progressively hypertonic as the seawater becomes more
dilute. When these partial osmotic regulators are subjected to reduced
salinities, additional ions are actively absorbed by their gills to compensate for the ions lost in their urine (figure 9.7). Thus, these and most
other estuarine crustaceans are osmotic conformers at or near normal
seawater salinities and are **osmoregulators** in dilute seawater.

Most estuarine animals are **stenohaline;** they tolerate exposure
to limited salinity ranges and therefore occupy only a limited portion
of the entire range of salinity regimes available within an estuary
(figure 9.8). A few opportunistic species of estuarine organisms are
euryhaline, capable of withstanding a wide range of salinities.
These species can be found throughout the range of estuarine salinities, with a limited number of euryhaline species also found in high-
salinity lagoons that fringe some of the world's arid coastlines. Lagoons
such as those along the coast of Texas and both sides of northern Mexico have shallow bottoms, high summer temperatures, excessive evaporation, and high salinities. The osmotic problems experienced by animal species in these high-salinity lagoon populations are similar to

Figure 9.8

Differing salinity tolerances of five species of amphipods, *Gammarus).* Of these, only *G. duebeni* is euryhaline.

Adapted from Nicol 1967.

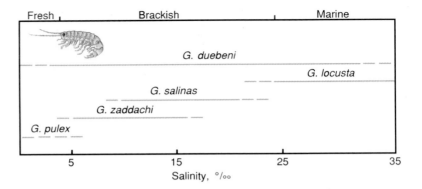

those encountered by bony fishes in seawater and are so severe that reproduction is seldom successful. Continued immigration of euryhaline species from nearby estuaries sustains these lagoon populations.

Species diversity and numbers of individuals usually decline considerably, from a maximum near the ocean to a minimum near the headwater of an estuary. The distributional patterns of estuarine animals are governed by salinity variations, patterns of food and sediment preferences, current action, water temperature variations, and competition between species. It is the collective interaction of all these factors that establishes and maintains the distribution of estuarine organisms.

Sediment Transport: Creating Habitats

Sediments are transported into estuaries from rivers that drain coastal watersheds and from coastal areas outside the estuary mouth. River sediment particles range in size from gravels and coarse sands to fine silts, clays, and organic detritus. They are derived from erosion of riverbanks denuded of their natural plant cover and from the scouring of meandering river channels. Fast-moving rivers carry large amounts of particles, but as the rivers widen and slow in coastal floodplains, they begin to meander and their sediments settle to the bottom. Estuaries thus serve as effective catch basins for much of the fine suspended sediments washed off the land. The current speed necessary to keep the sediment load suspended diminishes in the protected and quiet waters of estuaries, and the water slows to a point at which only the finest silts and clays remain suspended in the water.

Storms and near-shore currents of the open ocean can also move coastal sand and detritus materials into the mouth of an estuary and add to the complex mix of estuarine sediments. Typically, these deposits show a characteristic distribution of different sediment types; coarse particles are deposited at the heads of estuaries and in shallow water, and finer particles settle nearer the mouth and in deeper water. These graded and sorted sediment deposits transported down from rivers and in from the sea provide a rich and varying substrate to support the estuarine communities.

Estuarine Habitats and Communities

Estuarine communities include wetlands, mudflats, and channels. The areas of highest elevation are the wetlands; they are periodically covered by estuarine water at high tides and consist of dense plant communities

(a)

(b)

that tolerate contact with seawater. Mudflats, or tideflats, are lower in elevation than the wetlands and are alternately submerged and exposed by changing tides. Channels, those areas that are under water even at the lowest tides, are prevented from filling with sediments by the scouring action of tides or river flow.

Temperate Wetlands: Salt Marshes

Temperate wetlands are essentially wet grasslands, or **salt marshes,** that grow along estuarine shores. The dominant members of these marshes are **halophytes,** a few species of plants that require, or at least are tolerant to, saline waters (figure 9.9). Salt marshes develop in the muddy deposits around the edges of temperate and subpolar

Figure 9.9

Two types of emergent salt marsh plants: *(a)* a dense stand of marsh grass, *Spartina; (b)* pickleweed, *Salicornia.*

249

Figure 9.10

Food particle production and
utilization in a typical estuary.
Adapted from Correll 1978.

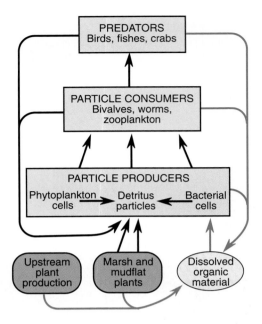

estuaries, creating a transition zone between land and estuarine
plant communities. Salt marshes may be inhabited by several plant
species, each with its own specific set of sediment, water, and exposure requirements. The lowest parts of a marsh, submerged for
longer periods of time, may be dominated by pickleweed, which
stores excess salt in its fleshy leaves, and marsh grass, which has special glands that enable it to rid itself of excess salt. In higher marsh
zones, grasses, rushes, and sedges that cannot tolerate prolonged sub-
mersion by the tides dominate the landscape.

Salt marshes form an important part of the base of estuarine food
webs. Some of the plants in the estuary are eaten directly by marsh
herbivores, but most of the vegetation decays and enters the food web
as detritus. The flooding and ebbing of the tides transport the detritus
from the marsh into the estuary and surrounding tidal creeks, where
the detritus sinks and decomposes further. Each winter, salt marsh
plants die back, and tides (or, in cold climates, the shearing action of
rising and falling tidal ice) harvest this grass and put it into the detrital
food web. There, it becomes the target of decomposing bacteria. The
microbial activities of the bacteria further break down the plant matter, especially the cellulose cell walls, which are undigestible by most
animals, convert some of it to bacterial cells, and release dissolved organic materials and inorganic nutrients to be reused by other plants
into the estuary.

Thus, bacteria, as well as the phytoplankton of the overlying waters, contribute heavily to the production of small, energy-rich detrital food particles. These microorganisms then become a major
source of food for large populations of estuarine particle consumers
(figure 9.10). The particle consumers themselves produce still more
food particles in the form of feces and rejected food items. These
particles are eventually recolonized by bacteria and recycled into the

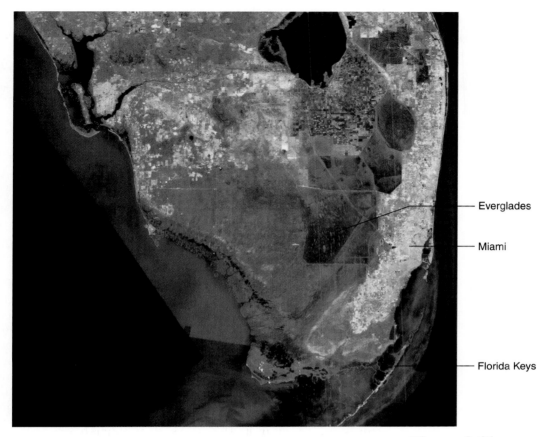

— Everglades

— Miami

— Florida Keys

Figure 9.11

Satellite image of the Everglades
National Park and the keys of
southern Florida.
Courtesy M. Miller, Earthsat.

particle pool of the estuary. In this manner, estuarine bacteria and
other microorganisms play a central role in transforming the productivity of the estuary margins into small detrital food particles available to numerous other species of estuarine animals.

Tropical Wetlands: Mangals

Dominating large expanses of rich estuarine muds in warmer climates
are excellent examples of both emergent and submergent plant-based
communities. Several species of mangroves (see figure 4.13) inhabit
the shores of protected coastal lagoons and estuaries in tropical and
subtropical latitudes. Mangroves range in size from small shrubs to
10-m-tall trees that are tolerant to seawater and capable of establishing
their roots in black anaerobic muds. Collectively, mangrove plants, the
major component of mangal communities, line about two-thirds of the
tropical coastlines of the world. In the United States, the distribution
of mangals reflects their need for warm waters protected from wave
action; they are found only along portions of the Gulf of Mexico and
the Atlantic Coast of Florida (figure 9.11).

   The southern coast of Florida is dominated by extensive interconnected shallow bays, waterways, and mangals. These mangals
form a nearly continuous narrow band along the coast, with smaller
fingers extending inland along creeks. Inland, toward the freshwater

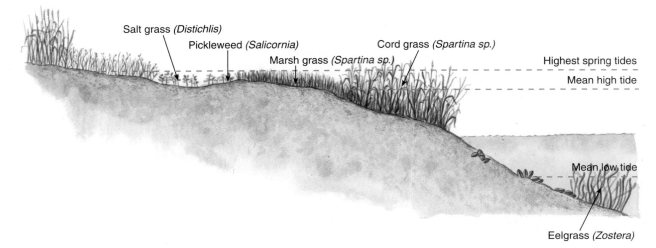

Salt grass *(Distichlis)*

Pickleweed *(Salicornia)*

Cord grass *(Spartina sp.)*

Marsh grass *(Spartina sp.)*

Highest spring tides

Mean high tide

Mean low tide

Eelgrass *(Zostera)*

Figure 9.12

Plant-dominated salt marsh, mudflat, and channel habitats of Chesapeake Bay and other East Coast estuaries; their vertical position relative to high tide is indicated.

Everglades, the mangroves are not high, but tree height of all three species (red, black, and white mangroves) increases to as much as 10 m at the coast. It is these taller coastal members of mangal communities that are especially prone to hurricane damage. In 1992, Hurricane Andrew cut a swath of destruction across southern Florida with sustained winds up to 242 km/h. The accompanying storm surge lifted the sea surface over 5 m above normal levels. Some of the more exposed coastal mangal communities experienced over 80% mortality, due mostly to wind effects and lingering problems of coastal erosion.

Mudflats

Mudflats are large estuarine expanses composed primarily of rich muds that are exposed at low tide. Where marine waters and rivers mix and the salinity gradients are large, dissolved ions interact with the sediments and flocculate (bind together to form larger particles) and add to the accumulating richness of the bottom muds. These unstable soft anaerobic mud deposits serve as the principal structural foundation of soft-bottom communities that thrive in estuaries.

A sometimes all-too-obvious feature of mudflats is the anaerobic condition that exists just below the surface. There, oxygen-depleted muds serve as an important habitat for many species of anaerobic bacteria. Some of these species produce hydrogen sulfide, the gas responsible for the rotten egg smell so characteristic of exposed mudflats. Other types of sediment dwellers, such as clams and mud shrimp, burrow in the muds to be protected from predators, to be sheltered from the drying effects of the sun at low tide, and to be in an environment where the salinity of the mud is more constant than that of the overlying water (figures 9.6 and 9.12).

Animals living in the mudflats have developed a variety of feeding habits to cope with the environment. Many of the near-surface infauna are suspension feeders, gleaning small food particles from water currents above their burrows. Others are deposit feeders, such as lug worms, which digest the bacterial and organic coatings from

Figure 9.13

A bed of eel grass, *Zostera,* at low tide.
Courtesy P. Flannigan.

NUTRIENT PUMP

sediment particles passing through their digestive system. Other types of burrowing animals, such as the arrow goby, leave their burrows at high tide and forage for food over the mudflat.

Three major groups of primary producers are found on mudflats: diatoms, larger algae, and sea grasses. Microscopic benthic algae, including abundant diatoms, coat the muds with a golden brown film. These photosynthesizers are a rich and important food source for benthic invertebrates. Green mats of macroscopic algae, such as sea lettuce, commonly cover rocks, shells, and pieces of wood debris on mudflats. These are important food sources for herbivores, especially certain worms, amphipods, and crabs.

Sea grasses, found at lower levels of the intertidal region, grow on substrates that range from clean sands to muds. Eel grass and a few other sea grasses constitute one of the few types of flowering plants that can survive completely submerged in saltwater (figure 9.13). Eel grass gets its name from the long, thin, straplike leaves (which can grow up to 2 m in length) that weave back and forth in the currents like eels. Eel grass production contributes greatly to the pool of detrital particles within an estuary. As a source of detritus, it is a valuable food source and is fed upon by some species of ducks (such as widgeon and brant), invertebrates, fishes, and insect larval stages. Algae and diatoms grow on its leaves, as do many types of hydroids, clam larvae, tunicates, ectoprocts, and crustaceans. In addition to its role in the detritus food web, eel grass is usually the first rooted plant to take hold in shallow water. Its roots and long leaves trap fine particulate materials and produce an ideal, food-rich, protected habitat. Finally, it also serves as a nutrient pump by taking nutrients from the sediment for growth and later releasing them to the water when it dies at the end of the growing season.

Channels

A channel is that area of the estuary where water is always present under all tidal conditions (figure 9.12). It may be as broad as the entire estuary, or it may be restricted to a narrow creeklike feature between mudflats. Numerous species of planktonic organisms inhabit channels, relying on the action of currents to move them around. Crabs, oysters, starry flounders, sculpins, anchovies, and killifish are also abundant in channels during certain times of the year.

Channel areas are also used as spawning and nursery areas for many animals. Herring and sole are ocean fishes that move into the protected areas of the estuary to spawn and to allow the juveniles to feed on the large amounts of available food. Crabs also use the estuaries as nursery areas. Anadromous fishes such as salmon use the estuary to get from the ocean to their spawning areas in freshwater streams, and then they return to the sea. Before continuing their migration upstream, they may linger for a time in the estuary to feed while adjusting to the seawater conditions.

Economic Uses of Estuaries

Estuaries on both coasts of North America and in other areas of the world are ecologically critical areas that support a wide variety of biological communities and serve as vital resting and feeding stops within the migratory paths of ducks, geese, Bald Eagles, and many species of shorebirds. Salt marshes, in particular, serve as important natural filters to trap some pollutants as resident bacteria convert some of them to less harmful substances. They also play a role in moderating flooding and sedimentation processes.

Most salt marshes and virtually all estuaries have been altered to some extent by human activity. These modifications have affected estuarine productivity, species diversity, and water quality. The environmental quality of estuaries depends on the types and intensity of human activities throughout the coastal drainage basin. The effects of development and industrialization in an estuary, combined with effects from pollutants carried downstream from watersheds, create conditions that have contributed to worldwide environmental degradation. Estuaries have always been important economic resources for coastal communities, and balancing the many conflicting uses and managing pollution sources are critical to protecting the economic values and natural resources of estuaries.

Estuarine shorelands are used for a wide variety of commercial and recreational industries (figure 9.14): agricultural and forest production; residential, commercial, industrial, and shipping facilities; and disposal sites for dredged materials. Most of the world's major seaports are situated in estuaries, so they become centers for industrialization and extensive population growth. Commercial and industrial activities lead to modification of estuaries. Estuary channels are often dredged to facilitate shipping and to accommodate large ships. Docks, pilings, piers, and jetties alter the natural flushing and circulation patterns of the estuary. Many wetlands have been dredged and filled to create additional flat acreage for development. This activity disturbs the bottom, reduces the number of species that live in the estuary, and affects

Agricultural runoff

Crop spraying

ACME

Industrial wastes

Logging-induced erosion

Urban and industrial wastes

Shipping wastes

Figure 9.14

Common sources of pollutants entering estuaries.

the primary productivity of the estuarine ecosystem. Over 75% of the estuarine wetlands in the United States have been lost because of draining and diking to create agricultural lands or areas for commercial development.

Estuaries have also been used extensively as dumping grounds for municipal and industrial wastes generated by activities associated with coastal developments. Since one-third of the population in the United States lives near estuaries, this is a serious pollution load, especially in estuaries with long flushing times. Solids dumped into the estuaries often smother benthic communities. Toxic chemicals from wastewaters accumulate in water, sediments, and animal tissues; excessive amounts of pollutants such as organic materials or fertilizers create a high biochemical oxygen demand (BOD) and can reduce the estuary of life-supporting oxygen. Human encroachment into estuaries reduces their water quality and productivity.

In addition to easily defined **point sources** of pollutants that discharge directly into estuaries (such as municipal outfalls), pollutants from upstream sources in the watershed can wash long distances downstream and accumulate in estuaries. These **non-point sources** of pollutants may change in toxicity after their first, intended use because of the complex physical and chemical processes at work in estuaries. Pesticides such as DDT and industrial contaminants such as PCBs and dioxin have all been found in high concentrations in estuarine sediments and in several fish and wildlife species (see Research in Progress: Estuaries and Eagles: The Columbia River, p. 256). Water quality in estuaries directly affects the quality and size of sport and

Research in Progress

Estuaries and Eagles: The Columbia River

The Columbia River drains the second largest watershed in the United States. It starts in the ice fields of British Columbia and flows through a drainage basin that extends over parts of five states and two provinces. It is rich in natural resources, supporting some of the largest salmon runs in the world and providing special habitat for sensitive, rare, and endangered species. One of these is our national symbol, the Bald Eagle.

The Columbia River also supports a diverse economy and serves as a major transportation route for world commerce. It provides a fairly typical example of the conflict that often exists between utilization and conservation of estuarine natural resources. The Columbia River drainage basin, home to over 8 million people, receives heavy waste loads associated with populated and industrially developed areas. Industrial discharges from pulp and paper mills and aluminum plants, wastewater from cities, and runoff from agricultural fields, forest harvest areas, and city streets all contribute fertilizers, pesticides, petroleum products, heavy metals, oxygen-consuming organic materials, silt, and soil bacteria to the Columbia River's drainage.

The estuary of the Columbia River extends from the river mouth upstream over 200 km to Bonneville Dam, located between the states of Oregon and Washington. Most pollutants that run off the land into the river or are discharged through pipes directly into the river eventually move downstream and end up in the estuary. Recent studies have shown that the introduction of two toxic pollutants, PCB and DDE (a degradation product of DDT), and the loss of critical habitat have had an adverse effect on Bald Eagles.

Refrigerants, wastes from plastics manufacture, and leaking electrical

generators and transformers are common sources of PCB in the environment. The insecticide DDT was banned in 1972 (see chapter 16), but its metabolite DDE is still biologically available. The sources of DDE and DDT are not known, but it is thought that they enter the river in storm runoff carrying sediments contaminated with agricultural pesticides or in sediments disturbed during river-dredging operations. Both compounds are stable, are slow to chemically degrade, and are biologically persistent, thus enhancing their ability to reenter the river and become magnified in food chains of the estuary.

The Columbia River Estuary is also home to 24 pairs of breeding Bald Eagles and as many as 150 that visit seasonally. The Oregon Cooperative Wildlife Research Unit at Oregon State University recently completed a study on the effects of DDE and PCB on Bald Eagles in the Columbia River Estuary. The study revealed high concentrations of DDE (4 to 16 ppm) and PCB (4.8 to 26.7 ppm) in Bald Eagle eggs and carcasses found in the estuary. Associated with these high concentrations of DDE and PCB was significant thinning of eggshells (mean eggshell thickness was 14% thinner than the shells of normal eggs) and consequent high reproductive failure. Only about 40% of the healthy eagle pairs around the Columbia River Estuary successfully raised chicks each year, the lowest reproductive success of any breeding Bald Eagle population in the Pacific Northwest. Detectable levels of DDE and PCB were also found in the blood of nestlings, and elevated blood levels of both DDE and PCB were common in subadults and adults.

For more information on this topic, see the following:
Dennison, W. C., et al. 1993. Assessing water quality with submerged aquatic vegetation. Bioscience 43:86–93.

commercial harvests of shellfishes and finfishes, oysters, salmon, and other food species. Dredging, harbor development, and discharge of untreated pollutants from agricultural runoff or waste discharges can seriously degrade habitats for these species, restrict industries based on their harvesting, and threaten human health.

Although multiple shoreland uses are an integral part of all coastal economies, they contribute contaminants to the coastal river and estuary systems that can cause habitat loss and a change in the ecological integrity that cumulatively affects public health, fish and wildlife habitat, and recreational resources. Despite past regulatory and management efforts, the water quality and habitat within estuaries have been seriously degraded around the world, and they continue to decline. To better understand some of the conflicts that surround the use of estuarine resources, we will examine in detail one major and well-known estuary, the Chesapeake Bay system on the East Coast of the United States.

The Chesapeake Bay System

The variety of estuary types that exist in North America makes the task of characterizing even the major estuaries difficult. Therefore, the focus of the remainder of this chapter will be one of the world's largest estuaries, the Chesapeake Bay. This system exemplifies the physical, chemical, and biological features of estuaries as well as the substantial social and political problems generated by conflicts between natural estuarine processes and the many additional uses imposed on these coastal habitats by humans.

The Chesapeake Bay system is a cascading series of five major and numerous smaller estuaries (figure 9.15). These estuaries were linked together when the ancestral Susquehanna River valley flooded after the LGM. The bed of the ancient Susquehanna is deep, but most of the bay is sufficiently shallow to allow sunlight to penetrate to the bottom.

The Chesapeake Bay drains a very large (166,000 km^2), heavily populated, and agriculturally rich land area of the central Atlantic coastal plain. Human activities have imposed some serious stresses on the bay in the form of increasing loads of heavy metals, fertilizers, pesticides, and incompletely treated sewage. A reduced oxygen level in the waters of the bay is only one of several complications resulting from this input. Yet, the Chesapeake Bay has been referred to as a protein factory. Millions of blue crabs, oysters, striped bass, and other finfishes are harvested from the Chesapeake Bay waters each year. In 1990, fisheries had a $1 billion impact on the economies of Maryland and Virginia. When the recreational, military, shipping, and other uses of Chesapeake Bay are added to the mix of conflicting uses, we have in microcosm a picture of some of the same problems confronting the larger world ocean.

In the Chesapeake Bay, the existing salinity gradient creates an up-bay low-salinity zone, a mid-bay brackish zone, and a lower-bay marine zone (figure 9.15). Although the tides in the Chesapeake Bay have an average vertical range of only 1 to 2 m, they are the major short-term mixing influence in the bay. On longer time scales, seasonal flooding and storms have dramatic effects on the salinity distribution patterns in the bay.

Figure 9.15

Chesapeake Bay and its numerous smaller side estuaries, showing mean surface salinity zones.

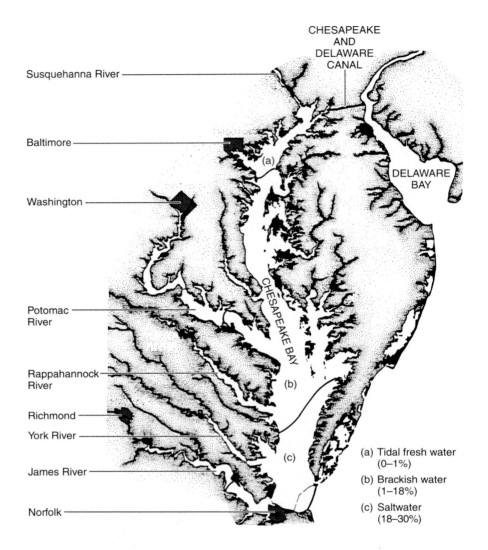

Susquehanna River

CHESAPEAKE AND DELAWARE CANAL

Baltimore

(a)

DELAWARE BAY

Washington

CHESAPEAKE BAY

Potomac River

Rappahannock River

(b)

Richmond

York River

James River

(c)

Norfolk

(a) Tidal fresh water (0–1%)

(b) Brackish water (1–18%)

(c) Saltwater (18–30%)

These habitats offer nearly ideal nursery conditions for the young of animals capable of tolerating the dynamic fluctuations of the Chesapeake Bay and other estuaries. Estuaries provide some protection against the physical stresses of nearby open coasts, and there is an abundance of food available in a large range of particle sizes. The majority of fish species commercially exploited along the Atlantic and Gulf coasts of the United States use estuaries as spawning or juvenile feeding areas. Some species occupy estuaries throughout their lives; others occupy estuaries for only a particularly crucial stage of their development. Figure 9.16 illustrates this range of utilization patterns for a few commercially important or otherwise notable Chesapeake Bay species.

At one extreme are oysters, which typically spawn, mature, and die within the confines of the bay (although some larvae occasionally may be transported to other nearby estuaries). Oysters are broadcast spawners, with fertilization and a two-week larval development period occurring in the moving water above the benthic habitat of the adults.

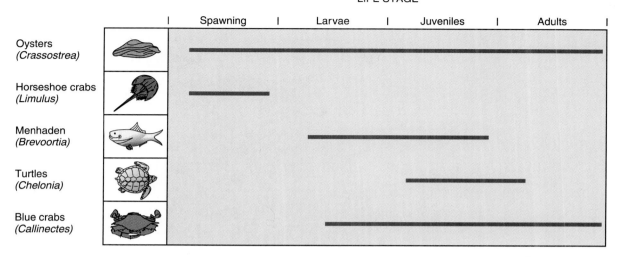

<figure>## Figure 9.16

Utilization of estuaries by differing life stages of five common inhabitants of Chesapeake Bay.</figure>

The double-layered circulation pattern of the bay is used by oyster larvae to avoid being washed out of the estuary. During ebb tides, when most of the tidal outflow is in the surface layers, the larvae remain in the deeper inflowing saltwater. During slack or incoming tides, the larvae venture into the shallower portions of the larger bay or its smaller side estuaries, where high levels of larval settling and retention occur.

Horseshoe crabs and menhaden are marine species that use Chesapeake Bay for early life stages only. Adult horseshoe crabs move into the bay only to spawn. During high tides in the spring, these animals crawl into salt marshes at the water's edge. There, the female digs a depression to deposit her eggs. The smaller male, who hitches a ride on the female's back (see figure 6.28), sheds sperm to fertilize the newly deposited eggs. The eggs are then covered to await hatching two weeks later, when they are again flooded by the next series of spring tides. After hatching, the larvae swim and feed near the surface while currents carry them out to sea, where development continues through as many as 13 successive larval stages.

Menhaden is a commercially valuable fish species of the middle Atlantic Coast. Although the adults live and spawn in coastal waters, menhaden larvae drift into estuaries such as Chesapeake Bay to continue their development. Juvenile marine turtles (loggerheads and Atlantic ridleys are the most common) also graze on the slowly shrinking beds of eel grass found along the bay edges. The 200 to 300 Atlantic ridleys seen each year in the Chesapeake Bay may represent nearly all the existing juveniles of this seriously endangered species.

For blue crabs, the pattern of bay utilization is different still. Adults live in estuaries along most of the United States Atlantic and Gulf coasts. After mating, females seek higher salinities in the open sea before releasing their larvae (figure 9.17). Larval development continues in coastal waters outside the bay, where winds and coastal currents combine to keep blue crab larvae close to shore until they return to the bay as young crabs. It is at this time that exchange of individuals between neighboring estuaries sometimes occurs, preventing genetic

Figure 9.17

Spawning migration of adult female blue crabs and return routes of planktonic larval stages.

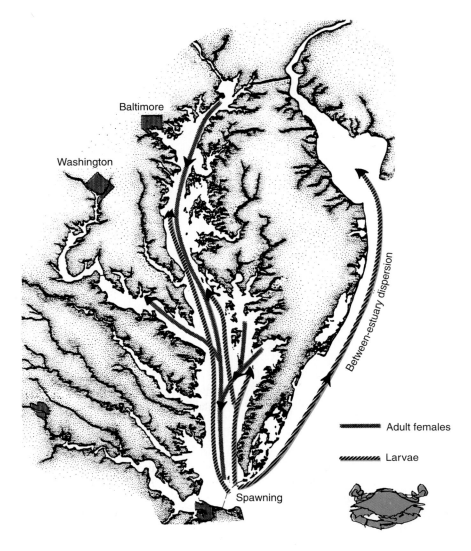

isolation of the crabs that occupy any one estuary. Within the estuary, young blue crabs seek eel grass beds and salt marshes as winter nursery areas for food and protection until they grow sufficiently to exploit other estuarine habitats.

Through their roles as nursery areas and feeding grounds, the value of estuaries such as the Chesapeake Bay extends far beyond the bounds of the estuary itself. Yet, the health of the Chesapeake Bay, as well as that of most of the earth's other major estuaries, has gradually, but seriously, deteriorated in the past few decades. Since 1960, submerged vegetation in the Chesapeake Bay, especially eel grass beds, has declined in size and abundance. Most of the decline has been in the upper and western parts of the bay, but the problem is moving down the bay as well. Dissolved nutrient loads draining into the bay have increased, causing changes in the species composition of the water. In the upper reaches of the Chesapeake Bay, concentrations of cyanobacteria and dinophytes have increased 250-fold since 1950. During the same period, submerged eel grass and marsh grass beds have declined dramatically. These changes in patterns of primary productivity have resulted in extremely high concentrations of algae, which die and sink to the bottom. Sinking of algal cells is part of the normal cycle of estuarine processes. When it occurs in such high densities, however, bacterial decomposers of this algal detritus consume oxygen that is also needed by other members of estuarine communities. Presently, most of the Chesapeake Bay water deeper than 13 m (downstream from near the mouth to the Rappahannock River) has little or no dissolved oxygen. As a consequence of these biological changes, recent catches of alewives, shad, striped bass, and oysters also have dropped precipitously.

Serious and sustained efforts on the part of state regulatory agencies of the six states with rivers that empty into the Chesapeake (Maryland, Delaware, Virginia, West Virginia, New York, and Pennsylvania), the District of Columbia, and several cooperating federal agencies are required to stem these adverse changes and reduce the nutrient and toxic substance loads presently carried into the bay. Only when these efforts are successful can we be assured of the continuing health of this hardy and beautiful estuary system that is the Chesapeake Bay.

Summary Points

- Estuaries are partially enclosed coastal embayments where fresh water and seawater mix. A large variety of estuary types exist, but most owe their present configuration to river or glacial erosion since the last glacial maximum (LGM).
- General patterns of water circulation in estuaries result from interactions between the outward flow of low-density fresh water over a higher density layer of seawater beneath. Tides, winds, and the Coriolis effect add additional complexity to the circulation patterns and salinity gradients within estuaries.
- Estuarine inhabitants exhibit varied adaptations to deal with the range of salinities found in most estuaries. Osmotic conformers are unable to regulate their internal fluid environments. Other inhabitants actively regulate the exchange of dissolved ions and water across their body tissue surfaces. Both approaches result in some species that are stenohaline, tolerating only narrow salinity ranges, and other species that are euryhaline, tolerating wider salinity ranges.
- Tides and river flow transport a mix of sediment particles that are sorted and deposited within estuaries, creating varied and changing substrates on which estuarine communities eventually develop. Salt marshes become established along the edges of many temperate-latitude estuaries, creating transition zones between land and marine communities. Nearer the equator, estuarine fringes are occupied by mangals, which are composed principally of low-growing mangrove trees. Salt marsh grasses, mangroves, and sea grasses of mudflats and deeper channels are the macroscopic primary producers in these estuaries. Much of their production is decomposed to detritus by bacteria, and, with contributions from phytoplankton and additional detritus carried into estuaries by tides, they maintain a rich pool of small food particles for estuarine deposit and suspension feeders.

- Alterations of estuaries and their upstream watersheds have led to a reduction in size and a degradation in the water quality of most of the world's estuaries. The conversion of estuarine marshland to agricultural or industrial uses and the contamination of estuarine waters with commercial, residential, and agricultural wastes have contributed to sometimes dramatic alterations of existing estuarine communities and serious reductions in fish and shellfish harvests.
- In North America, the Chesapeake Bay system is the largest and most productive estuary. Because of its long history of use, it exemplifies the problems created when resource use and habitat protection come into conflict.

Review Questions

1. Describe the pattern of water circulation in a typical estuary during periods of high river runoff. Of low river runoff.
2. Describe the resulting general vertical distribution of salinity values from the estuary mouth upstream to its head.
3. Draw a food web representative of the Chesapeake Bay ecosystem.
4. Discuss the advantages and disadvantages of being an osmotic conformer in a typical estuary.

Challenge Questions

1. Why is it critical to nonestuarine animals, such as migratory ducks and geese, that some estuaries remain relatively undisturbed by human activities?
2. Describe how you imagine the Chesapeake Bay appeared during the last glacial maximum.

Suggestions for Further Reading

Books

Cloern, J., and F. Nichols. 1985. *Temporal dynamics of an estuary.* Boston: Kluwer Academic.

National Wildlife Foundation. 1989. *A citizen's guide to protecting wetlands.* Washington, D.C.: National Wildlife Foundation.

Teal, J., and M. Teal. 1983. *Life and death of the salt marsh.* Boston: Little, Brown.

Warner, W. W. 1976. *Beautiful swimmers.* Boston: Little, Brown.

Articles

Bertness, M. D. 1992. The ecology of a New England salt marsh. *American Scientist* 80:260-68.

Botton, M. L., and H. H. Haskin. 1984. Distribution and feeding of the horseshoe crab, *Limulus polyphemus,* on the continental shelf off New Jersey. *Fishery Bulletin* 82:383-89.

Coutant, C. C. 1986. Thermal niches of striped bass. *Scientific American* (May) 255:98-104.

Dennison, W. C., et al. 1993. Assessing water quality with submerged aquatic vegetation. *Bioscience* 43:86-93.

Durbin, A., and E. Durbin. 1974. Grazing rates of the Atlantic menhaden (*Brevoortia tyrannus*) as a function of particle size concentration. *Marine Biology* 33:265-77.

Heinle, D. R., R. P. Harris, J. F. Ustach, and D. A. Flemer. 1977. Detritus as food for estuarine copepods. *Marine Biology* 40:341-53.

Johnson, D. R. 1985. Wind-forced dispersion of blue crab larvae in the Middle Atlantic Bight. *Continental Shelf Research* 3:425-38.

Kusler, J. A., W. J. Mitsch, and J. S. Larson. 1994. Wetlands. *Scientific American* (January) 270:64-70.

Miller, J. M., and M. L. Dunn. 1980. Feeding strategies and patterns of movement in juvenile estuarine fishes. In: *Estuarine perspectives,* ed. V. S. Kennedy. New York: Academic Press.

Milliman, J. 1989. Sea levels: Past, present, and future. *Oceanus* 32(2):40-43.

Nichols, F., et al. Temporal dynamics of an estuary: San Francisco Bay. *Science* 231:567-73.

Phillps, R. C. 1978. Seagrasses and the coastal marine environment. *Oceanus* 21(2):3-40.

Estuaries as Nurseries
http://www.harborside.com/home/s/ssnerr/nursery1.html
Describes how some life forms remain in estuaries while other life
   forms begin life in estuaries then move elsewhere.
Classification of Estuaries
http://www.harborsie.com/home/s/ssnerr/class1.html
Explains how and why different types of estuaries are classified.
The Estuary as Pollution Trap and Filter
http://www.harborside.com/home/s/ssnerr/pollute6.htm
Shows how plants and water motion filter pollutants. Tom J. Gaskill,
   Education Program Coordinator, South Slough National Estuarine
   Research Reserve. South Slough National Estuarine Research
   Reserve Estuaries Feature Series **http://www.harborside.com/
   home/s/ssnerr/artindx.html** Links to tides, types of estuaries,
   estuarine dependence, pollution, sturgeon, clams, food chains and
   food webs, geology, continental drift, and more.

The Estuary At Work
http://www.wa.gov/puget_sound/aboutps/estuary.html
Puget Sound's circulation patterns and pattern influences. Puget
   Sound Water Quality Action Team—Office of the Governor,
   Washington State. Puget Sound On-Line
http://www.wa.gov/puget_sound/index.html Many links to
   virtually every aspect of Puget Sound.

Hatfield Marine Science Center Sites
Where the River Meets the Sea
http://www.hmsc.orst.edu/education/programs/estuary.html
Full description with photos in, under and around Yaquina Bay,
   Oregon; eel grass beds, mudflats, more. Oregon State University.
   Hatfield Marine Science Center **http://www.hmsc.orst.edu/**

River Mouths, Brackish & Estuarine Wetlands
http://bonita.mbnms.nos.noaa.gov/sitechar/river.html
Geographical comparisons, regional patterns, Elkhorn slough, plants
   and animals. Monterey Bay National Marine Sanctuary. Monterey Bay
   National Marine Sanctuary Site Characterization
http://bonita.mbnms.nos.noaa.gov/sitechar/index.html

Arcata Marsh Wildlife Sanctuary Saltwater Vegetation
http://sorrel.humboldt.edu/~ere_dept/marsh/saltveg.html
Background, producers and growth patterns; the salt marsh is a
   dynamic system constantly receiving and giving. Greg McCormick,
   Mary Grace Tecson, Elizabeth A. Eschenbach. The Arcata Marsh and
   Wildlife Sanctuary **http://sorrel.humboldt.edu/~ere_dept/
   marsh/index.html**

An Introduction to Estuaries
http://inlet.geol.sc.educ/nerrsintro.html
Describes in pictures and words different types of estuaries and their
   plant and animal life.
Estuarine Ecology
http://inlet.geol.scarolina.edu/estecohp.html

Discusses the ecology of estuaries. National Estuarine Research Reserve System, National Oceanic and Atmospheric Administration (NOAA), Susan Lovelace, Education Coordinator. Welcome to the National Estuarine Research Reserve's Estuary-Net Project

http://inlet.geol.scarolina.edu/estnet.html

National Estuary Program: Bringing Our Estuaries New Life

http://www.epa.gov/nep/nepbroc.html

Discusses the role of estuaries and their economic, recreational, and aesthetic value. Short discussions of many estuaries, bays, lagoons, etc., around the U.S. National Estuary Program **http://www.epa.gov/nep/** The NEP's mission is to protect and restore the health of estuaries while supporting economic and recreational activities, and was established by Congress in 1987 as part of the Clean Water Act. Home page provides a text search to documents on the site as well as many links.

Chesapeake Bay Observing System

http://cbos.hpel.umd.edu/

Real-time air and water temperature, temperature and salinity at different depths, wind direction and speed, and humidity.

Coral Reefs

For many people, thoughts of tropical islands conjure up images of a special type of marine ecosystem, coral reefs. Coral reefs are famous for a diversity of species that rivals rain forests, for the myriad colors exhibited by their inhabitants, and for the amazing biological interactions that have evolved there. For example, unlike the intertidal communities discussed in previous chapters, coral reefs are actually produced by the organisms that live on them. The entire reef, which may extend for hundreds of kilometers, is primarily composed of a veneer of tiny, sea anemone-like creatures called coral polyps. These small animals slowly produce the massive carbonate infrastructure of the reef itself, on, around, and in which a vast array of organisms live.

Therein lies a wonderful biological paradox. Think of any common terrestrial ecosystem, a temperate forest, a tropical jungle, a midwestern plain, or the field adjacent to your house. These areas are dominated by a great variety of plants (the producers) and contain just a handful of animal species, both herbivores and carnivores (the consumers). Conversely, a typical coral reef contains an impressive assemblage of consumers and just a few plants. The coral animals that create the reef feed by removing plankton from the water column, as do the many sponge species that decorate the reef, representing the second most important component of the benthic fauna on coral reefs. Yet everyone knows that tropical seas are virtually devoid of plankton! That is why azure tropical waters are so transparent. So a coral reef can be viewed as one giant animal that is inhabited by hundreds of other animals, such as sponges, snails and clams, squids and octopuses, sea anemones and jellyfishes, shrimps and crabs, worms and fishes. The questions remain: Where are the primary producers on a coral reef? Can an ecosystem violate the second law of thermodynamics by containing more consumers than producers? Why don't planktivorous reef creatures, such as corals and sponges, starve to death in the nearly plankton-free waters that surround them? This chapter will attempt to answer these fascinating biological riddles.

Coral Anatomy and Growth

This chapter is about coral reefs (and the organisms that inhabit them). Let's begin by examining the terms *coral* and *reef* in more detail. *Coral* is a general term used to describe a variety of cnidarian species (see pp. 155–156). Some grow as individual colonies; hence, not all corals produce reefs. Not all reefs are formed by corals; some modern reefs are formed by oysters, annelid worm tubes, red algae, or even cyanobacteria.

Reef-forming corals, the primary species that secrete the calcium carbonate ($CaCO_3$) matrix of coral reefs, are members of the class

Figure 10.1

Extended polyps of a coral colony.
The numerous light-colored spots on
the tentacles are nematocysts.
Photo by T. Phillipp.

Anthozoa (see p. 156). All anthozoans are radially symmetrical, a morphology that is adaptive for sessile organisms, such as corals and sea anemones. Anthozoans are divided into two subclasses. The soft corals, sea fans, and sea pens (subclass Alcyonaria) are characterized by the presence of eight pinnate, or featherlike, tentacles. Members of the subclass Zoantharia possess a mouth that is surrounded by multiples of six smooth tentacles and includes three orders of sea anemones (some exist as individuals, some are in colonies, and others are tube dwellers) and three orders of corals (stony corals, false corals, and black corals). One group, the stony corals (order Scleractinia) is responsible for creating coral reefs. Stony corals and most of their cnidarian relatives are carnivores that use tentacles armed with nematocysts that ring the mouth (figure 10.1) to capture prey and push it into their gastrovascular cavity, where it is digested.

Most corals are colonial, built of numerous basic structural units, or polyps (figure 10.2 and see p. 156), each usually just a few millimeters in diameter. Coral polyps sit in calcareous cups, or **corallites,** an exoskeleton secreted by their basal epithelium. Several bladelike **septa** radiate from the walls of each corallite, and a stalagmite-like **columella** extends upward from its floor. Periodically, the coral polyp grows upward by withdrawing itself up and secreting a basal plate, a partition that provides a new, elevated bottom in the corallite. In addition, the coral colony also increases in diameter by adding new, asexually cloned polyps to its periphery. These new polyps secrete their own $CaCO_3$ corallites, which share a wall with neighboring polyps. All polyps that make up the colony are interconnected over the lips of their corallites via a thin sheet of tissue called a **cenosarc.** Therefore, touching a living coral colony in any way can easily crush the cenosarc against its own $CaCO_3$ skeleton and leave the colony open to infection by pathogens.

The growth rate of corals is affected by light intensity (which is affected by water motion, depth, turbidity, and sedimentation), day

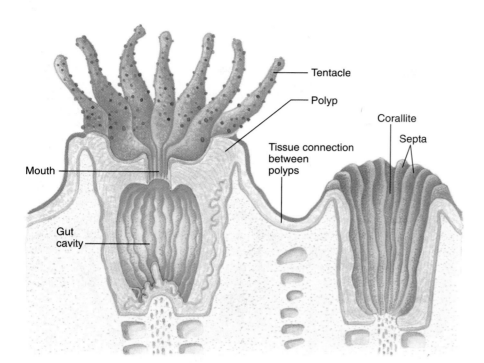

Figure 10.2

Cross section of a coral polyp and a calcareous corallite skeleton. The living coral tissue forms a thin interconnection, the cenosarc, over the surface of the reef.

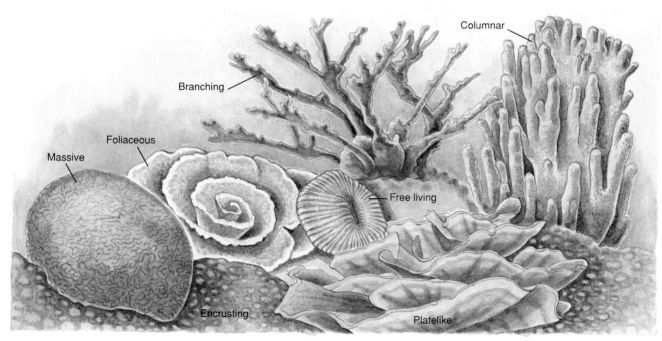

Figure 10.3

Corals exhibit a large variety of growth forms.

length, water temperature, plankton concentrations, predation, and competition with other corals. Stony corals exhibit a large variety of growth forms; they are typically described as encrusting, massive, branching, or foliaceous species (figure 10.3). In addition, many species are rather polymorphic because of morphological changes that occur owing to differences in exposure or depth. Therefore, growth may seem like a simple parameter to measure, but it is not. Suggestions for monitoring the growth of corals include measuring an increase in weight, or diameter, or surface area, or number of branches, or length of branches. Individual coral colonies may grow continually for centuries or even longer. Some species exceed several meters in size. In general, species with more porous skeletons grow more rapidly than species with denser skeletons, and branching species grow more quickly than massive species. Overall, the entire reef grows up to 1 mm/yr vertically and up to 8 mm/yr horizontally.

The growth of an individual coral or an entire reef is not simply a function of the local rate of calcification for that species. The persistence of a coral colony or reef depends on a balance between the deposition and removal of $CaCO_3$. The loss or erosion of carbonate is caused by grazers or scrapers, such as sea urchins and fishes (echinoids are the major grazers in the Atlantic); etchers, such as bacteria, fungi, and algae (especially *Ostreobium*) that penetrate coral substrates; and infaunal organisms, such as sponges, bivalves, sipunculans and polychaetes, that drill or bore into coral skeletons. Aside from carbonate deposition by corals, several types of encrusting and segmented calcareous red and green algae, calcareous colonial hydrozoans, skeletons of crustaceans, ectoprocts, and single-celled foraminiferans, mollusk shells, tests and spines of echinoderms, sponge spicules, serpulid polychaetes, and the calcareous remains of other reef inhabitants also contribute to the structure of coral reefs. From this encrusted, integrated base of skeletal remains, coral reef ecosystems have evolved as the most complex of all benthic associations.

Coral Distribution

Like sea anemones, corals are ubiquitous. Non-reef-forming corals can be found in the deep sea (e.g., black corals) and in temperate zones (e.g., *Astrangia* on shipwrecks off New England) as well as in the tropics. However, there are several restrictions to the distribution of reef-forming corals, which are more abundant and diverse in the Indo-Pacific (about 700 species) than in the Atlantic Ocean (about 145 species; figure 10.4). First, coral reefs are restricted to tropical and subtropical regions (usually below 30° latitude), where the water temperature never dips below 18° C. Second, coral reefs are much better developed on the eastern margins of continents than on the western margins. Third, coral reefs thrive only in normal-salinity seawater; hence, reefs are rare on the eastern coast of South America because of the enormous outflow of fresh water there from the Amazon River system. Fourth, reef-forming corals are usually found within 25–70 m of the surface, in clear water on exposed surfaces. The first two biogeographical restrictions suggest that reef-forming corals thrive only in warmer water (because only in warm waters can the high rates of $CaCO_3$ deposition needed for reef building be achieved). Hence, they are found in low latitudes and on eastern seaboards, where coastal upwelling of cold water is less common (see figure 5.15) and where the major ocean gyres direct warm, tropical currents (see figure 1.33). These latitudinal limits of coral reef development also are often set by competition with macroalgae, with

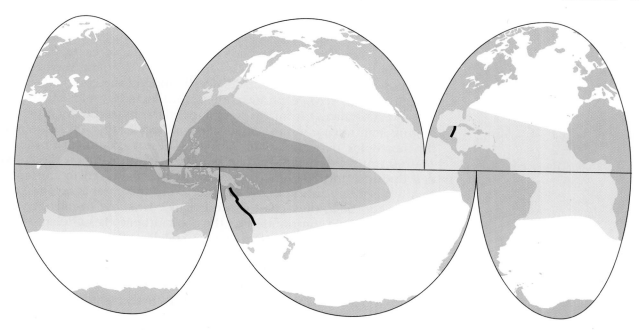

macroalgae being favored in higher latitudes because of increased nutrient concentrations, decreased water temperatures, and perhaps decreased grazing pressure. The third limit to the global distribution of coral reefs suggests that coral animals can not tolerate low-salinity seawater. Rivers and freshwater runoff also can increase sedimentation and nutrification, two additional coral killers (see Coral Diversity and Catastrophic Mortality; on page 281. The final biogeographical limitation seems more puzzling. Why would an animal such as a coral colony require sunlight, and why would its growth rate be affected by light intensity and day length, as described earlier? The answer to that question is also the answer to the apparent paradox posed at the beginning of this chapter.

Coral Ecology

Living intracellularly within the endodermal tissues of all reef-building, or **hermatypic,** corals are masses of symbiotic **zooxanthellae,** unicellular algae that, like all other photosynthetic organisms, require light. Solitary, non-reef-building corals, such as *Astrangia,* do not possess zooxanthellae and are termed **ahermatypic.** *Zooxanthellae* is a general term for a variety of photosynthetic cells that are symbiotic with a variety of invertebrate species. The most common species is *Symbiodinium microadriaticum,* a member of the protist division Dinophyta (see pp. 87–89). Unlike the dinophytes that appear in figure 3.17, zooxanthellae lose their flagellae and articulating cellulose cell walls. They occur in concentrations of up to 1 million cells per square centimeter of coral surface and often provide most of the color seen in corals. In fact, corals that grow in bright sunlight are often creamy white, whereas those in deep shade are nearly black. This difference is due to the concentrations of photosynthetic pigments in each cell rather than to differences in the densities of cells.

1 mc/cm²
Dinos lose flagella
+ cellulose wall

89

Zooxanthellae and coral derive several benefits from each other. Thus, this relationship usually is considered a mutualistic one (see figure 2.13). Corals provide the zooxanthellae with a constant, protected environment and an abundance of nutrients (CO_2 and nitrogenous and phosphate wastes from cellular respiration of the coral). In return, the host corals receive photosynthetic products (O_2 and energy-rich organic substances) from the symbiotic algae by stimulating or promoting their release with specific signal molecules that appear to alter the membrane permeability of the algae. These zooxanthellae produce 10–100 times more carbon than is necessary to maintain their own cell division, and 90–99% of the excess is transferred to the coral. Nearly all of the carbon that is transferred to the coral is respired and not used to build new coral tissue because it is low in nitrogen and phosphorus. This contribution by the zooxanthellae is sufficient to satisfy the daily energy needs of several species of corals (in fact, soft corals are obligate symbionts, having lost the ability to capture and ingest plankton). Moreover, the total contribution of symbiotic zooxanthellae to the energy budget of the reef is several times higher than the phytoplankton production occurring in the water above many reefs. Hence, corals are able to construct enormous reefs in plankton-poor waters because they receive a significant supply of food from their algal associates. The coral animals also avoid the necessity of excreting some of their cellular wastes (which the algae absorb and utilize) and enjoy greater calcification rates than hermatypic corals that have been experimentally separated from their algal symbionts (figure 10.5). Interestingly, one investigator, E. B. Marshall, reported that the calcification rate of a tropical, ahermatypic coral, *Tubastrea faulkneri,* without zooxanthellae was not significantly different from that of a hermatypic species *(Galaxea fascicularis)* with polyps of a similar size and shape. This important study demonstrated that, although light profoundly influences the calcification rate of hermatypic corals, that rate is not significantly faster than the rate of calcification in ahermatypic corals without zooxanthellae. Therefore, the widespread supposition that ahermatypic corals have lower calcification

Figure 10.5

Exchange of materials between zooxanthellae and their coral host.

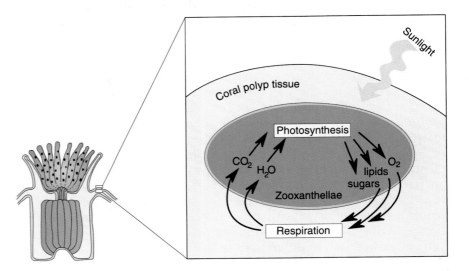

rates than hermatypic corals is false and rested on the observation of lower calcification rates of hermatypic corals that had their zooxanthellae removed experimentally. Hence, the success of hermatypic corals as reef builders may be more attributable to their autotrophic capabilities than to enhanced calcification rates. Additional primary production on coral reefs is provided by several types of rather cryptic plants, including encrusting calcareous red algae, filamentous green algae that invades dead corals, a brown algal turf, photosynthetic symbionts in other reef invertebrates, sand algae, sea grass, and phytoplankton cells in the water column over the reef.

Despite the nutritional contribution of zooxanthellae, coral polyps remain superbly equipped to prey upon external sources of food, and few, if any, depend solely on zooxanthellae. Corals with large polyps and tentacles, such as *Favia* and *Mussa,* feed exclusively on small fish and larger zooplankton, such as copepods, amphipods, and worms. Species with smaller polyps, such as *Porites* and *Siderastrea,* use ciliary currents to collect small plankton and detritus particles. Most coral polyps are capable of using mesentarial filaments to harvest particulate organic carbon from surrounding sediments. Corals also use their mucus-ciliary system (analogous to the ciliated epithelium in humans' windpipes) to trap and ingest organic particles as small as suspended bacteria, bits of drifting fish slime, and even organic substances dissolved in passing seawater. Finally, the controversial concept of endo-upwelling has been suggested as a possible source of additional dissolved nutrients; geothermal heat deep within island reefs drives the upwelling of nutrient-rich water through the reef structure from depths of several hundred meters.

Corals are not the only animals on the reef that possess photosynthetic symbionts. Zooxanthellae also occur in other anthrozoans, some medusae (such as *Cassiopea*), sponges, and giant clams (see page 280). In addition, it is well documented that sponges possess photosynthetic cyanobacteria. These photosynthetic symbionts are found in about 40% of the species from the Atlantic and Pacific oceans, although their contribution to sponge ecology in the two oceans differs dramatically. On the Great Barrier Reef in the Pacific Ocean, 90% of the sponges on the outer reefs are **phototrophic** (they are flattened and obtain up to half of their energy from cyanobacteria), with 6 of 10 species producing three times as much oxygen as they consume. Virtually none of the sponges studied in the Caribbean Sea are phototrophic, and consequently Caribbean sponges consume an order of magnitude more prey than their Pacific relatives. Perhaps this different reliance on energy from cyanobacteria is due to the fact that primary productivity in the western Atlantic is higher than in the western Pacific.

Finally, nitrogen fixation, an activity that is light dependent, has recently been found to be associated with bacteria living in the skeletons of various hermatypic corals. These nitrogen-fixing bacteria benefit from organic carbon excreted by the coral tissue. Perhaps this symbiotic association is as important to corals as their mutualism with zooxanthellae.

The living richness of coral reefs stands in obvious contrast to the generally unproductive nature of surrounding tropical oceans. The precise trophic relationships between producers and consumers on the reef are still largely unknown. Coral colonies seem to function as

Figure 10.6

Kayangel Atoll, capped with four
small low-lying islands, in the Belau
Islands, Micronesia.
Jeff Rotman Photography.

highly efficient trophic systems, each with its own photosynthetic, herbivorous, and carnivorous aspects. Critical nutrients are rapidly recycled between the producer and consumer components of the coral colony. Because much of the nutrient cycling is accomplished within the coral tissues, little opportunity exists for the nutrient to escape from the coral production system. Coral colonies, therefore, are able to rapidly recycle their limited supply of nutrients between internal producer and consumer components and keep productivity in coral reef communities relatively high (up to 5,000 gC/m^2/yr) compared with other regions of the ocean (see table 5.1).

Coral Reef Formation

Coral reefs occur as two general types: **Shelf reefs** grow on continental margins, and **oceanic reefs** surround islands. Oceanic reefs may be divided into three general subtypes: **fringing reefs, barrier reefs,** and **atolls.** The majority of shelf reefs are fringing reefs, which form borders along shorelines. Some of the Hawaiian reefs and other relatively young oceanic reefs are also of this type. The longest fringing reef known extends throughout the Red Sea, spanning some 400 km. Barrier reefs are farther offshore and are separated from the shoreline by a lagoon. The Great Barrier Reef of Australia is by far the largest single biological feature on the earth, bordering some 2,000 km of Australia's northeastern coast. Smaller barrier reefs occur in the Caribbean Sea. Atolls are generally ring-shaped reefs from which a few low islands project above the sea surface (figure 10.6). The largest atoll on the planet is Kwajalein Atoll in the Marshall Islands, which has a lagoon that is 100 km long and 55 m deep.

Charles Darwin is famous for his concept of natural selection, which he proposed as the mechanism by which biological evolution occurs. Darwin's propensity to view the world from a perspective of geological

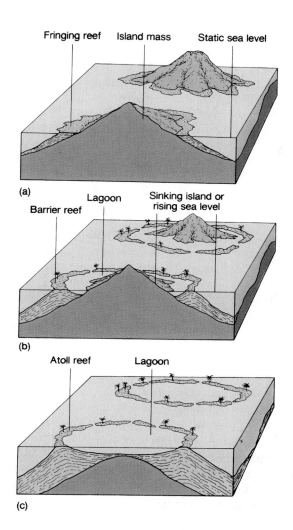

Figure 10.7

The developmental sequence of coral reefs, from young fringing reefs (*a*) barrier reefs (*b*), and finally to atolls (*c*). In the cross sections, volcanic island material is dark brown and reef formations are tan.

time certainly aided him in the development of his revolutionary hypothesis. This tendency also influenced his other studies. For example, before he published his *Origin of Species,* he published two works that required this perspective of deep time. One concerned the churning effect that earthworms had on topsoil by steadily eating and defecating while crawling their way through our lawns. The second concerned the development of the three types of coral reefs described above.

Darwin studied the morphology of coral reefs on several islands while serving as a naturalist aboard the H.M.S. *Beagle* during its circumglobal voyage from 1831 to 1836. His observations led him to propose that essentially all oceanic coral reefs were supported by volcanic mountains beneath their surfaces. Fringing reefs, barrier reefs, and atolls, he suggested, were sequential developmental stages in the life cycle of a single reef. Within the tropics, he argued, newly formed volcanic islands and submerged volcanoes that almost reach the sea surface are eventually populated by planktonic coral larvae from other nearby coral islands. The coral larvae settle and grow near the surface close to the shore, forming a fringing reef (figure 10.7a). The most rapid growth occurs on the outer sides of the reef, where food and oxygen-rich waters are most abundant. Waves break loose pieces of the reef and move them down the slopes of the volcano. More corals establish themselves on this debris and grow toward the surface. Darwin reasoned that the weight of the expanding reef and the increasing density of the cooling volcano cause the island to sink slowly. If the upward growth of the reef keeps pace with the sinking of the island, the coral maintains its position in the sunlit surface waters. If the upward growth of the reef does not keep pace with the sinking of the island, the reef sinks into the cold darkness below the photic zone and expires. Such a dead, sunken reef is called a **guyot.**

As the island sinks away from the growing reef, the top of the reef widens. Eventually, this reef crest or flat becomes so wide that many of the corals on the quiet inner edge of the reef die because the only water that reaches them is devoid of nutrients and oxygen and contains high concentrations of waste products. The dead corals are soon covered with reef debris and form a shallow lagoon. Delicate coral forms survive in the lagoon, protected from the waves by what is now a barrier reef (figure 10.7b). With further sinking, the volcanic core of the island may disappear completely beneath the surface of the lagoon and leave behind a ring of low-lying islands supported on a platform of coral debris, an atoll (figure 10.7c).

Darwin's concept of coral reef formation is, with a few modifications, widely accepted today. Test drilling on several atolls has revealed, as Darwin predicted, thick caps of carbonate reef material overlying submerged volcanoes. Two test holes drilled by Harry Ladd on Enewetak Atoll (the site of U.S. H-bomb tests in the 1950s) penetrated 1,268 and 1,405 m of shallow-water reef deposits before reaching the basalt rock of the volcano. For the past 60 million years, Enewetak apparently has been slowly subsiding as its surrounding reef has grown around it. Because this transition from a fringing morphology through a barrier morphology to an atoll requires a great deal of time, and because the Atlantic Ocean is much younger than the Pacific Ocean, atolls are virtually nonexistent in the Atlantic Ocean.

Some anecdotal information reinforces the scientific data that support Darwin's hypothesis of coral reef development. For example, British explorer Captain James Cook discovered Hawaii in January 1779, during Makahika, a festival to honor the god Lono. The Hawaiian natives initially thought that Captain Cook was Lono (who was said to come from the sea). After realizing their mistake, they killed him. A monument was soon built in the surf to commemorate his arrival and death. Today, that monument can be found under 20 m of water, yet the reef surrounding the island is still growing just under the surface of the sea!

For the past several hundred thousand years, the formation and melting of vast continental glaciers have produced extensive worldwide fluctuations in sea level. Darwin was aware of these fluctuations but he had no means of predicting their effects on coral reef development. Fifteen thousand years ago during the last glacial maximum (LGM), average sea level was about 150 m below its present level. As the ice melted, the sea level gradually rose (about 1 cm/yr) until it reached its present level nearly 6,000 years ago. Many coral reefs did not grow upward quickly enough and perished. Those that did keep

Figure 10.8

The Hawaiian Island–Emperor Seamount chains of volcanoes are carried, in a conveyer-belt fashion, north into deeper water by the movement of the Pacific Plate. Each volcano was formed over the "hot spot," a continuous source of new molten material presently under Hawaii, and is carried to its eventual destruction in the Aleutian Trench. Courtesy National Geophysical Data Center.

up with the rising sea are the living reefs we see today. Coral reefs in the Atlantic Ocean seem most susceptible to glacier-induced changes in their morphology, and barrier reefs are most common in the Atlantic Ocean.

Coral reefs have also been subjected to the effects of global plate tectonics. The Hawaiian Islands and the reefs they support have been transported to the northwest by the movement of the Pacific Plate. Atolls at the northern end of the chain appear to have drowned as they reached the "Darwin Point," a threshold beyond which coral atoll growth cannot keep pace with recent changes in sea level (figure 10.8). At the Darwin Point, only about 20% of the necessary $CaCO_3$ production is contributed by corals.

Reproduction in Corals

Corals reproduce in a variety of ways, both asexually and sexually. Most corals bud off new polyps along their margins asexually as they increase in diameter. Sometimes, these new polyps sever the cenosarc and initiate a new colony that is a clone of their neighbor. Storms or anchors frequently break branching species, such as *Acropora,* into clonal colonies by fragmentation, the production of new colonies from portions broken off of established colonies. Fragmentation decreases the risk of mortality of the genotype and avoids the risk of high mortality of larvae and juveniles during sexual reproduction. In addition, fragmentation by species with high growth rates

(a)

(b)

Figure 10.9

Spawning corals; (a) female staghorn
coral releasing eggs, (b) male
mushroom coral releasing sperm.
Courtesy Animals Animals/OSF.
Photos by P. Harrison

often results in that species' dominating certain reef zones (such as
the buttress zone discussed on page 278) as well as rapid recolonization after a disturbance. Researchers also have observed "polyp bail-
out" in the laboratory, when polyps crawl out of their corallites and
drift away. It is not known whether these polyps remain viable or if
this occurs naturally on coral reefs.

Corals also reproduce sexually, either by brooding fertilized eggs
internally or by spawning millions of gametes into the water column
for external fertilization. The eggs of brooding species remain in the
gastrovascular cavities of the adults, where they are fertilized by
motile sperm cells. The developing zygotes and resultant planula larvae are brooded before they are released to settle nearby. Some evidence suggests that coral species with small polyps have low numbers
of eggs combined with internal fertilization and brooding, whereas
large-cupped species spawn huge quantities of eggs that are fertilized
externally. In addition, the strategy of sexual reproduction used
(brooding larvae versus spawning gametes) is highly correlated with
taxonomic affiliation at the family level. Members of the families
Agariciidae, Dendrophylliidae, and Pocilloporidae commonly brood,
whereas broadcast spawning is predominant in the Acroporidae,
Caryophyllidae, Faviidae, and Rhizangidae. The family Poritidae includes both brooders and broadcasters.

Of 200 species of corals studied on the Great Barrier Reef, 131
were hermaphroditic spawners, 37 were diecious spawners, 11 were
hermaphroditic brooders, and 7 were diecious brooders. Hence,
spawning by hermaphrodites seems to be the most common method
of sexual reproduction among corals. Spawning is usually accomplished during a highly synchronous event known as mass spawning.
On the Great Barrier Reef of Australia, mass spawning by corals is a
spectacular sight. More than 100 of the 340 species of corals found
there synchronously spawn on only one night each year, just a few
days after the late spring full moon (figure 10.9). A similar episode of
mass spawning has been documented in the Gulf of Mexico, in the
evening eight days after the full August moon. Mass spawning by

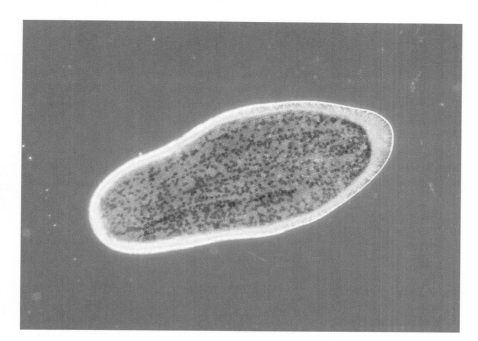

Figure 10.10

Micrograph of a planula larva of the coral *Acropora*.
Photo by P. Parks, Oxford Scientific Films.

corals seems to be induced by specific dark periods, and it can be delayed by experimentally extended light periods. Mass spawning also seems to be broadly influenced by temperature. Such highly seasonal spawning is surprising in the tropics, an area wherein reproduction throughout the year is said to be the norm because of relatively constant climatic conditions.

A few days after spawning, the fertilized eggs develop into a ciliated **planula** larva (figure 10.10). Because these larvae, each with its own supply of zooxanthellae, are initially positively phototactic, they remain near the sea surface, where maximal dispersal by surface currents is likely. Then, after a species-specific interval, they become negatively phototactic and attempt to settle on the seafloor. They thrive only if they encounter their preferred water and bottom conditions. From these planula larvae, new coral colonies develop and mature in 7–10 years. R. H. Richmond has reported that the larvae of *Pocillopora damicornis* are capable of reversible metamorphosis. The planula larva of this species settles and begins to metamorphose into a juvenile. It forms a $CaCO_3$ exoskeleton, a mouth, and tentacles. However, if it is stressed within the first three days of settling, it will sever its attachments to its carbonate exoskeleton, revert back into a planula larva, and reenter the water column to search for a better settlement site. During their planktonic phase, coral planula larvae are capable of settling new volcanic islands some distance from their island of origin. When they do, the form of reef they eventually create depends on existing environmental conditions and the developmental history of the reef.

It is unclear why corals spawn synchronously and why this event occurs just several nights after the full moon. One advantage to mass spawning is that the chance of fertilization will increase greatly for one species. It is unclear, however, why mass spawnings are multispecies events, in that simultaneous spawning by many species probably increases the risk of gamete loss via hybridization. Perhaps the egg and sperm cells, both of which are motile in corals, are chemotactic and cue on chemicals released from the gametes of their own species. It has also been suggested that such an epidemic spawning event would overwhelm (and satiate) active predators and filter feeders in the area, increasing the likelihood of gamete survival. However, these species also risk big losses by spawning on just a few nights each year. For example, a sudden rainstorm, and subsequent drop in salinity of surface waters, during a mass spawning event around Magnetic Island in November of 1981 negated the entire reproductive effort for those corals for that year. Mass spawning is not a universal behavior of reef corals; in the northern Red Sea, none of the major species of corals reproduce at the same time as any of the other major species.

Interestingly, calcareous green algae in the Caribbean also exhibit mass spawning. Nine Caribbean species in five genera participate in the predawn episode, with a total of 17 species of green algae exhibiting highly synchronous reproductive patterns. Unlike the coral phenomenon described previously, closely related algal species broadcast their gametes at different times, and the environmental or biological triggers of these events remain unknown. In all cases, gametes from both sexes remain motile for 40–60 minutes after release but sink quickly after combining to form a zygote.

Zonation on Coral Reefs

Environmental conditions that favor some coral reef inhabitants over others in a particular habitat depend a great deal on wave force, water depth, temperature, salinity, and a host of biological factors. These conditions vary greatly across a reef and provide for both horizontal and vertical zonation of the coral and algal species that form the reef. Figure 10.11, a cross section of an idealized Indo-Pacific atoll, includes the major features and zones of the reef.

The living base of a coral reef begins as deep as 150 m below sea level. Between 150 and 50 m on outer reef slopes, a few small, fragile species, such as *Leptoseris,* exist despite the fact that little sunlight penetrates at these depths. Above 50 m, and extending up to the base of vigorous wave action (approximately 20 m), is a transition zone between deep and shallow water associations. In this zone, the corals and algae receive adequate sunlight yet are sufficiently deep to avoid the adverse effects of surface waves. Several of the delicately branched species commonly found in the protected lagoon waters also occur in this transition zone.

From a depth of about 20 m to just below the low tide line is a rugged zone of spurs, or **buttresses,** radiating out from the reef. Interspersed between the buttresses are grooves that slope down the reef face. This windward profile of alternating buttresses and grooves is useful in dissipating some of the energy of waves that crash into the face of the reef, but damage to the reef and its inhabitants is inevitable. The grooves drain debris and sediment produced by wave impacts off the reef and into deeper water. Continual heavy surf has limited detailed studies of the buttress zone, but it is known to be

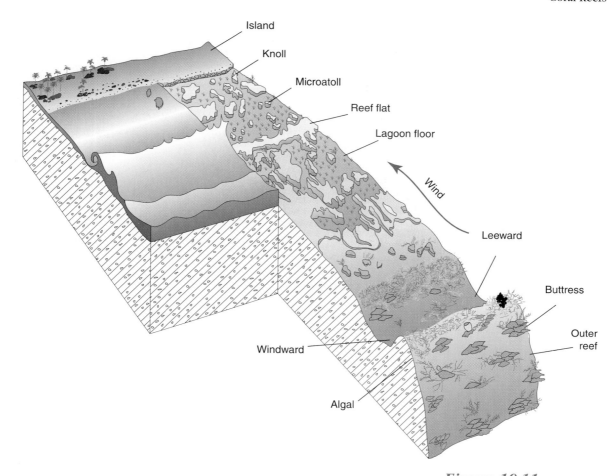

Labels: Island, Knoll, Microatoll, Reef flat, Lagoon floor, Wind, Leeward, Buttress, Outer reef, Windward, Algal

Figure 10.11

Cross-sectional structure of an atoll.

dominated by several species of encrusting coralline algae and by rapidly growing, branching coral species (such as some *Acropora*) that repair damage quickly and thrive when fragmented. Small fishes seem to be in every hole and crevice on the reef, and many of the larger fishes of the reef—sharks, jacks, and barracudas—patrol the buttresses in search of food.

A large portion of the geographical range of coral reefs is swept by the broad reaches of the trade winds. The waves generated by these winds crash as thundering breakers on the windward sides of reefs. Windward reefs are usually characterized by a low, jagged **algal ridge.** The algal ridge suffers the full fury of incoming waves. In this severe habitat, a few species of calcareous red algae, especially *Porolithon, Hydrolithon, Goniolithon,* and *Lithothamnion,* flourish and produce the ridge, producing new reef material as rapidly as the waves erode it. A few snails, limpets, and urchins (figure 10.12) can also be found wedged into surface irregularities. Slicing across the algal ridge are surge channels that flush bits and fragments of reef material off the reef and down the seaward slope.

Extending behind the algal ridge to the island (or, if the island is absent, to the lagoon) is a **reef flat,** a nearly level surface barely covered by water at low tide. The reef flat may be narrow or very wide,

Figure 10.12

Echinometra, a common tropical sea urchin.
Courtesy T. Ebert.

consists of several subzones, and has an immense variety of coral species and growth forms. In places where the water deepens to a meter or so, small raised **microatolls** occur. Microatolls are produced by a half-dozen different genera of corals and, with other coral growth forms, provide the framework for the richest and most varied habitat on the reef. Burrowing sea urchins are common, and calcareous green algae and several species of large foraminiferans thrive and add their skeletons to the sand-sized deposits on the reef flat. The sand, in turn, provides shelter for other urchins, sea cucumbers, and burrowing worms and mollusks.

Possibly the most spectacular animal of the reef flat is the giant clam *Tridacna.* The largest species of this genus occasionally exceeds a meter in length and weighs over 100 kg. Some tridacnids sit exposed atop the reef platform; others rock slowly to work themselves into the growing coral structure beneath (figure 10.13). Like corals and many other invertebrates, tridacnids house dense concentrations of zooxanthellae in specialized tissues, particularly the enlarged mantle that lines the edges of its shell. When the shell is open, the pigmented mantle tissues with their zooxanthellae are fully exposed to the energy of the tropical sun.

Tridacnid clams were long thought to "farm" their zooxanthellae in blood sinuses within the mantle and then transport them to the digestive glands, where they were thought to be digested by single-celled **amebocytes.** Using elaborate staining and electron microscope techniques, P. V. Fankboner demonstrated that the digestive amebocytes of *Tridacna* selectively cull and destroy old or degenerate zooxanthellae. Healthy zooxanthellae are maintained to provide photosynthetic products to their hosts in dissolved rather than cellular form.

Siphon

Mantle tissue

Figure 10.13

A giant clam, *Tridacna,* amid mixed corals from Fiji. Note the siphon opening and extended mantle tissue between open shell edges.
Photo by T. Phillipp.

The tranquil waters of the lagoon protect two general life zones: the lagoon reef and the lagoon floor. The lagoon reef forms the shallow margin of the lagoon proper. It is a leeward reef, usually free of severe wave action. It lacks the algal ridge characteristic of the windward reef and, in its place, has a more luxuriant stand of corals (figure 10.14). Other algae, some specialized to burrow into coral, and uncountable species of crustaceans, echinoderms, mollusks, anemones, gorgonians, and representatives of many other animal phyla flourish in the lagoon reef. In this gentle, protected environment, single coral colonies of *Porites* and *Acropora* may achieve gigantic proportions. Branching bush and treelike forms extend several meters from their bases. The plating, branching, and overtopping structures common in the protected lagoon are most likely structural adaptations evolved in response to competition for particles of food and sunlight, two resources vital to the survival of reef-forming corals.

Coral Diversity and Catastrophic Mortality

The great diversity of species on coral reefs is legendary and rivals that of tropical rain forests. The classic explanation for this high diversity began with the assumption that the species composition of coral reefs is maintained near equilibrium (hence, it would persist if there were no perturbation). The working hypothesis was that the uniform and predictable conditions on coral reefs promoted high diversity by enabling species to become increasingly specialized. Recently, that view has been challenged by an argument that suggests that the high diversity of coral reefs, like that of rocky intertidal communities, is a nonequilibrium state whose diversity can persist only if it is disturbed. Like some rocky intertidal communities discussed in chapter 8, coral reefs are subject to severe disturbances often enough that equilibrium, or climax stage, may never be reached, and high diversity is maintained by frequent catastrophic mortality.

(a)

(b)   (c)

Figure 10.14

Variation in coral growth forms from the Solomon Islands: (*a*) plate coral, *Acropora;* (*b*) brain coral, *Diploria;* and (*c*) elkhorn coral, *Acropora.*

Photos by T. Phillipp.

Catastrophic mortality of corals and coral reef species is caused by a variety of agents, both natural and anthropogenic. One common natural cause of catastrophic coral mortality is storm waves from hurricanes and typhoons. At Heron Island on the Great Barrier Reef, for example, the highest number of species of corals occurs on the crests and outer slopes, which are constantly exposed to damaging waves. In fact, it has been reported that the most significant factor determining the spatial and temporal organization of Hawaiian coral reef communities is physical disturbance from waves.

Outbreaks of predators also result in massive mortality of corals. For example, there have been two outbreaks of the coral-eating crown-of-thorns sea star, *Acanthaster planci,* in the western Pacific Ocean: the first from 1962 to 1977 and the second from 1979 to 1986. In the 1960s, during the first outbreak, ecologists speculated that these sudden occurrences of large populations of this damaging sea star were the result of human activities, specifically the disappearance of its major predator, a large and beautiful snail called *Charonia tritonus* that had been nearly exterminated by shell collectors. Others suspected that the population increase was due to natural causes, such as unusually frequent storms that caused nutrient washout and subsequent increases in successful larval settlement. However, recent evidence from fossil skeletal remains of *Acanthaster* retrieved from cores of reef sediments suggests that large populations of *Acanthaster* have played a role in the ecosystem of the Great Barrier Reef for at least 8,000 years. Thus, outbreaks of *Acanthaster* may be an enduring ecological phenomenon that simply has escaped notice until recently. Interestingly, an outbreak of *Acanthaster* appeared in the eastern Indonesian Archipelago in February 1996, with coral mortality adjacent to aggregations of the sea stars approaching 100%. Such recurrent mortality supports the model of coral reefs as a high-diversity community in a nonequilibrium state.

Sudden outbreaks of pathogens also result in high mortalities of corals, either directly or indirectly. For example, a water-borne pathogen killed large numbers of a ubiquitous, long-spined black sea urchin, *Diadema antillarum,* in the Caribbean Sea beginning in January 1983 and continuing for 13 months. The rapid extermination of this urchin, an important grazer of algae, enabled algal populations to overgrow corals in their competition for space on reefs. Some Caribbean reefs, including many along the northern coast of Jamaica, still have not recovered and remain green, fuzzy reminders of their former beauty. More recently, **blackband disease** is causing high mortalities in susceptible corals in the Caribbean. This disease, characterized by a band of blackened, necrotic tissue that slowly advances around coral colonies (figure 10.15), is caused by *Phormidium corallyticum.* This cyanobacterium grows as a densely interwoven mat that separates the cenosarc from the coral's skeleton and consequently results in the death of the coral by enabling opportunistic pathogens to invade the coral's tissue. Occasional outbreaks of these pathogens also would maintain coral species in a high-diversity, nonequilibrium state.

Perhaps the best-studied cause of episodic coral mortalities is **bleaching,** a recently characterized phenomenon wherein

Figure 10.15

External symptoms of blackband disease on coral.
Photo by John Morrissey.

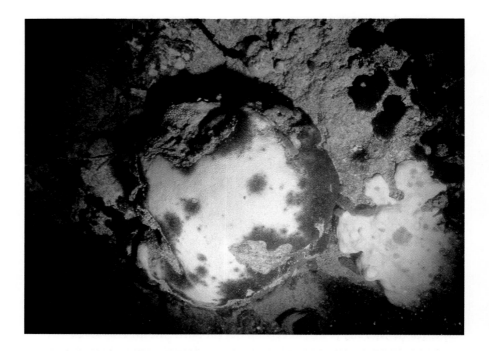

Figure 10.16

Effect of coral bleaching on a Caribbean coral head.
Photo by John Morrissey.

physiologically stressed corals expel their mutualistic zooxanthellae (figure 10.16). Expulsion of the zooxanthellae results in a whitening of the colony (due to the appearance of the $CaCO_3$ skeleton of the coral as seen through its now pigment-free cenosarc) and perhaps its death. Bleaching events have been correlated with increased sea-surface temperatures, such as those that occur in the tropical eastern Pacific Ocean during El Niño–Southern Oscillation events (ENSO; see pp. 37–38 and 136–137). For example, 95–99% of the living coral in the Galápagos Islands bleached and died after the severe 1982–1983 ENSO episode, and another even more intense ENSO event is occurring as this book goes to press. Similarly, ENSO-related coral mortality and a subsequent increase in encrusting coralline algae and sea urchins have resulted in profound changes to the coral reef ecosystem off western Panama. Before ENSO, the reef was depositional; now it is eroding at a rate of $0.2 \text{ kg/m}^2/\text{yr}$. Recent experiments and observations have demonstrated that corals exposed to high temperatures exhibit a loss of zooxanthellae, cellular abnormalities, and coral mortalities similar to those observed during the 1982–1983 ENSO. Elsewhere, coral bleaching has been correlated with the times of maximum seasonal sea-surface temperatures.

The first Caribbean bleaching event occurred in 1987–1988 and affected all species living from 0 to 30 m depth (only *Madracis* and *Acropora* seem to have been minimally affected). A second event occurred in 1990. Unlike Pacific episodes, which are usually attributed to increased water temperatures, neither mass bleaching event in the Caribbean Sea was readily explained by temperature alone. Recent studies have suggested that increased levels of ultraviolet radiation may play an additional role in Caribbean bleaching episodes during periods of calm, clear water, which occur during ENSO events.

More worrisome are anthropogenic causes of coral mortality, especially increased sedimentation and nutrification of surrounding waters (similar to the problems experienced by many estuaries discussed in the previous chapter). A coating of sediment on a coral colony is harmful for several reasons. First, it will increase the colony's energy expenditure by causing its mucus-ciliary system to work overtime to rid its surface of sediment particles. Second, it will decrease the photosynthetic output of zooxanthellae by shading them and reducing their light absorption. Finally, although some corals are adept at cleaning their surfaces of sediment particles, they may be growth-limited in turbid waters because the resulting muddy seafloor does not provide a suitable foundation for successful larval settlement. Consequently, corals do not thrive near river mouths with high discharges of fresh water and heavy loads of suspended sediments. Local rates of sedimentation are increased dramatically by deforestation or removal of mangrove stands (see pp. 251–252) during shoreline development. With these botanical sediment traps removed, runoff from land contains enough sediment to quickly smother and kill nearby reefs.

Enhanced nutrification occurs when runoff from agricultural areas injects enormous quantities of fertilizer into the waters that bathe coral reefs. These increased concentrations of nitrogen and phosphorus enhance algal growth and enable algae to dominate corals in their quest for space on the reef. These two common man-made causes of coral mortality, increased sedimentation and nutrification, are thought to be responsible for the devastation of corals in the Florida Keys, in parts of Hawaii, and elsewhere that has occurred in recent years.

Coral Reef Fishes

Associated with the reef and lagoon, but with the mobility to escape the limitations of a benthic existence, are thousands of species of reef

Reef Face · Algal Ridge · Reef Flat

Figure 10.17a

(a) General habitats of some common reef fishes on an exposed tropical Pacific reef: 1 (*Triaenodon*), 2 and 3 (*Carcharbinus*)—requiem sharks; 4 (*Zebrastoma*), 6, and 13 (*Acantburus*)—surgeonfishes; 5 (*Chaetodon*)—butterfly fishes; 7 (*Paracirrbites*)—hawk fishes; 8 (*Epinepbelus*)—groupers; 9 (*Gymnotborax*)—moray eels; 10 (*Scarus*)—parrotfishes; 11 (*Lutjanus*)—snappers; 12 (*Pempberis*)—sweepers.

fishes (figure 10.17). These fishes find protection on the reef, prey on the plants and animals living there, and sometimes nibble at the reef itself. Several groups are common to all the major regions characterized by coral reefs. These include grunts, snappers, cardinalfishes, moray eels, porcupinefishes, butterfly fishes, squirrelfishes, groupers, triggerfishes, gobies, wrasses, parrotfishes, surgeonfishes, sea horses, sharks, and rays. Many of these fishes are thought to be major importers of important limiting nutrients to local reef systems by foraging on pelagic prey during the day, then defecating at night while resting on the reef. Others feed in surrounding seagrass meadows at night and defecate on the reef while resting during the day. The results of this off-reef predation are converted through detritus food webs to dissolved nutrients usable by plants, phytoplankton, and the coral-based zooxanthellae.

These assemblages of shallow-water coral reef fish species are easily observed by divers and have been intensively studied for decades. Less well known are the fish of the deeper portions of coral reef communities (below 100 m). Submersible-based studies have recently

Figure 10.17b

Close-up of a small coral head with some of its many associated fish species.

Photo by permission from Castro and Huber, 1997

demonstrated that, as one works down the reef face into deep water, the same general assemblages are present, but individual numbers and species diversity both diminish.

Symbiotic Relationships

Excellent examples of all the types of symbiosis (summarized in figure 2.13) can be found in many of the abundant animal groups of the coral reef. Our discussions will be limited to some of the better-known symbiotic relationships involving coral reef fish. The relationships span the entire spectrum of symbiosis, from very casual commensal associations to highly evolved parasitic relationships.

Remoras (figure 10.18) associate with sharks, parrotfishes, sea turtles, and even the occasional dolphin in a mutualistic symbiosis. The remora's first dorsal fin is modified as a sucking disc and is used to attach itself to its host. From its attached position, it feeds on scraps from the host and often cleans the host of external parasites. Thus, the remora gains food, a free ride, and protection via proximity (a special benefit of symbiosis called **inquilinism**), whereas the host is rid of many ectoparasites. A similar association in the open ocean exists between sharks and pilotfishes *(Naucrates)*. The pilotfishes swim below and in front of their hosts and scavenge bits of food from the shark's meal. It has been speculated that pilotfishes may attract prey species to the shark.

It is common for smaller, defenseless fishes to live on or near better-defended species of reef invertebrates. For example, shrimpfishes often hover vertically in a head-down position among the long, sharp spines of sea urchins in a commensal symbiosis (figure 10.19). The shrimpfishes acquire protection from that sea urchin without affecting it. Brightly colored clownfishes and anemonefishes find equally effective shelter by nestling among the stinging tentacles of several species of sea

Figure 10.18

Two remoras with modified dorsal fins accompanying a nurse shark. Photo by John Morrissey.

Figure 10.19

Shrimpfish, *Aeloiscus,* seeking shelter amid the spines of a sea urchin.

anemones (figure 10.20). This relationship, somewhat more complex than those just described, is also probably a mutualistic one. In return for the protection they obtain, clownfishes assume the role of "bait" and lure other fishes within reach of the anemone. They occasionally collect morsels of food and, in at least one observed instance, catch other fishes and feed them to the host anemone. Clownfishes, however, are not immune to the venomous nematocysts of the host. Instead, they mouth and nibble at the anemone's tentacles, and substances contained in the mucus provide the mechanism to inhibit nematocyst discharge.

The increased popularity of skin diving and SCUBA diving has revealed some remarkable cleaning associations involving a surprising

Figure 10.20

Clownfishes, *Amphiprion,* hovering
near their host anemone.
Photo by T. Phillipp.

number of animals. **Cleaning symbiosis** is a form of mutualism; one
partner picks external parasites and damaged tissue from the other.
The first partner gets the parasites to eat; the other partner has an irri-
tation removed.

The behavioral and structural adaptations of cleaners are well developed in a half-dozen species of shrimps and several groups of small
fishes. Tropical cleaning fishes include butterfly fishes, young stages of
angelfish, gobies, and several wrasses, including all known species of
Labroides (figure 10.21). All tropical cleaning fishes are brightly marked,
are equipped with pointed pincerlike snouts and beaks, and occupy a
cleaning station around an obvious rock outcrop or coral head. Most are
solitary; a few species, however, live in pairs or larger breeding groups.

Host fishes approach cleaning stations, frequently queuing up and
jockeying for position near the cleaner. Often, they assume unnatural
and awkward poses similar to courtship displays. As the cleaner fish
moves toward the host, it inspects the host's fins, skin, mouth, and gill
chambers and then picks away parasites, slime, and infected tissue.

In the Bahamas, C. Limbaugh tested the cleaner's role in subduing
parasites and the infections of other reef fishes. Two weeks after he
had removed all known cleaners from two small reefs, the areas were
vacated by nearly all but territorial fish species. Those species that remained had an overall ratty appearance and showed signs of increased
parasitism, frayed fins, and ulcerated skin. Limbaugh concluded that
symbiotic cleaners were essential in maintaining healthy fish populations in his study area.

G. S. Losey conducted similar studies on a Hawaiian reef. In this
situation, the small cleaning wrasse *Labroides phthirophagus* (the
major cleaner on the reef) was excluded from the study site for more
than six months. During that time, no increase in the level of parasite
infestation was observed. This result suggests that for some cleaner-

Figure 10.21

A small wrasse, *Labroides,* cleaning external parasites from a turkey fish.
Courtesy C. Farwell, Scripps Institution of Oceanography.

host associations the role of the cleaner is not crucial. The cleaner may be dependent on the host for food, but the host's need for the cleaner seems to be variable.

The fine line separating mutualistic cleaning of external parasites and actual parasitism of the host fish is occasionally crossed by cleaning fishes such as *Labroides.* In addition to unwanted parasites and diseased tissue, some cleaners take a little extra healthy tissue or scales or graze on the skin mucus secreted by the host. Thus, the total range of associations displayed by cleaning fishes encompasses mutualism, commensalism, and parasitism.

Because parasitism is such a widespread way of life in the sea, few fishes avoid contact with parasites at some time in their lives. The groups notorious for creating parasitic problems in humans—viruses, bacteria, flatworms, roundworms, and leeches—also plague marine fish. Despite the bewildering array of parasites that infest fish, very few fishes become full-time parasites themselves. A remarkable exception are pearlfishes. They find refuge in the intestinal tracts of sea cucumbers and the stomachs of certain sea stars. When seeking a host, pearlfishes detect a chemical substance from the cucumber and then orient themselves toward the respiratory current coming from the cucumber's cloaca. (Sea cucumbers draw in and expel water through their cloacae for gas exchange.) The fish enters the digestive tract tail first via the anus. The hosts are not willing participants in this relationship. They sometimes eject their digestive and respiratory organs in an attempt to rid themselves of the symbiont. In fact, sea cucumbers of the genus *Actinopyga* have evolved five teeth on their anal margin, perhaps as an antipearlfish mechanism. But once this association is

(a)

(b)

Figure 10.22

A well-camouflaged fish (*a*), with magnified chromatophores from a section of skin (*b*). The black and brown pigments of some are expanded and diffused; others are densely concentrated in small spots. (*a*) Photo by T. Phillip. (*b*) Courtesy C. Stepien. Scripps Institution of Oceanography.

established, the pearlfishes assume a parasitic existence, feeding on and seriously damaging the host's respiratory structures and gonads.

Coloration

Against the colorful background of their coral environment, reef fishes have evolved equally brilliant hues and color patterns. The colors are derived from skin or internal pigments and from iridescent surface features (like those of a bird's feathers) with optical properties that produce color effects. Most fishes form accurate visual color images of what they see. Like humans, they are susceptible to misleading visual images and camouflage.

Our interpretation of the adaptive significance of color in fishes falls into three general categories: concealment, disguise, and advertisement. Some seemingly conspicuous fishes resemble their coral environment so well that they are nearly invisible when in their natural setting. Extensive color changes often supplement their basic camouflage when they are moving to different surroundings. These rapid color changes are accomplished by expanding and contracting the colored granules of pigmented cells **(chromatophores)** in the skin and are governed by the direct action of light on the skin, by hormones, and by nerves connected to each chromatophore. As the chromatophore pigments disperse, the color changes become more obvious (figure 10.22). When the grannules are contracted, the pigment retreats to the center of the cell,

Research in Progress

Marine Sanctuaries

In 1972, a century after the United States established the first national park at Yellowstone, legislation was passed to create the National Marine Sanctuaries Program. The intent of this legislation was to provide similar protection to selected coastal habitats as we have for land areas designated as national parks. The designation of an area as a marine sanctuary says to all that, like our national parks, this is a safe refuge where people can observe but where organisms and their environment may not be harmed or removed.

The National Marine Sanctuaries Program is administered by the National Oceanic and Atmospheric Administration, a branch of the Department of Commerce. Initially, 70 sites were proposed as candidates for sanctuary status. Two and a half decades later, only 15 sanctuaries have been designated (see map), with half of these established after 1987. These range in size from the tiny (less than 1 km²) Fagatele Bay National Marine Sanctuary in American Samoa to the Monterey Bay National Marine Sanctuary, extending over 15,744 km².

The National Marine Sanctuary Program is a crucial part of new management practices in which whole communities, and not just individual species, are offered some degree of protection from habitat degradation and overexploitation. Only in this way can a reasonable degree of marine species diversity be maintained in a setting that also maintains the natural interrelationships that exist among those species.

Several other types of marine protected areas exist in the United States and other countries. A federally managed National Estuarine Research Reserve System (see map) includes 23 designated and protected coastal estuaries. In addition, most coastal states, partially funded by the Federal Coastal Zone Management Act, have developed their own coastal management programs.

Abroad, marine protected area programs exist as marine parks, reserves, and preserves. Over 100 designated areas exist around the periphery of the Caribbean Sea. Others range from the well-known Australian Great Barrier Reef Marine Park to little-known parks in developing countries such as Thailand and Indonesia, where tourism is placing growing pressures on fragile coral reef systems.

Agencies at state, national, and international levels are slowly recognizing the importance of conserving marine biodiversity, and marine protected areas, whether as sanctuaries, parks, or estuarine reserves, will play an important future role in conserving that diversity.

For more information on this topic, see the following:
Marine Protected Areas. 1993. Oceanus 36(3). Entire issue

and little of it is visible. Other cells, called **iridocytes,** contain reflecting crystals of guanine. Iridocytes can produce an entire spectrum of colors within a few seconds.

Several distinctive fishes conceal themselves with color displays reminiscent of disruptive coloration, or dazzle camouflage. Bold, contrasting lines, blotches, and bands tend to disrupt the fish's image and draw attention away from recognizable features such as eyes. Eyes are common targets for attack by predators, so a disguised eye is a protected eye. One common strategy masks the eye with a dark band across the black staring pupil so that it appears to be continuous with some other part of the body (figure 10.23). To carry the deception even further, masks around the real eyes are sometimes accompanied by fake eyespots on other parts of the body or fins. Eyespots, intended as visual attention getters, are usually set off by concentric rings to form a bull's-eye. Presumably, predators are drawn away from the eyes and head and drawn to less vital parts of the body.

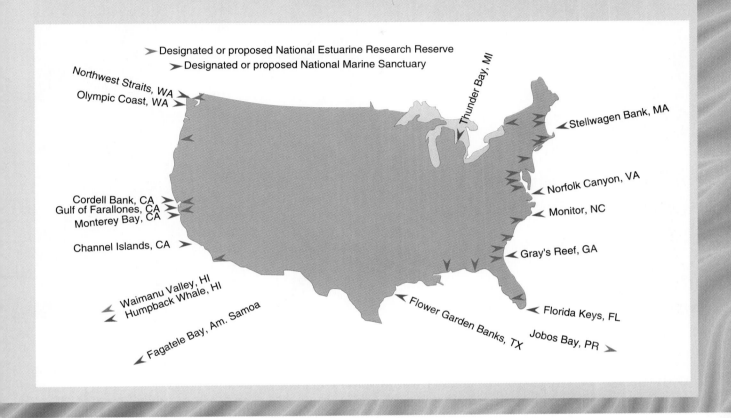

The flashy color patterns of cleaning fishes serve different functions. If the fishes are to attract any business, they must be conspicuous. So they advertise themselves and their location with bright, startling color combinations. These bold advertisement displays are also useful for sexual recognition. One or both sexes of certain species assume bright color patterns during the breeding period. The colors play a prominent role in the courtship displays, which lead to spawning. During this period, the positive value gained from sexual displays must offset the adverse impact of attracting hungry predators. Between breeding periods, these fishes usually assume a drab, less conspicuous appearance.

Advertisement displays are also employed to warn potential predators that their prey carry sharp or venomous spines, poisonous flesh, or other features that would be painful or dangerous if eaten. Predatory fishes recognize the color patterns of unpalatable fishes and learn to avoid them.

Figure 10.23

Disruptive coloration patterns of a
butterfly fish, *Chaetodon.*
Photo by T. Phillipp.

Occasionally, a species capitalizes on the advertisement displays of
another fish by closely mimicking its appearance. The cleaner wrasse
(*Labroides,* the small fish in figure 10.21) is nearly immune to predation because of the cleaning role it performs for its potential predators. Over much of its range, *Labroides* lives close to a small blenny
(Aspidenotus). The blenny so closely resembles *Labroides* in size,
shape, and coloration (figure 10.24) that it fools many of the predatory
fishes that approach the wrasse's cleaning station. Not content to
share *Labroides'* immunity to predation, the blenny also uses its disguise to prey on fishes that mistakenly approach it for cleaning. This
ability to disguise and thereby be protected is known as **mimicry.**

Only in the clear waters of the tropics and subtropics does color
play such a significant role in the lives of shallow-water animals. In the
more productive and turbid waters of temperate and colder latitudes,
light does not penetrate as deeply nor is the range of colors available.
In coastal waters and kelp beds, monotony and drabness of appearance, not brilliance, are the keys to camouflage. In the deep ocean,
color is even less important. Without light to illuminate their pigments, it matters little whether deep-water organisms appear red,
blue, black, or chartreuse when viewed at the surface. In the abyss,
they would all assume the uniform blackness of their surroundings
were it not for bioluminescence (discussed in the next chapter).

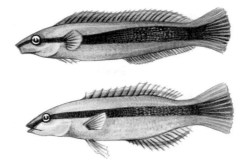

Figure 10.24

The small cleaner *Labroides* (above) and its mimic, *Aspidenotus* (below).

Summary Points

- Corals are colonial, sea anemone-like members of the phylum Cnidaria that live inside a $CaCO_3$ exoskeleton.
- Coral reefs are limited to the photic zone of tropical and warm subtropical coasts where clean, hard substrates are available.
- Reef-forming corals contain photosynthetic zooxanthellae that nourish the coral polyps and dramatically increase the productivity of the reef. These corals form the structural foundation for very complex shallow-water communities.
- Fringing reefs, barrier reefs, and atolls are sequential ontogenetic stages in the life cycle of a single coral reef.
- Corals reproduce both asexually (via budding and fragmentation) and sexually (by brooding fertilized eggs internally and by spawning millions of gametes into the water column for external fertilization).
- Wave force, water depth, temperature, salinity, and a host of biological factors favor some coral reef inhabitants over others. These conditions vary greatly across a reef and provide for both horizontal and vertical zonation of the coral and algal species that form the reef.
- Coral reefs are subject to catastrophic mortality often enough that high species diversity is maintained. This catastrophic mortality of corals and coral reef species is caused by both natural and anthropogenic agents.
- Thousands of species of reef fishes are associated with coral reefs. These fishes find protection on the reef, prey on the plants and animals living there, nibble at the reef itself, and are major importers of important limiting nutrients to local reef systems.
- Symbiotic associations are common, and numerous examples of commensalism, mutualism, and parasitism exist.
- The brightly colored patterns of coral reef fishes illustrate the advertisement, disguise, and concealment roles of brilliant coloration in a color reef environment.

Review Questions

1. Describe the growth of a coral head from its inception as a single polyp that has recently settled on a coral reef.
2. Describe all physical limitations to the biogeographical distribution of coral reefs.

3. List, in sequence, the names of the three stages of coral reef development around a new oceanic volcanic island.
4. Describe the phenomenon of mass spawning of coral reef organisms.
5. What is the most diverse zone on an atoll? What factors contribute to this high diversity?
6. What is coral bleaching? What are its potential causes? What is its great significance?
7. What is cleaning symbiosis?
8. List two specific functions of bright coloration in coral reef fishes.

Challenge Question

1. Coral reefs give the appearance of highly productive ecosystems, yet they exist in oceanic regions with very low production rates and standing crops. Why do coral reefs appear to be so productive?

Suggestions for Further Reading

Books

Darwin, C. (1842) 1962. *The structure and distribution of coral reefs.* Berkeley: University of California Press.

Endean, R. 1983. *Australia's Great Barrier Reef.* New York: The University of Queensland Press.

Fox, D. L. 1979. *Biochromy: Natural coloration of living things.* Berkeley: University of California Press.

Halstead, B. W., P. S. Auerbach, and D. R. Campbell. 1990. *A colour atlas of dangerous marine animals.* Boca Raton, FL: CRC Press.

Thresher, R. E. 1984. *Reproduction in reef fishes.* Neptune City, NJ: T. H. F. Publications.

Vernon, J. E. N. 1993. *Corals of Australia and the Indo-Pacific.* Honolulu: University of Hawaii Press.

Articles

Aeby, G. S. 1991. Costs and benefits of parasitism in a coral reef system. *Pacific Science* 45:85–86.

Babcock, R. C., P. L. Bull, A. J. Heyward, J. K. Oliver, C. C. Wallace, and B. L. Willis. 1986. Synchronous spawnings of 105 scleractinian coral species on the Great Barrier Reef. *Marine Biology* 90:379–94.

Birkeland, C. 1989. The Faustian traits of the crown of thorns starfish. *American Scientist* 77:154–63.

Blair, S. M., T. L. McIntosh, and B. J. Mostkoff. 1994. Impacts of Hurricane Andrew on the offshore reef systems of the central and northern Dade County, Florida. *Bulletin of Marine Science* 54:961–73.

Clifton, K. E. 1997. Mass spawning by green algae on coral reefs. *Science* 275:1116–18.

Connell, J. H. 1978. Diversity in tropical rain forests and coral reefs. *Science* 199:1302–10.

Fadlallah, Y. H. 1983. Sexual reproduction, development and larval biology in scleractinian corals: A review. *Coral Reefs* 2:129–50.

Falkowski, P. G., Z. Dubinsky, L. Muscatine, and L. McCloskey. 1993. Population control in symbiotic corals. *Bioscience* 43:606–11.

Goreau, T. F., N. L. Goreau, and C. M. Yonge. 1971. Reef corals: Autotrophs or heterotrophs? *Biological Bulletin* 141:247–60.

Hatcher, B. G. 1990. Coral reef primary productivity: A hierarchy of pattern and process. *Trends in Ecology and Evolution* 5:149–55.

Hawkins, J. P., C. M. Roberts, and T. Adamson. 1991. Effects of a phosphate ship grounding on a Red Sea coral reef. *Marine Pollution Bulletin* 22:538–42.

Highsmith, R. C. 1982. Reproduction by fragmentation in corals. *Marine Ecology Progress Series* 7:207–26.

Hutchings, P. A. 1986. Biological destruction of coral reefs: A review. *Coral Reefs* 4:239–52.

Jackson, J. B. C., and T. P. Hughes. 1985. Adaptative strategies of coral-reef invertebrates. *American Scientist* 73:265–74.

Jones, G. P. 1990. The importance of recruitment to the dynamics of a coral reef fish population. *Ecology* 71:1691–98.

Muscatine, L., and J. W. Porter. 1977. Reef corals: Mutualistic symbiosis adapted to nutrient-poor environments. *Bioscience* 27:454–60.

Richmond, R. H. 1985. Reversible metamorphosis in coral planula larvae. *Marine Ecology Progress Series* 22:181–85.

Schener, P. J. 1977. Chemical communication of marine invertebrates. *Bioscience* 27:644–68.

Scot, R. D., and H. R. Jitts. 1977. Photosynthesis of phytoplankton and zooxanthellae on a coral reef. *Marine Biology* 41:307–15.

Stoddart, D. R. 1973. Coral reefs: The last two million years. *Geography* 58:313–23.

Walbran, P. D., R. A. Henderson, A. J. T. Jull, and M. J. Head. 1989. Evidence from sediments of long-term *Acanthaster planci* predation on corals of the Great Barrier Reef. *Science* 245:847–50.

Warner, R. R. 1990. Male versus female influences on mating-site determination in a coral reef fish. *Animal Behavior* 39:540–48.

Wilkinson, C. R. 1987. Interocean differences in size and nutrition of coral reef sponge populations. *Science* 236:1654–57.

Yamamuro, M., H. Kayanne, M. Minagawa. 1995. Carbon and nitrogen stable isotopes of primary producers in coral reef ecosystems. *Limnology and Oceanography* 40:617–21.

Internet Addresses

CREST (Coral Reef Education for Students and Teachers)
http://www.petsforum.com/IMA/CRAA.html
An 80+ page on-line manual about coral reefs; excellent supplemental resource to text. International Marinelife Alliance **http://www.petsforum.com/IMA/**

Coral Reef Information Page
http://planet-hawaii.com/sos/coralreef.html
Links to over 20 sites related to conserving and saving coral reefs and to other marine topics. Save Our Seas, Carl Stepath, Executive Director, e-mail sos@aloha.net. Save Our Seas **http://planet-hawaii.com/sos/index.html**

The Hawaii Coral Reef Network
http://www.coralreefs.hawaii.edu/ReefNetwork/default.html
Comprehensive site that includes a wide variety of information on marine biology and ecology of coral reefs. Brian N. Tissot, Univ. of Hawaii at Hilo; Eric Brown, Pacific Whale Foundation; Carl Stepath, Save Our Seas; Dave Raney, Sierra Club.

Corals and Coral Reefs
http://www.seaworld.org/coral_reefs/introcr.html
Links to pages with Classification, Habitat and Distribution, Physical Characteristics, and much more. Busch Entertainment Corporation. The Sea World/Busch Gardens Animal Information Database **http://www.seaworld.org**

Reef Resource Page
http://www.indiana.edu/~reefpage/
"The site's principle objective is to provide an integrated review of what is known about reefs, from the basics to the forefront of our understanding." Click on home page links to connect with subpage topics such as Reef Geology. Dr. Paul Blanchon (Indiana University).

Coral Reef Ecosystems—Tropical Rain Forest of the Sea
http://geosun1.sjsu.edu/~dreed/105/coral.html
Many links to other sites with pictures, projects, information and issues. Donald L. Reed, Dept. of Geology, San Jose State University.

Learn About Marine Biology
http://www.odysseyexpeditions.org/course.html
Click on topics like these: Coral Reefs—biology, distribution, and evolution; Coral Reef Bleaching—what causes the whitening of the reefs, is global climate change possibly going to cause the destruction of coral reefs worldwide; Coral Reef Fish Ecology—an in-depth look at the challenges faced by fishes on coral reefs.

Jason Buchheim, Odyssey Expeditions, Inc. Odyssey Expeditions Tropical Marine Biology Voyages
http://www.odysseyexpeditions.org

Main links page for International Year of the Reef 1997
http://www.aza.org/yearofthereef/

International Year of the Reef—Photo Section
http://www.aza.org/yearofthereef/photos/

Seventeen high-quality photos of fish in reefs.
International Year of the Reef—Fun Facts
http://www.aza.org/yearofthereef/ffreef.html

International Year of the Reef—Top Ten Coral Reefs
http://www.aza.org/yearofthereef/topten.html

American Zoo and Aquarium Association
http://www.aza.org/

Below the Tides

L iving conditions on the sea bottom change gradually as wave action, light intensity, and water temperature diminish with increasing depth. Sand and mud deposits become more extensive and dominate the seafloor below the low tide line. The soft bottoms of the continental shelves are comparatively level and are characterized by a uniformity of environmental conditions generally not found along the intertidal shoreline. Benthic plants are largely restricted to occasional small reefs, rocky outcrops, and kelp forests on the inner shelf. Nearly all the bottom fauna are dependent on the slow rain of plankton and detritus from the sunlit photic zone above. Filter feeders, deposit feeders, and their predators are the dominant macrofauna.

Sedimentary materials found in the deep ocean basins away from continental margins are composed of the mineralized skeletal remains of planktonic organisms and of fine particles blown or washed off land. These skeletal deposits, known as oozes, are characterized by their chemical composition. Siliceous oozes contain cell walls of diatoms and the internal silicate skeletons of planktonic radiolarians (see figure 6.2). The skeletons of other planktonic protozoans, the foraminiferans, constitute most of the extensive calcareous oozes found on the ocean floor. These oceanic oozes accumulate very slowly: approximately 1 cm of new sediment every 1,000 years. In the deep sea, oceanic oozes exhibit distributional patterns that reflect the surface abundance of their biological sources. These patterns are shown in figure 11.1.

Shallow Subtidal Communities

Extensive studies of the life in shallow-water soft sea bottoms were initiated by the Danish biologist C. G. J. Petersen, in the early part of the twentieth century. His intent was to evaluate the quantity of food available for flounders and other commercially useful bottom fish. After sorting and analyzing thousands of bottom samples from Danish seas, Petersen concluded that large areas of the level sea bottom are inhabited by recurring associations, or communities, of infaunal species (figure 11.2). Each community has a few very conspicuous or abundant macrofauna as well as several less obvious forms. On other bottom types, distinct communities of other species can be found. When exposed to similar combinations of environmental conditions, widely separated shallow-bottom communities in temperate waters closely resemble each other in structure and species composition (figure 11.3). Although Petersen considered his communities statistical units only, these recurring and predictable links between certain animal species and particular environmental features led G. Thorson to argue for the recognition of Petersen's communities as biological realities.

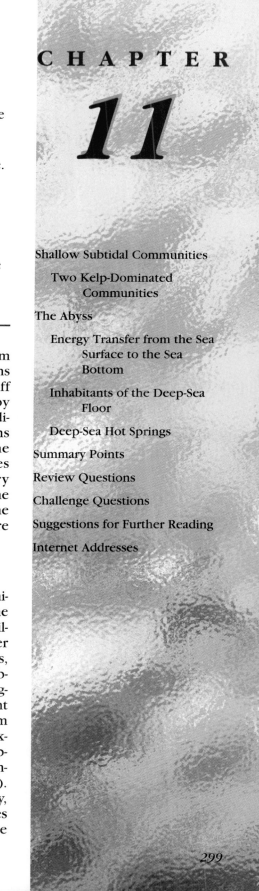

CHAPTER

11

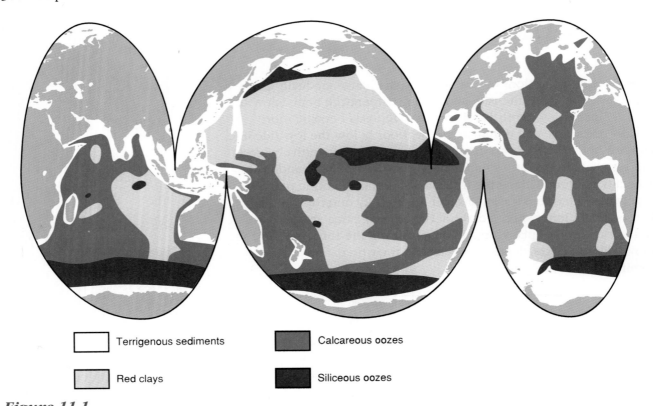

☐ Terrigenous sediments	▨ Calcareous oozes
▨ Red clays	■ Siliceous oozes

Figure 11.1

Distribution of ocean-bottom sediment deposits.
Adapted from Tait, 1963, and Sverdrup, Johnson, and Fleming, 1942.

Since Petersen's time, a prime objective of many benthic ecologists has been to describe the manner in which the benthos is distributed on the seafloor and to explain how this distribution is related to the sediments, the overlying water, and the influence of community members on each other. Using diving gear for direct observations in shallow water and an assortment of dredges, trawls, and grabs to obtain bottom samples in deeper water, benthic ecologists have found parallel shallow-water communities in much of the cold and temperate regions of the world ocean. This parallel community concept has been extended beyond obvious animal associations to include bottom type, depth, and water temperature as additional key factors in shaping benthic community structures. However, Petersen's concept of parallel communities breaks down in tropical and subtropical waters. Here, large numbers of species exist, and seldom does a single species dominate a community.

Two Kelp-Dominated Communities

The infaunal communities just discussed exist only in soft, muddy marine sediments that dominate the continental shelves of the world. Where rocky outcrops do exist on the shallow ocean floor in temperate areas, very different communities (introduced on p. 219) develop, dominated by assemblages of brown algae. The large size (up to 30 m) of some kelp plants adds an important three-dimensional structure to kelp "forests" analogous to the canopy structure of terrestrial forests.

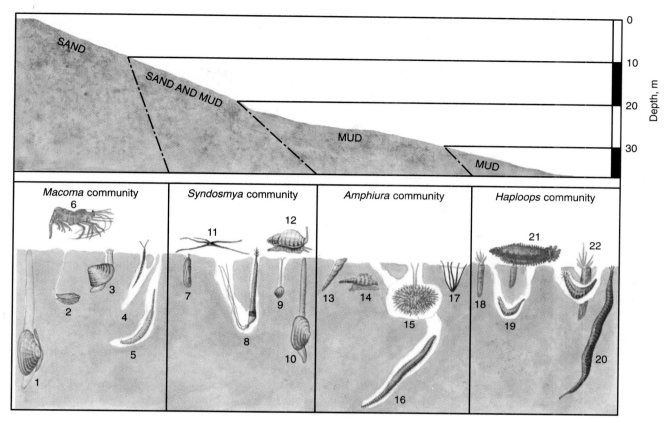

Figure 11.2

A series of soft-bottom benthic communities in the Danish seas, including bivalve mollusks (1, 2, 3, 7, 9, 10), gastropod mollusks (12, 13, 14), crustaceans (4, 6, 18, 22), polychaete worms (5, 8, 16, 19, 20), and echinoderms (11, 15, 17, 21).
Adapted from Thorson 1969.

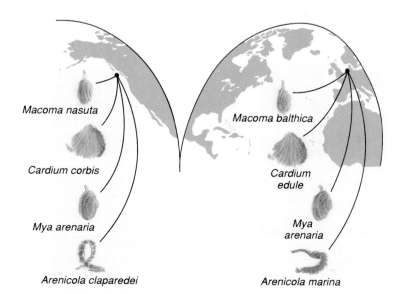

Figure 11.3

Diagram showing the close similarity in the dominant members of soft-bottom communities in the northeastern Pacific and the northeastern Atlantic. The upper three species in each community are bivalve mollusks; the lower one is a polychaete annelid.
Adapted from Thorson 1957.

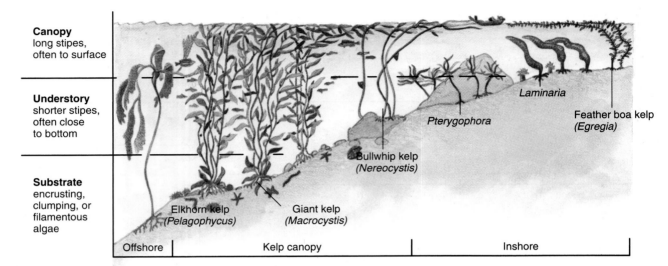

Canopy
long stipes,
often to surface

Understory
shorter stipes,
often close
to bottom

Substrate
encrusting,
clumping, or
filamentous
algae

Laminaria

Pterygophora

Feather boa kelp
(*Egregia*)

Bullwhip kelp
(*Nereocystis*)

Elkhorn kelp
(*Pelagophycus*)

Giant kelp
(*Macrocystis*)

Offshore   Kelp canopy   Inshore

Figure 11.4

General structure of a West Coast
kelp forest, with a complex
understory of plants beneath the
dominant *Macrocystis* or
Nereocystis.

Consequently, in the more complex ecological terrain of kelp forests, more niches exist than do on nearby flat, soft sea bottoms.

Most kelp plants are perennial. Although they may be battered down to their holdfasts by winter waves, their stipes will regrow from the holdfast for several successive seasons. Thus, the extent of the kelp canopy and the overall three-dimensional structure of the kelp forest are quite variable over annual cycles. Occasionally, herbivore grazing or the pull of strong waves frees the holdfast of its attachment and causes the upper parts of the plant to wash ashore. More commonly, small fragments of blades and stipes are continually being cast ashore to decompose into food for detritus feeders.

Along most of the North American West Coast, subtidal rocky outcrops are cloaked with massive growths of several species of brown algae, dominated by either *Macrocystis* or *Nereocystis*. In the dimmer light below the canopy of these large kelps exists a shorter understory of mixed brown and red algae (figure 11.4). Together, these large and small kelp plants accomplish very high rates of primary production and support a complex community of grazers, suspension feeders, scavengers, and predators (figure 11.5).

Compared with the richness of species observed in northeastern Pacific kelp forests, the kelp beds of the northwestern Atlantic Coast exhibit low diversity in most taxonomic groups. Unlike the shores of the western United States, the rocky intertidal and subtidal shores of the New England states and neighboring Canadian Maritime provinces were scoured to bare rock (in places to several hundred meters below sea level) by several episodes of continental glaciation. Only since the retreat of the most recent glacial episode 8,000–10,000 years ago have these shores been recolonized, and that recolonization is not yet complete.

The lower species diversity of northwestern Atlantic kelp beds leads to somewhat simpler trophic interactions than those occurring in U.S. West Coast kelp forests; still, similar species perform the same major trophic roles (figure 11.6). The macroscopic primary producers are dominated by the kelp *Laminaria*, with an understory of mixed

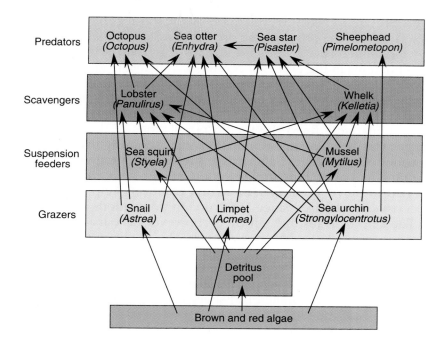

Figure 11.5

Trophic relationships of some dominant members of a southern California kelp community.

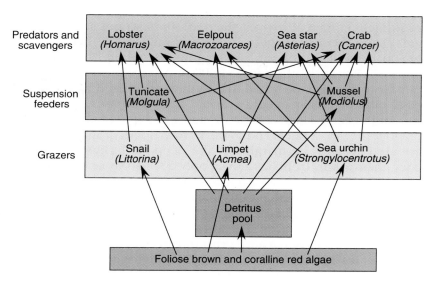

Figure 11.6

Trophic relationships of the common members of a New England kelp community.

red and brown foliose algae. In clear patches, encrusting coralline red algae cover rock surfaces with a bright pink pavement of $CaCO_3$.

On both the Atlantic and Pacific coasts of the United States, kelp beds exist in a delicate balance with their major grazers, sea urchins. Since World War II, numerous kelp beds on both coasts have been devastated by dense aggregations of sea urchins grazing on the holdfasts, causing the remainder of the plant to break free and wash onto the shore. These large urchin populations, capable of completely eliminating local kelp beds, seem free of the usual population regulatory mechanisms—predation and starvation. A major predator of West Coast kelp bed urchins is the sea otter (*Enhydra lutris,* see figure 7.15). East Coast sea urchins are similarly preyed upon by the lobster (*Homarus americanus*). However, both of these predators have been subjected to intensive commercial harvesting and have experienced major population reductions in the past two centuries. Available evidence indicates that the effects of this reduced predation have been magnified by the increased concentrations of dissolved organic materials in coastal waters (mostly from urban sewage outfalls). These energy-rich substances apparently freed the urchin populations of the usual consequences that befall animal populations that seriously overgraze their plant food sources. These alternative sources of energy ensured that large numbers of urchins survived long enough after eliminating one kelp bed to move to another. In central California kelp beds that sea otters have recolonized since 1950, urchin populations are now kept low. The kelp beds just off San Diego, however, have made a dramatic recovery since 1960 without sea otters. The recovery there is likely due to improved urban sewage treatment and the activities of other urchin predators, particularly sea stars and the sheephead fish. Currently, the recovered urchin populations are themselves the target of a rapidly expanding commercial fishery to supply the local and international sushi market.

The Abyss

Beyond the continental shelves, the seafloor descends sharply on continental slopes to the perpetual dark and cold of the deep-sea floor. Three-quarters of the ocean bottom (the abyssal and hadal zones of figure 1.37) lies at depths below 3,000 m. A large portion of the deep ocean basin consists of broad, flat, sediment-covered abyssal plains. Abyssal plains typically extend seaward from the bases of continental slopes and oceanic ridge and rise systems at depths between 3 and 5 km below the sea surface. The long, narrow floors of marine trenches distributed around the margins of the deep-sea basins are generally deeper than 6 km. Trenches cover only 2% of the seafloor and are characterized by the most extreme pressure regimes experienced by living organisms anywhere on this planet.

Separating abyssal plains are mountainous linear ridge and rise systems encircling the globe. The axes of these ridge and rise systems are the source of new oceanic crust, created as the global forces of plate tectonics rift apart plates of the earth's crust. Growth of new crust is slow, about as fast as your fingernails grow, yet it causes about 60% of the earth's surface (or 85% of the seafloor) to be recycled every 200 million years. Several remarkable discoveries associated with ridge and rise spreading centers have been made in the past two decades. These discoveries are radically changing our understanding of life in the deep sea and are clarifying how metal-rich mineral deposits are formed from chemical interaction between seawater and the earth's crust.

The deep-sea bottom is one of the most inaccessible environments on the earth. Studies of deep-sea life have been fragmentary at best, inhibited by the high costs of sampling so far below the surface and by the difficulties of bringing healthy deep-sea animals to the surface for study. Captured animals encounter extreme temperature and pressure

Figure 11.7

A group of large crustacean amphipods feeding on bait in the Peru-Chile Trench at a depth of 7,000 m. For scale, the cable at the right is just over 6 cm in diameter.
Courtesy Scripps Institution of Oceanography, Marine Life Research Group.

changes when hauled from the bottom. By the time they reach the surface, they are usually dead or seriously damaged. Consequently, we presently know more about the back side of the earth's moon than we do about organisms in vast areas of the deep-sea floor.

Below 3,000 m, the water is cold, averaging 2° C and dipping slightly below 0° C in polar regions. Water temperatures at these depths vary little on time scales of years to decades, creating an extremely constant thermal environment for the inhabitants of the deep sea. Pressures created by the overlying water are tremendous, ranging from 300 to 600 (atm) atmospheres on the abyssal seafloor and exceeding 1,000 atm in the deepest trenches. Laboratory studies have confirmed that metabolic rates of deep-sea bacteria are lower at pressures normally experienced on the seafloor than they are at sea surface pressures. Less clear is the response of multicellular organisms to high pressures. Several studies have suggested that pressure-induced reductions in metabolic rates may lead to lowered growth rates, lowered reproductive rates, and increased life spans in the deep sea, culminating in occasional examples of deep-sea gigantism (figure 11.7). Other recent studies have found that the depth-related decline in metabolic rates of crustaceans can be explained as metabolic adjustments to temperature declines with increasing depth and not to a separate depth or pressure effect. Because the effects of high pressures on growth rates and maximum sizes of deep-sea animals are not very well understood yet, this question will not be resolved without more study.

Dissolved O_2 needed to support the metabolic activities of deep-sea animals must come from the surface waters above. Diffusion and sinking of cold, dense water masses are the chief mechanisms of O_2 transport into the deep-ocean basin, with diffusion providing O_2 from the surface downward and sinking water masses filling the deeper ocean basins with oxygen-rich water. As dissolved O_2 diffuses downward in areas with a permanent pycnocline, it is slowly consumed by animals and bacteria, leaving an O_2 minimum zone at intermediate depths (usually between 1,000 and 2,000 m; figure 1.24). Below this zone, available O_2 just above the sea bottom gradually increases. The presence of this O_2 minimum zone reduces the abundance and activity of mid-depth consumers. In one study of animal assemblages on the steep sides of a Pacific volcanic seamount that penetrated the O_2 minimum zone, the number and abundance of benthic species were much greater just below the O_2 minimum zone than within the zone a few tens of meters higher on the seamount.

Energy Transfer from the Sea Surface to the Sea Bottom

A critical and variable resource in the deep sea is food. Because there is no sunlight, there are no photosynthesizers. Food for deep-sea benthic communities, by necessity, comes from above. Only recently have studies been carried out to determine the rate and condition of the food's arrival at the sea bottom. Several studies have demonstrated that, contrary to earlier assumptions, a tight coupling exists between near-surface primary productivity and the eventual consumption of that production by inhabitants of the deep-sea floor (figure 11.8). In most areas of the deep ocean, seasonal variations in the organic content of surface sediments reflect the seasonality of primary production occurring in the photic zone directly above. Such seasonal pulses of organic matter are often detectable several centimeters below the sediment surface as well, as infaunal animals rework this material through their burrowing and feeding activities.

Sinking rates of food-rich particles are largely a function of their size, with smaller particles sinking more slowly. As they sink slowly through the photic zone, much of the near-surface phytoplankton are aggregated into larger gelatinous blobs or are compacted into zooplankton fecal pellets, which settle rapidly to the bottom. This fallout of fecal particles and gelatinous aggregates accelerates the transport of organic material to the abyss, falling from the surface waters in a few days rather than the weeks or months of settling time necessary for smaller phytoplankton particles. Thus, the time lag between the initial production of organic material in the photic zone and its arrival on the deep-sea floor ranges from several days to a few weeks. On the way down, sinking particles of organic material are repeatedly consumed by pelagic scavengers and colonized by bacteria. Consequently, sinking particles, especially the very small ones, lose much of their nutritive value by the time they settle to the bottom. The amount of photic zone productivity that actually reaches the sea bottom, typically only a few percent, is depth dependent, decreasing at greater depths.

(a)

(b)

Figure 11.8

Sea-floor photographs showing the deposition of phytodetritus before (*a*) and (*b*) two months later after a phytoplankton bloom in the photic zone above.

Courtesy R. Lampitt, Southampton Oceanography Centre, U.K.

Inhabitants of the Deep-Sea Floor

It is not possible to establish a precise global depth boundary between the animals of the deep sea and the shallow-water fauna of the continental shelves. In general, the boundary exists as a vague region of transition on the continental slopes bordering the deep-sea basin. However, animals of the "deep sea" commonly extend into shallower water in polar seas and, on occasion, extend into the continental borderlands of southern California and even into the inner portions of high-latitude continental shelves.

The inhabitants of the deep-sea floor are a distinctive group, although few exhibit structural adaptations that make them appear notably different from their shallow-water relatives. Instead, a shift in dominant taxonomic groups occurs in deeper water. Echinoderms (especially sea cucumbers and crinoids), polychaete worms, pycnogonids, and isopod and amphipod crustaceans become abundant, whereas bivalves and other mollusks and sea stars decline in number. Some taxonomic groups are virtually absent until relatively great depths are reached. Most species of pogonophorans (see figure 11.15), for instance, are found below 3,000 m, and 30% of them are restricted to trenches below 5,000 m.

Low temperatures, high pressures, and a limited food supply led to an early and widespread belief that the rigorous and specialized climate of the deep sea would not support a highly diversified assemblage of animals. It was assumed that only a few highly adapted animals could succeed in the abyss. Repeated sampling efforts did reveal a marked decline in both density and biomass of organisms at greater depths, most likely in response to the limiting effect of decreasing food and O_2 with depth. Because of recent improvements in sampling equipment and sample analysis techniques, deep-sea assemblages have been found to contain a diversity of species comparable to or even exceeding the diversity of soft-bottom communities in shallow inshore waters. To explain the relatively high species diversity of deep-sea life, researchers have proposed several complementary explanations.

First, the enormous extent of the deep sea (about 300 million km^2) provides large areas with few barriers to block dispersal. When large areas of the deep sea are sampled at the same depth, approximately one additional species is added for each additional square kilometer sampled. When different depths are sampled, the rate at which additional species are acquired is even greater. These trends suggest very high species diversity in the deep sea and also point out one of the basic difficulties in clarifying the pattern of that diversity: Enormous areas must be sampled (at enormous costs) before we have an accurate picture of the number and distribution of species in the deep sea.

Second, within large areas of nearly constant deep-sea environments, small-scale variations in food availability and physical or biological disturbance create new opportunities for colonizing by additional species. Events such as food-falls of large animal carcasses or the sorting and modification of sediment deposits when animals form tubes and compact sediments to produce fecal pellets and castings may serve to maintain some spatial diversity in what otherwise seems a very uniform environment.

A portion of the smaller particles of organic material settling to the bottom is immediately claimed by suspension feeders. These are often epifauna clinging to rocky outcrops, shells, or manganese nodules. Other epifauna employ stiltlike appendages to keep their bodies from sinking into the sediments (see figures 8.6 and 11.11). The majority of benthic animals in the deep sea, however, are infaunal deposit feeders. Analyses of their stomach contents indicate that many infauna indiscriminately engulf sediments containing smaller infauna and bacteria as well as dead organic material. It has been estimated that 30 to 40% of the organic material available at the bottom is first absorbed by benthic bacteria, which in turn are consumed by larger animals. An important component of the organic material is the chitinous exoskeletons shed by planktonic crustaceans; it cannot be digested by most animals but is decomposed and utilized by bacteria.

Large, rapidly sinking particles, including dead squids, fishes, and an occasional whale or seal, may provide a significant, although unpredictable supply of food to the inhabitants of the deep-sea floor. When large food parcels do contribute significantly to the energy budget of the deep sea, they are rapidly consumed by various wide-ranging scavenger-predators. Most scavenger-predators seem swift enough to evade bottom trawls and are best known from photographs taken by underwater cameras baited with food (figure 11.9). They seem to be food generalists, capable of rapidly locating and consuming large food items as they arrive on the bottom. Some of the food is eventually dispersed as detritus and fecal wastes to other benthic consumers.

(a)

(b)

Figure 11.9

A bait can lowered to the sea floor at a depth of 1390 m off the northern Baja California coast quickly attracts several types of scavenging fishes. (*a*) Several hours later (*b*), the same scene is dominated by slower-moving invertebrates, including tanner crabs.

Photo courtesy Scripps Institution of Oceanography, Marine Life Research Group

In deep-ocean basins, as in shallow-water soft-bottom communities, deposit feeding is common. A variety of worms, mollusks, echinoderms, and crustaceans obtain their nourishment by ingesting accumulated detritus and digesting its organic material. These deposit feeders engulf sediments and process them through their digestive tracts (figure 11.10). They extract nourishment from the organic material in the sediment in much the same manner as earthworms do. Some deposit feeders indiscriminately ingest any available sediments; others select organically rich substrates for consumption.

Several animal species are capable of extracting sufficient nourishment from sediments by conducting digestive processes outside their bodies. These animals absorb the products of digestion either through specialized organs or across the general body wall. Deep-sea pogonophorans, small worms related to the large tube worms recently discovered in association with deep-sea hot springs (see figure 11.15), depend on this type of absorptive feeding, as do numerous echinoderms. Sea stars are usually carnivorous, but a few species are opportunistic foragers. When feeding on bottom sediments, some sea

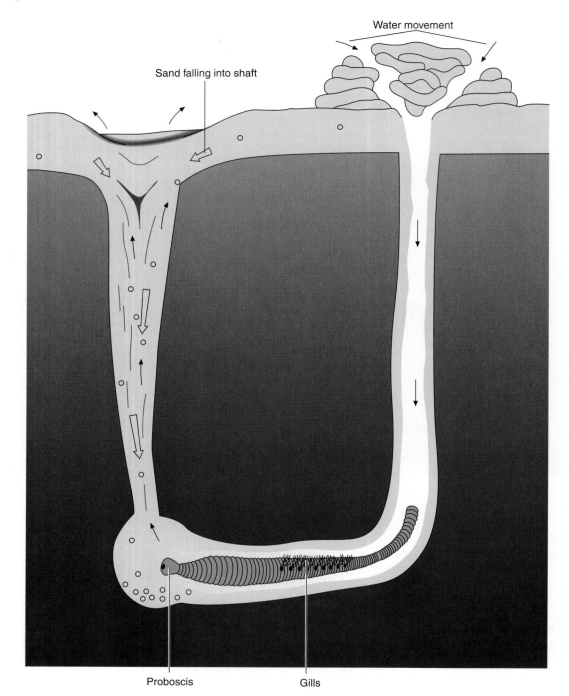

Sand falling into shaft

Water movement

Proboscis

Gills

Figure 11.10

Funnel feeding by the polychaete, *Arenicola Marina*. Black arrows show water flow, open arrows show path of sediments. Sediments are drawn down into anoxic layer, selectively ingested, and deposited above the tail shaft.

Figure 11.11

An aggregation of deep-sea
cucumbers, *Scotoplanes*, feeding on
bottom deposits at a depth of
approximately 1,000 m.
Courtesy E. Barham, National Marine
Fisheries Service.

stars extrude their stomachs outside their bodies to digest and absorb
organic matter from the sediments. The omnipresent bacteria also depend on extracellular digestion. As they absorb nutrients and their
population grows, they in turn become a significant source of particulate food for deposit feeders.

The term **cropper** has been applied to deep-sea animals that
have merged the roles of predator and deposit feeder (figure 11.11).
These croppers, by preying heavily on populations of smaller deposit feeders and bacteria, may be responsible for reducing competition for food and for permitting coexistence between several
species sharing the same food resource. Thus, even though a deep-
sea community as a whole is food limited, populations within the
community need not be as long as they are heavily preyed upon.
With competition for food reduced by cropping, fewer species are
pushed to extinction (on a local scale), and species diversity remains high. The disruptive effects of large croppers on smaller ones
can be compared to predators, ice, or logs as disturbance mechanisms in rocky intertidal communities. By nonselectively reducing
competition, they lessen the possibility of a species being excluded
by resource competition.

In the physically stable environment of the deep sea, patterns of
reproduction are thought to differ appreciably from the reproduction patterns of shallow-water benthic animals. Few deep-water benthic species produce planktonic larvae because the larvae's chances
of reaching the food-rich photic zone several kilometers above and

successfully returning to the ocean floor for permanent settlement are extremely remote. To compensate for the absence of a dependable external food supply, fewer and larger (leithotrophic) eggs are produced. Leithotrophic eggs supply the larvae with adequate yolks so that they can develop to more advanced stages before hatching. Brood pouches and other similar adaptations further enhance the chances of survival by protecting the eggs until they hatch.

Deep-Sea Hot Springs

Tectonic spreading centers along seafloor ridge and rise systems are sites of active vulcanism. Several remarkable discoveries associated with these centers have been made in the past two decades. These discoveries are radically changing our understanding of life in the deep sea and are clarifying how metal-rich mineral deposits are formed from chemical interaction between seawater and the earth's crust. At seafloor ridge and rise axes, crustal plates are spreading apart (see figure 1.5), and molten magma is rising to fill the gap. The molten magma cools, solidifies, and creates a new oceanic crust. This oceanic crust is basalt, a black volcanic rock. It was predicted that hot springs like those found in Yellowstone National Park also occur in the deep sea at areas of comparable volcanic activity.

Deep-sea hot springs were first directly observed in 1977 when a team of geologists studying the Galápagos Rift Zone west of Ecuador discovered several unusual assemblages of animals behaving in unusual ways nearly 3 km deep. Dense aggregations of large mussels, clams, giant pogonophoran tube worms, and crabs were found clustered in small areas of shimmering warm water pouring from the seafloor.

The Galápagos Rift Zone is a site of active seafloor spreading, with extensive lava flows so young that very little sediment has had a chance to accumulate. In the few years since this discovery, enough other vent communities have been found along the axes of both the Mid-Atlantic Ridge and the East Pacific Rise (figure 11.12) to suggest that such communities may be a normal feature of hot, actively spreading rift zones in other oceans as well.

Water temperature sometimes exceeds 100° C at the hot spring sites, in sharp contrast to the near-uniform 2° C abyssal water just a few meters away. These hot springs are the end product of seawater's circulating through the many cracks and fissures of the new crust as it forms along the axes of the rise system. It has been estimated that a volume of water equivalent to the volume of all the earth's oceans circulates through these ridge crack systems approximately every 8 million years. This hydrothermal activity is limited to the upper 1 km of the seafloor crust, yet it is driven by the heat and circulating mechanism of the entire underlying lithosphere (see figure 1.5).

As seawater percolates through these cracks, it is heated to about 350° C, and some complex chemistry occurs between the circulating seawater and the hot basaltic rock of the ocean crust. Sulfate (SO_4^{-2}) is the third most abundant ion dissolved in seawater. When heated, sulfate reacts with water to form hydrogen sulfide (H_2S) which, despite its being toxic to most animals, then plays a central role in subsequent processes as the heated water emerges from the seafloor vent as a hot spring.

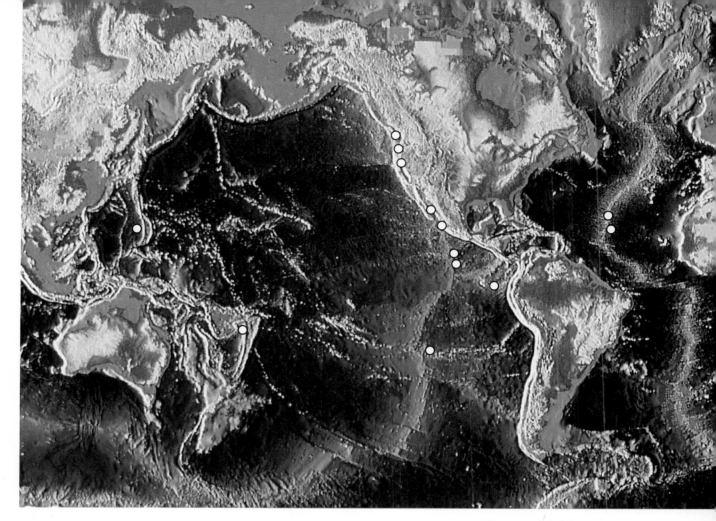

Figure 11.12

Locations (white dots) of confirmed
deep-sea vent communities. All are
associated with actively spreading
ridge systems.
Courtesy National Geophysical Data Center.

Some of the H_2S reacts with iron and other metal ions within the
crust to form metal sulfides in concentrations several million times
greater than those found in average seawater. At the vent opening, this
super-enriched metal sulfide solution quickly mixes with cold surrounding seawater, and a dense black "smoke" of iron sulfide particles
is produced (figure 11.13). These particles either settle immediately
around the vent or are oxidized by dissolved oxygen to produce thick
deposits of metal oxides in the vicinity of the hot spring.

The massive quantities of H_2S spewing from the chimneys atop
these hot springs (known as black smokers; figure 11.14) are utilized as
a primary energy source by perpetual dense blooms of bacteria. Bacteria use dissolved O_2 from the surrounding water to oxidize the abundant H_2S back to SO_4^{-2}, releasing energy in the process. These bacteria
are the primary producers of the hot spring communities, relying on
chemosynthesis rather than photosynthesis to fuel their metabolic requirements. The bacteria in turn are consumed by dense populations of
clams and mussels that are found nowhere else in the sea. These bacterial communities also experience large cyclical variations in productivity rates. Rather than being seasonally driven as are phytoplankton
communities near the surface, vent bacteria periodically thrive and
bloom in response to release of H_2S and CO_2 by earthquakes.

Figure 11.13

A simplified drawing of a deep-sea hot spring community. Some of the organisms shown here can also be seen in figure 11.15.

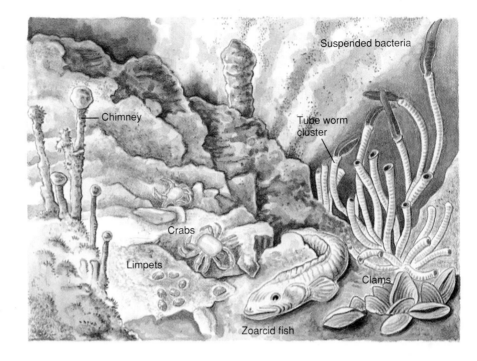

Figure 11.14

A black smoker on the Galápagos Rift Zone. Particles in the "smoke" may be the major source of minerals for the formation of manganese nodules (see figure 8.3).
Photo by D. Foster.

Figure 11.15

Red-plumed tube worms, *Riftia,* with a galatheid crab and a few other members of this unusual deep-sea community.
Courtesy Scripps Institution of Oceanography

The clams and mussels are suspension feeders well adapted to remove the extremely small bacterial cells from the water. The red-plumed giant vestimentiferid worm *Riftia* (figure 11.15), however, completely lacks a digestive tract. Instead, it maintains internal bacterial symbionts that oxidize H_2S within a specialized organ, the **trophosome.** Major blood vessels link the trophosome to the bright red obturacular plume, where gas exchange occurs. The blood of *Riftia* contains two proteins to deal with the high concentrations of dissolved H_2S and small amounts of dissolved O_2 in the deep-sea hot spring environment. The first protein is a rare sulfide-binding protein used to transport large quantities of H_2S from the obturacular plume, where it is absorbed, to the trophosome, where it is used by bacterial symbionts. **Hemoglobin,** the other blood protein, is used to bind and transport O_2 within these animals.

In these deep-sea hot spring communities, bacterial use of O_2 from the water in the immediate vicinity of the vents is extensive. To compete effectively with bacteria for the limited available O_2, larger animals need some type of blood pigment with a strong affinity for O_2. Hemoglobin is abundant in *Riftia* and in large bivalve mollusks, including the giant white clam *Calyptogena*. In the crabs most commonly found in deep-sea hot spring communities (*Bythograea*), the blue pigment **hemocyanin** is used instead. The blood pigments of these hot spring animals are typically quite insensitive to changing temperatures. Such temperature insensitivity is unusual and likely has evolved to accommodate the extremely wide range of water temperatures experienced over very short distances around vents.

These pigments thus assist the larger deep-sea hot spring animals to extract the little O_2 left over by the abundant bacteria. With these adaptations, these populations of benthic animals have tapped into an unusual energy source—the heat and chemicals of the earth's crust—and have achieved a nutritional emancipation from the fall-out products of the photic zone above.

How these hot springs are initially colonized is still a puzzle. Hot spring communities are often a thousand kilometers or more apart, and several "bridging" mechanisms to explain the presence of the same species at different hot springs have been proposed. An intriguing one is the outgrowth of a modest observation made off the California coast in 1987 from the deep-sea research submersible *Alvin*. A decomposing, but still identifiable, whale carcass on the seafloor was surrounded by numerous clams of two species previously seen only at hot spring sites. Dense mats of sulfide bacteria were also found. This observation suggests that hot spring community members may "hop" from carcass to carcass, eventually moving long distances along the sea bottom. Such larval hops, if they occur, may be assisted by the vertical rise and eventual horizontal drift of warm plume water from an active vent. It is not yet known if such "whale-falls" occur frequently enough to provide the larval-hopping network of food-rich material necessary to bridge the distances that exist between these remarkable deep-sea communities. However, observations of a rift site on the East Pacific Rise demonstrated that, once established, vent communities develop rapidly. Barren of larvae in 1991, the site was transformed to a community resembling that shown in figure 11.15 only two years later.

A scattering of other types of chemosynthetic communities have been found in the deep sea always associated with a rich localized energy source. In addition to whale-falls, cold spring seeps near Japan, methane seeps in the Gulf of Mexico, and the earthquake-disturbed sediments of the Laurentian fans in the eastern North Atlantic support small dense communities of benthic animals and chemosynthetic bacteria similar to those found at deep-sea hot spring sites. The unexpected discoveries of these communities suggest that we have only begun to understand the biology of this fascinating environment.

Summary Points

- Below the effects of waves and tides, the sea bottom is depositional, with most of it covered with a blanket of soft sediments.
- Subtidal kelp communities inhabit temperate, shallow seafloors. They are more complex and extensive on the West Coast of North America than they are on the East Coast.
- The benthic animals of the abyss are not well known. They experience little in the way of fluctuations in their physical environment and are limited by the absence of a constant and abundant food supply. Most are thought to have very generalized feeding habits and to function either as wide-ranging predator-scavengers or as sedentary croppers.
- Food for animals of the abyss is transported from sunlit surface layers by sinking and by feeding actions of midwater animals.

- Deep-sea hot springs, recently discovered along the axes of ridge and rise systems, support unique communities of deep-sea animals. Dissolved H_2S emerging from seafloor cracks is used as an energy source by chemosynthetic bacteria. The bacteria in turn are the source of nutrition for dense animal populations clustered around these springs.

Review Questions

1. Describe the seafloor conditions that lead to more complex kelp communities on the U.S. West Coast than on the East Coast.
2. Describe the different sources of food available to animals of the deep sea.

Challenge Questions

1. Where in the Atlantic Ocean would you predict that additional deep-sea hot springs and their associated chemosynthesis-powered communities might be found? Why?
2. Why is broadcast spawning less common in deep-sea animals than in shallow-water animals?

Suggestions for Further Reading

Books

Ernst, W. G., and J. G. Morin, eds. 1982. *The environment of the deep sea.* Englewood Cliffs, NJ: Prentice Hall.

Articles

Arp, A. J., and J. J. Childress. 1983. Sulfide binding by the blood of the hydrothermal vent tube worm *Riftia pachyptila. Science* 219:295–97.

Childress, J., H. Felback, and G. Somero. 1987. Symbiosis in the deep sea. *Scientific American* (May) 256:114–20.

Corliss, J. B., and R. D. Ballard. 1977. Oases of life in the cold abyss. *National Geographic* 152:441–53.

Edmond, J. M., and K. Von Damm. 1983. Hot springs on the sea floor. *Scientific American,* (April) 252:78–93.

Grassle, J. F. 1991. Deep-sea benthic biodiversity. *Bioscience* 41:464–69.

Hollister, C. D., A. R. M. Nowell, and P. A. Jumars. 1984. The dynamic abyss. *Scientific American* (March) 253:42–53.

Jannasch, H. W., and C. O. Wirsen. 1977. Microbial life in the deep sea. *Scientific American,* (June) 236:42–52.

Lutz, R. A., 1991. The biology of deep-sea vents and seeps. *Oceanus* 34(3):75–83.

Sebens, K. P. 1985. The ecology of the rocky subtidal zone. *American Scientist* 73:548–57.

Snelgrove, P. V. R., and J. F. Grassle, 1995. The deep sea: Desert and rainforest. Debunking the desert analogy. *Oceanus* 38(2):25–29.

Tunnicliffe, V. 1992. Hydrothermal-vent communities of the deep sea. *American Scientist* 80:336–49.

Internet Addresses

Kelp Forest and Rocky Subtidal Habitats
http://bonita.mbnms.nos.noaa.gov/sitechar/kelp.html
Kelp forest distribution and ecology, algal assemblages, invertebrate and vertebrate assemblages.
Shallow Soft Bottom Habitats
http://bonita.mbnms.nos.noaa.gov/sitechar/shallow.html
Physical features, zonation and disruption of zonation pattern.
Deeper Bottom Habitats
http://bonita.mbnms.nos.noaa.gov/sitechar/deep.html
Faunal Assemblages on the Continental Shelf and Deep Sea Sedimentary Megafaunal Communities
Cold Seep Communities
http://bonita.mbnms.nos.noaa.gov/sitechar/cold.html

Discovery of Hydrothermal Vents and Cold Seeps and Cold Seeps in Monterey Bay. Monterey Bay National Marine Sanctuary. Monterey Bay National Marine Sanctuary Site Characterization
http://bonita.mbnms.nos.noaa.gov/sitechar/index.html

Creatures of the Thermal Vents by Dawn Stover
http://seawifs.gsfc.nasa.gov/OCEAN_PLANET/HTML/ ps_vents.html

Review of investigation of a recently formed hydrothermic vent and the discovery of life in a vent during a dive by the submersible, Alvin. Links to tube worms and "black smokers." Smithsonian Institution. **Ocean Planet http://seawifs.gsfc.nasa.gov/ocean_planet.html**

Deep-sea Hydrothermal Vents
http://www.ocean.washington.edu/people/grads/summit/ overview.html

Discussion of deep-sea vents with many clickable references. Melanie Summit, University of Washington Oceanography. University of Washington School of Oceanography
http://www.ocean.washington.edu/

Seaweek '97 "Explore the Deep Sea"
http://www.education.monash.edu.au/peninsula/seaweek/ deepsea.html

Take an imaginary journey to the depths of the seas. Descriptions of environment at lower and lower levels, sketches and explanations of life at each level, strategies, adaptations, reproduction, and more. Thorough, but not overwhelming. Peter Biro, Ed Online, Marine Education Society of Australia, Inc. Seaweek "Celebrate the Sea"
http://www.education.molnash.edu.au/peninsula/seaweek/ deepsea.html

Giant Bladder Kelp
http://www.catalinas.net/seer/marine/algae/kelp_mp.html
Discussion of Macrocystis pyrifera with photos, diagrams and sketches.

Other Kelp Species
http://www.catalinas.net/seer/marine/algae/kelpmisc.html
Brief discussion and two photos of Pelagophycus porra and Nereocystisluetkeana. © Dr. William W. Bushing and Santa Catalina Island Conservancy. Catalina Island Conservancy Intranet with links to many other interesting pages
http://www.catalinas.net/seer/menu_mar.html

Vents Video Clips
http://www.pmel.noaa.gov/vents/geology/video.html
A selection of video highlights from VENTS research activities, including Alvin and ROV dives, and more. MPEG player required to view videos, but stills from each video are also available. NOAA Pacific Marine Environmental Laboratory VENTS Program. Vents Geology Program with several topic-related links and pages
http://www.pmel.noaa.gov/vents/geology/

The Pelagic Realm

Chapter 12

Chapter 13

Chapter 14

ECOLOGICAL PERSPECTIVES
The Pelagic Realm-The Zooplankton

Chapter 12 discusses the interesting phenomenon of diel vertical migration of the zooplankton community. There are more copepods than any other kind of zooplankton; they dominate nearly all oceans. Therefore, most investigations of diel vertical migration use copepods as the experimental subject. In chapter 12, you will learn about the size of copepods (less than 1 cm in total length), their swimming speed during migrations, and the hypothetical advantages of diel vertical migration. But first, stop to think for a moment about the journey from a copepod's point of view. If you do, this phenomenon will become even more fascinating! First, you must consider the relative distance that the copepod swims each day. Suppose we are dealing with a copepod that is 1 mm in total length. Assume that this individual spends its days at a depth of 450 m and migrates to within 150 m of the surface at night. Hence, this copepod makes a daily trip of 600 m (300 m up and 300 m back down) each day. This is a total distance equivalent to 600,000 body lengths! To put this incredible journey into human terms, if an average human 5.5 feet tall were to swim a round-trip of 600,000 body lengths in a day, she or he would need to swim 625 miles! Perhaps this perspective amazes you, and it should! But that's only a third of the story. Not only does a 1-mm copepod swim the equivalent of 625 miles each day, it does so at an enormous rate of speed, in relative terms. For example, as discussed in chapter 13, a 1-m barracuda is capable of swimming at 40 km/h, or about 11 body lengths per second. That speed seems incredible when you consider that Olympic-class swimmers attain a speed of just 1.6 body lengths per second. However, a 1-mm copepod swimming at 50 m/h is moving at a relative speed of about 14 body lengths per second! Third, our little copepod not only swims an enormous distance at a great rate of speed, it does so through a medium (seawater) that seems much "stickier" to it than to you. In any fluid, there are two competing forces: viscosity and inertia. Small objects, such as copepods, are dominated by viscous forces and possess very little inertia. The instant that a copepod stops swimming, it stops. It does not drift after becoming motionless because it is living in what amounts to a sticky medium. Therefore, a copepod must continuously exert a force against water to swim. In summary, copepods swim an enormous distance each day, at a great rate of speed, through a medium that must seem like molasses to them! Diel vertical migration is indeed a fantastic accomplishment.

The Zooplankton

L ife in the pelagic division of the marine environment exists in a three-dimensional, nutritionally dilute medium. Microscopic protists and the major groups of small herbivores and many of their larger predators live in near-surface waters. The distribution of pelagic animals reflects their nutritional dependency on the primary producers of the sea. Like primary producers, pelagic animals concentrate in regions of upwelling, over continental shelves and other shallows, and elsewhere in or near the photic zone. The upper few hundred meters of the sea teem with animal life. At greater depths, population densities diminish rapidly, but animal life never completely disappears.

The animals of the pelagic division include the zooplankton and nekton. Most nektonic animals begin life as members of the zooplankton community. As the zooplankton grow and improve their swimming capabilities, they eventually graduate to the status of nekton. This chapter examines some general aspects of the biology of zooplankton; the nekton will be considered in the following two chapters.

Zooplankton Groups

Zooplankton are represented by temporary **meroplankton,** including larval stages of shallow-water invertebrates and fishes, and a variety of permanent planktonic forms, the **holoplankton.** Meroplankton are concentrated in near-shore neritic provinces over continental shelves and near shallow banks, reefs, and estuaries. Their distribution and abundance are related to the seasonal distribution and productivity cycles of local phytoplankton communities (see chapter 5).

More than 5,000 species of holoplankton have been described from numerous phyla in two kingdoms. Prominent among these are protozoans, cnidarians, ctenophores, mollusks, chaetognaths, crustacean arthropods, and invertebrate chordates, introduced in chapter 6. The microscopic tintinnids, flagellates, and other protists seldom move far under their own power. Because of their very small cell sizes, they have great difficulty trying to overcome the viscous forces between water molecules by swimming. As a result, there is almost no glide in this microscopic world. A cell must swim continuously to move. When it stops working, it stops swimming. Holoplankton employ flotation and buoyancy devices similar to those found in phytoplankton. Because most holoplankton are characteristically small, they increase their frictional resistance to the water by having high surface-area-to-body-volume ratios. A profusion of spines, hairs, wings, and other surface extensions also increases frictional resistance to sinking (figure 12.1).

A few genera of siphonophores (a type of colonial cnidarian) maintain neutral or even positive buoyancy by secreting gases into a float, or **pneumatophore.** *Velella* (sometimes called by-the-wind

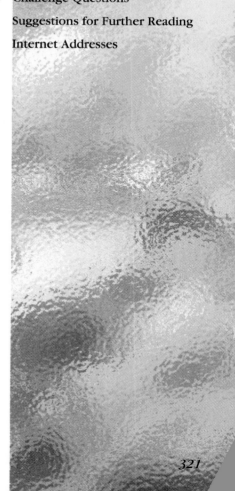

Figure 12.1

Some planktonic copepods exhibiting structural adaptations for flotation: (*a*) *Aegilsthus,* (*b*) *Oithona,* (*c*) side and (*d*) *top views of Sapphirina.*

From H. U. Sverdrup, Martin W. Johnson, and Richard H. Fleming, *The Oceans: Their Physics, Chemistry, and General Biology.* © 1942, renewed 1970. By permission of Prentice Hall, Inc. Upper Saddle River, New Jersey.

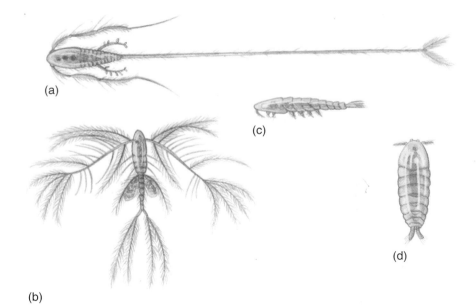

(a)

(c)

(d)

(b)

Figure 12.2

The Portuguese man-of-war, *Physalia,* floating at the sea surface. The trailing tentacles may reach 50 m in length.

Courtesy J. Trent, Scripps Institution of Oceanography.

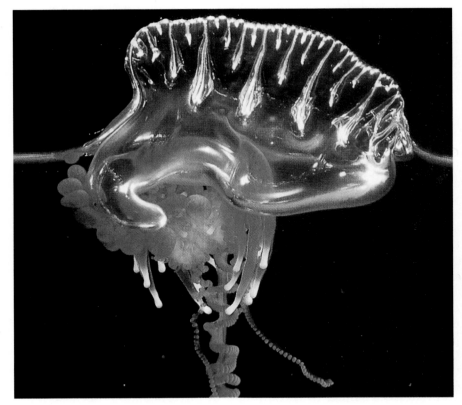

sailor, Jack-by-the-wind, or simply, purple sail) and the larger Portuguese man-of-war, *Physalia* (figure 12.2), have large pneumatophores and float at the sea surface. The pneumatophore, acting as a sail, catches surface breezes and transports the colony long distances. Both *Velella* and *Physalia,* with only one species in each genus, have worldwide distributions.

Figure 12.3

A midwater siphonophore with a
small gas-filled pneumatophore at the
upper end.
Courtesy A. Alldridge.

Other siphonophores with gas floats are neutrally buoyant and
can easily change their vertical position in the water column by swimming (figure 12.3). A gas gland within the pneumatophore secretes gas
into the float. Excess gases are vented through a small pore that is
opened and closed by a sphincter muscle, and the siphonophore's
buoyancy is adjusted accordingly.

Gas is also used for buoyancy by a small planktonic nudibranch,
Glaucus. It produces and stores intestinal gases to offset the weight of
its shell. Another planktonic snail, *Janthina,* forms a cluster of bubbles
at the surface and clings to it. This adaptation is apparently related to
the snail's preference for feeding on the soft parts of *Velella* (which
also floats at the surface). These and other zooplankton adapted to live
at the air-seawater interface are known as **neuston.** (Another example
of a neustonic animal is the water strider shown in figure 1.12.)

A large variety of gelatinous zooplankton exists, including numerous medusae (or jellyfishes), siphonophores, pelagic mollusks (figure
12.4a), ctenophores (figure 12.4b), and tunicates. The tunicates include barrel-shaped salps, with life cycles alternating between solitary
individuals and colonial clones (figure 12.4c), as well as the smaller appendicularians. Because the bodies of all these gelatinous species contain at least 95% water and their body densities are very close to that
of seawater, buoyancy is not a problem. Their small proportion of organic material accounts for their low metabolic rates and for their
body sizes, which range from about 1 mm to several meters. Their relatively large sizes (in comparison with zooplankton) and nearly transparent appearance in water confer some protection from larger
pelagic predators, such as sea turtles.

Crustaceans are the most numerous and widespread species of
holoplankton. Copepods, euphausiids, amphipods, decapods, chaetognaths, salps, and ctenophores (figures 12.1 and 6.30) all contribute
greatly to near-surface plankton communities. Calanoid copepods,

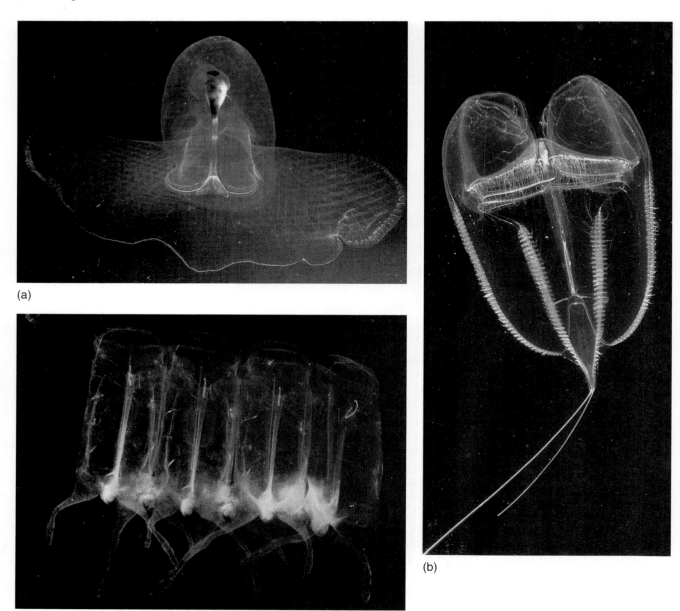

(a)

(b)

(c)

Figure 12.4

Some large gelatinous zooplankton.
(*a*) A pelagic mollusk, *Corolla.* (*b*) A
ctenophore, *Eurhamphea*, swimming
with eight rows of ciliated combs.
(*c*) A short chain of seven salps
(Pegea) cloned from a single parent.
All photos approximately life size.
Courtesy J. Trent, Scripps Institution of
Oceanography.

such as *Calanus,* account for the bulk of herbivorous zooplankton in
the 1 to 5 mm size range. Euphausiids, the giants of the planktonic
crustaceans, seldom exceed 5 cm.

Life cycles of planktonic crustaceans involve several definable stages,
each punctuated by one or more molts of the exoskeleton and some accompanied by structural metamorphosis. Planktonic mysids exhibit a
fairly generalized crustacean life cycle, which includes four distinct
stages: the nauplius, protozoea, zoea, and adult (figure 12.5). These stages
are seen in the development of many other crustaceans, including crabs,
shrimps, and euphausiids (figure 12.5, middle). Planktonic copepods (figure 12.5, bottom) have life cycles that differ somewhat from those of
mysids and euphausiids. Their eggs hatch into nauplius larvae that move

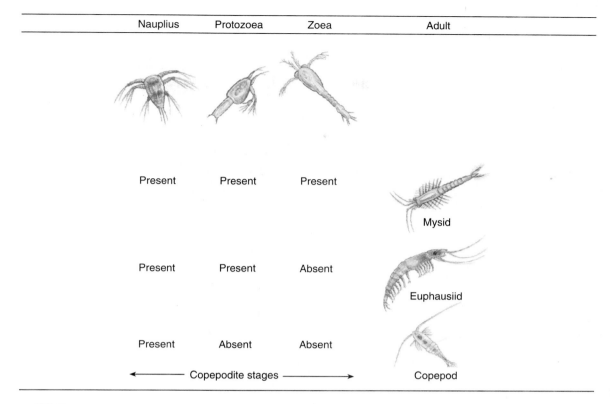

Nauplius	Protozoea	Zoea	Adult
Present	Present	Present	Mysid
Present	Present	Absent	Euphausiid
Present	Absent	Absent	Copepod

← ——— Copepodite stages ——— →

Figure 12.5

Developmental stages of three groups of planktonic crustaceans.

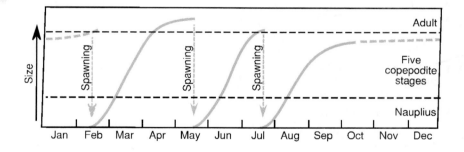

Figure 12.6

Growth and reproductive cycles of a North Atlantic copepod, *Calanus finmarchicus*. The adults of each brood produce eggs for the next brood (arrows). Dashed lines indicate overwintering of copepodites in deep water that experience little growth.

Adapted from Russell, 1935.

with their antennae. After six molts, the larvae enter the copepodite stage, a sexually immature form resembling adults. Five more molts lead to the adult stage; thereafter no more molts occur. The yearly growth and reproductive cycles for a copepod are shown in figure 12.6.

The Pelagic Environment

The pelagic division of the world ocean is a realm that presents few obvious ecological niches for its inhabitants. With little local variation in temperature and chemical characteristics smoothed out by turbulent mixing and diffusive processes, patterns of light intensity, water temperature, and food availability change only on horizontal scales of several kilometers.

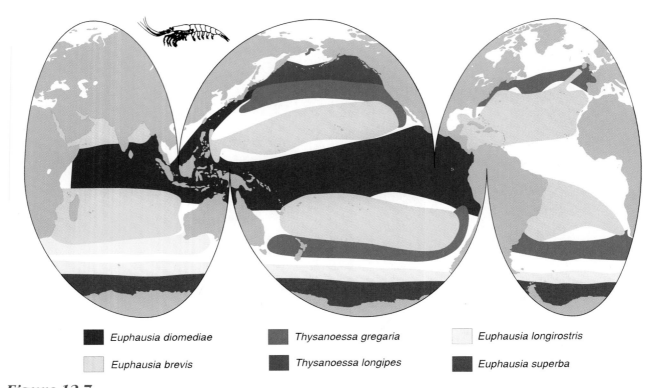

| ■ *Euphausia diomediae* | ■ *Thysanoessa gregaria* | ☐ *Euphausia longirostris* |
| ☐ *Euphausia brevis* | ■ *Thysanoessa longipes* | ■ *Euphausia superba* |

Figure 12.7

The global distribution of six species of epipelagic euphausiids.

Away from the influences of continental borders, zooplankton in the upper 200 m of the world ocean slowly drift along in large, semi-enclosed current gyres (see figure 1.33). This upper layer of the oceanic province, the **epipelagic zone,** is approximately coincident with the photic zone. In marked contrast to the numerous life zones available to animals on the sea bottom, the epipelagic zone can be partitioned into only a few major habitats, each broadly defined by its own unique combination of temperature and salinity characteristics. These major epipelagic habitats reflect the major marine climatic zones shown in figure 1.21. Each habitat is occupied by a suite of zooplankton species that, over a long period of time, have adapted to a special set of environmental conditions.

Figure 12.7 maps the distribution patterns of six species of planktonic euphausiids. These patterns resemble the general large-scale distribution of many other epipelagic animal species. As noted in figure 12.7, there is a tendency for each species to be distributed in a broad latitudinal band across one or more oceans. Well-defined patterns of tropical *(Euphausia diomediae)*, subtropical *(E. brevis)*, and south polar *(E. superba)* distribution are evident. Some species, such as *E. diomediae* and *E. brevis,* are broadly tolerant to their environmental regimes and occupy wide latitudinal bands. Other species *(E. longirostris, Thysanoessa gregaria,* and *E. superba),* occupy narrower latitudinal ranges. Similar regimes in both the Northern and Southern hemispheres are frequently inhabited by the same species. The subtropical *E. brevis* and the temperate-water *T. gregaria* exhibit this **antitropical distribution.** Often these lower-latitude species extend into all three major ocean basins *(E. brevis* and *T. gregaria),* but occasionally they do not *(E. diomediae* is conspicuously absent

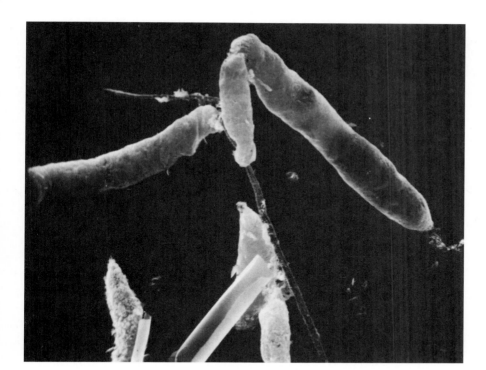

Figure 12.8

A scanning electron micrograph of a copepod fecal pellet containing fragments of phytoplankton cells.
Courtesy P. Azam, Scripps Institution of Oceanography.

from the tropical Atlantic). The distribution of *E. diomediae* suggests that, although comparable latitudes in the Atlantic, Pacific, and Indian oceans provide similar environmental conditions, each ocean still retains significant individual differences.

High-latitude species in the Southern Hemisphere (*E. longirostris* and *E. superba*) also extend around the globe, aided by extensive oceanic connections between Antarctica and the other southern continents. Similar circumglobal distributions are less common in the higher latitudes of the Northern Hemisphere (*T. longipes* is restricted to the North Pacific).

The boundaries of these zones overlap slightly, but analyses of the distribution patterns of numerous other species of zooplankton and nekton confirm that these boundaries define, in a very real way, the major epipelagic habitats of the oceanic province. However, smaller scale variations of environmental features do exist and do influence the structure of pelagic communities by contributing additional texture to the physical-chemical terrain and by creating potential niches for occupation. Zooplankton, like phytoplankton, have patchy distributions at virtually all levels of sampling, from several kilometers down to microscopic distances. Some of this patchiness develops from the local effects of grazing or predation (see figure 5.20), some from responses to chemical and physical gradients, and some from the social responses of planktonic species (such as aggregation or avoidance).

Even finer scale patchiness develops around macroscopic particles. These particles are produced in epipelagic waters by two associated processes that package extremely small, abundant particles into a few larger particles. One of these processes is the production of fecal pellets by herbivorous grazers (figure 12.8). The other process is the incorporation of living and dead material into variously shaped organic aggregates sometimes known as "marine snow." These aggregates, ranging from 1 mm to several millimeters in size, occur in concentrations up to 30,000/m^3 in neritic waters and are composed of living and dead phytoplankton cells, abundant bacteria, exoskeletons shed by crustaceans, and detrital material. (Sometimes these aggregates also include fecal pellets, so the distinction between the two types of particles blurs somewhat.) Both types of particles serve as sites of additional aggregation by other members of the plankton community. Bacteria inoculate the particles and initiate processes of decomposition. Dinophytes exploit the nutrients released by the activity of the bacteria, and grazing herbivores are attracted by the concentration of energy-rich organic material, both living and dead.

Vertical Migration

Zooplankton, by definition, cannot make directed, long-distance horizontal movements. They can, however, experience major environmental changes by vertically moving modest distances (a few tens of meters). Water temperature, light intensity, pressure, and food availability all change markedly as the distance from the sea surface increases.

Below the sunlit waters of the epipelagic zone lies the **mesopelagic zone,** a world where animals live in very dim light and depend on primary production from within the photic zone above. The mesopelagic zone extends from the bottom of the epipelagic zone down to about 1,000 m. Some permanent members of the mesopelagic zone rely totally on the flux of particles from above for food. This downward transport is accelerated by the conversion of dispersed microscopic cells to fecal pellets and organic aggregates. The fecal pellets of calanoid copepods, for instance, sink about 10 times faster than do the individual phytoplankton cells constituting the pellet.

The mesopelagic zone offers some distinct advantages when compared with life nearer the sea surface. Predators find it more difficult to see their prey in the dim light. Decreased water temperatures at mid-depths lower the metabolic rates and the food and oxygen requirements of mesopelagic animals. The cold water, with its increased density and viscosity, also slows the sinking rates of food particles.

Large numbers of mesopelagic animals (including many fish species) periodically migrate upward to feed in near-surface waters. The most common pattern of vertical migration occurs on a daily cycle. At dusk, these midwater animals ascend to the photic zone and feed throughout the night. Before daybreak, they begin migrating to deeper, darker waters to spend the day. The following evening, the pattern is repeated. In Antarctic waters, the daily pattern of vertical migration often breaks down; the animals generally remain in the photic zone during the summer and in deeper waters during the winter.

The general pattern of daily, or diurnal, **vertical migration** has been deduced from numerous sources of information. Net collections of animals from several depths at different times throughout the day have shown that more animals are near the surface at night than during the day (figure 12.9). Direct observations from submersible vehicles support these conclusions.

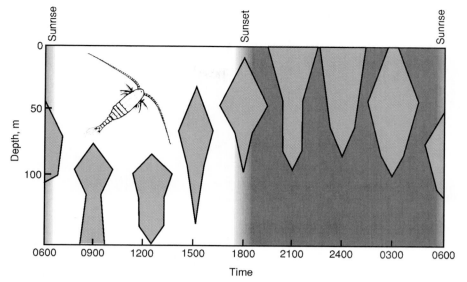

Figure 12.9

A generalized kite diagram of net collections of adult female copepods, *Calanus finmarchicus,* during a complete one-day vertical migration cycle. The relative width of each part of each "kite" represents relative numbers of animals. Night hours are shaded.

Underwater sound pulses from ship-mounted echo sounders have also been used extensively to study the behavior of vertically migrating animals. The pulses are partially reflected by concentrations or layers of midwater animals. These **sound-scattering layers (SSL)** ascend nearly to the surface at dusk and then break up (figure 12.10). At daybreak, the layers reform and descend to their usual daytime depths (200–600 m). Often, three or more distinct layers are discernible over broad oceanic areas.

The composition of the SSL is still an unsettled question. Most inhabitants of the mesopelagic zone are too small or too sparsely distributed to strongly reflect sound signals. Net tows and observations from manned submersibles suggest that there are three groups of animals that cause the deep scattering layers: euphausiids, small fishes (primarily lantern fishes; figure 12.11), and siphonophores (figure 12.3). Euphausiids and small fishes are often abundant members at depths where the SSLs occur. The strong echoes of sound pulses may be due, in part, to the resonating qualities of the gas-filled swim bladders of lantern fishes and air floats of siphonophores. Relatively few swim bladders or air floats are necessary to produce strong echoes at certain sound frequencies.

Whatever the composition of the SSLs, they are merely sound-reflecting indicators of a much more extensive vertically migrating assemblage of animals not detected by echolocation. Undoubtedly, members of the SSLs graze on smaller vertical migrators and are, in turn, preyed upon by larger fishes and squids.

These vertical migrations, only a few hundred meters in extent, occur in a short period of time. The copepod *Calanus,* for example, is only a few millimeters long, yet it swims upward at 15 m/h and descends at 50 m/h. Larger euphausiids (2 cm long) move in excess of 100 m/h. If diurnal vertical migrations are foraging trips from below into the productive photic zone, why do these animals descend after feeding? Why don't they remain in the photic zone? Many explanations for the adaptive value of vertical migration have been offered.

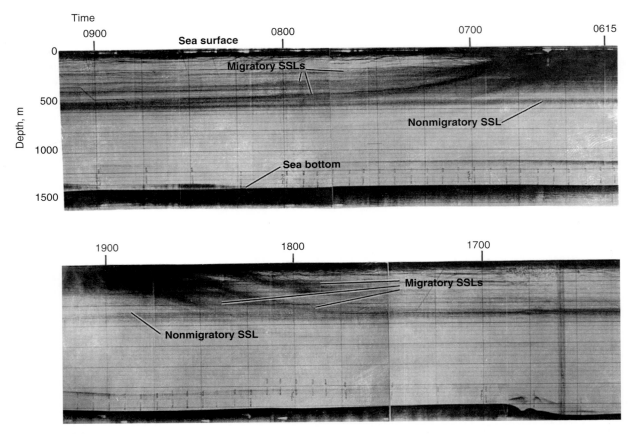

Figure 12.10

A SONAR record of several distinct sound-scattering layers at a station near the Canary Islands. Except for slight drift corrections, the recording ship remained stationary. The time scale reads from right to left. Multiple layers begin migrating downward at daylight (0700 hours), remain near 500 m throughout the day, then migrate upward in the evening (1800–1900 hours).
Courtesy E. Kampa, Scripps Institution of Oceanography and the National Institute of Oceanography.

Figure 12.11

A midwater lantern fish, *Bolinichthys.* Small white dots are light-producing photophores.

One explanation is that diurnal vertical migration enables animals to capitalize on the more abundant food resources of the photic zone in the dark of night and to escape visual detection by predators in the refuge of the dimly lit mesopelagic zone during the day. But vertical migration is useful in other ways. Lower water temperatures at the deeper depths reduce an animal's metabolic rate and its energy requirements. The energy conserved may be sufficient to offset the lack of food during the day and the energy expenditures incurred during the actual migration. Each of these explanations offers plausible

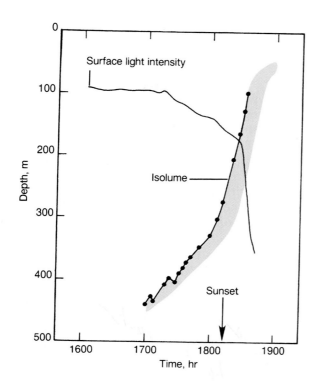

Figure 12.12

The upward migration of a sound-scattering layer (colored portion of the graph) at sunset, November 4, near the Canary Islands. Note the very close correspondence between the isolume and the top of the scattering layer, with both rising as the surface light intensity diminishes. Redrawn from Boden and Kampa 1967.

mechanisms favoring the selection of individuals possessing the genetic information required to accomplish vertical migration.

Vertical migrations of zooplankton and small nekton also can occur on seasonal time scales. While in the late copepodite stage, *Calanus finmarchicus* spends the winter months of low primary productivity in the North Atlantic at depths near 1,000 m. When the spring diatom bloom develops, it molts to the adult form and begins to rise into the photic zone. In polar seas, diurnal vertical migration is often completely suppressed during summer months of continuous daylight and high plant productivity. Major herbivores, such as *Euphausia superba,* rise from winter depths of 250–500 m to the surface in late spring. In autumn, they descend once more to their wintertime depths.

Daily or seasonal changes in light intensity seem to be the most likely stimulus for vertical migrations. Large-scale experiments with mixed coastal zooplankton populations have demonstrated that, under constant light and temperature conditions, some species of copepods maintained their diurnal migratory behavior as a circadian rhythm for several days without relying on external cues such as light intensity. Other species did not migrate and apparently required light or another external stimulus to induce vertical migration. Electric lights lowered into the water at night and bright moonlight can drive the SSL downward from 60 m to 300 m. Solar eclipses cause the opposite to occur; vertical migrators move toward the surface during an eclipse, even at midday.

These responses demonstrate a sensitivity to light intensity by natural populations of vertical migrators. Each SSL follows an isolume (a constant light intensity) characteristic of the top of the layer at its normal daytime depth (figure 12.12). As the sunlight intensity decreases

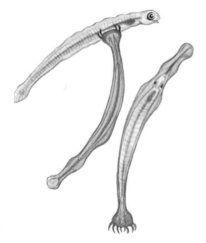

Figure 12.13

A small chaetognath, *Sagitta,* capturing and consuming a fish larva its own size.

Figure 12.14

(*a*) A scanning electron micrograph of the thorax and filtering mechanism of *Calanus,* shown in side view. (*b*) Higher magnification of *Calanus,* showing the ventral view of the two mandibles (far left) and the filtering basket formed by the second maxillae.
Courtesy J. R. Strickler.

in late afternoon, the isolume moves toward the sea surface, and the SSL follows with a precision seldom seen in natural populations.

Feeding

Zooplankton employ all conceivable methods of capturing food, from suspension feeding of bacteria and phytoplankton to direct predation on other zooplankton and small nekton. The small, soft-bodied chaetognaths, a small phylum with about 50 species, are voracious predators of other zooplankton. They are found throughout the world ocean in numbers sufficient to affect whole broods of young fishes (figure 12.13). In addition to their significant role in pelagic food webs, some species of chaetognaths are well-known biological indicators of distinctive types of surface ocean water.

Copepods also are common prey of chaetognaths. All adult calanoid copepods species are similar in body form and general feeding behavior, suggesting a very successful functional form. Copepods and other small pelagic particle grazers are typically exposed to a wide spectrum of food particle sizes. Food particles range from abundant minute bacteria through the common types of phytoplankton to organic aggregates and large centric diatoms. This size spectrum presents an opportunity for small, versatile particle grazers to adopt a feeding strategy that involves selectivity of optimal-sized food items.

Although copepods prey on large phytoplankton cells when they are available, calanoid species can capture the smaller, more abundant microplankton with a basketlike filtering mechanism derived from their complex feathery feeding appendages (figure 12.14). The hairlike setae on the appendages of *Calanus* are fine enough to retain food particles larger than 10 μm. Laboratory observations have revealed that food particles are carried into the filter basket by currents generated by the feeding appendages and the five posteriorly positioned

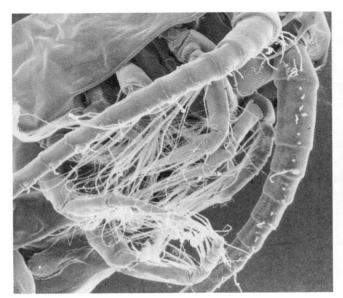

(a)

(b)

pairs of thoracic swimming legs. Special long setae on the feeding appendages remove the trapped particles and direct them to the mouth. With this filtering mechanism, *Calanus* species are capable of exploiting a wide size range of food particles.

Even though *Calanus* and similar copepods can shift rapidly from one food particle size to the other, they prefer larger food particles to smaller ones. Studies using high-speed photomicrography techniques have shown that calanoid copepods efficiently capture sparsely scattered, single, large food items such as protozoans, small fish eggs, and large diatoms. As the copepods' feeding actions drive them forward in the water, their antennae extend laterally to function as an array of flow sensors that detect minute disturbances surrounding larger food items (figure 12.15). If the food particle is detected near the end of an antenna, the animal quickly adjusts its swimming direction to bring the particle within reach of an extended feeding appendage. The mouthparts then seize and manipulate the particles before eating them. Because filtering and large-particle seizure cannot operate simultaneously, this mode of feeding is interrupted when the copepods are filtering small particles.

Such particle-size selectivity by copepods is likely an important factor in stabilizing phytoplankton populations. When phytoplankton populations of a particular cell size become more abundant through growth and reproduction, they attract increased grazing pressures as more copepods shift feeding strategies to concentrate on them. It is unlikely, however, that the phytoplankton population would be grazed to extinction. Several species exhibit ingestion rates that are dependent on the concentration of phytoplankton cells (figure 12.16). For a certain food particle size, ingestion rates increase with increasing particle density to some critical maximum. Beyond the maximum particle density, some aspect of the copepod's food-processing system appears to become saturated, and no further increase in ingestion rates occurs. Conversely, ingestion rates decrease with decreasing phytoplankton densities; copepods most likely will shift to another nearly optimal concentration of food particles before the first population is exhausted completely.

In contrast to the rigid filter devices of crustaceans, many gelatinous herbivores rely on nets or webs of mucus to ensnare food particles. One highly evolved example is *Gleba*, a planktonic pteropod mollusk. When feeding, *Gleba* secretes a mucous web that often exceeds 2 m in diameter. The free-floating web spreads horizontally as it

Figure 12.15

Copepod detection and capture behavior of individual diatoms (green). Sensory cells arranged in arrays on large antennae provide information to detect prey and guide the response until the capture is made (right).

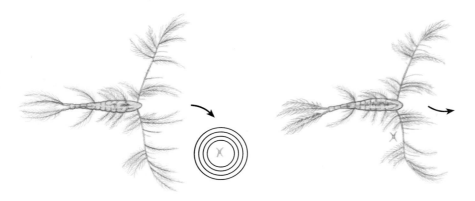

Figure 12.16

The ingestion rate of a copepod, *Calanus,* as a function of the concentration of its food (in this case, the diatom *Thalassiosira,* shown in figure 3.9). The ingestion rate peaks near 3,000 diatom cells per milliliter, and no further increase is seen even at much higher concentrations.
Redrawn from Frost 1972.

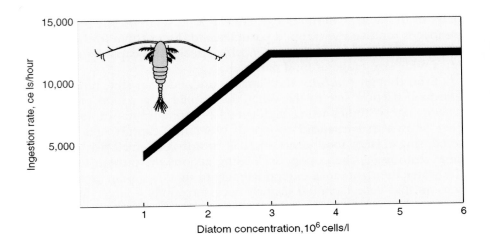

Figure 12.17

The appendicularian *Oikopleura,* within its mucous bubble. Arrows indicate path of water flow.

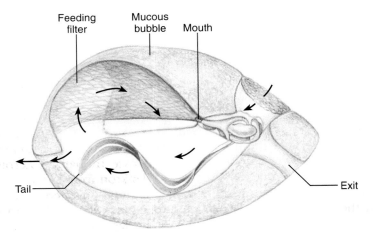

is produced, maintaining a single point of attachment at the mouth. As the animal and web slowly sink, bacteria and small phytoplankton become trapped in the mucus. The web, with its load of food, is then formed into a mucous string, directed to the mouth, and ingested. *Gleba* then swims upward to repeat the behavior.

Another elaborate mucous feeding system is found in appendicularians, small, tadpole-shaped invertebrate chordates. Most appendicularians live enclosed within delicate, transparent mucous bubbles (figure 12.17). Food-laden water, pumped by the tail beat of the occupant, enters the bubble through openings at one end. These openings are screened with fine-meshed grills to exclude large phytoplankton cells. Smaller cells enter the bubble and are trapped on a complex internal mucous feeding filter. Every few seconds, the animal sucks the particles off the filter and into its mouth. When the incurrent filter becomes clogged or the interior is fouled with feces, the entire bubble is abandoned and a new one is constructed, sometimes in as little as 10 minutes.

With this feeding mechanism, even bacteria-sized particles can be harvested with high efficiency, for these animals achieve filtering rates several times higher than those of calanoid copepods. The larger salps are even better; a small chain of colonial salps (such as the seven

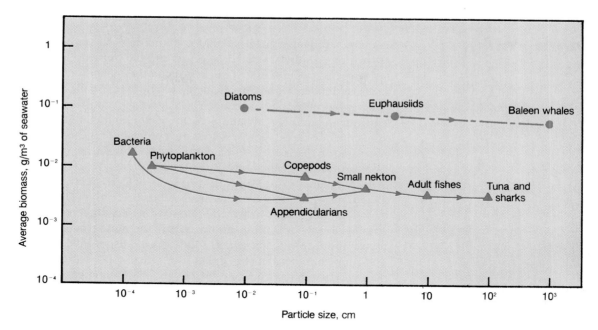

shown in figure 12.4c) is estimated to be capable of filtering more water than 3,000 copepods could in the same amount of time.

The mechanisms employed by zooplankton to glean small, diffuse food particles from the water reflect the crucial roles these animals play in pelagic food webs. Planktonic tunicates are the largest herbivorous grazers to exploit the very small phytoplankton found in tropical and subtropical areas of low productivity. Because their filters clog quickly when they encounter the high phytoplankton densities of upwelling areas or temperate spring blooms, they are less successful competitors in these areas. It is in these regions of high productivity that planktonic crustaceans, particularly the calanoid copepods and euphausiids, thrive.

Two generalized food webs are shown in figure 12.18 to illustrate some fundamental differences in trophic linkages between oceanic regions of high and low productivity. Food webs of subtropical waters are low in biomass, begin with very small phytoplankton, and include numerous trophic levels; leading to tuna-sized predators. The Antarctic upwelling system, in contrast, is characterized by relatively large (though still microscopic) primary producers, larger filter-feeding herbivores, and fewer trophic levels; this system produces the largest carnivores on the earth, the baleen whales.

In the productive waters of the Antarctic (which correspond to the area occupied by *Euphausia superba* in figure 12.7), the average biomass at all trophic levels is approximately 10 times as great as that in subtropical gyres. Within each region, the biomass is about the same at each trophic level. In other words, during summer, a square kilometer of Antarctic ocean surface includes as much diatom material as it does baleen whale biomass. The difference in life span, from hours in diatoms to decades in whales, permits the whales to compensate for their relatively small population with very large individual body size. These pelagic animals of higher trophic levels are the subjects of the next two chapters.

Figure 12.18

The relationship between food particle size and biomass in two pelagic food webs. Note that the biomass in the Antarctic (dashed lines) is about 10 times as high at all trophic levels as those in subtropical gyres (solid lines) and that the biomass of each food web is approximately the same at all trophic levels.

Redrawn from Steele 1980.

Summary Points

- Permanent members of zooplankton communities are drawn from several animal phyla. Small crustaceans are the most abundant, with fewer species of medusae, tunicates, and other gelatinous forms. All live in a vast fluid environment characterized by latitudinally defined geographical life zones.
- Vertically, conditions of light intensity, water temperature, and food availability decrease rapidly below the epipelagic zone. Herbivorous zooplankton, like the phytoplankton they consume, are characterized by large- and small-scale patchy distribution patterns.
- Many zooplankton and small nekton make daily round-trip vertical migrations between the productive waters near the surface and the darker, cooler waters below the photic zone. Feeding mechanisms employed by zooplankton are varied, but the most common mechanisms utilize filters or mucous nets to collect small, dispersed food particles.
- The zooplankton's ability to collect very small food particles and package them into larger ones places zooplankton in critical positions in marine food webs.

Review Questions

1. Describe mechanisms that establish and maintain patchy distributions of zooplankton.
2. Describe the role of the sound-scattering layer (SSL) in the exchange of nutrients and energy between the epipelagic zone and deeper water masses.
3. List three types of animals commonly found in the SSL.
4. Describe two plausible benefits of daily vertical migration for members of the SSL.

Challenge Questions

1. Describe how temperature and light intensity change along the sea surface from the equator poleward. From the sea surface at the equator downward to 1,000 m.
2. Discuss what competitive advantages gelatinous bodies might confer on zooplankton such as medusae and tunicates.

Suggestions for Further Reading

Books

Briggs, J.C. 1974. *Marine zoogeography.* New York: McGraw-Hill.

Hardy, A. 1971. *The open sea: Its natural history. Part I: The world of plankton.* Boston: Houghton Mifflin.

Kerfoot, W. C., ed. 1981. *Evolution and ecology of zooplankton communities.* Halstead, NH: University Press of New England.

Marshall, S. M., and A. P. Orr. 1972. *The biology of a marine copepod.* New York: Springer-Verlag.

Steele, J. S., ed. 1978. *Spatial pattern in plankton communities.* New York: Plenum Press.

Wickstead, J. H. 1976. *Marine zooplankton.* London: E. Arnold.

Articles

Allan, J. D. 1976. Life history patterns in zooplankton. *American Naturalist* 110:165–80.

Alldredge, A. L., and L. P. Madin. 1982. Pelagic tunicates: Unique herbivores in the marine plankton. *Bioscience* 32:655–63.

Boyd, C. M. 1976. Selection of particle sizes by filter-feeding copepods: A plea for reason. *Limnology and Oceanography* 21:175–79.

Dietz, R. S. 1962. The sea's deep scattering layers. *Scientific American* (August) 207:44–50.

Isaacs, J. D. 1977. The life of the open sea. *Nature* 267:778–85.

Lam, R. K., and B. W. Frost. 1976. Model of copepod filtering response to changes in size and concentration of food. *Limnology and Oceanography* 21:490–500.

Porter, K. G., and J. W. Porter. 1979. Bioluminescence in marine plankton: A coevolved antipredation system. *American Naturalist* 114:458–61.

Rubenstein, D. I., and M. A. R. Koehl. 1977. The mechanisms of filter feeding: Some theoretical considerations. *American Naturalist* 111:981–94.

Sheldon, R. W., A. Prakash, and W. H. Sutcliffe Jr. 1972. The size distribution of particles in the ocean. *Limnology and Oceanography* 17:327–40.

Silver, M. W., A. L. Shanks, and J. D. Trent. 1978. Marine snow: Microplankton habitat and source of small-scale patchiness in pelagic populations. *Science* 201:371–73.

Steele, J. 1980. Patterns in plankton. *Oceanus* 25(3):3–8.

Stoecker, D. K., and J. M. Capuzzo. 1990. Predation on protozoa: Its importance to zooplankton. *Journal of Plankton Research* 12:891–908.

Strickler, J. R. 1985. Feeding currents in calanoids: Two new hypotheses. In: *Physiological adaptations of marine animals,* M S. Laverack, ed. Symposium, *Society of Experimental Biology* 39:459–85.

Turner, J. T., P. A. Tester, and W. F. Hettler. 1985. Zooplankton feeding ecology. *Marine Biology* 90:1–8.

Zaret, T. M., and J. S. Suffern. 1976. Vertical migration in zooplankton as a predator avoidance mechanism. *Limnology and Oceanography* 21:804–13.

Internet Addresses

Marine Ecology
http://darter.ocps.k12.fl.us/classroom/klenk/Ecology.html
Abiotic factors and biotic interactions, zonation, feeding relationships, energy transfer and population cycles. Ted Klenk (Apopka High School, Apopka, FL). Apopka High Marine Science
http://darter.ocps.k12.fl.us/classroom/klenk/index.html

Zooplankton
http://bonita.mbnms.nos.noaa.gov/sitechar/pelagic4.html
Good discussion with photos of zooplankton and comparisons between shallow and deep-water zooplankton. Monterey Bay National Marine Sanctuary. Monterey Bay National Marine Sanctuary Site Characterization http://bonita.mbnms.nos.noaa.gov/sitechar/index.html

Copepods
http://www.mov.vic.gov.au/crust/copbiol.html
Brief description and drawings of anatomy, biology and distribution. Joanne Taylor & Gary Poore, Museum of Victoria. Marine Crustaceans of Southern Australia
http://www.mov.vic.gov.au/crust/page1a.html

Paul Yancey's Deep-Sea Pages
http://www.bmi.net/yancey/
Photos and material about deep-sea marine biology. Paul H. Yancey, Whitman College.

Jellyfish
http://www.ios.bc.ca/ios/plankton/ios_tour/zoop_lab/jelly.html
Several photos and descriptions. Watch a short QuickTime or MPEG movie of a Jellyfish in action. Stephen Romaine.

Nekton—Distribution, Locomotion, and Feeding

More than 5,000 species of nektonic animals roam the pelagic province of the world ocean. These animals represent some of the taxonomic groups that have achieved the large body sizes and great swimming powers needed to exploit the pelagic realm of the world ocean. Absolute body size is crucial; once a well-muscled animal exceeds a few centimeters in body length, the viscous forces of water that limit continuous swimming by zooplankton begin to diminish, and more efficient patterns of swimming are possible.

Most nektonic animals are vertebrates, and most vertebrates are fishes. Thus, this chapter is, to a large extent, a chapter about fishes. Of the numerous groups of marine invertebrates that live in the sea, only the squids and a few species of shrimps are truly nektonic. In some regions, vast numbers of small squids (less than 1 m in length) form important intermediate links in epipelagic food webs. The giant squid (*Architeuthis*) lives at greater depths. These large animals occasionally reach 18 m in length and weigh a ton. Various parts of the body are scaled proportionally, with tentacles approximately 10 m long and eyes as large as soccer balls.

Vertical Distribution of Nekton

Even though two-thirds of the ocean's volume lies below the epipelagic and mesopelagic zones, the majority of nektonic species are found in those zones. Considerably less is known about the biology of mesopelagic animals than about those animals living in the epipelagic zone, and less still is known about animals living below 1,000 m. Most epipelagic nekton are carnivorous predators of the higher trophic levels of pelagic food webs. They are typically large in size when compared with zooplankton, are effective swimmers, and have a variety of well-developed sensory capabilities for prey detection, orientation, and navigation. Some accomplish impressive feats of migration to locate food or to improve their chances of successful reproduction.

Epipelagic animals of the open ocean seldom exhibit the bright coloration so common in coral reef fishes and invertebrates. Instead, **countershading** is a common pattern of coloration (figure 13.1). Many abundant fishes, whales, and squids have dark, often green or blue, pigmentation on their dorsal surfaces, and they have silvery or white pigmentation on their ventral surfaces. When viewed from above, the pigmented upper surfaces of countershaded fishes blend with the darker background below. From beneath, the silvery undersides are difficult to distinguish from the ambient light coming from

Vertically oriented light-reflecting
crystals in skin scales reflect
surface light downward to make
fish appear dark from above.

Crystals reflect light sideways to balance
horizontal light blocked by fish's body.

Crystals reflect surface light downward
to make fish appear light from below.

Figure 13.1

Patterns of light reflection from the
skin of a countershaded fish.

the sea surface. From either view, these fishes tend to blend visually into, rather than stand out against, their watery background. Not only does countershading protect animals against predators, but the flashing of silvery bellies or dark backs during abrupt turns may alert individuals in the school to the maneuvers of their immediate neighbors.

Fishes living in the mesopelagic zone are typically much smaller than fishes of the epipelagic zone. This group includes lantern fishes (figure 12.11) and many other vertical migrators. Mesopelagic fishes seldom exceed 10 cm in length and many are equipped with well-developed teeth and large mouths (figure 13.2a). Because only dim light penetrates from above, many species have evolved large, light-sensitive eyes (figure 13.2b) to detect prey and predators alike. Regardless of their color at the sea surface, they appear uniformly black at these depths.

Correlated with large eyes is the presence of **photophores,** light-producing organs most commonly arranged on the ventral surface of the body (figures 12.11 and 13.2). The position and arrangement of photophores suggest two likely functions. The light produced by the ventral photophores approximates the intensity of the background light found at the normal daytime depths of these fishes. The light from the photophores may disrupt the visual silhouette of the fish when observed from below and causes the fish's silhouette to visually blend with the background light from above. The effect of the photophores is similar to countershading in near-surface fishes. Elaborate arrangements of photophores are unique to single species and suggest that photophores are also used for species identification. With little to be seen at these depths except the pattern of photophores, appropriate mate selection may depend on the existence of species-specific patterns of photophores.

Below the mesopelagic zone, light from the surface is so dim that it cannot be detected with human eyes, and it does not stimulate the visual systems of deep-sea fishes. The light seen at depths below 1,000 m comes largely from photophores. At these depths, photophores may be employed as lures for prey, as species recognition signals, and possibly even as lanterns to illuminate small patches of

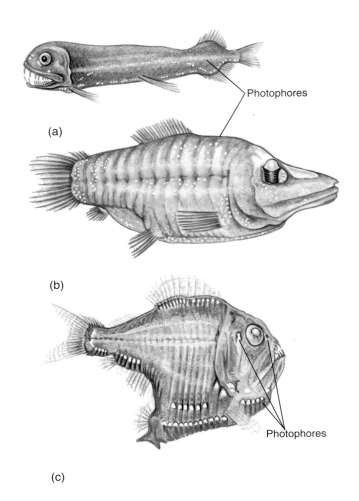

Figure 13.2

Some mesopelagic fishes:
(*a*) loosejaw, *Aristostomias;*
(*b*) spookfish, *Opisthoproctus;*
(*c*) hatchetfish, *Argyropelecus.* All
are less than 5 cm in length.

the surrounding blackness. Most fishes found at these depths are not
vertical migrators. Instead, some depend on the unpredictable sinking of
food particles from the more heavily populated waters above. These
fishes are typically small and have flabby, soft, nearly transparent flesh
supported by very thin bones (figure 13.3). Others feed on mesopelagic
fishes, often engulfing fishes that are nearly their own size.

Getting Oxygen

Marine vertebrates are characteristically active animals. Their activity is
fueled by the oxidation of lipids and other energy-rich foods. The oxygen used in the mitochondria of active cells necessarily either comes
from the atmosphere (for tetrapods) or is dissolved in seawater (for
fishes). For air-breathing tetrapods that swim below the sea surface, air
is rich in O_2 (21%) but is available only when they return to the surface to breathe. Fishes, on the other hand, use their gills to extract constantly accessible O_2 from seawater, but water has a very low solubility
for O_2 (just a few parts per million).

However they acquire their O_2, nearly all vertebrates use hemoglobin to store and transport it in the blood. Hemoglobin is found in
other animal phyla (including several inhabitants of deep-sea, hot
vent communities (figure 11.12), but only vertebrates package their

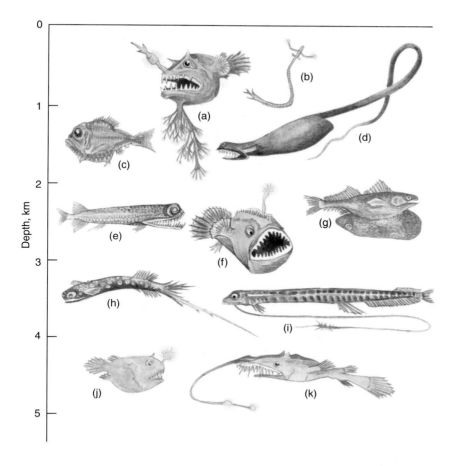

Figure 13.3

A few fishes of the deep sea, shown at their usual depths. Most have reduced bodies, large mouths, and lures to attract prey. (*a*) An angler, *Linophyrne*. (*b*) Young *Idiacanthus* with eyes on stalks. (*c*) A hatchetfish, *Argyropelecus*. (*d*) A gulper, *Saccopharynx*, (*e*) A widemouth, *Malacosteus*. (*f*) Another angler, *Melanocetus*. (*g*) The "great swallower," *Chiasmodon,* with a larger fish in its stomach. (*h*) A giant tail, *Gigantura*. Three more anglers, (*i*) *Eustomias*, (*j*) *Borophyrne*, (*k*) *Lasiognathus*.

hemoglobin inside red blood cells. Hemoglobin is a pH-sensitive protein that has a high chemical affinity for O_2. Each hemoglobin molecule can bind with up to four O_2 molecules. In marine vertebrates, it matters little whether O_2 is obtained with gills or with lungs; the function of hemoglobin is essentially the same.

Breath-Hold Diving in Marine Mammals

All marine tetrapods (reptiles, birds, and mammals) breathe air to obtain O_2. The length of time that they can spend under water is determined by their capacity for storing O_2 and by their metabolic rates. Marine reptiles are poikilothermic, and their relatively low body temperatures drive correspondingly low metabolic rates. The homeothermic birds and mammals have much higher metabolic rates and activity levels that require more O_2 to support. The mechanisms for prolonged breath-hold diving in birds and mammals are similar, and only marine mammals are emphasized in the following discussion.

Aristotle recognized more than 20 centuries ago, that dolphins were air-breathing mammals. Yet, it was not until the classic studies conducted by Irving and Scholander nearly halfway into the twentieth century that the physiological basis for the deep and prolonged breath-holding dives by marine birds and mammals basis was defined. The diving capabilities of the two groups vary immensely. Some are

Table 13.1
Diving and Breath-Holding Capabilities of Humans and a Few Marine Mammals

Animal	Maximal Depth (m)	Maximal Time of Breath-Hold (min)	Resting Breathing Rate (breaths/min)
Human (Homo)	66.5	6	15
Dolphin (Tursiops)	305	6	2–3
Sea lion (Zalophus)	168	30	6
Fin whale (Balaenoptera)	500	30	1–2
Weddell seal (Leptonychotes)	600	75	?
Elephant seal (Mirounga)			
Female	1,250	120	?
Male	1,530	77	?
Sperm whale (Physeter)	2,250	90	?

Compiled from Kooyman and Andersen, in Andersen 1969; Norris and Harvey 1972; Kooyman et al. 1981; Delong and Stewart 1991.

little better than the Ama pearl divers of Japan who, without the aid of supplementary air supplies, repeatedly dive to 30 m and remain under water for 30 to 60 seconds. The maximal free-diving depth for humans is about 60 m; breath-holds lasting as long as six minutes have been independently achieved, although not while diving. But even the best efforts of humans pale in comparison to the spectacular dives of some whales and pinnipeds (table 13.1). With dive times often exceeding 30 minutes, these exceptional divers are no longer closely tied to the surface by their need for air.

When diving, marine tetrapods experience a triad of worsening physiological conditions: Stored O_2 is diminished while CO_2 and lactic acid become more concentrated at the very time that activity is increasing.

Several respiratory adjustments are necessary to achieve prolonged dives such as those listed in table 13.1. As the last column of the table indicates, breathing rates of marine mammals are decidedly lower than those of humans and other terrestrial mammals. The pattern of breathing is also quite different. In general, marine mammals exhale and inhale very rapidly, even when resting at the sea surface, then hold their breaths for prolonged periods before exhaling again. Smaller dolphins, for instance, exhale and inhale in a fraction of a second, then hold their breaths for 20 to 30 seconds before repeating the pattern. Even the larger baleen whales can empty their lungs of 1,500 liters of air and refill them in as little as two seconds. In the larger species of whales, dives of several minutes' duration are commonly followed by several blows 20 to 30 seconds apart before another prolonged dive is attempted. This **apneustic breathing** pattern (figure 13.4) is also exhibited by pinnipeds both in and out of the water.

Extensive elastic tissue in the lungs and diaphragms of these animals is stretched during inspiration and recoils during expiration to rapidly and nearly completely empty the lungs. Apneustic breathing provides time for the lungs to extract additional O_2 from the air held in the lungs. Dolphins can remove nearly 90% of the O_2 contained in each breath. (Humans use only about 20% of the O_2 inspired.) Oxygen uptake within the **alveoli** (air sacs) of the lungs may be enhanced as lung air is moved into contact with the walls of the alveoli by the

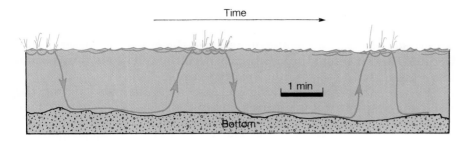

Figure 13.4

Apneustic breathing pattern of a gray whale, observed while the whale was feeding. Blows at the surface represent individual breaths.

Figure 13.5

A self-portrait of Tuffy, a bottle-nosed dolphin, taken at a depth of 300 m. The water pressure at that depth caused the thoracic collapse apparent behind the left flipper. From Sam H. Ridgeway, *Mammals of the Sea, Biology and Medicine,* 1972.

Courtesy S. Ridgeway.

kneading action of small muscles scattered throughout the lungs. In some species, an extra capillary bed surrounds each alveolus and may also contribute to the exceptionally high uptake of O_2.

Each of these features may seem insignificant by itself, but taken together they represent a style of breathing that permits marine mammals increased freedom to explore and exploit their environment some distance from the sea surface. Still, apneustic breathing alone cannot explain how some seals and whales are capable of achieving extremely long dive times.

Cetaceans typically dive with full lungs, whereas pinnipeds often exhale before diving. These differences suggest that the volume of lung air at the beginning of a dive is adjusted to achieve neutral buoyancy and is of little value in supplying O_2 during a dive. Moreover, the lungs and their protective rib cage are sufficiently resilient to enable the lungs to collapse as the water pressure increases with depth (figure 13.5). For a dive from the sea surface to 10 m, the external pressure is doubled, causing the air volume of the lungs to be compressed by half and the air pressure within the lungs to double.

Complete lung collapse probably occurs in the upper 100 m; any air remaining in the lungs below that depth is squeezed by increasing water pressure out of the alveoli and into the larger air passages (the **bronchi** and **trachea,** or windpipe). Even the trachea is flexible and undergoes partial collapse during deep dives.

By tolerating complete lung collapse, these animals sidestep the need for respiratory structures capable of resisting the extreme water pressures experienced during deep dives (over 200 atm for a sperm whale at 2,250 m). And they receive an additional bonus. As the air is forced out of their collapsing alveoli during a dive, the compressed air still within the larger air passages is blocked from contact with the walls of the alveoli. Consequently, little of these compressed gases are absorbed by the blood, and marine mammals avoid the serious diving problems (decompression sickness and nitrogen narcosis) sometimes experienced by humans when they breathe compressed air at moderate depths while under water. After prolonged breathing of air under pressure (with hard hat or SCUBA gear), large quantities of compressed lung gases (particularly N_2) are absorbed by the blood and distributed to the body. As the external water pressure decreases during rapid ascents to the surface, these excess gases are frequently not discharged quickly enough by the lungs. Instead, they form bubbles in the body tissues and blood, causing excruciating pain, paralysis, or even death. Excess N_2 dissolved in the blood also has a narcotic effect on human divers and seriously restricts the time within which they can function effectively at depth. Deep-diving marine mammals avoid both of these problems simply because the air within their lungs is forced away from the walls of the alveoli as the lungs collapse during a dive, thereby preventing excess N_2 from diffusing into the blood.

Because the collapsed lungs of deep-diving marine mammals are not effective stores for O_2, it must be stored elsewhere in the body or its use must be seriously curtailed during a prolonged dive. Both options are exercised by diving mammals. Additional stores of O_2 are maintained in chemical combination with hemoglobin of the blood or with myoglobin in muscle cells. Red blood cells (which contain the hemoglobin) are about the same diameter in diving mammals and nondiving mammals; however, there are more red blood cells in diving mammals, and each cell tends to be somewhat inflated by its extra load of hemoglobin. The blood volume of diving mammals is also significantly higher than that of nondiving mammals. About 21% of the total body weight of sperm whales, for instance, is blood. Much of the additional blood volume is accommodated in an extensive network of capillaries, a **rete mirabilia,** such as the extensive rete found along the dorsal side of the thoracic cavity (figure 13.6). The **vena cava** (the major vein returning blood to the heart) in some species is baglike and elastic. In the elephant seal, it alone can accommodate 20% of the animal's total blood volume. These features all contribute to the total reserve of stored O_2 for use during a dive.

The swimming muscles of marine mammals are highly tolerant to anaerobic conditions during a dive, so they and other nonessential organs (such as the kidneys and digestive tract) may be deprived of the reserve O_2 stored in the blood. The arteries leading to these peripheral

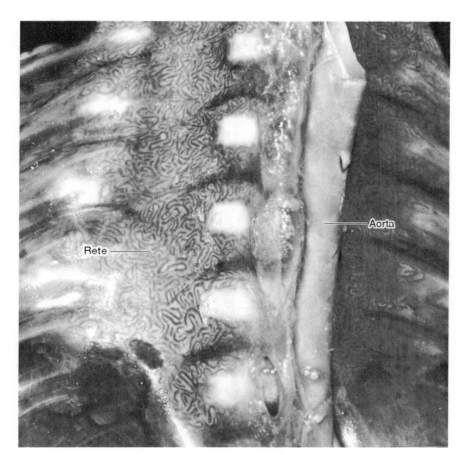

Figure 13.6

The right thoracic rete mirabilia of a small dolphin, *Stenella*.

muscles and organs constrict, and most of the circulating blood is shunted to a few vital organs, primarily the heart and brain. Simultaneously, the heart rate slows dramatically to accommodate pressure changes in a much-reduced circulatory system comprising the heart, the brain, and connecting blood vessels. Other circulatory structures also help to smooth out and moderate fluctuations in the pressure of blood going to the brain. An elastic bulbous "natural aneurism" in the **aorta** (the large artery leaving the heart) and another rete in the smaller arteries at the base of the brain both help to dampen blood pressure surges each time the heart beats.

Bradycardia (the marked slowing of the heart rate that accompanies a dive) probably occurs in all diving vertebrates, including birds, reptiles, and mammals. Even grunion (see figure 14.18) experience bradycardia when they come out of the water to spawn and are deprived of a continuous supply of O_2. The intensity of bradycardia varies widely among marine mammal groups. During experimental dives in laboratory conditions, heart rates of restrained cetaceans are reduced to 20 to 50% of their predive rates. Many seals in similar conditions drop their resting heart rates of 100 to 150 beats/min to 10 beats/min when diving. The triggering mechanism for bradycardia is not completely understood, but it seems to involve sensors in the face and possibly in the respiratory system. The combined

Figure 13.7

Weddell seal, *Leptonychotes.*
Courtesy G. Kooyman, Scripps Institution of
Oceanography.

Table 13.2
A Summary of Dive Responses of Weddell Seals

1. Cessation of breathing
2. Variable bradycardia depending on dive duration
3. Variable peripheral and central vasoconstriction
4. Reduced aerobic metabolism in most organs
5. Rapid depletion of muscle O_2
6. Lactic acid accumulation in muscles after 20 minutes
7. Variety of blood chemistry changes during and immediately after dive, depending on dive duration
8. Voluntary reduction of core body temperature

response of bradycardia and peripheral circulation shutdown has been referred to as the **dive reflex.**

Recent studies of Weddell seals (figure 13.7) in Antarctic waters suggest a very different picture of the diving responses of unrestrained mammals in their natural habitat. G. Kooyman equipped numerous seals with instrument packages to record dive time, depth, heart rate, and other physiological responses. He monitored lactic acid buildup by taking blood samples before and immediately after the dive. Weddell seals were ideal subjects for this type of study because they breathe by surfacing at holes maintained in the fast sea ice. To breathe, each seal must return to its own hole after a dive. The instrument pack can then be retrieved.

Kooyman and his associates found that Weddell seals can dive for 25 minutes without using the mechanisms previously described. Only during dives lasting longer than 25 minutes are peripheral circulation shutdown and bradycardia apparent. These results suggest that Weddell seals have enough stored O_2 at the initiation of a dive to last about 25 minutes. If the dive is to be shorter, none of the oxygen-conserving mechanisms are employed because these seals store large volumes of O_2 due to their high hematocrit and blood volume. For longer dives, the magnitude of the diving reflex is a function of the length of the dive. Together, these responses (summarized in table 13.2) enable Weddell seals to accomplish some of the longest breath-holds known for mammals.

Although not as well studied, elephant seals may surpass Weddell seals in their breath-holding ability. Elephant seals spend months at sea foraging for squids and fishes at depths between 300 and 1,500 m (see table 13.1). Their feeding dives are typically 20 to 25 minutes long, with females usually going to depths of about 400 m and males to depths of 750 to 800 m. Both sexes dive night and day for weeks on end without sleeping and usually spend only two to four minutes at the surface between dives. These short surface times between long, deep dives suggest that these are not unusual dives but are the norm for this species. Further studies may show that the dive responses of Weddell seals, as outlined in table 13.2, are essentially what all marine tetrapods do to varying degrees.

Fish Gills

All cartilaginous and bony fishes take water and dissolved gases into their mouths and pump them over their gills. Each **gill arch** supports a double row of bladelike gill filaments (figure 13.8). Each flat filament bears numerous smaller secondary lamellae to further increase the gill surface available for gas exchange. Active fishes, such as mackerel, may have up to 10 times as much gill surface as body surface. The gill surfaces of sedentary bottom fishes, on the other hand, are not as extensive because their O_2 requirements are not as great.

Microscopic capillaries circulate blood very near the inner surface of the secondary lamellae. As long as the O_2 concentration of the blood is less than that of the water passing over the gills, O_2 continues to diffuse across the very thin walls of the lamellae and into the bloodstream. In fish gills, the efficiency of O_2 absorption into the blood is enhanced by the direction of water flow over the gill lamellae (arrows, figure 13.8), a direction reverse that of the blood flow within the lamellae. Oxygen-rich water moving opposite to the flow of oxygen-depleted blood establishes a very effective **countercurrent** system for gas exchange (figure 13.9). Blood returning from the body with a low concentration of O_2 enters the lamellae adjacent to water that has already given up much of its O_2 to blood in other parts of the lamellae. As the blood moves across the lamellae, it continually encounters water with greater O_2 concentrations; as a result, the blood picks up more O_2 as it goes. Thus, O_2 continually diffuses from the water into the blood along the entire length of the capillary bed within the lamellae. With such a countercurrent O_2 exchanger, some fishes are capable of extracting up to 85% of the dissolved O_2 present in the water passing over the gills. In contrast, air-breathing vertebrates, such as humans, generally use less than 25% of the O_2 that enters their lungs.

Oxygen is transported in vertebrate blood not as dissolved O_2 but rather in chemical combination with the red pigment hemoglobin contained within red blood cells. In such combination, O_2 does not generate diffusion gradients, so it is osmotically invisible as long as it remains bound with hemoglobin. Although nitrogen is also absorbed by the gills, much less is transported because it remains dissolved in the fluids of the blood. Hemoglobin functions by combining with O_2 at the high levels found at the gills (or lungs in other vertebrates) and

Figure 13.8

Cutaway drawing of a fish showing the position of the gills (*a*) Broad arrows in (*b*) and (*c*) indicate the flow of water over the gill filaments of a single gill arch. Small arrows in (*c*) indicate the direction of blood flow through the capillaries of the gill filament in a direction opposite that of incoming water.

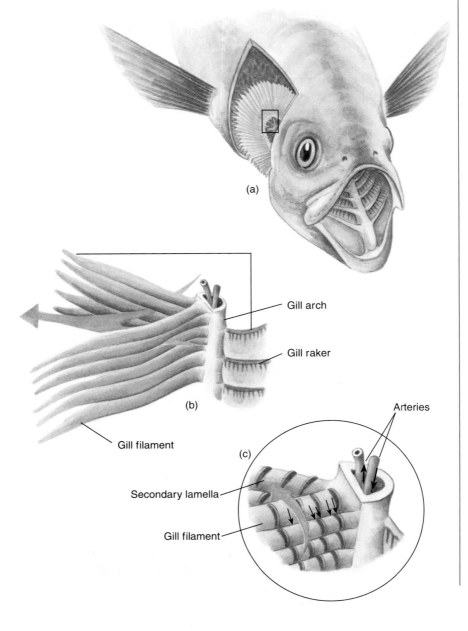

Gill arch

Gill raker

(b)

Gill filament

Arteries

(c)

Secondary lamella

Gill filament

Figure 13.9

A countercurrent gas exchange system of fish gills. Nearly all the oxygen from water flowing right to left diffuses across the gill membrane into the blood flowing in the opposite direction. Numbers represent arbitrary oxygen units.

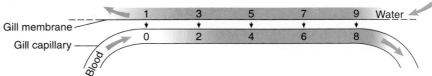

Gill membrane

Gill capillary

Blood

Water

by releasing O_2 to the body tissues at low O_2 concentrations. The quantity of O_2 carried by hemoglobin depends on the O_2 concentration of the water flowing over the gills and on the demands made by the tissues where it is used. Antarctic ice fish and a few types of eel larvae are among the very few fishes that lack hemoglobin.

Buoyancy

Living and moving in three dimensions above the seafloor presents some buoyancy problems for pelagic animals because most of the tissues of these animals are denser than seawater (which has a specific gravity of 1.02–1.03). The specific gravity of muscle is near 1.05; of bone, scale, and shell, 2.0; of cartilage, 1.1; and of fat, wax, and oil, 0.8–0.9.

Like many of the planktonic animals already mentioned, some deep-water nekton offset the weight of heavy bone and muscle tissue by reducing body fluid densities and by storing fats or oils. The giant squid *Architeuthis* excludes divalent ions from its body fluids and replaces them with less dense ammonium ions derived from metabolic wastes. Another squid, *Chiroteuthis,* has one of its four pairs of arms filled with low-density body fluids. These arms, which have less muscle than the other arms and appear swollen, are located on the ventral side of the animal. Because these arms are lighter than seawater, they may cause *Chiroteuthis* to swim and float upside down.

Stored fats and oils, which are less dense than water, are also common buoyancy devices in pelagic marine animals. Whales, elephant seals, and other large marine animals maintain thick blubber layers just under the skin. The average blue whale is about 18% blubber. Approximately 80% of that blubber is fat, and the remainder is connective tissue and blood vessels. Many sharks and a variety of bony fishes store great quantities of oils in their livers and muscle tissues. In fact, in some species of sharks, the liver accounts for more than one-quarter of the body weight.

Gas Inclusions

Fats, oils, and body fluids of reduced densities, although widely employed for buoyancy, are still only slightly less dense than seawater. This fact poses a serious challenge for many small but active nektonic species. They cannot energetically afford to pack around a huge oily liver or a thick blubber layer, nor can they sacrifice muscle and bone to lighten the load. The solution for many marine animals is an internal gas-filled flotation organ. At sea level, air is only about 0.1% as dense as seawater. Thus, a small gas volume provides a relatively large amount of lift.

The amount of lift derived from a volume of gas depends on the volume of seawater the gas displaces. Unlike water, gases are compressible; they occupy different volumes at different pressures and depths. At sea level, the pressure created by the earth's envelope of air is about 1 kg/cm^2 or 15 lb/in^2 or 1 atm. Below the sea surface, the water pressure increases about 1 atm for each 10 m increase in depth. Thus, the total pressure experienced by a fish at 5,000 m is 501 atm (more than 3.5 tons/in^2).

The gas-filled buoyancy organs of some marine animals are rigid and strong and can structurally resist the increased water pressures found at great depths. Other organs maintain their buoyancy in flexible, compressible containers.

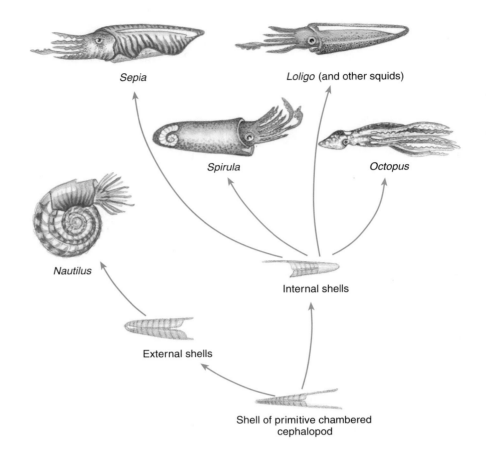

Rigid Buoyancy Tanks

Rigid-walled gas containers are found in only a few types of cephalopods. All cephalopods are believed to have evolved from an ancestral stock that had an external shell (figure 13.10). *Nautilus* is the only living cephalopod that has retained its external shell. The shells of other living cephalopods are either reduced to an internal chambered structure, as in the cuttlefish *(Sepia)* and *Spirula* (a deep-water squid), or are absent entirely, as in *Octopus.* In squids other than *Spirula*, a thin chitinous structure (the **pen**) extends the length of the mantle tissue and represents the last vestige of what was once an internal shell.

Nautilus, Sepia, and *Spirula* all have numerous hard transverse septa, partitions that separate adjacent chambers of the shell. In *Nautilus,* only the last and largest chamber is occupied by the animal. As *Nautilus* grows, it moves forward in its shell and adds a new chamber by secreting another transverse septum across the area it just vacated. The chambers are connected by a central tubelike tissue, the **siphuncle.** The siphuncle of *Nautilus* removes salts from the fluids left behind when a new chamber is formed, and water from within the chamber diffuses into the siphuncle. As water leaves, it is replaced by gases (mostly N_2) from tissue fluids. The gases diffuse in, and the total pressure of the gases dissolved within the chambers never exceeds 1 atm. The lift obtained from the gas inclusion offsets the

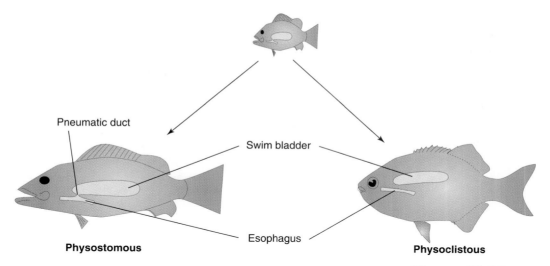

Pneumatic duct

Swim bladder

Esophagus

Physostomous

Physoclistous

Figure 13.11

The development and relative positions of physostomous and physoclistous swim bladders.

weight of the shell in water, and *Nautilus* becomes neutrally buoyant. *Sepia* and presumably *Spirula* evacuate fluids from their chambered shells in a similar manner.

These chambered cephalopods are confronted with the same depth-limiting factor that plagues submarines. Their depth ranges are limited by the resistance of their shells to increased water pressure. Each species has a critical implosion depth at which the external water pressure becomes too great for the design and strength of its shell and the shell collapses. The implosion depth of *Nautilus* shells, for example, is somewhat below 500 m, yet this animal does not normally live below 240 m.

Swim Bladders

Many bony fishes, especially active species with extensive muscle and skeletal tissue, have body densities about 5% greater than that of seawater. To achieve neutral buoyancy, many of these fishes have an internal swim bladder filled with gases (mostly N_2 and O_2). The swim bladders of bony fishes develop embryonically from an outpouching of the esophagus (figure 13.11). The densely woven fibers that make up the bladder wall are embedded with a layer of overlapping crystals of guanine to make the bladder wall nearly impermeable to O_2 and N_2 gases.

The connection between the esophagus and swim bladder, called the **pneumatic duct,** is present during the larval or juvenile stages of all bony fishes. In some species, the pneumatic duct remains unchanged in the adult. This is known as a **physostomous** condition and is found in salmon and other relatively primitive fishes. In other primarily marine species, the duct disappears as the fish matures to create a **physoclistous** swim bladder (figure 13.11). Nearly half of the more than 23,000 species of bony fishes, however, lose not only the pneumatic duct but also the swim bladder when they mature. Swim bladders are notably lacking in some bottom fishes and in active, continuously swimming fishes such as tuna.

Swim bladders are not rigid structures; the volume of water they displace is subject to changing water pressures at different depths.

Figure 13.12

A physoclistous swim bladder and associated blood vessels. The area of the gas gland is diagrammed in greater detail in figure 13.14.

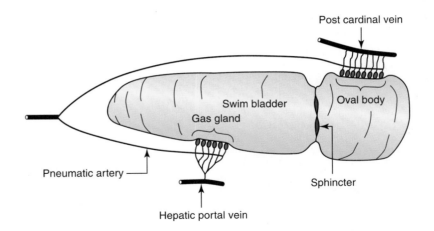

To maintain neutral buoyancy at different depths, a fish's swim bladder volume must remain constant. A fish that swims downward experiences greater external water pressure, which compresses its swim bladder and reduces the bladder's volume. The quantity of gas in the bladder must then be increased to compensate for the pressure change. Conversely, an ascending fish must get rid of swim bladder gases as rapidly as they expand. Some shallow-water physostomous fishes fill their swim bladders simply by gulping air at the sea surface. Excess gases from physostomous swim bladders are also expelled through the pneumatic duct and eventually out the mouth, anus, or gills. Lacking a pneumatic duct, a fish with a physoclistous swim bladder must reabsorb excess gases into the bloodstream. Physoclistous swim bladders have a specialized region, the oval body (figure 13.12), that is richly supplied with blood vessels for resorption of gases. The oval body is isolated from the remainder of the swim bladder by a muscle ring, or sphincter, that restricts access of the bladder gases to the oval body.

In most cases, both types of swim bladders have a gas gland that secretes gas from the blood into the bladder when these fishes have no access to air. But again there are exceptions. Herring, for example, lack gas glands and are restricted to reasonably shallow waters. The capacity of fishes with physoclistous swim bladders to quickly add or remove bladder gases to compensate for a rapid depth change is limited. If a deep-water fish rapidly ascends, the decreased water pressure allows the gases within the somewhat elastic swim bladder to expand and reduce the overall density of the fish. The density decrease may be so great that the fish is unable to descend for some time. This fact is well illustrated by the appearance of many fishes brought to the surface (unwillingly, of course) from deep water on fishing lines or in trawls. It is not unusual for the swim bladders of such fishes to expand (figure 13.13), causing severe internal organ damage.

In shallow water, the gas composition of swim bladders resembles the gas composition of air, about 20% O_2 and 80% N_2. At greater depths, the pressure of both gases increases to match the increasing water pressure. Fishes with gas-filled swim bladders have been taken from depths as great as 7,000 m. The gas pressure needed within the swim bladder to balance the water pressure at that depth (5 tons/in²) is about 700 atm. Such extreme gas pressures are achieved by a dramatic

Figure 13.13

Two deep-sea fishes on the deck of a ship after being hauled up from a depth of 800 m. Both fishes were seriously damaged and distorted by the rapid expansion of gases in their swim bladders as they were brought to the surface.

increase in the O_2 concentration of the bladder gases. Oxygen commonly accounts for more than 50%, and occasionally exceeds 90%, of the gas mixture of the swim bladders of deep-ocean fishes.

The general picture of swim bladder gas composition poses two crucial questions concerning the mechanism for filling the swim bladder in deep water. First, how are O_2 and N_2, which are dissolved in seawater at pressures no greater than 1 atm, concentrated in swim bladders at pressures as great as 700 atm? Second, why is O_2 so much more abundant than N_2 within swim bladders at great depths when N_2 is more abundant than O_2 in seawater? In some instances, O_2 is at least 1,000 times as concentrated within the swim bladders of deep-water fishes as in the water in which they are swimming. Nitrogen is generally concentrated by no more than 10–20 times.

Deep-water fishes fill their swim bladders with O_2, N_2, and traces of other gases absorbed from seawater by their gills. These gases are transported in the blood to the gas gland of the swim bladder, then are secreted into the bladder at pressures equal to external water pressures.

When hemoglobin loaded with O_2 reaches the gas gland of the swim bladder (see figure 13.12), the O_2 must be induced to leave the hemoglobin and diffuse into the swim bladder, often in the face of high O_2 pressures within the bladder. The role of the gas gland in this process is simple, but crucial. As oxygenated blood enters the gas gland, lactic acid is produced. The lactic acid diffuses into the blood vessels and lowers the pH of the blood. Lower pH conditions reduce the oxygen-carrying capacity of hemoglobin and induce it to dump part of its O_2 load. The unloaded O_2, which has not yet left the blood, is now no longer associated with the hemoglobin. Additional lactic acid production creates lower blood pH conditions and may cause hemoglobin to release up to 50% of its O_2 load. The total effects of lactic acid on hemoglobin are sufficient to produce about 2 atm of O_2 pressure at the gas gland of the swim bladder.

Eventually, the O_2 will diffuse into the swim bladder if the O_2 pressure there is not greater than 2 atm. This mechanism alone,

Figure 13.14

A simplified diagram of the rete mirabilia and gas gland associated with the swim bladders of many bony fishes. Inset illustrates the countercurrent arrangement of blood flow (small arrows) and the diffusion of oxygen from outgoing to incoming blood vessels (broad arrows).

Adapted from Hoar 1966.

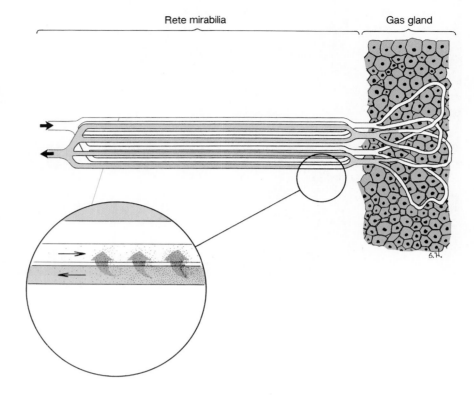

however, is not capable of producing the very high gas pressures found in the swim bladders of deep-water fishes. All deep-water fishes with gas-filled swim bladders have an extensive rete mirabilia, leading to and going away from the gas gland (figures 13.12 and 13.14). The rete capillaries that approach the gas gland carrying oxygen-rich hemoglobin are situated adjacent to and parallel with capillaries leaving the gas gland. A rete system may contain a few hundred or as many as 200,000 such capillary channels, depending on the species. These complex rete systems form another counter-current exchange system that operates on the same principle as that described for the gills to concentrate O_2.

Large amounts of O_2 forced to dissociate from hemoglobin by lactic acid at the gas gland may be blocked from diffusing into the swim bladder because of the higher gas pressures there. If the O_2 leaves the region of the gas gland via a capillary of the rete system, it diffuses across the capillary walls and back into the incoming blood of adjacent capillaries. The O_2 forced off the hemoglobin is thus trapped in this recycling system as it leaves the gas gland. Eventually, the pressure of O_2 in the capillaries surpasses even the very high pressures of the swim bladder, and O_2 diffuses from the gas gland into the bladder.

As one might expect, a long rete is capable of concentrating more O_2 at the gas gland than is a short one. Still the rete need not be un-manageably long. It has been estimated that a rete only 1 cm long could secrete O_2 at pressures up to 2,000 atm, well in excess of the swim bladder pressures needed in the deepest parts of the sea. The

Research in Progress

Swimming

In a 1975 article, D. Tucker summarized and calculated the minimum costs of locomotion for a large variety of animals spanning several orders of magnitude in body size (see figure 13.15). The minimum cost of transport, referred to here as the COT_{min}, is defined as the minimum power required to transport an animal's weight over some distance; it usually occurs at some particular speed. This calculation is analogous to determining the minimum amount of fuel necessary to drive an automobile between two cities. For both animals and automobiles, the power:speed curve is U-shaped, with the power requirements reaching a minimum at some intermediate speed. The COT_{min} occurs at the speed at which the power requirements are minimum. COT, then, is power/(mass × distance).

It is apparent from Tucker's figure that, within any one mode of locomotion (running, flying, or swimming), the COT_{min} decreases with increasing body size and is essentially independent of taxonomic affiliation. Of the three general modes of transport, swimming is the least costly, for swimmers need not support their weight against the constant tug of gravity.

The relationship between body size and COT_{min} suggested to several researchers that large swimming animals should have exceedingly low COT_{min}, but experimental evidence to test that prediction was not available until recently. Measuring power output rates requires the measuring of the metabolic rates of a subject animal must be measured, and that is difficult to do with a large, unrestrained swimming animal. T. A. Williams and co-workers trained two Atlantic bottlenose dolphins to swim in open water beside a pace boat and to match its speed. Heart and breathing rates, previously calibrated to oxygen consumption rates, were monitored and recorded continuously during each 20–25 minute test session to estimate metabolic rate. Blood samples for lactate (a product of anaerobic respiration) analysis were collected immediately after each session.

The results of this study indicate that the COT is minimum for these swimmers at speeds of 2.1 m/s and that COT is doubled at 2.9 m/s. However, when speeds were increased above approximately 3 m/s, the dolphins invariably switched to wave riding, a behavior that is best described as surfing the stern wake of the pace boat. When the dolphins were wave riding at 3.8 m/s, their COT was only 13% higher than the minimum at 2.1 m/s. The large energy savings that accompanies wave riding at higher speeds explains dolphins' common practice of riding the bow or stern waves of ships and even of large whales, apparently with little effort.

How does the COT of dolphins compare with the COT of other swimmers? Williams and colleagues demonstrated that dolphins are efficient swimmers, with COT_{min} about an order of magnitude lower than that of humans or of other surface swimmers (see figure 13.15). Yet the dolphin COT_{min} is still several times higher than a hypothetical fish of comparable size. The additional costs incurred by dolphins are presumably associated with the mammalian requirement of maintaining high body temperatures. In essence, this is the overhead cost of keeping the motor warmed up and running, a cost not experienced by poikilothermic fish. The COT_{min} of dolphins is substantially lower than that of pinnipeds and is comparable (when body mass differences are accommodated) to the estimated COT_{min} of migrating adult gray whales. Interestingly, the swimming speed at which COT was minimum is identical, 2 m/s, for both dolphins and gray whales.

For more information on this topic, see the following:

Sumich, J. L. 1983. *Swimming speeds, breathing rates, and estimated costs of transport in gray whales,* Eschrichtius robustus. Canadian Journal of Zoology 61:647-52.

Williams, T. A., W. A. Friedl, M. L. Fong, R. M. Yamada, P. Sedivy, and J. E. Haun. 1992. *Travel at low energetic cost by swimming and wave-riding bottlenose dolphins.* Nature 355:821-23.

rete mirabilia concentrates N_2 as well as O_2. The lack of a specialized transport system for N_2 (as hemoglobin is for O_2), however, relegates N_2 to the role of a minor gas in swim bladders at great depths.

As the pressure of gases inside swim bladders increases, so do their densities. At 7,000 m, the gas within a swim bladder is so compressed that its specific gravity is about 0.7, or similar to that of fat. For some fishes at great depths, the constant energy costs necessary to maintain a full swim bladder become unrealistic. At these depths, many fishes have fat-filled swim bladders. Fat-filled swim bladders provide almost as much buoyancy as gases do at 7,000 m but have few of the attendant maintenance problems.

Fat-filled swim bladders are also found in vertically migrating fish species, such as lantern fishes (figure 12.11), that face the problems of moving through pressure changes of 10 to 40 atm twice daily.

Locomotion

Animals move to improve their conditions for survival. They move to reproduce, migrate, find food, avoid predators, obtain lift, aerate the gills, and for a host of other reasons. Structural or behavioral adaptations that permit animals to swim with reduced energy expenditures enable them to divert more energy to growth and reproduction and contribute to the potential success of an individual.

Some generalizations about the energy costs associated with locomotion can be made. A comparison of the cost of transport (COT) can be made between different modes of locomotion and for animals of different sizes (see Research in Progress: Swimming, p. 355). All species in all modes of locomotion have some preferred speed at which their COT is minimum. When the COT_{min}, in joules per kilogram body mass per kilometer traveled, is calculated and plotted as a function of body mass, it is clear that both flying and swimming impose lower COT_{min} than does walking or running (figure 13.15). Flying is energetically demanding but covers large distances in little time, so the COT_{min} is low. Swimmers, regardless of size, need not support the

Figure 13.15

Relationship between COT_{min} and body weight for different modes of locomotion. M, COT_{min} values for a variety of machines.
Adapted from Tucker, 1975.

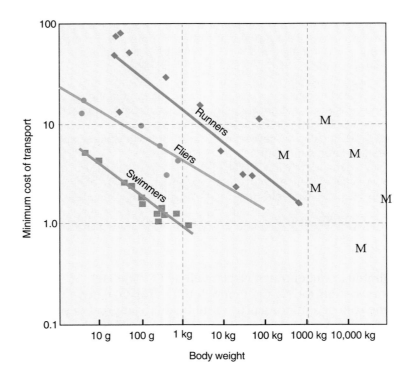

weight of their bodies, so their COT_{min} is lower still. In contrast to fliers, though, which cannot be larger than about 40 kg, swimmers can be extremely large and still move efficiently. Because flying is accomplished above the sea surface, the following discussion of aquatic locomotion focuses on swimming.

Different marine organisms swim in varying ways. Whereas most are generalists (such as the surfperch in figure 13.16), others are specialized for one of three modes of swimming: sprinting (barracuda), fine maneuvering (butterfly-fish), or nearly continuous high-speed cruising (tuna). These specialized approaches to swimming, with appropriate adaptations of body shape, fins, and muscle, reflect the variety of ecological niches available to nekton.

Body Shape

Some fishes live in situations, especially near the ocean bottom or in coral reefs or kelp beds, where speed is not crucial for survival but camouflage is. In these situations, body shape is often quite variable (figures 13.2 and 13.3).

The streamlined shape of most fast nekton is actually a compromise between various possible body forms that enables the animal to slip through the water with as little resistance, or drag, as possible. **Frictional drag** results from the interaction of the animal's surface

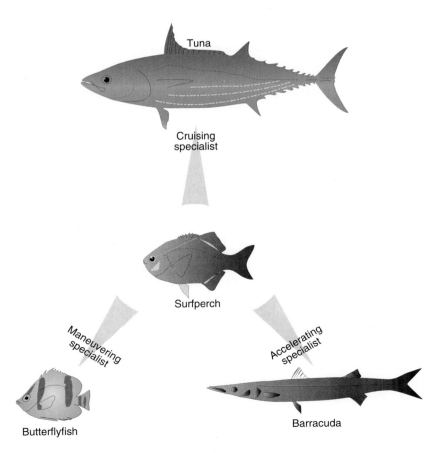

Figure 13.16

Examples of body shape specialization for three different swimming modes.
Adapted from Webb 1984.

Figure 13.17

The combination of shapes needed to minimize frictional, form, and turbulent resistance to a body moving through a fluid results in the streamlined shape.

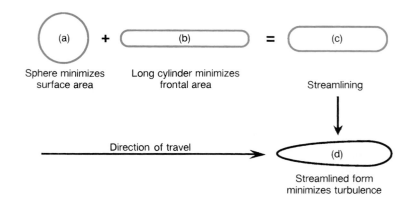

with the water surrounding its body. If frictional drag alone were to be reduced, the ideal shape would be a sphere, which has a minimal surface area for the animal's volume (figure 13.17a). However, an animal swimming through water must overcome more than just frictional resistance. As it swims forward, an amount of water equal to the size of the animal's largest cross-sectional area (from a head-on view) must be displaced to permit the animal to progress. **Form drag,** by itself, can be minimized with a shape that has a small cross-sectional area, a body shaped like a long, thin cylinder (figure 13.17b). The actual shape of a fast swimmer such as a tuna or a porpoise is neither spherical nor cylindrical; it is a compromise form (figure 13.17c). One additional drag factor, **turbulence,** must also be considered. Wind tunnel tests have demonstrated that the ideal shape of a high-speed body in a fluid medium, be it fish, missile, or torpedo, is one that has a length about 4.5 times its greatest diameter. In addition, it should be roundly blunt at the front end, tapered to a point in the rear, and round in cross section (figure 13.17d). The form shown in figure 13.17d is the optimal overall shape to minimize the total drag resulting from friction (a function of surface area), form (a function of cross-sectional area), and turbulence (a function of streamlining). Most fast marine animals, excluding their fins, closely approximate this ideal shape.

In contrast, rapidly accelerating fishes, such as barracudas, tend to have thinner, more elongated bodies, possibly to reduce their chances of being seen and recognized as they rush their prey. On the other hand, maneuverers such as butterfly-fishes are tall and elliptical in cross section, with large fins extending even greater distances from the body. The increased amount of body surface, while adding to the overall drag, produces additional thrust and also serves as a control surface for making fine position adjustments.

Fins

The push, or thrust, needed for swimming generally comes from the sides and fins of the animal's body (figure 13.18). The bending motion of the anterior part of the body, initiated by the contraction of a few muscle segments (myomeres) on one side, throws the body into a curve. This curve, or wave, passes backward over the body by sequential contraction and relaxation of the myomeres (figure 13.19). The

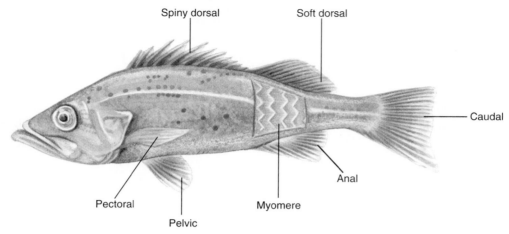

Figure 13.18

Position and name of fish fins. A portion of the musculature has been exposed to show the arrangement of myomeres.

Figure 13.19

Progression of a body wave as an eel swims from left to right.

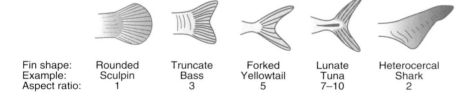

Fin shape:	Rounded	Truncate	Forked	Lunate	Heterocercal
Example:	Sculpin	Bass	Yellowtail	Tuna	Shark
Aspect ratio:	1	3	5	7–10	2

Figure 13.20

Examples of shapes and aspect ratios for caudal fins.

contraction of each myomere in succession reinforces the wave form as it passes toward the tail. Immediately after one wave has passed, another starts near the head on the opposite side of the body, and the entire sequence is repeated in rapid succession. Forward thrust is developed almost entirely by the backward component of the pressure of the animal's body and fins against the water.

Caudal Fins

Caudal fins typically flare dorsally and ventrally to provide additional surface area to develop thrust. Increased fin size increases the total frictional drag of the fish. The ratio of thrust to drag changes with the shape of the caudal fin. One index of the propulsive efficiency of the fin, based on its shape, is the **aspect ratio:**

$$\text{Aspect ratio} = \frac{(\text{fin height})^2}{\text{fin area}}$$

The caudal fins of most pelagic fishes fit into five profile categories: rounded, truncate, forked, lunate, and heterocercal (figure 13.20). Each has a different aspect ratio.

Figure 13.21

A pelagic white shark, *Carcharodon*. Lift is obtained from its heterocercal tail and the large pectorals extending from the flattened underside of the body.

Courtesy M. Snyderman.

The angelfish has a rounded caudal fin that is soft and flexible and it has a low aspect ratio. When the fin moves laterally, it bends and allows water to "slosh" past it. This flexibility permits the caudal fin to be used effectively for accelerating and maneuvering. Truncate and forked fins have intermediate aspect ratios, produce less drag, and are generally found on faster fishes. These fins are also flexible for maneuverability.

The lunate caudal fin characteristic of the tuna, sailfish, marlin, and swordfish has a high aspect ratio (up to 10 in swordfish) for reduced drag at high speeds. The shape closely resembles the swept-wing design of high-speed aircraft. These fishes are among the fastest marine animals. The caudal fin is quite rigid for high propulsive efficiency but is poorly adapted for slow speeds and maneuvering. Fish with high aspect ratio caudal fins (especially forked and lunate types) are capable of long-distance, continuous swimming.

The heterocercal tail (figure 13.20) characteristic of sharks has a shape very different from that of most bony fishes (which are homocercal, or symmetrical about the long axis of the body). The heterocercal tail is asymmetrical. When the caudal fin is moved from side to side, a forward thrust develops. Because of the angle of the trailing edge of the tail, however, it produces some lift as well (figure 13.21). The caudal fin asymmetry (and the lift it produces) is reduced in several species of fast-swimming pelagic sharks; at higher speeds a smaller fraction of their total swimming power output is needed to create lift. The paired pectoral fins of sharks are flat and large and extend horizontally from the body like wings of an aircraft (figure 13.21). The ventral side is nearly flat in front and, with the flat extended pectoral fins, produces a

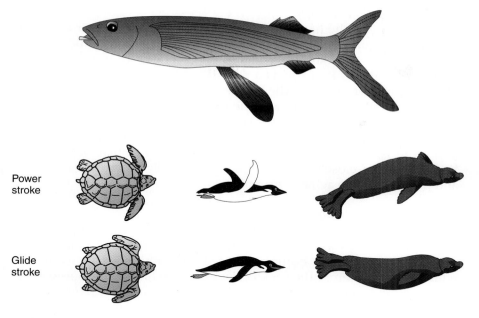

Power
stroke

Glide
stroke

Figure 13.22

A flying fish, *Exocoetus,* uses enlarged pectoral fins for gliding.

Figure 13.23

Power and glide portions of pectoral swimming strokes of three marine tetrapods.

large hydrofoil surface. This hydrofoil meets the water at an angle and produces lift for the front part of the body to balance the lift produced by the tail. Pelagic sharks and other cartilaginous fishes lack swim bladders and need this lift to maintain their position in the water column. This mechanism for achieving lift, however, does have its disadvantages. These fishes cannot stop or hover in midwater. To do so would cause them to settle to the bottom, a bottom that may be some distance away in the open ocean. Maneuverability is also reduced; the large and rigid paired fins that function as hydrofoils are not well suited for making fine position adjustments.

Paired Fins

Unlike sharks, bony fishes equipped with swim bladders have their pectoral and pelvic fins free for other uses. In most bony fishes, the paired fins are used solely for turning, braking, balancing, or other fine maneuvers. When the fishes are swimming rapidly, these fins are folded back against their bodies. Wrasses and surgeonfishes, however, swim with a jerky, fanning motion of their pectorals and hold the remainder of their bodies straight. Some skates and rays swim by gracefully undulating the edges of their flattened pectoral fins or, in the cases of manta and eagle rays, by flapping their pectorals like large wings.

   The greatly enlarged pectoral fins of the flying fish in figure 13.22 do not enable this animal to actually fly; yet it can glide in air for long distances. Flying fishes build up considerable speed while just under the sea surface and then leap upward with their pectorals extended. The pectorals do not flap during flight, so the length of the glide is dependent on wind conditions and the initial speed of the fish as it leaves the water. These "flights" are apparently a means of escaping predators; glides up to 400 m have been reported. Tetrapods, too, use paddling (turtle; figure 13.23) or underwater flying motions (penguin and sea lion; figure 13.23), with their pectoral flippers doing most of the work.

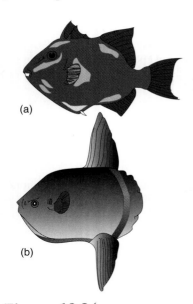

Figure 13.24

Two fishes that use their dorsal and anal fins for propulsion: (*a*) triggerfish, *Balistes;* (*b*) ocean sunfish, *Mola.*

Figure 13.25

The sea horse, *Hippocampus,* swims vertically, using its dorsal fin for propulsion.

Anal and Dorsal Fins

Triggerfishes and the ocean sunfish swim by undulating their anal and dorsal fins (figure 13.24). These fins extend along much of the triggerfish's body. The large sunfish, which reaches lengths of nearly 3 m and attains weights up to a ton, is a sluggish fish and is often seen "sunning" at the surface. The little swimming it does is accomplished by the long dorsal and anal fins and the foreshortened caudal fin.

The sea horse usually swims vertically with its head at right angles to the rest of its body (figure 13.25). The prehensile tail tapers to a point and is used to cling to coral branches and similar objects. Sea horses and the closely related pipefishes rapidly vibrate their dorsal and pectoral fins to achieve propulsion.

Propulsion by Other Nekton

Examples of fish counterparts can be found in the swimming patterns of several types of nonfish vertebrates, from sea snakes to whales, and generalizations made concerning fish swimming patterns apply equally well to them. Other nekton lack midline fins for propulsion yet are nonetheless effective swimmers. Most marine invertebrates are not well known for their speed, but a few are fast and can maneuver well enough to be successful nekton. Shrimps and prawns use their abdominal paired appendages **(pleopods)** and their tail fin for swimming. Squids and other cephalopods take water into their mantle cavities and then expel it at high speeds through a nozzle-like **siphon.** The siphon can be aimed in any direction for rapid course corrections and for maneuvering purposes. Squids and cuttlefishes also use their undulating lateral fins in much the same manner as skates and rays.

Speed

Several species of nekton are noted for their amazing swimming speeds. The oceanic dolphin *Stenella* has been clocked in controlled tank situations at better than 40 km/h (approximately 25 mi/h). Top speeds of killer whales are estimated to be 40 to 55 km/h. A barracuda only 1 m long has been clocked at 40 km/h. For a comparison, consider that human Olympic-class swimmers achieve sprint speeds of only 4 to 5 km/h. To clock fishes that can easily outdistance a speeding ship, specially designed fishing poles have been developed to measure the speed at which the fishing line is stripped from the reel. When a fish takes the bait and flees, its speed is measured and recorded. In this manner, investigators clocked a yellowfin tuna less than 1 m long at a maximum speed of 74.6 km/h (45 mi/h) for 0.19 second. A tunalike wahoo slightly more than 1 m in length was clocked at 77 km/h for about 0.1 second. It has been suggested that large tuna are capable of speeds in excess of 110 km/h (70 mi/h). This estimate may not be as farfetched as it seems because some species of tuna achieve lengths of 4 m and presumably would be much faster than a fish only 1 m long, but it has yet to be confirmed.

What enables tuna to swim so fast? Most fast marine animals exhibit nearly optimal streamlined body shapes (figure 13.26). Yet the exceptional swimming abilities of tuna and tunalike fishes go beyond simply having a streamlined body form and an efficient caudal fin. The

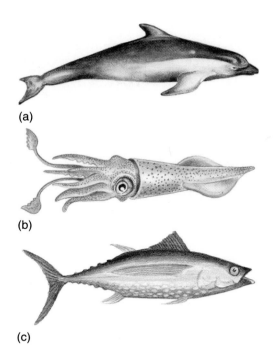

(a)

(b)

(c)

Figure 13.26

Three swift pelagic animals with streamlined body forms: (*a*) bottle-nosed dolphin, *Tursiops;* (*b*) squid, *Loligo;* (*c*) tuna, *Thunnus.*

streamlined body form is complemented by other friction-reducing features. The first dorsal fin can be retracted into a slot and out of the path of water flow when not needed for maneuvering. Tuna scales are small and minimize friction with the water. Their eyes do not bulge beyond the profile of their head and are covered with adipose eyelids to further reduce turbulence. Numerous small median **finlets** on the dorsal and ventral sides of the rear part of the body function to reduce turbulence in that region. Their body is quite rigid and provides little of the forward thrust.

Most of the caudal flexing is localized in the region of the **caudal peduncle,** the region where the caudal fin joins the rest of the body. The caudal peduncle, flattened in cross section, produces little resistance to lateral movements. Several small finlets just anterior to the peduncle guide the water posteriorly toward the caudal fin rather than allowing it to slosh over the peduncle. The rigid caudal fin is lunate and has a high aspect ratio (usually greater than 7). The tail beats rapidly with relatively short strokes. This type of caudal fin creates little drag but also provides very little maneuverability.

Nearly 75% of the total body weight of a tuna is composed of swimming muscles. In tuna, each myomere overlaps several body segments and is anchored securely to the vertebral column. Tendons extend from the myomeres across the caudal peduncle and attach directly to the caudal fin.

Tuna swimming muscles consist of segregated masses of red and white muscle fibers. Structurally, red muscle fibers are much smaller in diameter (25 to 45 μm) than white muscle fibers (135 μm) and are rich in myoglobin (a red pigment with a strong chemical affinity for O_2 similar to that of hemoglobin). The small size of the red muscle cells provides additional surface area, which, in conjunction with

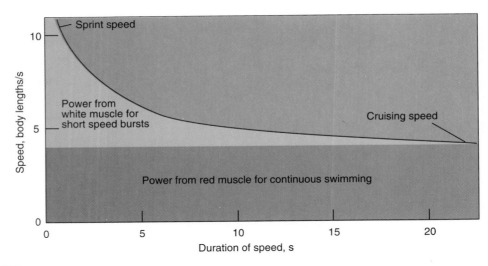

Figure 13.27

Duration of swimming speeds for white and red muscles. White muscle is used for short bursts at top speed and fatigues rapidly; red muscle maintains continuous cruising speeds.

After Bainbridge 1960.

myoglobin, greatly facilitates O_2 transfer to the red muscle cells. Physiologically, red muscle cells respire aerobically and white muscle cells respire anaerobically, converting glycogen to lactic acid.

The metabolic rate (and power output) of tuna red muscle, and probably of red muscle in other fishes, is about six times as great as that of white muscle. The relative amount of red and white muscle a fish has is related to the general level of activity the fish experiences. A slow-moving grouper has almost no red muscle, but its large mass of white muscle fibers can power short, fast lunges to capture prey or elude predators. At the other extreme are tuna, with over 50% of their swimming muscles composed of red muscle fibers. The red muscles of sculpins, which operate their pectoral fins, and of puffers, which swim by fanning their anal and dorsal fins, are concentrated at the bases of their swimming fins.

Electrodes have been inserted into red and white muscle tissue of small sharks and some tuna to measure muscle activity. At slow, normal cruising speeds, only red muscles contract. White muscles come into play only at above-normal speeds. Top speeds of about 10 body lengths per second can be maintained for about 1 second, but cruising speeds of two to four body lengths per second can be maintained indefinitely (figure 13.27). Apparently, the power for continuous swimming comes from the red muscle, with white muscle being held in reserve for peak power demands. White muscle does not require an immediate O_2 supply; it can operate anaerobically and accumulate lactic acid during stress situations. The lactic acid can be converted back into glycogen or some other substance when the demand for O_2 has diminished. Tuna, with a greater proportion of red muscle, are able to indefinitely maintain a faster cruising speed than most other fishes.

Fishes are generally thought to be poikilothermic animals. The heat generated by metabolic processes within the body may elevate the body temperature slightly above the ambient water temperature, but the heat gain is quickly lost to the surrounding seawater (table 13.3, left column). A few exceptionally fast fishes, however, have red muscle masses that are much warmer than the surrounding water

Table 13.3
Elevation of Red Muscle Temperatures above Seawater Temperatures
for Some Marine Fishes

Fish with Slightly Elevated Temperatures		Fish with Dramatically Elevated Temperatures	
Yellowtail (*Seriola*)	+1.4° C	Porbeagle shark (*Lamna*)	+7.8° C
Mackerel (*Scomber*)	+1.3° C	Mako shark (*Isurus*)	+4.5° C
Bonito (*Sarda*)	+1.8° C	Tuna (*Thunnus*)	+5 to +13° C
			occasionally to +23° C

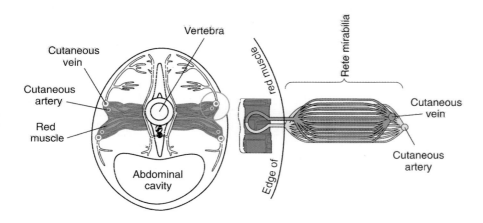

Figure 13.28

Cross section of a tuna, showing the position of the red muscles (shaded) and the countercurrent system of small arteries and veins serving the red muscles.
After Carey 1973.

(table 13.3, right column). The magnitude of muscle temperature elevation above the water temperature is usually consistent for each species. The one well-studied exception is the bluefin tuna *(Thunnus thynnus)*, which has a consistently high red muscle temperature regardless of water temperature. In water of 25° C, for example, the core muscle temperature of the bluefin tuna is near 32° C and declines only slightly to 30° C when the animal is moved to seawater with a temperature of 7° C.

Within certain limits, metabolic processes, including muscle contractions, occur more rapidly at higher temperatures. Consequently, the power output of a warm muscle can be greater than that of a cold muscle. Tuna and porbeagle sharks exhibit some behavioral characteristics that elevate and control their internal temperatures to some extent. These fishes are most abundant in tropical and subtropical regions, where differences between body and water temperatures are not great. More important, though, are their heat-conserving anatomical features. The swimming muscles of most bony fishes receive blood from the dorsal aorta just under the vertebral column. The major blood source for the red muscle masses of porbeagle sharks and most tuna is a cutaneous artery under the skin on either side of the body (figure 13.28). The blood flows from the cutaneous artery to the red muscle and then returns to the cutaneous vein. Between the cutaneous vessels and the red muscles are extensive countercurrent heat exchangers that facilitate heat retention within the red muscle. Cold blood enters the countercurrent system and is

Table 13.4

Functional Comparison of Some Features That Influence the Swimming Speeds of a Noncruising Fish (Rockfish) and a Specialized Cruiser (Tuna)

Characteristic	Rockfish	Tuna
Body feature		
Shape		
Front view		
Rigidity	Flexible body	Rigid body
Scales	Abundant large scales	Small scales
Eyes	Bulging eyes	Nonprotuding eyes covered with adipose lid
% of thrust by body	50%	Almost none
Dorsal fin	Broad-based and high	Small, fits into slot
Caudal peduncle		
Form	Compressed	Depressed
Cross-sectional shape		
Keels	Absent	Present
Finlets	Absent	Present
Caudal fin		
Aspect ratio	Low, 3	High, 7–10
Rigidity	Flexible	Rigid
Maneuverability	Good	Poor
Tail beat frequency	Low	High
Tail beat amplitude	Large	Small
Swimming muscles		
% of body weight	50–65%	75%
% red muscle	20%	50% or more
Body temperature	Ambient	Elevated

Modified from Fierstine and Walters 1968.

warmed by the blood leaving the warm red muscle. As a result, little of the heat generated in the red muscles is lost.

All the previously described features collectively function to provide tuna and other similar fishes with the capability of cruising continually at moderate speeds and with the opportunity to be the efficient pelagic predators they are. Table 13.4 summarizes these features and compares tuna with a normally noncruising fish (a rockfish) that typically lies in wait for its prey.

Schooling

The successful use of filter-feeding techniques by large whales, numerous fishes, and even a few birds and seals is dependent on the presence of abundant and dense aggregations of smaller animals. In addition to the patchiness of zooplankton described in the previous chapter, hundreds of species of smaller fishes and a few types of squids and larger crustaceans create well-defined social organizations called **schools.** Fish schools vary in size from a few fishes to enormous populations extending over several square kilometers. Schools

Figure 13.29

A skipjack, *Katsuwonus*, in a school of baitfish.

Courtesy Honolulu Laboratory, National Marine Fisheries Service, NOAA, Department of Commerce.

usually consist of a single species, with all members similar in size or age. Larger fishes swim faster than smaller ones, and mixed populations quickly sort themselves out according to their size. The spatial organization of individuals within a school remains remarkably constant as the school moves or changes direction. Individual fishes line up parallel to each other, swim in the same direction, and maintain fixed spacings between individuals. When the school turns, it turns abruptly, and the animals on one flank assume the lead. The spatial arrangement within schools seems to be maintained with the use of visual or vibrational cues.

Why do small fishes band together to be so conveniently eaten by larger predators (as indicated in figure 13.29)? Ironically, part of the answer seems to be that for small animals with no other means of individual defense, schooling behavior provides a degree of protection.

Most of our present understanding of the survival value of schooling behavior is based on conjecture because experiments with natural populations are exceedingly difficult to conduct and evaluate. Predatory fishes have less chance of encountering prey if the prey are members of a school because the individuals of the prey species are concentrated in compact units rather than dispersed over a much larger area. Large numbers of fishes in a school may achieve additional survival advantages by confusing predators with continually shifting and changing positions; they might even discourage hungry predators with the illusion of an impressively large and formidable opponent. Schooling may act as a drag-reducing behavior and enable closely spaced individuals to capitalize on the turbulence generated by their neighbors. Laboratory studies with fishes that instinctively school also indicate that if these fishes are isolated at an early age and prevented from schooling, they learn more slowly, begin feeding later, grow more slowly, and are more prone to predation than are their siblings who are allowed to school. It is also thought that schooling serves as a mechanism to keep reproductively active members of a population together. Schooling species typically reproduce by broadcast spawning. Dense concentrations of mature individuals spawning simultaneously ensure a high proportion of egg fertilization and probably greater larval survival (for the same reasons that large numbers of their parents survived to produce them).

Two Specialized Approaches to Feeding

Most nekton are carnivores, feeding at the second trophic level or higher. In marine communities, individual prey typically become larger in size but fewer in number at successively higher trophic levels, yielding approximately the same biomass at each trophic level (see figure 12.18). To secure adequate diets, large and active nekton either must obtain large but relatively rare prey items or be able to efficiently consume very large numbers of smaller, more abundant prey.

White Sharks

White sharks are examples of the first approach. In contrast to the image portrayed by the movie *Jaws* and its sequels, this shark is not a capricious or mindless killer. Instead, it is a stealthy, skilled predator of pinnipeds and may actually consider human flesh unpalatable. About 1 m long at birth, young white sharks initially feed on small bony fishes. By the time they are about 3 m long, they begin to shift to the adult preference for pinnipeds, especially young seals (see figure 13.30). This shift in prey coincides with a movement to higher latitudes (in both hemispheres), where aggregations of seals are abundant.

The predatory behavior of white sharks on seals involves a stealthy approach along the sea bottom in shallow areas where seals enter the water. White sharks exhibit strong countershading patterns (see figure 13.21), and their approach must be difficult for seals to visually detect from above. When the prey is located, the shark swims upward from below and bites the prey, often causing profuse bleeding (figure 13.30). The bleeding seal is carried underwater in the shark's jaws or is left at the surface until it dies.

Large white sharks seem to prefer seals and other pinnipeds or whales over other kinds of prey, such as birds, sea otters, or fishes. Seabirds, fishes, and otters are rarely found in the stomachs of large white sharks. This selective preference for marine mammals that use fat-rich blubber for insulation over leaner birds, otters, and fishes may be related to the metabolic demands of maintaining elevated muscle temperatures and high growth rates. Although white sharks occupy cool, temperate waters, their growth rates may be greater than those of tropical lemon sharks, and a fat-rich diet may be essential to maintain these rates.

Figure 13.30

Typical attack pattern of a white shark on a seal. The shark approaches the seal from beneath to attack, then may carry the seal underwater. The shark often bites, lets the seal float free to bleed profusely, then returns to feed.
Adapted from Klimley 1994.

Baleen Whales

Baleen whales are among the largest animals on the earth, and all are filter feeders. All except the gray whale feed on planktonic crustaceans or small shoaling fishes. The character of the baleen, as well as the size and shape of the head, mouth, and body, differ markedly between species of baleen whales (see figure 7.18). Bowhead whales have very fine, long baleens well adapted to collect *Calanus* and other small copepods less than 1 cm in size. Most of the rorquals and the humpback whales have coarser baleen fibers and greatly distensible throats. They feed by engulfing entire shoals of euphausiids, sand lances, or capelin. The gray whale has the coarsest and shortest baleen of all mysticetes (see figure 7.20).

Several distinct types of feeding behaviors have been described for baleen whales, depending on the type of whale as well as on its prey. The large, slow right whales and bowhead whales use their very long and fine baleen plates to trap copepods and other small planktonic crustaceans (figure 13.31). The larger and faster blue whales and fin whales are equipped with 70 to 80 throat pleats that permit the floor of their mouths to expand enormously. These whales engulf tens of tons of water with the contained zooplankton in each mouthful. Their gigantic muscular tongues act as huge pistons (sometimes in concert with surfacing behavior) to force the water out through the baleen and to assist in swallowing trapped zooplankton. Humpback whales frequently lunge open-mouthed into shoals of prey. When prey are too dispersed for lunge feeding, humpback whales sometimes produce a curtain or net of ascending bubbles to concentrate the prey into tight food balls for more efficient feeding.

These distinctive feeding patterns are best observed in the Northern Hemisphere, where competition for food encourages specializations for and partitioning of available food resources. In the Southern Hemisphere, all of the large baleen whale species exploit a single prey, the enormously abundant krill, *Euphausia superba,* of the Antarctic upwelling region.

Gray whales exhibit the most unusual feeding behavior of all mysticetes. In their shallow summer feeding grounds of the Bering and

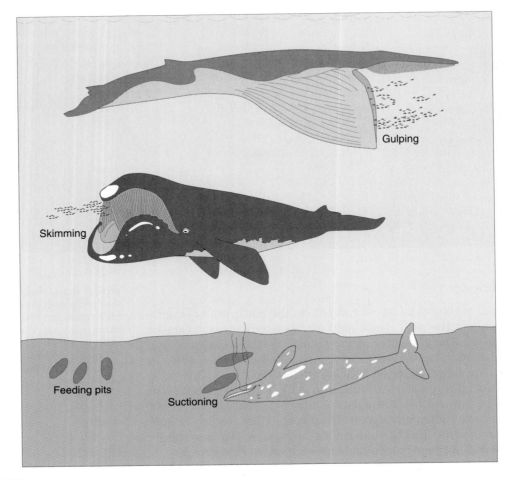

Figure 13.31

Three feeding styles of baleen whales: (top) fin whale, (middle) right whale, (bottom) gray whale. Benthic feeding pits are created by the feeding activities of gray whales.

Adapted from Pivorunas 1979 and Nerini 1984.

Arctic seas, these medium-sized whales feed on bottom invertebrates, especially amphipod crustaceans. It was thought that gray whales fed by dredging up mouthfuls of soft sediment and the resident invertebrates, flushing the mud out through the coarse baleen. But recent direct observations on the feeding behavior of gray whales demonstrate that these animals roll to one side and suck their prey into the side of the mouth and expel water out the other side (figure 13.31).

Summary Points

- The nekton of the world ocean principally comprise three groups of vertebrates—the sharks, bony fishes, and whales—although other tetrapod vertebrates and a few species of swimming invertebrates are also included. Most nekton are found in or just below the epipelagic zone. Those living in deeper waters exhibit adaptations to sparse food supplies, stable environmental conditions, and low-light intensities.
- Fat or oil deposits, gas-filled flotation devices, or lift from swimming movements provide buoyancy for animals with tissues denser than seawater. Air-filled floats are reasonably simple devices for animals that

do not change depths or that have rigid gas containers. Air and other gas mixtures are compressible, however, and if the container is also compressible (as are the swim bladders of bony fishes), some pressure-compensating mechanism is needed. The gas gland and associated countercurrent rete mirabilia of some bony fishes are capable of concentrating gases from the blood into their swim bladders at high pressures.

- With their buoyancy problems solved, nekton have evolved a large variety of swimming patterns and associated body forms and fin shapes. These are forms and shapes related to some combination of specializations for fine maneuvering, rapid sprinting, or prolonged cruising.
- Some species of nekton find protection from predation by schooling. Others take advantage of these dense aggregations to harvest small prey items efficiently.

Review Questions

1. What advantages do cetaceans derive from apneustic breathing patterns besides those directly associated with prolonged breath holding?
2. List two anatomical locations in adult tuna where countercurrent blood vessel systems are found.
3. How do fish achieve extremely high concentrations of gases in their swim bladders? At great depths, why is nitrogen, the most abundant gas in air, of relatively minor importance in swim bladders?
4. Explain the structural and physiological adaptations that account for the high swimming speeds achieved by tuna and similar fishes. Compare these high-speed fishes with a typical lunger, such as a bass or a grouper.
5. Describe the structural and physiological differences that allow a greater work output from the red muscle of fish than from a similar amount of white muscle.

Challenge Questions

1. Discuss the adaptive advantages of schooling by fish.
2. Suggest a relationship between the evolution of the swim bladder in bony fishes and their high species diversity and widespread distribution.
3. Is your own breathing a voluntary or involuntary response? Can you stop breathing for 1 minute? For 2 minutes? For 5 minutes? Can you voluntarily alter your heart rate or body temperature? Compare your responses with what must occur in a Weddell seal when it is preparing for a long dive.

Suggestions for Further Reading

Books

Alexander, R. McN. 1970. *Functional design in fishes.* London: Hutchinson.

———. 1988. *Elastic mechanisms in animal movement.* New York: Cambridge University Press.

Blake, R. W. 1983. *Fish locomotion.* New York: Cambridge University Press.

Bond, C. E. 1979. *Biology of fishes.* Philadelphia: W.B. Saunders.

Elsner, R., and B. Gooden. 1983. *Diving and asphyxia: A comparative study of animals and men.* New York: Cambridge University Press.

Keenleyside, M. H. A. 1979. *Diversity and adaptation in fish behavior.* New York: Springer-Verlag.

Kooyman, G. 1989. *Diverse divers: Physiology and behavior.* New York: Springer-Verlag.

Articles

Carey, F. G., et al. 1971. Warm-bodied fish. *American Zoologist* 11:137–45.

Eastman, J. T., and A. L. DeVries. 1986. Antarctic fishes. *Scientific American* (November) 255:106–14.

Nelson, C., and K. Johnson. 1987. Whales and walruses as tillers of the seafloor. *Scientific American* 256:112–17.

Oceanus. 1980. Special issue on sensory reception in marine organisms. 23(3).

Oceanus. 1982. Special issue on sharks. 25(4).

Partridge, B. L. 1982. The structure and function of fish schools. *Scientific American* 246:114–23.

Perutz, M. F. 1978. Hemoglobin structure and respiratory transport. *Scientific American* (December) 239:92–125.

Sanderson, J. L., and R. Wassersug. 1990. Suspension-feeding vertebrates. *Scientific American* (January) 262:96–101.

Scholander, P. F. 1957. The wonderful net. *Scientific American,* (April) 196:96–107.

Webb, P. W. 1984. Form and function in fish swimming. *Scientific American* (April) 251:72–82.

Internet Addresses

Nekton

http://www.ma.org/classes/oceanography/boconnor/home.html

General description of the different groups, describing movement, buoyancy, respiration, defense and predators. Author: Brendan O'Connor, Student, Marin Academy. Site Owner: Don Alexander, Oceanography Instructor, Marin Academy, San Rafael, CA 94901, e-mail: dalexander@ma.org. Marin Academy's Oceanography Home Site

http://www.ma.org/classes/oceanography/home.html

Marine Birds and Mammals

http://darter.ocps.k12.fl.us/classroom/klenk/MMAM.html

Advantages, disadvantages, adaptations, feeding, and moving are all covered. Ted Klenk (Apopka High School, Apopka, FL). Apopka High Marine Science

http://darter.ocps.k12.fl.us/classroom/klenk/index.html

Gray Whale Feeding Method

http://www.slocs.k12.ca.us/whale/whale4.html

Brief description of gray whale feeding method and favorite food; with sketch. San Luis Obispo County Schools World Wide Web Server

http://www.slocs.k12.ca.us/default.html

Marine Mammals

http://bonita.mbnms.nos.noaa.gov/sitechar/mamm.html

Marine mammals: seals and sea lions; whales, dolphins and porpoises; sea otter. Monterey Bay National Marine Sanctuary. Monterey Bay National Marine Sanctuary Site Characterization

http://bonita.mbnms.nos.noaa.gov/sitechar/index.html

Respiration in Fishes

http://www.csuchico.edu/~pmaslin/ichthy/fshrsp.html

Comprehensive discussion including sketches and diagrams. Biol. 261, Ichthyology, Paul Maslin e-mail pmaslin@oavax.csuchico.edu.

http://www.csuchico.edu/~pmaslin/ichthy/ichthy.html

Chiroteuthis

http://www.soest.hawaii.edu/tree/cephalopoda/coleoidea/decapoda/chiroteuthididae/chiroteuthis/chiroteuthis.html

Brief description, photo with close-ups of photophores. Cephalopoda

http://www.soest.hawaii.edu/tree/cephalopoda The Tree of Life

http://phylogeny.arizona.edu/tree/phylogeny.html

The Sea World/Busch Gardens Animal Information Database
http://www.seaworld.org
Fairly complete descriptions of these marine animals' lives. Click on
   links for Bony Fishes, Bottlenose Dolphins—Adaptations for an
   Aquatic Environment, Sharks and Their Relatives, Baleen Whales,
   Beluga Whales, and Killer Whales. © Busch Entertainment
   Corporation.

Learn About Marine Biology
http://www.odysseyexpeditions.org/course.html
Click on links for "Fish, a Quick Course in Ichthyology" and for "Sharks,
   rays, and bony fish: Evolution, Ecology, Physiology, Reproduction and
   Sensory Biology." Jason Buchheim, Odyssey Expeditions, Inc.
   Odyssey Expeditions Tropical Marine Biology Voyages
http://www.odysseyexpeditions.org

CHAPTER 14

Nekton—Migration, Sensory Reception, and Reproduction

In the sea, only the larger and faster swimming nekton are capable of accomplishing long-distance periodic migrations. These migrations often serve to integrate the reproductive cycles of adults into local and seasonal variations in the patterns of primary productivity. In this chapter, the importance of the relationships between migration and successful reproduction and the role of sensory perception of some nekton are examined.

Migration

Many species of nekton participate in well-defined migratory movements that are larger in both time and space scales than the patterns of vertical migration described in chapter 12. Some migrators travel oceanic distances. In general, these migrations are adaptations to better exploit a greater range of resources for feeding or for reproduction. For example, the food available in spawning areas may be appropriate for larval and juvenile stages, but it might not support the mature members of the population. So the adults congregate for part of the year in productive feeding areas elsewhere that may be unsuitable for the survival of the younger stages. The regular and directed movements between feeding areas and regions used for reproduction are called migrations.

Migratory patterns of marine animals often exhibit a strong similarity to patterns of ocean surface currents. Juvenile stages of some species may be carried long distances from spawning and hatching areas by ocean currents. Although adults may use currents for a free ride, many types of larvae and juvenile fishes are absolutely dependent on current drift for their migratory movements. The downstream drift of these young may require the adults to make an active, compensatory return migration upstream against the current flow to return to the spawning grounds.

Pelagic animals typically move below the sea surface and well away from the coast, making it difficult or even impossible for us to observe directly their migratory behavior. Most of our understanding of oceanic migrations has been inferred from studies employing visual or electronic tags and from distributional patterns of eggs and subsequent developmental stages. Animals marked with visual tags can yield valuable information about their migratory routes and speeds, but only if the tags are recovered. The application of tagging programs is thus limited to animals that can be recaptured in large numbers (usually commercially important species) or to animals whose tags can be observed frequently at the sea surface. Newer techniques, such as continuous tracking of individual animals fitted with radio or sound transmitters, have added considerably to our knowledge of oceanic migration patterns.

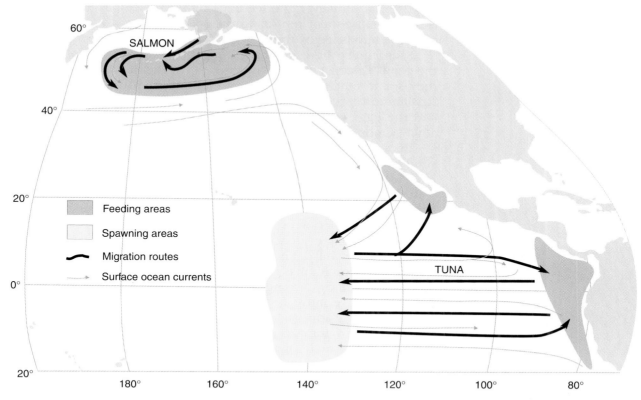

60°

SALMON

40°

20°

	Feeding areas
	Spawning areas
	Migration routes
	Surface ocean currents

TUNA

0°

20°

180°   160°   140°   120°   100°   80°

Figure 14.1

The general oceanic migratory patterns of the eastern Pacific skipjack tuna and the Bristol Bay sockeye salmon. Note the apparent relationship between these migratory patterns and surface ocean currents. These currents are identified in figure 1.33.

Adapted from Royce, Smith, and Hartt 1968 and Williams 1972.

Migratory patterns may also be determined by analyzing the distribution of eggs, larvae, young individuals, and adults of a species. When a general progression of developmental stages from egg to adult can be found extending from one oceanic area to another, a migratory route between those areas may be inferred. Some specific examples are described in the next section.

Some Examples of Extensive Oceanic Migrations

The skipjack tuna is widely distributed in the warm waters of the world ocean. Several genetically distinct populations probably exist, but we will examine only the eastern Pacific population.

Skipjack tuna spawn during the summer in surface equatorial waters west of 130° W longitude (figure 14.1). For several months, the young fishes remain in the central Pacific spawning grounds. After reaching lengths of approximately 30 cm, they either actively migrate or are passively carried to the east in the North and South Equatorial countercurrents. These adolescent fishes remain in the eastern Pacific for about one year while they mature. Two feeding grounds, one off Baja California and another off Central America and Ecuador, are the major centers of skipjack concentrations in the eastern Pacific.

As the skipjack approach sexual maturity, they leave the Mexican and Central-South American feeding grounds and follow the west-flowing Equatorial currents back to the spawning area. After spawning, the adults follow the Equatorial countercurrents they followed as adolescents. However, the feeding adults are seldom found as far to the east. Subsequent returns to the spawning area follow the general pattern established by the first spawning migration.

Salmon also have extensive migrations. Six species of salmon in the genus *Oncorhynchus* live in the North Pacific. All are anadromous; they spend much of their lives at sea and then return to freshwater streams and lakes to spawn. They deposit their eggs in beds of gravel, and the eggs remain there through the winter. After spawning, the adult salmon die.

Because the migratory patterns of the various types of salmon are similar, only the patterns of the sockeye salmon will be described here. After hatching in the spring, the young sockeye remain in freshwater streams and lakes for about two years as they develop to a stage known as **smolts.** The smolts then migrate downstream and into the sea and enter a period of heavy feeding and rapid growth.

Accumulating evidence indicates that the sockeye, as well as other salmon, follow well-defined migratory routes, usually 10 to 20 m deep, during the oceanic phase of their migrations. These migrations closely follow the surface current pattern in the North Pacific (figure 14.1), but the sockeye move faster than the currents. After several years at sea, the sockeye approach sexual maturity, move toward the coast, and seek out freshwater streams. Strong evidence supports a home-stream hypothesis that each salmon returns to precisely the same stream and tributary in which it was spawned. There it spawns for its only time and then dies.

The Atlantic eel *Anguilla* exhibits a migratory pattern just the reverse of the Pacific salmon. This eel also migrates between fresh and saltwater. But, in complete contrast to salmon, Atlantic eels are **catadromous.** They hatch at sea and then migrate into lakes and streams, where they grow to maturity.

Two species of the Atlantic eel exist—the European eel and the American eel. The distinction between the species is based on geographical distribution and the anatomical and genetic differences of the adults (figure 14.2). Both species spawn deep beneath the Sargasso Sea region of the North Atlantic. Their eggs hatch in the spring to produce a leaf-shaped, transparent **leptocephalus larva** about 5 cm long (figure 14.3). The leptocephalus larvae, drifting near the surface, float out of the Sargasso Sea and move to the north and east in the Gulf Stream. After one year of drifting, American eel larvae metamorphose into young **elvers** that move into rivers along the eastern coast of North America. The European eel larvae continue to drift for another year across the North Atlantic to the European coast (figure 14.4). There, most enter rivers and move upstream. The remainder of the European population requires still another year to cross the Mediterranean Sea before entering fresh water.

After several years (sometimes as many as 10) in fresh water, the mature eels (now called yellow eels) undergo physical and physiological changes in preparation for their return to the sea as silver eels. Their eyes enlarge, and they assume a silvery and dark countershaded pattern characteristic of midwater marine fishes. Then they migrate downstream and, presumably, return to the Sargasso Sea, where they spawn and die.

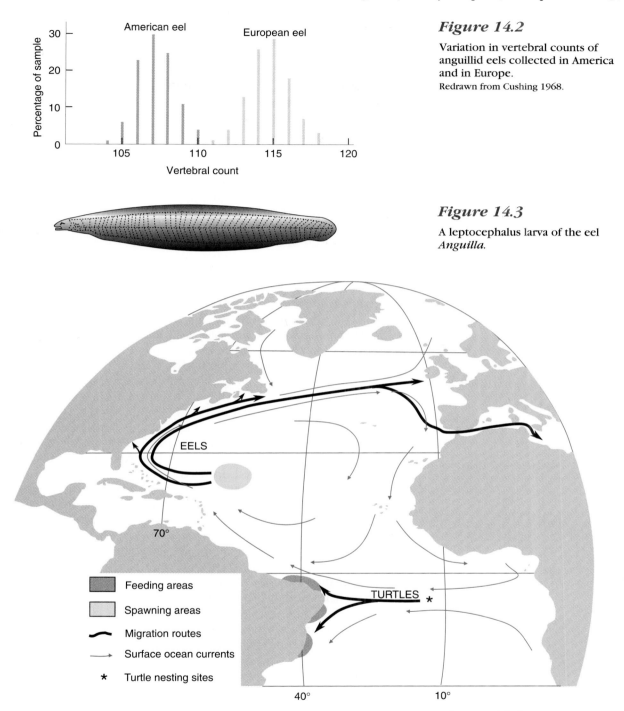

Figure 14.2

Variation in vertebral counts of anguillid eels collected in America and in Europe.
Redrawn from Cushing 1968.

Figure 14.3

A leptocephalus larva of the eel *Anguilla*.

Legend:
- Feeding areas
- Spawning areas
- Migration routes
- Surface ocean currents
- * Turtle nesting sites

Figure 14.4

Migratory routes of the larvae and young of anguillid eels and green turtles. The return migrations of the respective adults have been omitted for clarity.
Adapted from Carr 1965.

Very few adult silver eels have been captured in the open sea, and none have been taken from the spawning area itself. Thus, the spawning migration back to the Sargasso Sea is still a matter of supposition. The European eels may backtrack the path they followed as leptocephalus larvae. However, swimming against the substantial current of the Gulf Stream would require greater energy expenditures. A more likely route would take the eels into the south-flowing Canary Current after leaving European rivers, then west in the North Atlantic Equatorial Current, and eventually to the region of the Sargasso Sea (figure 14.4). The American eels apparently swim across the Gulf Stream to their spawning area.

Studies of variation in the structure of mitochondrial DNA confirm the contention that American and European eels are genetically isolated and should be treated as separate species. These studies also revealed a hybrid population of eels inhabiting streams in Iceland. But, despite repeated attempts with sophisticated SONAR, underwater video cameras, and high-speed nets, no adult eels of either species (nor of their hybrids) have yet been observed or captured in the presumed spawning areas of the Sargasso Sea.

South of the Sargasso Sea, several species of sea turtles lay their eggs in nests dug in sandy beaches above the high tide lines of tropical and subtropical shores. Green sea turtles, *Chelonia,* have a strong tendency to migrate for nesting from coastal feeding grounds to remote, isolated islands. These islands apparently lack many of the predators that would harass the turtles and raid their nests on mainland beaches. The best-documented feats of island finding by green turtles are migrations between the east coast of Brazil and Ascension Island. Ascension Island is a tiny piece of land only 8 km wide in the Atlantic Ocean midway between Brazil and Africa (figure 14.4).

These turtles lay their eggs in the warm, sandy beaches along the north and west coasts of Ascension Island. Immediately after hatching, the young turtles instinctively dig themselves out of the sand, scurry into the water, and head directly out to sea. During this very short period, they are heavily preyed upon by seabirds and large fishes. Once they are beyond the hazards of shoreline and surf, they presumably are picked up by the South Atlantic Equatorial Current and are carried toward Brazil at speeds of 1 to 2 km/h. Less than two months is needed to passively drift to Brazil, yet nothing is known of the young turtles' whereabouts or activities during their first year.

As they mature, the turtles congregate along the mainland coast of Brazil, where they graze on turtle grass and other sea grasses in shallow flats. Features of the nesting migration back to Ascension Island are not well known, but the adult turtles do show up there in great numbers during the nesting season.

Mating apparently occurs only near the nesting ground. Either the males accompany the females on their migration from Brazil to Ascension, or they make a precisely timed, but independent, trip on their own. Either way, the males get to the nesting area and can be seen just outside the surf zone splashing and fighting for the attentions of the females.

The females go ashore several times during the nesting season and deposit about a hundred eggs each time. These egg-laying episodes provide the only opportunity researchers have to capture and tag large numbers of green turtles at their nesting sites. (Because males do not leave the water, almost nothing is known about their migratory behavior.) The tagging results indicate that the females leave Ascension Island after laying the eggs and return to the Brazilian

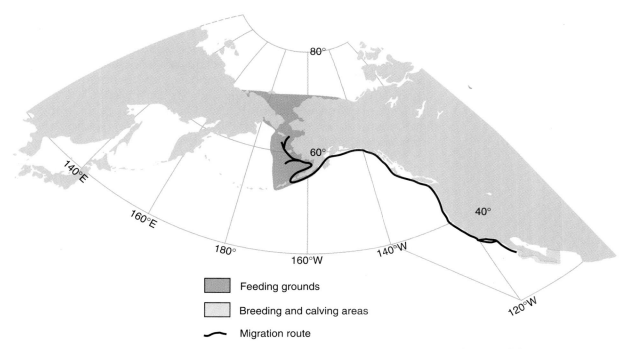

- Feeding grounds
- Breeding and calving areas
- —— Migration route

Figure 14.5

The migratory route of the California gray whale.

coast. Two or three years later, they return to the tiny island of Ascension again to mate and lay their eggs.

Like marine turtles, several of the larger marine mammals undertake impressive seasonal migrations. The large whales alternate between cold-water summer feeding grounds and warm-water winter breeding and calving grounds. The annual migration of the California gray whale has been extensively studied and is the best known of the large whale migrations. These whales migrate an impressive 18,000 km (11,000 mi) round-trip each year. Most gray whales spend the summer in the Bering Sea and adjacent areas of the Arctic Ocean as far north as the edge of the pack ice. Their habit of feeding on bottom invertebrates and their limited capacity to hold their breath (4–5 minutes) restrict their feeding activities to the shallow portions of these seas (usually less than 70 m).

The southward migration is initiated in autumn possibly in response to shortening days or to the formation of sea ice in Arctic waters. The migration is a procession of gray whales segregated according to age and sex. Pregnant females leave first and are followed by nonpregnant females, immature females, adult males, and finally immature males. Recent observations have shown that, after they pass through the Aleutian Islands, gray whales follow the long, curving shoreline of Alaska (figure 14.5).

South of British Columbia, the whales can be observed traveling reasonably close to the shoreline (in water usually less than 200 m deep). However, a few travel well offshore in water over 1,000 m deep. The average speed of southbound gray whales is 7 km/h. At that speed, most of the whales reach the warm, protected coastal lagoons of Baja California by late January.

It is in these lagoons that the pregnant females give birth and the males and nonpregnant females mate. The new mothers remain with their calves in the lagoons for about two months. During that time, the nursing calves rapidly put on weight to face the rigors of a long migration back to the chilly waters of the Bering Sea.

In early spring, the northward migration begins and is much the reverse of the previous southbound trip, with nursing females and their calves the last to leave the lagoons. Traveling at a more leisurely pace than when going south, the whales reach their Arctic feeding grounds in the late spring or early summer. They spend the summer rapidly restoring their depleted fat and blubber reserves in preparation for their next migratory performance a few months later.

A few generalizations can be made from these examples of long-distance swimming migrations of marine animals. Timing and precise routes are not known for many migrating species. Despite these gaps in our information, we do know that numerous species of marine animals undertake and successfully accomplish long and sometimes complex migrations. In general, the migrations link areas that ensure reproductive success with other areas that provide an abundance of food. And, quite often, these migratory paths follow ocean current patterns.

Orientation

How do migratory species know where they are and where they are going? Before an animal can successfully accomplish a directed movement from one place to another, it must orient itself both in time and in space. Biological clocks operating on circadian and longer period rhythms (see chapter 8) are important factors in the orientation process. A variety of environmental factors serve as cues to adjust or reset the timing of these rhythms. Well known among these timing factors is the day length, which changes with predictable regularity through the seasons. Day length, water temperature, and food availability might serve as useful cues for following the passage of the seasons. These and other factors have been suggested as cues that trigger the seasonal migrations of gray whales and other marine animals.

Orientation in space is somewhat more complex than time orientation. Terrestrial animals and birds are known to use recognizable landmarks to orient themselves. Much of the gray whale migration occurs within sight of land. These whales frequently thrust their heads vertically out of the water; it has been suggested that this behavior is a means of getting visual bearings on coastal headlands and other recognizable landmarks. Because they usually stay inside the 200 m depth contour, gray whales might also follow ocean bottom contours of the continental shelf.

Several species of birds are capable of accurately navigating over completely unfamiliar terrain by sensing the direction of the earth's magnetic field. When researchers attached small magnets to the birds' necks to disrupt the earth's magnetic field around the birds' heads, they lost their homing ability. Control birds with nonmagnetic bars on their necks homed correctly under the same conditions. Sharks, rays, and green turtles also have a demonstrated ability to sense and respond to the earth's magnetic field.

Some species of birds are known to navigate at night using only a few stars for guidance. Directional information derived from the apparent position of the sun, moon, and stars might also be useful to migrating marine animals, but only if they can see the sky. Whales and turtles are frequently at the surface to breathe and might get their bearings and make course corrections using celestial cues. However, present evidence indicates that these air-breathing marine tetrapods have quite myopic vision in air and, thus, may have difficulty seeing stars or coastlines clearly.

It has been hypothesized that adult green turtles use a straightforward navigation system using the sun on their spawning migration back to their Ascension Island nesting sites. Ascension Island lies due east of Brazil at 8° S latitude. The adult turtles could conceivably use the height of the noonday sun to judge latitude, swim to the east at 8° S latitude, and eventually make landfall on Ascension Island. Though still hypothetical, this system would enable the turtles to make course corrections if they wandered or drifted to the north or south of the 8° S latitude line. Island finding by the turtles might be improved if, once they were within about 50 km of Ascension Island on the down-current side, they detected a characteristic chemical given off by the island. No one is sure how well green turtles can use celestial cues (if at all) or how well they can detect chemicals dissolved in water. Until these aspects of turtle biology are studied further, the guidance system of green turtles must remain hypothetical.

It is known that eels, salmon, sharks, and many other fishes have extremely keen olfactory senses. Since the 1970s, an impressive body of evidence has been gathered to support the idea that salmon use olfactory cues to guide them to their home stream. Rather than assuming that each stream and tributary has its own characteristic odor detectable some distance out to sea, F. R. Hardin-Jones postulated a slightly different sequential odor hypothesis for migrating salmon. He suggested that young salmon smolt are imprinted with a sequence of stream odors during their downstream trip to the ocean. When returning as adults, the remembered cues are played back in reverse. Hardin-Jones also suggested that the odors in the river are not "homed" on as they are encountered, but rather they act as a sequence of sign stimuli that release a positive response to swim upstream.

The chemical nature of the characteristic odors in stream water is still unidentified. However, these or similar odors are not likely to be concentrated sufficiently in the open ocean to guide the oceanic phase of the salmon's migration. What then are the guideposts available to marine fishes migrating across huge expanses of open ocean well below the sea surface?

Currents are among the most stable regional features of the oceans. The migratory patterns of many fishes and other marine animals seem closely related to surface current patterns. But how can a fish detect the direction or speed of an ocean current if it can see neither the surface nor the bottom? The sharp temperature and salinity gradients sometimes found at the edges of ocean currents might be detected by some fishes, but only if they leave the current. Available evidence suggests, however, that salmon, tuna, and possibly adult eels migrate within currents and not along their edges.

Current speeds and directions are difficult to detect from the surface if the observer is being carried by the current. But fishes

Figure 14.6

Possible speed and direction cues for fishes in an ocean current. To a drifting fish above an accumulation of debris and plankton, the debris appears to move backward. From below, the debris appears to be carried forward in the direction of the current. Numbers indicate current speed.

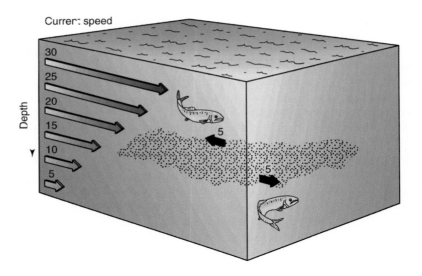

below the surface may be able to detect ocean surface currents by visually observing the speeds and direction of horizontally moving debris and plankton (figure 14.6). In the ocean, the water velocity generally decreases with depth. Fishes near the bottom of the current should see particles above them moving in the direction of the current. Slower-moving particles below the fishes would appear to move backward as the fishes were carried forward by the current. In this manner, the fishes could determine the current direction and orient their swimming motions either in the same direction or directly against it.

When charged ions of seawater are moved by the ocean's currents through the magnetic field of the earth, a weak electrical potential is generated. (A process that is similar to the operation of an electrical generator.) These ocean current potentials have been measured with ship-towed electrodes and are used to compute current speeds. Some preliminary laboratory evidence suggests that both the Atlantic eel *Anguilla* and the Atlantic salmon, *Salmo* are sensitive to electrical potentials of the same magnitude as those generated by ocean currents. In addition, they are most sensitive to the electrical potential when the long axes of their bodies are aligned with the direction of the current. If fishes can detect these potentials in the ocean, some, such as sharks, have an extremely accurate system for locating and responding to the directionality of ocean currents.

Sensory Reception

To participate successfully in migration, feeding, mating, or any other important life events, animals must be able to evaluate their immediate surroundings and to update those evaluations repeatedly. This evaluation is accomplished by employing a variety of sensory devices that serve to connect their internal nervous systems with chemical, mechanical, or electromagnetic stimuli from their external world. These sensory devices are receptor cells or organs specialized to convert stimuli into a nerve impulse; the nerve impulse is then conducted to the brain, where perception of the stimulus occurs and a response is initiated. This section will focus on the more obvious or important sensory devices of nekton, and most of that attention will be directed to fishes. The equally important internal visceral sensory systems used to maintain homeostatic conditions by monitoring internal states such as temperature, blood pressure, pH, and skeletal muscle tension are not discussed.

To begin, the concept that vertebrates possess five basic sensory capabilities (taste, smell, touch, vision, and hearing) is outdated. As mentioned in the previous section, some vertebrates exhibit electroreceptive and magnetoreceptive abilities that have no known counterparts in most terrestrial vertebrates, including humans. In addition, when animals submerged in seawater are considered, even comfortable human notions like the differences between taste and smell become confusing. Our ability to smell depends on tens of millions of ciliated sensory cells located in the nose that detect thousands of different chemicals carried to us by air. Taste, on the other hand, responds to a limited range of substances (sugars, acids, and salts) that must be dissolved in water and delivered to a few hundred taste buds on the tongue, mouth, and lips. How are these distinctions of taste and smell to be applied to invertebrates lacking noses and tongues or to vertebrates that live constantly under water?

Chemoreception

Both taste and smell are chemoreceptive senses. For nekton, they are used to detect and identify chemical substances dissolved in seawater. For marine fishes swimming in an aquatic medium of near-uniform salinity, with a buffered pH, and a nearly complete absence of sugar, an ability comparable to human taste may be of limited use. But **olfaction,** the detection with olfactory sensory cells of chemicals dissolved in water, is highly evolved in fishes. Salmon and some species of large, predatory sharks respond to very low concentrations of odor molecules. Just a few parts per billion of chemicals in the water of their olfactory sacs are sufficient for recognition. With such olfactory capabilities, it is possible for predatory sharks to locate odor sources using very dilute chemical trails left by injured prey or for a migrating salmon to locate and identify the stream of its birth.

Vision

Most animals rely on ambient light from the sun, moon, or stars to illuminate their visual fields and provide the energy needed to stimulate their photoreceptor cells. (A few animals, especially mid- and deep-water fishes, have light-producing photophores to illuminate their own very small visual fields.) Although sunlight travels several kilometers through our atmosphere with little loss in intensity, an additional few hundred meters through the clearest ocean water so reduces the intensity that photosynthesis is impossible and vision is very limited. As light intensity is reduced, visual fields shrink to a few meters, and the range of colors available narrows to the green and blue portions of the visible spectrum. (See the section on light in the sea in chapter 1.)

Figure 14.7

Cross section of a fish eye. Note the solid, round lens that is focused by being moved near to or away from the retina by the retractor muscle.

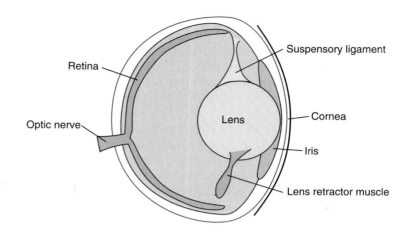

Marine vertebrates, except hagfishes, and all cephalopods, except *Nautilus,* obtain visual images of their surroundings with a remarkable organ, the camera eye (figure 14.7). The basic structure is similar to a human eye, with light focused by a lens through a light-tight and nearly spherical eye cavity to the light-sensitive receptor cells of the **retina** at the back of the eye. In contrast to eyes fit for vision in air, fish and squid eyes must accommodate the higher refractive power of water. Fish and squid eyes are typically flattened in front and are fitted with a round and rigid lens that focuses by moving nearer to or away from the retina rather than by changing shape.

Rods, in vertebrates, are a type of light-sensitive retinal cell specialized for low-intensity light detection. They can be triggered with lower light energies than can cones, the other light-sensitive retinal cells. Typically, cones serve as high-light and color receptors, and rods serve as low-light receptors. Fishes living below the photic zone usually have fewer cones than rods, and deep-sea fishes often lack cones. For all fishes with only one type of cone (or no cone cells at all), vision is limited to detecting variations in light intensity; they see their world in varying shades of gray. Many fish species nearer the surface and in better-illuminated marine environments such as coral reefs, are capable of varying degrees of color vision. The retinas of these fishes contain either two or three different types of cone cells, each type sensitive to a particular and different range of wavelengths. In bright light and clear water, those fishes with three types of cones apparently possess a visual color acuity that may exceed that of humans.

Equilibrium

A critical aspect of orientation in space is knowing which way is up or down. For human divers away from surface light or diving at night, the absence of clear notions of up and down can be very disorienting. Equilibrium organs in a range of marine animals from jellyfishes to whales detect the tug of gravity. From that, they obtain enough information to sense and control their orientation in space. The equilibrium organs are small fluid-filled structures called **statocysts;** they contain a calcareous particle (the statolith) suspended by cilia extending from sensory cells (figure 14.8) lining the statocyst. Statocysts respond only to the pull of gravity.

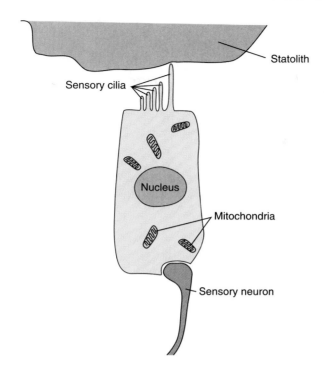

Figure 14.8

General structure of a mechanosensory hair cell. When the sensory cilia of the hair cell are bent, a nerve impulse is initiated and passed to the associated sensory neuron.

It is useful for actively moving animals to also know their rate of speed and any changes in their direction. With this information, body positions and orientations can be updated repeatedly. To do this, almost all nekton have two types of receptors within their organs of equilibrium; one detects gravity, and the other detects acceleration forces.

In squids and cuttlefishes, the organs of equilibrium are embedded in the capsule surrounding the brain. Each organ contains extensive plates of sensory hair cells (like those shown in figure 14.8) to detect the movement of the statolith or the surrounding fluid.

Vertebrates use the basic sensory **hair cell** design to serve a variety of mechanoreceptive chores. Fishes and other vertebrates have an equilibrium organ similar in both structure and function to that of squids. It is the **labyrinth organ,** located on either side of the head (figure 14.9). Each labyrinth organ consists of three semicircular canals and two smaller sac-shaped chambers. These sac-shaped chambers are the gravity detectors, with small stony secretions suspended by hair cells known as **neuromasts.** The canals are the acceleration detectors. Each canal is filled with endolymph and has an enlarged ampulla that is lined with hair cells supporting a cupola. Acceleration in any direction moves the endolymph against the cupola and stimulates the neuromasts of at least one canal.

Sound Reception

Sound is transmitted through air or water as spreading patterns of vibrational energy. This energy travels at 1,500 m/s in water, about five times faster than in air. Although the sea may at times seem surprisingly silent to human divers, animals with sensory organs attuned to that medium must find it a somewhat noisy place. Swimming animals

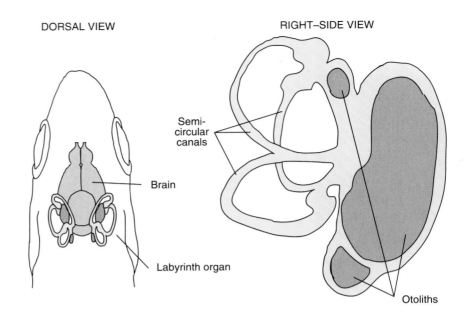

Figure 14.9

Anatomical location (left) and
general structure (right) of a
labyrinth/otolith organ of a bony fish.

produce unintentional noises as they move, feed, and bend their bodies. Others make intentional grunts, groans, chirps, and other noises. Add those to surface wave noises and vocalizations from whales and seals, and the ocean becomes a cacophony of sounds within, below, and well above the range of human hearing.

Fishes detect sounds with mechanoreceptive sensory hair cells nearly identical to those in their organs of equilibrium. The organs associated with sound detection in fishes are the otoliths (figure 14.9). Otoliths are small calcareous stones embedded in and associated with part of the labyrinth organ. Together they constitute the inner ear. On each side of the head, two or three otoliths are suspended in fluid-filled sacs, where they contact neuromasts. Arriving sound waves move the fish very slightly, and the denser otoliths lag behind, bending the neuromast cilia and stimulating a nerve impulse. Hair cells with different orientations relative to the otoliths may provide some directional information about the sound source.

The lateral line system of fishes consists of canals contained in scales and extending along each flank and in complex patterns over their heads (figure 14.10). Within the canals are neuromasts. The cilia of the neuromasts are stimulated by water movement and by pressure differences at the fish's body surface. These are communicated to the lateral line canals through pores in the skin surface. In this way, the lateral line systems function to detect disturbances caused by prey or by predators, by swimming movements of nearby schooling companions, and sometimes by sound vibrations.

Echolocation

In most tetrapod species vision is well developed. Yet several groups function well in conditions where the lack of light renders vision nearly useless. About 20% of all mammal (and even a few bird) species have overcome the problems of orienting themselves and locating objects in

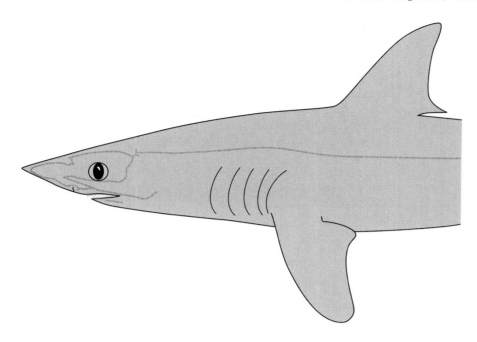

Figure 14.10
The major branches of the left lateral line system (blue) of a shark. Pores scattered in clusters over the snout are the surface openings to the ampullae of Lorenzini.

the dark by producing sharp sounds and listening for reflected echoes as the sounds bounce off objects. Bats are well-known echolocators, but so too are some shrews, golden hamsters, flying lemurs, and many marine mammals.

Soon after the first hydrophone was lowered into the sea, it became apparent that whales and pinnipeds could generate a tremendous repertoire of underwater vocalizations. Many of the moans, squeals, and wails are evidently for communication. Bottle-nose dolphins, *Tursiops,* also produce a large variety of whistle-like sounds, and captive individuals have been shown to understand complex linguistic subtleties. Other sounds, especially those of the humpback whale, *Megaptera,* have a fascinating musical quality. The songs of each humpback whale population are identifiably different from the songs of other populations, are probably produced exclusively by breeding males, and are culturally transmitted from one individual to another within each population. Each song is composed of numerous phrases, some of which are repeated several times. During each breeding season, the songs evolve; some phrases are modified and others are added or deleted.

The sounds most useful for echolocation are neither squeals nor songs but trains or pulses of clicks of very short duration. Much more is known about the echolocating capabilities of the smaller whales, such as *Tursiops,* for they are easily and frequently maintained in captivity. *Tursiops* uses clicks consisting of sound frequencies audible to humans as well as higher-frequency clicks well beyond the upper range of human hearing. Each click lasts only a fraction of a millisecond and is repeated as often as 800 times each second. Click repetition rates are adjusted to allow the click echo to return to the animal during the very short lull between outgoing clicks. As each click strikes a target, a portion of it is reflected back to the source (figure 14.11). The time required for a click to travel from an animal to the

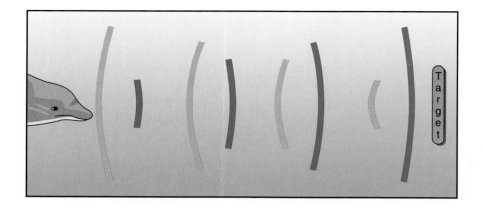

Figure 14.11

Pattern of click production for echolocation. Outgoing clicks (dark pink areas) are spaced so that echoes returning from the target (light pink areas) can be received with little interference.

reflecting target and back again is a measure of the distance to the target. As that distance varies, so will the time necessary for the echo to return. Continued evaluation of returning echoes from a moving target can indicate the target's speed and direction of travel.

Low-frequency clicks usually serve as orientation or scanning clicks for surveying an animal's general surroundings. Higher-frequency clicks always occur in situations in which fine discriminations must be made. Relying solely on their echolocating abilities, blindfolded bottle-nosed dolphins have repeatedly demonstrated an aptitude for discriminating between objects of a similar nature: two fishes of the same general size and shape, plates of different metals, and pieces of metal differing only slightly in thickness. In the wild, these animals must acoustically survey their surroundings, while simultaneously distinguishing their own echolocation clicks from the cacophony of other sounds so frequently present in large herds of wild dolphins (figure 14.12).

How do whales produce the sounds involved in echolocation, and how do they receive and process the echoes? The larynx of toothed whales is well muscled and complicated in structure, yet it lacks vocal cords. The elongated tip of the larynx extends across the esophagus into a common tube leading to the blowhole at the top of the head. This arrangement completely separates the pathways for food and air; consequently, underwater feeding and sound production can occur simultaneously.

At the blowhole are a pair of heavily muscled valves, the nasal plugs. These plugs, with an associated complex of **air sacs** branching from the nasal passage, are the sites of click production in the smaller toothed whales (figure 14.13). High intensities of emitted clicks measured over the surface of the head tend to be centered above the margins of the upper jaw and suggest a sound production site somewhere in the vestibular sac region. Clicks produced here are directed forward by the concave front of the skull and then focused by the fatty lens-shaped **melon** (the rounded forehead structure so characteristic of toothed whales (figure 14.14), to concentrate the clicks into a narrow, directional beam. Recent research indicates that some species of toothed whales may also stun fish prey with intense blasts of sound energy, presumably using the same sound production system used for echolocation.

Figure 14.12

A pod of dolphins at sea.

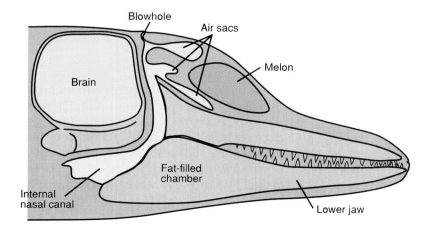

Figure 14.13

Midsection of a dolphin head, showing the bones of the head, the air passages, and the structures associated with sound production.

Figure 14.14

A bottlenose dolphin, *Tursiops,* with a prominent melon.

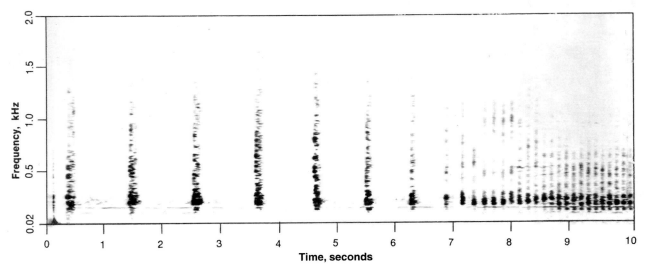

Figure 14.15

A sonogram of the echolocation click pulses of a sperm whale approaching a recording hydrophone. Each pulse shown (starting about 1 s apart) consists of several separate clicks not resolved in this illustration. Note the accelerating pulse repetition rate.

Courtesy J. Fish, Naval Undersea Center, San Diego.

The echolocation clicks of the much larger sperm whales are lower in frequency and generally have a slower repetition rate. They can be quite powerful and may carry for several kilometers in the sea. Each click lasts about 24 milliseconds and is composed of a pulse or burst of up to nine separate clicks. Figure 14.15 is a sonogram of a portion of a train or sequence of click pulses emitted by a sperm whale at sea. The boat from which the recording hydrophone was suspended was apparently the target of the whale's echolocation efforts. As the whale swam toward the boat to investigate, the time required for successive pulses to travel from the whale to the boat and back to the whale decreased. The whale compensated by increasing the repetition rate of click pulses (middle of figure 14.15) to keep the echoes returning between the outgoing sound pulses. Near the boat, the click pulses were being

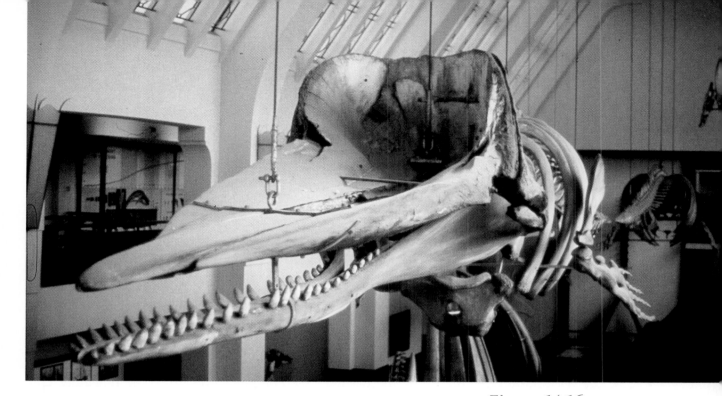

Figure 14.16

A sperm whale skeleton. Note the concave shape of the skull, which in life is filled by the spermaceti organ.

emitted very rapidly and then abruptly ceased as the whale passed beneath the boat and presumably came within visual range.

The powerful, long-range echolocation systems of sperm whales may partially explain their success as efficient predators of the larger squid of midwater depths. Visualize these whales cruising along at the sea surface with all the air they need, periodically scanning the unseen depths below with a short burst of echolocation click pulses. Only when a target worthy of pursuit is detected and its location pinpointed does the whale depart from its air supply and go after its meal. In addition to their likely function in echolocation, there is some evidence to suggest that the click trains of sperm whales also serve as a means of communication between individual whales during dives. These sounds travel well underwater, and the patterns of clicks produced serve as recognition codes for individual whales so that they can keep track of other pod members while diving.

A complex sound production system has been proposed for the compound click pulses of sperm whales. These whales are noted for their massive and very distinctive foreheads. Inside the forehead is a highly specialized melon, the **spermaceti organ,** which may occupy 40% of the whale's total length and 20% of its weight. This organ is filled with waxy spermaceti oil, a fine-quality liquid once prized by whalers for candlemaking and for burning in lanterns. The spermaceti organ is encased within a wall of extremely tough ligaments, and the entire structure sits in the hollow of the rostrum and the amphitheater-like front of the skull (figure 14.16).

At either end of the spermaceti organ are two large flattened air sacs. These sacs are connected to each other and to the remainder of the respiratory system by the left and right nasal passages (figure 14.17). The large left nasal passage penetrates the spermaceti organ and leads directly to the blowhole at the tip of the snout. One branch

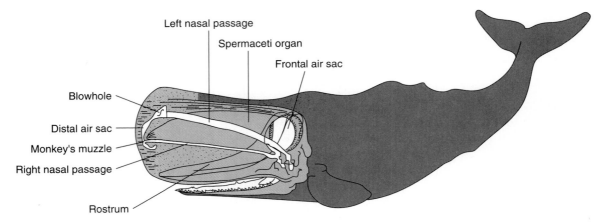

Left nasal passage
Spermaceti organ
Frontal air sac
Blowhole
Distal air sac
Monkey's muzzle
Right nasal passage
Rostrum

Figure 14.17

A cutaway view of the complex melon structure of the sperm whale. Adapted from Norris and Harvey 1972.

of the much smaller right nasal passage extends to the frontal air sac at the posterior end of the spermaceti organ; the other branch is directed anteriorly along the base of the spermaceti organ and ends at the monkey's muzzle. The monkey's muzzle is a structure that consists of a pair of hard, well-matched, and tightly compressed lips. In front of these lips, the distal air sac continues upward to the blowhole, connects with the left nasal passage, and completes the loop of air passages associated with the spermaceti organ.

These structures are responsible for the production of the multiple-click pulses of sperm whales. The clicks are produced as air is forced between the hard lips of the monkey's muzzle; the hard lips part and abruptly snap back together to create a sharp report or click. A portion of this sound signal is emitted directly into the water ahead of the whale and probably represents the first click of each click pulse. Subsequent clicks of declining intensity within a click pulse may be derived from reverberations of that initial signal as it bounces between two sound mirrors, the distal and frontal air sacs, at each end of the spermaceti organ. All of these rapidly reverberated clicks together produce one of the click pulses shown in figure 14.15. The spermaceti organ itself may function as an effective sound channel to guide the reflected click between the two air sacs. The air used to activate the lips of the monkey's muzzle can be recycled back through the left nasal passage and can be used repeatedly without loss during a dive.

Behavioral studies suggest that all marine mammals have good hearing. Experimental evidence, however, is largely restricted to studies of captive small toothed whales. Humans are sensitive to sound frequencies ranging from 16 to 20,000 vibrations per second (1 vibration/s = 1 Hertz [Hz]). The bottlenose dolphin, and presumably some other toothed whales, respond to sound frequencies in excess of 150,000 Hz. Their sound detection systems must be attuned to very weak echoes of their own clicks but must simultaneously withstand the powerful blast of outgoing clicks generated in adjacent regions of the head. The sound-processing structures of the middle ear are enclosed in a bony case, the **tympanic bulla.** In toothed whales, the bulla is separated from adjacent bones of the skull by air sinuses filled with an insulating emulsion of mucus, oil, and air. The bulla is suspended in this emulsion, supported only by a few wisps of connective tissue. Thus, each middle ear functions as a separate sound receiver to pinpoint sound sources better.

The **external auditory canal** is the usual mammalian sound channel connecting the external and middle ears. The auditory canal of mysticetes is commonly blocked by a plug of earwax; in toothed whales, the canal is reduced to a tiny pore or is completely covered by skin. Mapping of acoustically sensitive areas of dolphins' heads has shown the external auditory canal to be about six times less sensitive to sound than the lower jaw. These results support the hypothesis of a unique sound reception system in toothed whales. The bones of the lower jaw are flared toward the rear and are extremely thin. Within each half of the lower jaw is a fat body (or, in some cases, liquid oil) that directly connects with the wall of the bulla of the middle ear. The fat or oil bodies, like the oil of the sperm whale's spermaceti organ, act as a sound channel to transfer sounds from the flared portions of the lower jaw directly to the middle ear. An area on either side of the forehead is nearly as sensitive as the lower jaw, providing multiple hearing channels with four very sensitive centers for sound reception.

How common is echolocation in marine mammals? Presently, it is uncertain because it is difficult to establish whether wild populations are indeed using echolocation-like clicks for the purposes of orientation and location. If judgments can be made from the types of sounds produced, then echolocation should be suspected in all toothed whales, some pinnipeds (the Weddell seal, California sea lion, and possibly the walrus), and at least a few baleen whales. Click trains have been recorded in the presence of gray whales in the North Pacific and blue whales and minke whales in the North Atlantic. It is not unreasonable to assume that these animals use these sounds, as well as any other sensory means they possess, to find food, locate the bottom, and evaluate the nonvisible portion of their surroundings.

Electroreception and Magnetoreception

Humans are totally oblivious to the weak electrical and nonvisible electromagnetic energy fields generated by contractions of muscles in swimming animals, by water currents moving past inanimate objects, and even by the earth's own magnetic field. Yet organisms as small as bacteria and as large as sharks detect and respond to some of these signals. These specialized senses are known or suspected to exist in several classes of vertebrates, but the best studied examples are the cartilaginous fishes. Sharks, skates, rays, and chimaerns all exhibit an extensive network of tiny pores or pits arranged on their snouts and pectoral fins. Each pit connects, via a short jelly-filled canal, to a flask-shaped ampulla of Lorenzini. These ampullae are associated with the lateral line system of cartilaginous fishes (figure 14.10) and of at least one marine bony fish, the marine catfish. Electroreception is accomplished by sensory cells located at the bottom of each ampulla, possibly evolved from the basic lateral-line sensory hair cell. With this sensory system, some sharks and rays are able to detect (at distances of a meter or so) bioelectrical fields equivalent to those generated by the muscle contractions of typical prey species. Similar electrical fields are also produced by some metal objects in seawater. The seemingly erratic responses by some sharks to these objects may be explained as the sharks' normal response to electrical fields mimicking those produced by their usual prey.

The study of geomagnetic reception in animals is still in its infancy. It is confirmed or suspected in cartilaginous fishes, some bony fishes such as tuna and salmon, some birds, and possibly some whales. The ampullae of Lorenzini are thought to be the organs of detection in sharks and rays; in other vertebrate groups, the organs of detection have not yet been identified.

Reproduction

Reproduction in pelagic animals usually proceeds as in most other animal groups. The eggs of the female are fertilized by the male in the water or within the female's reproductive tract, and embryonic development leads to a new generation. Breeding, spawning, and other reproductive activities are generally periodic and are most often associated with higher water temperatures or the greater primary production of spring and summer months.

Nonseasonal Reproduction

The reproductive cycles of some species deviate from annual cycles for a couple of reasons. Deep-water species experience little seasonal change in their environment and may breed or spawn irregularly. Still, little is known about the reproductive features of most deep-ocean pelagic animals.

Other species vary from seasonal reproductive patterns because small size permits them to reproduce more frequently or because their very large size prohibits them from meeting an annual reproductive schedule. At one extreme are the large whales, which, after reaching maturity, breed only every second or third year. Their **gestation period** (the time between fertilization and delivery) is slightly over one year, and the energy demands that a year of pregnancy followed by six months of nursing make on the mother are enormous. Consequently, most of the larger species of whales include at least a year of rest and recovery between pregnancies.

The grunion, a small fish found in coastal waters of southern California, exhibits a curious and unusual spawning behavior (figure 14.18). On the second, third, and fourth nights after each full or new moon of the spring and summer spawning season, the grunion move up on the beach by the thousands to deposit their eggs in the sand and away from water. Even more remarkable is their precise timing; they spawn only during the first three hours immediately after the highest part of the highest spring tides. During the spring and summer, these tides occur only at night.

As the highest spring tides occur at the time of full and new moons, the grunion spawn immediately after high tides, but they spawn on successively lower tides each night (figure 14.19). Thus, the eggs are buried by sand tossed up on the beach by the succeeding lower tides, and they are not washed out of the sand until the next series of spring tides. Nine or ten days after the last spawning, tides of increasing height reach the

Figure 14.18

Grunion, *Leuresthes,* spawning
in the sands of a southern California beach.

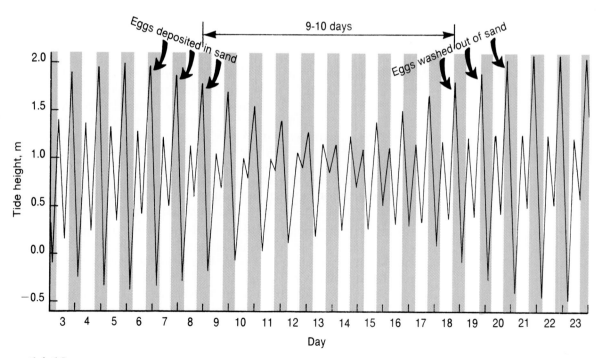

Figure 14.19

Predicted tide heights for a three-week period at San Diego, California. Spring tides appropriate for grunion spawning occur on
days 6, 7, and 8 (arrows at left). Nine to ten days later, the next set of spring tides (arrows at right) wash the eggs from the sand,
and the eggs hatch. Shaded portions indicate night hours.

area where the lowest eggs were buried (figure 14.19). Wave action erodes the sand away and bathes the eggs with seawater. Almost immediately after being agitated and wetted by the waves, the eggs hatch and the young grunion swim out to deeper water. There they feed and grow, reaching sexual maturity about one year later.

From Yolk Sac to Placenta

Viviparity is often considered a trait characteristic of mammals. Fishes, on the other hand, are generally thought to be egg layers (oviparity). Most fishes are oviparous, and some are quite prolific. A mature female cod may lay as many as 15 million eggs in a single season. These eggs are small and hatch into meroplanktonic larval stages that experience very high mortality rates. In contrast, skates and benthic sharks usually produce only a few large eggs (figure 14.20). The developing embryos are protected by a durable outer case and are nourished by the abundant yolk inside. When the eggs hatch (sometimes several months after laying), the young fishes are well developed and quite capable of surviving on their own.

Other sharks and a few bony fishes produce eggs that are maintained within the reproductive tract of the female until they hatch. A single egg of a whale shark can measure 30 cm in length. They are thus delivered alive into the world, and they obtain their nourishment from the yolk of their own eggs, from eating unfertilized eggs, from drinking uterine milk, or even from eating other embryos (figure 14.21). This intermediate condition between viviparity (live birth) and oviparity (egg laying) is known as **ovoviviparity.** Ovoviviparity is a method for incubating eggs internally and differs little functionally from the pouch-brooding habit of sea horses and pipefishes. Only it is the male sea horse and pipefishes that are equipped with abdominal brood pouches. The eggs are deposited in these pouches by the females and remain for the eight- to ten-day incubation period.

The several species of ovoviviparous fishes that provide embryonic nutrition in addition to the nutrition contained in the yolk, such as some pelagic sharks and rays, possess uteri and oviducts that are lined with numerous small projections called **villi.** The villi secrete a highly nutritive uterine milk for the embryos. In one butterfly ray, *Gymnura,* the secreting villi of the uterus extend down into the esophagus of the embryo. Thus nourished, the young of *Gymnura* at birth are 50 times larger than the initial size of their yolk sacs. Some sharks such as white-tip and hammerhead sharks, absorb nutrients through a placenta-like connection between the yolk sac and the uterine wall. The embryos of surfperches are equipped with large, vascularized fins to absorb additional nutrients from the mothers' uterine walls.

Along the continuum from ovoviviparous fishes (such as *Squalus*) to those fishes that are obviously viviparous exists a variety of reproductive conditions, some of which are rather exotic. The embryos of the porbeagle shark, *Lamna,* for instance, have no structures with which to absorb nutrients from the reproductive tract of the female. When the oldest embryo within a female *Lamna* has used its own yolk, it simply turns on the other eggs within the oviduct and consumes them. With the nutrition gained from its potential siblings, the

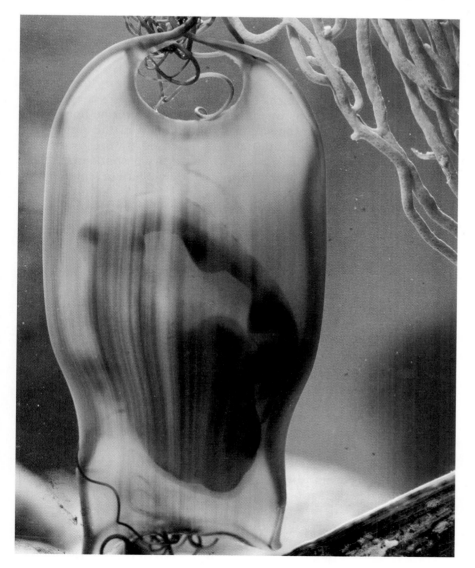

Figure 14.20

Developing swell-shark embryo, *Cephaloscyllium,* enclosed in a tough, protective egg case.
Courtesy R. E. Williams.

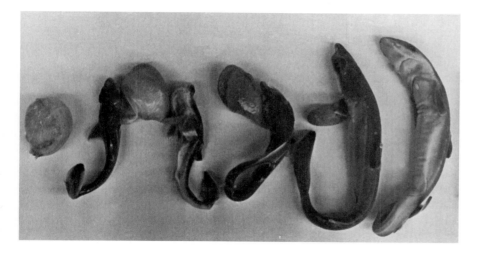

Figure 14.21

A developmental series of the dogfish shark, *Squalus,* from an egg (left) to a completely formed embryo ready for birth (right). Note the twins (second from left) joined to a common yolk.

single embryo is developmentally much better prepared for a pelagic existence before leaving the protective confines of its mother.

An extension of the strategy employed by *Lamna* is that of the sand tiger shark, *Corcharias,* found off the U.S. East Coast, South Africa, and southern Australia. The single enormous ovary found in females of this species produces large numbers of pea-sized eggs. After the two surviving embryos, one in each of the two uterine horns, have consumed their developing siblings, they remain in the uterus and consume thousands of additional eggs released by the ovary. This process may continue for a year, producing two shark pups each about 1-m long (a notable feat for a mother only 2.5 m in length).

Some Alternatives to Conventional Gender Ratios

Most species of sexually reproducing animals include approximately equal numbers of females and males. The maleness or femaleness of many animals is determined by their complement of **sex chromosomes.** In humans, the nucleus of each cell houses 23 pairs of chromosomes; one pair is the sex chromosomes, the other 22 pairs are **autosomes** not directly involved in sex determination. A human female has two large, similar sex chromosomes (an XX condition); human males have one large X and one small Y chromosome. The same is true for all other mammals. Sex in birds is also established by a pair of sex chromosomes, but the pattern is opposite that of mammals. Male birds have two similar sex chromosomes and female birds have chromosomes that differ in size and shape.

The influence of sex chromosomes on the gender of fishes is less straightforward and, in fact, is quite variable. Some guppies, for instance, reflect the mammalian pattern of sex chromosomes; XX is female and XY is male. Occasionally, though, these genders are reversed. The sex chromosomes of bony fishes lack the absolute control over gender determination found in birds and mammals. The genes involved in gender determination of fishes, unlike those of mammals and birds, are also carried on the autosomes. In some fishes, these autosomal sex genes apparently influence and even regulate the production of sex hormones, especially **androgen,** a male hormone, and **estrogen,** a female hormone. These hormones, in turn, influence the expression of several sexual characteristics and the determination of gender. Fishery scientists have found that a high percentage of salmon eggs treated with estrogen will hatch as females and that most salmon eggs treated with androgen will hatch as males.

The fluid and unfixed nature of gender determination in bony fishes has been effectively exploited through the evolution of a broad range of sex ratios and reproductive habits not common in other vertebrate groups. Part of this sexual diversity is due to the separation of sexes. Separate sexes housed in different individuals eliminates the possibility of self-fertilization and its accompanying reduction in genetic variation. Even in hermaphroditic fishes such as the sea basses *Serranus* and *Serranelus* behavioral interactions with others of the same species ensure that cross-fertilization will occur. Some deep-sea fishes also function simultaneously as both males and females (simultaneous hermaphrodites). The paths of these fish cross infrequently in the deep

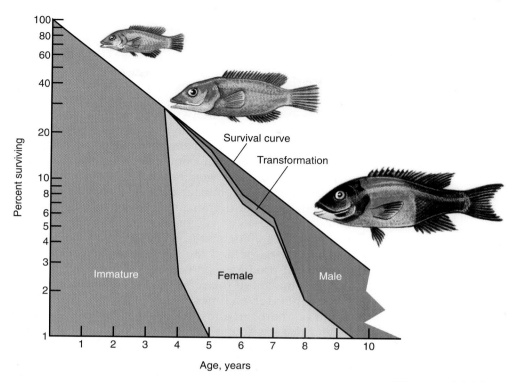

Figure 14.22

Distribution of sexes according to age of the Catalina Island sheephead population. The survival curve is based on an assumed 30% annual mortality rate. The ages of sexual maturity and sexual transformation determine the relative numbers of females and males in the population at any one time.
Data from Warner 1973.

sea, and hermaphroditism ensures that encounters between two individuals will be successful, whereas meetings between individuals belonging to the same gender would not be successful.

Because males of many species are capable of fertilizing the eggs of several females, fewer males than females are needed to accomplish reproduction. Thus, in a reproductive sense, the males of some fish populations are sexual excesses, and the reproductive potential of the individual would be better served were it a female instead.

A few fish species are exceptions to the 1:1 gender ratio. These species produce offspring that all clearly show the functional characteristics of one gender as they mature; then, at some point in their lives, some or all of them undergo a complete and functional transformation to the opposite gender. These fishes are hermaphroditic, but, unlike *Serranus,* they are sequential hermaphrodites. The entire gonad functions for one gender when it first matures and then changes to the other gender. In the California sheephead, the fishes become sexually mature as females at about four years of age. Those that survive to seven or eight years of age undergo gender transformation, become functional males, and mate with the younger females. The actual ratio of females to males depends on the survival curve of the population (figure 14.22) and on the age at which gender transformation occurs. For the sheephead, it is approximately five females to one male. In *Labrus* (belonging to the same family, Labridae, as the sheephead), gender transformation is size dependent. All *Labrus* mature as females and remain female until they reach a length of about 27 cm. Beyond that size about 50% of the surviving individuals change to males. A few species of sea basses do the reverse; they begin life as males and then change to females.

The ultimate example of manipulating the gender ratio for increased reproductive potential is found in the tropical cleaner fish *Labroides* (also in the family Labridae). This inhabitant of the Great Barrier Reef of Australia occurs in small social groups of about 10 individuals. Each group consists of one dominant functional male and several females existing in a hierarchical social group. The single male accommodates the reproductive needs of all the females. This type of social and breeding organization is termed **polygyny.** In polygynous populations, only the dominant, most aggressive individual functions as the male and, by himself, contributes half the genetic information to be passed on to the next generation.

In the event the dominant male of a *Labriodes* population dies or is removed, the most dominant of the remaining females immediately assumes the behavioral role of the male. Within two weeks, the dominant individual's color patterns change, the ovarian tissue is replaced with testicular tissue, and the population has a new male. In this manner, males are produced only as they are needed, and then only from the most dominant of the remaining members of the population.

These gender changes seem to be controlled by the relative amounts of androgen and estrogen produced by the gonads as fishes grow and mature. Young female sheepheads, when artificially injected with the male hormone androgen, change to males at a younger age than normal. Injections of estrogen delay gender transformation and maintain the individual in a prolonged state of femaleness. Conditions of social stress imposed by the dominant male *Labriodes* may induce estrogen production in the females and inhibit gender transformation. Removal of the male may eliminate that imposed stress and permit the dominant female to transform.

These examples are but a few of the vast array of reproductive patterns occurring in marine fishes. In the course of evolution, the selective advantage of each is being continually tested and retested. Whether the reproductive strategy of a particular species relies on millions of small eggs or a few large ones, oviparity or viviparity, separate sexes or hermaphroditism, each in its own way contributes to the biological success of that species.

Reproduction in Marine Mammals

Most of our knowledge concerning the reproductive patterns of marine mammals has been gleaned from observations of captive animals in oceanariums, from carcasses on board whaling ships, and from field studies of pinnipeds in breeding rookeries. Marine mammals, like their terrestrial kin, give birth to live young (figure 14.23). The young of cetaceans, capable swimmers at birth, instinctively surface to breathe. Most pinnipeds are unable to swim at birth, so the pups are invariably delivered on land or on ice floes.

The newborn of some marine mammals are relatively large. Gray whale calves weigh in at approximately 1 ton; blue whales weigh closer to 3 tons at birth. Still, these newborn mammals are smaller than their parents, and their insulating layers of blubber or fur are not usually well developed. Several factors compensate for the high surface-area-to-volume ratios and the potentially serious problem of

Figure 14.23

A killer whale birth.
Redrawn from a photo by Sea World, Inc.,
San Diego.

heat loss and body temperature maintenance in newborn marine mammals. Terrestrial pupping in pinnipeds provides some time for growth before the pups must face their first winter at sea. The larger cetaceans, including the gray whale described in the previous chapter, spend their summers feeding in cold polar and subpolar waters and then undertake long migrations to their calving grounds in tropical and subtropical seas (figure 14.24). In these warm waters, their calves have an opportunity to gain considerable weight before migrating back to their frigid feeding grounds.

The growth rates of the young of some marine mammal species are truly astounding. Weddell seal pups gain 3 kg each day, and elephant seals gain as much as 7 kg a day. Pups of both of these species double their weights within two weeks after birth. Nursing blue whales grow from 3 tons at birth to 23 tons when weaned a scant seven months later (an average weight gain of almost 100 kg a day). These prodigious growth rates are supported by an abundant supply of high-fat milk. Cetacean milk is 25 to 50% fat (cow's milk ranges from 3 to 5% fat). The daily milk yield of a large baleen whale has been estimated at nearly 600 liters (over a half ton). In smaller pinnipeds, 2 to 5 liters are more typical. Pinniped milk is also generally high in fat; however, it may be as low as 16% in the California sea lion. In both cetaceans and pinnipeds, the species occupying colder waters consistently produce milk with a high fat content.

The energy demands made on the female to produce a relatively large offspring and then to supply it with large quantities of fatty milk until it is weaned (usually a few weeks to several months) are exceedingly high. Even the water that goes into the milk imposes additional osmotic costs that must be paid with further energy expenditures. Marine mammals have a reasonably long gestation period (several months to a year) and tend to reproduce not more than once a year. It is typical for the larger whales and at least the walrus, among pinnipeds, to mate only once every two or even every three years.

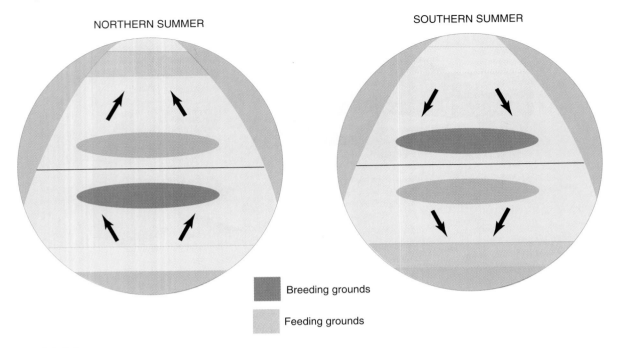

NORTHERN SUMMER

SOUTHERN SUMMER

Breeding grounds

Feeding grounds

Figure 14.24

Generalized migratory patterns of large whales between summer feeding and winter breeding grounds. Northern and southern populations follow the same migratory pattern but do so six months out of phase with each other. Consequently, northern and southern populations of the same whale species remain isolated from each other, even though both populations approach equatorial latitudes.
Adapted from Mackintosh 1966.

Figure 14.25

The reproductive cycle of female fin whales, *Balaenoptera physalus.*
Adapted from Mackintosh 1966.

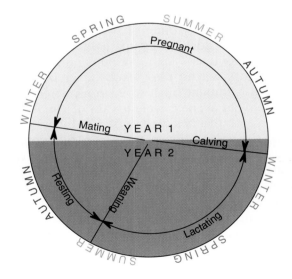

The breeding cycle of the larger baleen whales is typified by the fin whale. The cycle consists of three parts (figure 14.25). An 11-month gestation period is culminated by the birth, in tropical waters, of a 2-ton calf that is 6 m long. The calf nurses for six months. During that time, the calf and its mother migrate back to their polar feeding grounds. With food abundant there, the calf is weaned. The female

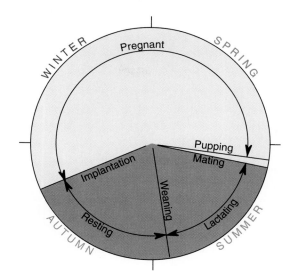

Figure 14.26

The reproductive cycle of the female Pribilof fur seal, *Callorbinus ursinus*. Light blue indicates the period of embryo dormancy; gray indicates the period of active fetal growth.

then enters a well-deserved six-month period of rest and recovery. During this feeding period, her fat and blubber reserves are replenished before she migrates back to the winter breeding grounds to mate and begin the cycle again.

The seasons and areas used for breeding by these migratory cetaceans tend to coincide with those used for calving. The same is true for many pinnipeds. The northern fur seal population, numbering about 1.5 million, disperses and forages over much of the North Pacific during the winter. Each summer they congregate in rookeries on the shores of North Pacific islands such as the tiny Pribilof Islands in the Bering Sea. Here both pupping and breeding occur. A few days after giving birth, female fur seals experience a short but intense period of **estrous** during which they are sexually receptive. This brief estrous is the only time during the year that the female ovulates and can become pregnant.

Here a problem arises. A fur seal fetus requires only seven months to develop. Yet seven months after estrous and mating, the pregnant female is far from land somewhere in the wintry North Pacific. These are impossible conditions for delivering a pup that cannot swim. This dilemma is solved by a uniquely mammalian reproductive feature: **delayed implantation.** After mating, the fertilized egg undergoes several cell divisions to form a small ball of cells, the **blastocyst.** Unlike the usual course of development for mammalian embryos, for the next four months, the blastocyst lies dormant in the female's uterus. After the four-month delay, the blastocyst becomes implanted on the inner wall of the uterus, a placental connection develops between the embryo and the uterine wall, and normal embryonic growth and development resume (figure 14.26). About seven months later, after the pregnant female has migrated back to the Pribilof Islands, she delivers a full-term fur seal pup. Delayed implantation is known to occur or is suspected in several other species of pinnipeds as well.

The breeding behavior of fur seals, elephant seals, and sea lions has several notable similarities. In the rookeries, these pinnipeds are extremely gregarious, assembling on the small pupping beaches in

Figure 14.27

Aerial photograph of an Alaskan
Stellar sea lion rookery beach in
1969, before this population began
a serious decline. Compare with
figure 14.29.
Courtesy National Marine Mammal
Laboratory, NMFS, NOAA.

unbelievable numbers (figure 14.27). These animals also exhibit a remarkable degree of sexual dimorphism (figure 14.28). The adult males
are three to six times larger than adult females; they have large canines, thick blubber and skin around the neck and special structures
for physical and vocal threats; and they are quite aggressive toward
other males. These exaggerated male characteristics serve one purpose: They secure and hold a group of females and deny other males
the opportunity to mate. This behavior imposes a polygynous social
order on breeding groups, with relatively few males monopolizing the
breeding activities of the population during any one breeding season.

The evolution and maintenance of polygyny and sexual dimorphism in pinnipeds is likely a consequence of the interplay between
conditions imposed by their mammalian heritage and their amphibi-
ous (terrestrial pupping and offshore feeding) life-styles. Unlike the
gender ratios of some bony fishes, the mammals cannot be manipulated to anything other than approximately 1:1. For females, lactation
places a strong selective premium on parental care and mate choice
behaviors, whereas males act to maximize the number of their mates
or mating opportunities.

In the water, these pinnipeds are quite mobile and tend to disperse some distance from the rookery in search of food. Offshore

Figure 14.28

Male and female elephant seals display obvious sexual dimorphism. The females lack the elongated nose, enlarged canines, thickened neck, and large size characteristic of sexually mature males.
Courtesy B. Maier, Sea Research.

feeding removes the competition for food from the rookery and promotes the fullest use of the few isolated sites that are appropriate as pupping and breeding rookeries. In congested breeding rookeries, males, even with their limited mobility on land, can easily contact and mate with several females. As with most mammals, a mature male is physically capable of fertilizing several females. Therefore, any male that mates with more than one female must compete for the females by excluding other males from the breeding activities. In fur seals, this competition revolves around the breeding territories. Only the most aggressive and vigorous males successfully establish and maintain breeding territories throughout the breeding season. The remaining males, although sexually mature, are excluded from the breeding activities and banished to bachelor groups around the fringes of the breeding population. Male aggressiveness (toward other males) is controlled in part by **testosterone,** a male hormone. The effects of testosterone reinforce the fertility of breeding males. Both male aggressiveness and sex drive are at a peak during the breeding season, a time when the populations in the rookeries are congested.

Successful territorial defense requires that the male fur seal become a permanent feature of the territory for the duration of the breeding season. If a male leaves to feed in the water, he gives up his territory to one of the many "bachelor," or "satellite," males. The males most capable of surviving these breeding fasts (which may last as long as two months) are the larger individuals with extensive fat reserves. The relationship between the large size of males and their reproductive success creates a positive genetic feedback to enhance

Research in Progress

Cetacean Intelligence

The actions of cetaceans (like those of many other animals) often seem purposeful and intelligent, as the following description of care-giving behavior directed to a captive Pacific common dolphin during birth demonstrates.

A female Pacific common dolphin, *Delphinus bairdi* Dall, arrived in Marineland . . . when, at approximately 11:50 A.M. on February 15, observers saw a small tail protruding from her birth canal. The birth progressed very rapidly and by 12:05 P.M. the entire posterior portion of the fetus had been expelled. The umbilical cord, which seemed stretched and taut, was clearly visible. The striped dolphins and false killer whale followed the laboring female. The dolphins showed particular interest and nosed the female's abdominal region on several occasions.

At 12:15 P.M., one of the striped dolphins grasped the fetal tail flukes in its mouth and withdrew the infant from the parental birth canal. A discharge of amniotic fluid and a little blood followed the delivery. The infant was stillborn, and delayed expulsion at a critical phase of parturition was no doubt incriminated in this fetal death. . . .

Our common dolphin, attended by the striped dolphins, carried her dead infant's body to the surface. These efforts were, however, terminated by the male pilot whale, who seized the body by its head. The pilot whale devoured the small cadaver, entirely, after carrying it to and from the surface for 38 minutes.

The common dolphin at first seemed little affected by the intervention of the pilot whale but appeared greatly distressed by its ingestion of the cadaver. Whistling constantly, she moved rapidly around the tank, swimming in an erratic manner, apparently searching for her calf. The animal quickly resumed a more normal swimming pattern, in the company of the striped dolphins, but she continued to vocalize intermittently for several hours.

. . . At 4:06 P.M., the common dolphin sought the company of the female false killer whale. She was observed at this time to deliberately avoid the company of the striped dolphins and begin to swim on the west side of the tank quite close to the surface.

The false killer whale swam to the little dolphin and, after an apparent deliberate examination of her genital area, gently grasped the umbilical remnant in her mouth and with a lateral movement of her head withdrew this tissue some six inches from the common dolphin's body. The dolphin rolled on her back and broke away from the larger animal but then returned and again waited for the false killer whale. Once more, the whale seized the placenta and repeated

---

sexual dimorphism generation after generation. Essentially, the only males that contribute genetic information to subsequent generations are the large, aggressive ones with physical and behavioral traits very different from the traits of females.

To illustrate the differential genetic contributions of male and female polygynous pinnipeds consider that a moderately successful fur seal bull maintains a breeding group of about 40 females for an average of five successive years. When unsuccessful matings are taken into account, this male will sire about 80 male and 80 female offspring. Each female during her reproductive lifetime will produce only about three males and three females. Thus, the total genetic contribution of a territorial bull to subsequent generations is about 25 times that of each female. This intense selective factor for exaggerated characteristics in male fur seals, elephant seals, and sea lions has led to the most extreme examples of sexual dimorphism of any mammal group.

The gregarious nature and relatively poor terrestrial locomotion of pinnipeds make them easy targets for sealers seeking skins and oil. In

the behavior. . . . At the third attempt, the female false killer whale was successful and withdrew the entire placental membrane from the smaller animal. This was released and immediately both animals resumed normal activity in the tank.[1]

The interpretation of such behavior as intentional or intelligent acts to provide assistance to another animal in distress, appealing though the concept is, is not presently supported by behavioral evidence.

The widespread belief that cetaceans are intelligent seems linked with an equally widespread assumption that cetaceans have large brains. When brain size is expressed as a ratio of brain size to body size, the smaller toothed whales compare well (many dolphins have brain-to-body ratios slightly greater than those of humans), but the larger baleen whales have ratios smaller than those of cows or rabbits. Numerous studies have rejected proposed relationships between such a simple index and intelligence. In several respects, the associative structures of the cetacean brain cortex have remained quite primitive while the absolute size of the entire brain has grown.

Then why the large brains? Some researchers suggest that most mammals use a process of reverse learning during REM sleep (rapid-eye-movement sleep, a sleep associated with dreaming) to erase useless memories and free neural networks for more-useful interactions. So far, the two types of nonprimate mammals shown to lack REM sleep, toothed whales and spiny anteaters (primitive egg-laying mammals), have brain-to-body ratios comparable to those of primates. Toothed whales and spiny anteaters may have large brains, not because they are intelligent, but because they are incapable of clearing out old and useless memories. If cetacean intelligence actually is on a level with that of dogs or parrots, we need to understand that, for they may not be intelligent enough to protect themselves in interactions with those "brainy" mammals known as humans.

For more information on this topic, see the following:
Bain, D. E. 1992. Multi-scale communication by vertebrates. In: Marine mammal sensory systems, *ed. J. A. Thomas, J. A., R. A. Kastelein, and A. Y. Supin. New York: Plenum Press.*

Eisenberg, J. F. 1986. Dolphin behavior and cognition: Evolutionary and ecological aspects. In: Dolphin cognition and behavior: A comparative approach, *ed. R. J. Schusterman, J. A. Thomas, and F. G. Wood. Hillsdale, NJ: Lawrence Erlbaum Associates, Publishers.*

[1]Brown, D. H., D. K. Caldwell, and M. C. Caldwell. 1966. By permission of Los Angeles County Museum of Natural History, *Contributions in Science*, pp. 7–12.

the past two centuries, several pinniped species have been severely decimated by commercial slaughters. The northern fur seal population numbered about 2.5 million when discovered by Russian sealers in the eighteenth century. By 1911, the population was reduced to about 100,000 animals. Protective regulations instituted at that time have allowed the northern fur seal herd to recover to a present population of approximately 1.5 million.

Through the early 1970s, 60,000 to 70,000 young Pribilof fur seal males (9% of the total male population) were killed annually for their furs. These three- and four-year-old males had not yet entered the competition for territories and females, so their furs were undamaged by fighting. Is the impact of harvesting 9% of the males from a population with plenty of excess males really significant? At that time, approximately 500,000 pups were born to the Pribilof fur seal herd each year; half were female, half were male. About 14% died from starvation, disease, or by being crushed by adult males before they left the rookery. Another 50% were lost at sea the first winter, leaving about 90,000 males alive at the end of their first year. By the time the young males reached three years of age, natural causes of mortality had further reduced their numbers to 70,000 to 80,000. When still another 60,000 to 70,000 were removed for commercial purposes, the number of males remaining for breeding was relatively small.

In the first half of the twentieth century, it was commonly assumed that "the killing of these bachelors does not affect the structure or breeding performance of the herd because of the animals' polygamous habits" (King 1964). That assumption completely ignores the significance of male aggressiveness and competitiveness in the evolution and maintenance of polygyny in pinnipeds. Fur seals have excess males because they are polygynous, but those males are not excess until their reproductive worth has been tested against other males. Potential breeders simply cannot be identified at three or four years of age, and many are slaughtered along with potential nonbreeding males. With competition reduced by the commercial take, the males that do survive have a greatly improved chance of obtaining and keeping a breeding territory regardless of their relative territorial and sexual capabilities. Might the long-term genetic consequences of continued intensive harvesting of the Pribilof fur seal population be too high a price to pay for a fur coat?

The Pribilof fur seal population has been in steady decline since the mid-1950s, despite intensive management and a cessation of commercial harvesting since 1972. The genetic effects of previous decades of harvesting have been blamed for the population decline. So has contamination of their insulating fur by crude oil. Another relatively recent hazard is entanglement in fishing nets. Fur seals are curious animals and will often investigate fixed nets or scraps of lost netting abandoned by expanding fisheries in the northern Pacific Ocean and Bering Sea. Entanglement often means drowning or slow starvation for the unlucky fur seal caught in this flotsam. These are the most likely culprits, but none provides a completely satisfactory explanation for the recent decline in the size of the fur seal population. Numbers increased in the late 1970s, then began to slip again. Presently, Pribilof fur seals use only 25% of the available rookery area used in 1955, and their numbers are continuing to decline. Other North Pacific pinniped species, including northern sea lions (figure 14.29), have suffered even greater declines.

The northern elephant seal, *Mirounga*, was even more seriously decimated by sealers than was the smaller fur seal. Once distributed from central California to the southern tip of Baja California, this species came under commercial hunting pressure in 1818. A scant half-century later, so few survived that they were not worth hunting. No elephant seals were sighted between 1884 and 1892. In 1892, C. Townsend discovered eight animals on Isla de Guadalupe, 240 km off the coast of Baja California. Seven were taken for museum specimens. Early census estimates suggest that in the 1890s as few as 20 individuals survived on a single inaccessible beach on Isla de Guadalupe. Beginning with protection afforded the northern elephant seal by Mexico and the United States around the turn of the century, that remnant population slowly recovered. Since then, the northern elephant seal has again spread throughout its former breeding range. The total population has swelled to more than 100,000 animals, and the future of this species now seems secure.

Figure 14.29

Aerial photograph of the same Stellar sea lion rookery beach shown in figure 14.27. This photograph was taken on the same day of the year in 1987, after this population was well into a serious decline that paralleled the fur seal decline and continues today.

Courtesy National Marine Mammal Laboratory, NMFS, NOAA.

Or does it? Is the present northern elephant seal population really as viable and hardy as the preexploitation population? Comparisons of 21 blood proteins from 159 animals of the "recovered" population suggests that they are not. In marked contrast to proteins of other vertebrate species, no structural differences were demonstrated either between individuals or between separate groups of northern elephant seals breeding on different islands. The lack of structural differences in these proteins points to a complete absence of variation in the genes controlling the synthesis of these proteins. M. L. Bonnell and R. K. Selander suggest that the absence of genetic variability in the existing northern elephant seal population is the result of a genetic bottleneck when the population was at its low point in the 1890s. It is quite conceivable that, on that isolated beach on Isla de Guadalupe, a lone elephant seal bull dominated the breeding of all the surviving sexually mature females for several years. If so, half the pool of genetic information possessed by the surviving representatives of this species was funneled through a single animal, and the genetic variability presumably inherent in the predecimation population was lost. The rapid recovery of the protected population indicates that genetic variability may not be essential to the short-term survival of this species, possibly because their existence has been cushioned by the relatively uniform and predictable marine environment of the past century. However, this species remains vulnerable to environmental changes occurring in an extended time frame for it may lack the genetic variability necessary to cope with such changing conditions.

Summary Points

- Many species of large nektonic animals make long-distance migrations to optimize conditions for survival. These migrations generally link reproductive areas with the feeding areas of adults. These are often thousands of kilometers apart. Several factors, from coastal landmarks and ocean currents to stars and the sun, may be employed as navigational cues by the animals that undertake these migrations.
- To find food and mates or to migrate in the ocean, nekton must be able to sense conditions and changes in their immediate surroundings. Fish exhibit several sophisticated sensory receptors to monitor dissolved chemicals, sound vibrations, light intensity, color, their own body orientations, and weak electrical and magnetic fields in the water.
- To compensate for reduced visibility and their inability to smell under water, toothed whales (and probably some other groups) have a sophisticated system of echolocation for target detection and orientation.
- Reproductive patterns vary immensely in pelagic animals. Some spawn frequently and produce large numbers of eggs. Others reproduce infrequently and conservatively. Some are oviparous, some ovoviviparous, others viviparous. A few even undergo gender transformation. Regardless of the different reproductive strategies, the goal is the same: to produce offspring that will achieve maturity and eventually reproduce.
- Marine mammals reproduce annually or at even longer intervals. Frequently, birth and mating occur in breeding areas or rookeries. In some pinnipeds that disperse offshore to feed and congregate on isolated rookeries for reproduction, polygyny and sexual dimorphism are extremely well developed.

Review Questions

1. List and discuss the adaptive advantages of sequential hermaphroditism in polygynous fishes.
2. Compare the reproductive cycles and migratory patterns of gray whales and Pribilof fur seals. What purpose does delayed implantation serve in these cycles?
3. Discuss and compare the life cycles and migratory patterns of European eels and sockeye salmon.
4. Describe the sexually dimorphic characteristics exhibited by mature elephant seals or fur seals.

Challenge Questions

1. Characterize the differences in the senses of olfaction and taste in mammals and sharks.
2. What special features of the life cycle of many pinnipeds probably have led to the evolution of polygyny in this group?

Suggestions for Further Reading

Books

Bonner, W. N. 1982. *Seals and man: A study of interactions.* Seattle: University of Washington Press.

Gaskin, D. E. 1982. *The ecology of whales and dolphins.* Portsmouth, NH: Heinemann.

Nicol, J. A. C. 1989. *The eyes of fishes.* New York: Oxford University Press.

Smith, R. J. F. 1985. *The control of fish migration.* New York: Springer-Verlag.

Articles

Blaxter, J. H. S. 1980. Fish hearing. *Oceanus* 23(3):27–33.

Jumper, G. Y., Jr., and R. C. Baird. 1991. Location by olfaction: A model and application to the mating problem in the deep-sea hatchetfish *Argyropelecus hemigymnus. American Naturalist* 138:1431–58.

Kalmijn, A. J. 1977. The electric and magnetic sense of sharks, skates, and rays. *Oceanus* 20(2):45–52.

Lohmann, K. J. 1992. How sea turtles navigate. *Scientific American* (January) 264:100–06.

Norris, K. S. 1968. Evolution of acoustic mechanisms in odontocetle cetaceans. *Evolution and Environment,* 297–324.

O'Brien, W. J., H. I. Browman, and B. I. Evans. 1990. Search strategies in foraging animals. *American Scientist* 78:152–60.

Oceanus. 1980. Special issue on sensory reception in marine organisms. 23(3).

Owens, D. W. 1980. The comparative reproductive physiology of sea turtles. *American Zoologist* 20:549–63.

Pennisi, E. 1989. Much ado about eels. *Bioscience* 39:594–98.

Robertson, D. R. 1972. Social control of sex reversal in a coral-reef fish. *Science* 177:1007–09.

Shapiro, D. Y. 1987. Differentiation and evolution of sex change in fishes. *Bioscience* 37:490–97.

Warner, R. R. 1984. Mating behavior and hermaphroditism in coral-reef fishes. *American Scientist* 72:128–36.

Internet Addresses

Marine Mammal Vocalizations: language or behavior?
http://www.umassed.edu/Public/People/kamaral/thesis/ marinemammalacoustics.html#echolocation
Thorough discussion of vocalizations and echolocation. Includes "clickables" to hear recordings of these sounds (for properly equipped computers). Kimberly Amaral, "The Web" and the Web of Life. Includes a series of articles on oceanography/coastal issues for Woods Hole Oceanographic Institution
http://www.umassed.edu/Public/People/kamaral/thesis/ Thesisexperiment.html

UF-Led Research Team Finds Sea Turtles Can Travel A Third Of The Planet
http://www.ufcn.aa.ufl.edu/UFCN/ufnews/turtles.html
Verification that sea turtle migration spans one-third of the planet; man-made dangers faced during their migration. University of Florida News Desk
http://www.ufcn.aa.ufl.edu/UFCN/ufnews/news.html

Report On Whale Migration by John Tocco
http://memorial.sdcs.k12.ca.us/LESSONS/Whales/Whale. migration.report html
"You are a Marine Biologist and have been asked to explore a whale species and write a paper on that whale species." Leads the student through the processes of investigating and writing. Memorial Academy School of Excellence, Principal; John Tocco, math/science teacher. Memorial Academy for International Baccalaureate Preparation Charter School
http://memorial.sdcs.k12.ca.us/default.html

Bluefin Tuna
http://www2.hp.com/abouthp/features/bluefin/taxonomy.html
Picture and complete description of bluefin.

Saving the Bluefin
http://www2.hp.com/abouthp/features/bluefin/
Two theories about bluefin migration. Computer-assisted studies and tracking help solve the questions. Photos and charts. © 1997 Hewlett-Packard Company **http://www2.hp.com/**

Northwest Salmon
http://kingfish.ssp.nmfs.gov/tmcintyr/fish/nwsalmon.html
Snake River Salmon: Life History; Why listed as Threatened and
  Endangered; Causes of Decline; additional information on
  Northwest Salmon. This site has been produced by the technical
  and program staff of the Office of Protected Resources, National
  Marine Fisheries Service
http:kingfish.ssp.nmfs.gov/tmcintyr/prot_res.html
Learn About Marine Biology
http://www.odysseyexpeditions.org/course.html
Click on links for Dolphins: Evolution, Distribution, Physiology,
  Sensory Biology, Behavior, and Training. Jason Buchheim, Odyssey
  Expeditions, Inc. Odyssey Expeditions Tropical Marine Biology
  Voyages **http://www.odysseyexpeditons.org**
Ecological Monitoring and Assessment Network
**http://www.cciw.ca/eman-temp/research/protocols/benthos/
  benthos1.html**
"Marine organisms can be conveniently categorized as benthic,
  nektonic, or planktonic according to the areas they customarily
  inhabit." Lengthy discussion in these areas. Introduction page for
  virtual book. Marine Biodiversity Monitoring
  **http://www.cciw.ca/eman-
  temp/research/protocols/benthos/intro.html#toc.**
By The Dark of the Moon, Eels Slither Out to the Sea
**http://washingtonpost.com/wp-srv/interact/longterm/
  horizon/100996/eels.html**
News story on the mysteries of the eels migration from freshwater
  streams to the Sargasso Sea, and on attempts to unravel those
  mysteries. Cheryl Lyn Dybas, The Washington Post—
  washingtonpost.com™ **http://washingtonpost.com/
  wp-srv/front.html**

Tuna entangled in drift net. Photo by Greenpeace.

Human Intervention in the Sea

Chapter 15

Chapter 16

ECOLOGICAL PERSPECTIVES

Human Intervention in the Sea: Food from the Sea

Most people are aware that the human population is increasing at an alarming rate. It is estimated that about 500 million people inhabited the earth in 1650. It took 200 years for our population to double to 1 billion people. Our population doubled again in just 80 years, reaching 2 billion in 1930. Just 45 years later, by 1975, our population had doubled yet again. If the present growth rate continues (which is actually a mosaic of various growth rates in different countries), the number of humans inhabiting the earth will double (to 8 billion) by the year 2017, just 43 years since the last doubling. Such explosive growth (and the resultant population size) is the primary cause of perhaps all environmental problems that currently confront humans, and their solution may lie in careful regulation of our numbers. But of more immediate concern is our desire to feed all humans that currently inhabit the earth. Because only 1% of the food consumed by living humans during the past several decades came from the sea, and because

estimates of the total potential production of fishes from the sea (see table 15.2) exceeds our current harvest by 3–20 times, it is commonly suggested that seafood represents a magic solution to our nutritional concerns. But there probably is little cause for optimism. After all, if the ocean is capable of producing 3–20 times as many fishes as we currently harvest, why are many of our fisheries, if not all, seriously overexploited? The answer is complex and hinges on our egocentric point of view. First, we cannot remove as much fish from the sea as we estimate that it produces because we must leave a relatively large percentage behind to produce a harvestable population in subsequent years. Also, the cultural bias inherent in our diet results in significant unexploited potential. For example, many abundant species are not harvested because they seem unpalatable, such as elasmobranchs, krill, jellyfishes, and plankton. In addition, some areas of the sea are unexploited because they are not near industrialized nations (e.g., the

Indian Ocean) or because of logistical difficulties (e.g., the mesopelagic zone). Moreover, our calculations are egocentric in that we forget that the sea must also feed our vertebrate competitors, such as marine mammals, sharks, and seabirds. Finally, we must reconsider shifting our harvest to a point that is lower down on the food web. Most terrestrial species that we consume are producers (i.e., plants) or herbivores (e.g., beef and chicken). Alternatively, most marine species that we harvest are consumers, many of which are carnivorous species, such as many fishes, lobsters, and cephalopods. This tradition of harvesting at the apex of the marine food web results in an enormous waste of primary production. In summary, the sea may be able to provide us with a great deal more carbon than we currently harvest. But it can do so only if we reconsider our dietary traditions and begin to eat a wider range of organisms, fish in underutilized areas, and target species at the base of the marine food web, such as phytoplankton.

Food from the Sea

For much of our history, the seas have been largely immune to the pressures created by the material needs of our land-based human population. Two thousand years ago, the human population of the earth was probably between 200 and 300 million people. The daily challenges of survival kept life expectancies short. High birthrates were balanced by high death rates, and the population grew slowly. Not until 1650, when the Industrial Revolution in Europe brought advances in medicine and technological improvements in food production, did human mortality rates decline and the population double to 500 million. By 1850, the population had doubled again to 1 billion people. Declining death rates continued to accelerate the rate of population growth through the twentieth century. Presently, the human population of the world is more than 6 billion people and is projected to surpass 7 billion by the end of this century. Of the present population, the United Nations Food and Agriculture Organization (UNFAO) estimates that at least one-half billion are seriously undernourished. The World Bank places that number nearer to 1 billion.

The sheer magnitude of the current human population creates enormous demands for food, fiber, minerals, and other commodities that support our modern social fabric. Today, parts of the world ocean are intensively exploited for recreation, military purposes, commercial shipping, fishing, dumping of waste materials, and extraction of gas, oil, and other mineral resources. Although the impact of these activities varies geographically, no portion of the world ocean is isolated from their effects. These effects often transcend national borders as well as the boundaries of many ecological and taxonomic groups. When the biological effects of these societal uses of the sea are large and measurable, they introduce a complex suite of important social, political, economic, aesthetic, and biological questions. Often such questions require value judgments that cannot be resolved by scientific means.

The demand for food made by the present human population is enormous and is increasing at a rapid rate. Uncounted millions of people starve to death each year, and hundreds of millions more are deprived of good health and vigor because of inadequate diets. Many people, therefore, regard marine sources of food as a critical part of the solution to present and future food problems.

Fishing is a multibillion dollar global industry. In 1995, the global import-export trade in fishery commodities was almost $55 billion. Most fishes are sold for human consumption, but an appreciable portion is also used for livestock fodder, pet foods, and industrial products.

The impact of fishing practices on the stability, and even the continued existence, of some of these marine populations has been severe. These practices have become an area of increasing concern since a relatively small number of species make up the bulk of the world fish harvest. These species have borne the brunt of the human population's demands for seafood. Some species have been eliminated from traditional fishing grounds; others face the imminent possibility of biological extinction.

A Brief Survey of Marine Food Species

The raw material of the fishing industry includes a number of species of bony and cartilaginous fishes, many mollusks and crustaceans (shellfishes), a variety of other aquatic animals (from worms to whales), and even some marine plants. Each year the UNFAO compiles and publishes global fishery catch statistics. Figure 15.1 summarizes catch results at decade intervals between 1960 and 1990. The annual catch size has varied somewhat; however, the relative ranking of most groups has remained reasonably constant from decade to decade.

The clupeoid fishes, including anchovies (figure 15.2), anchovetas, herrings, sardines, pilchards, and menhaden are very abundant and account for about one-quarter to one-third of the total commercial catch. A single species, the Peruvian anchoveta (*Engraulis ringens*), provided nearly 19%, or more than 13 million tonnes (metric tons; 1 metric ton = 1,000,000 g) of the total 1970 catch, but extensive overfishing, and more recently the occurrence of ENSO, caused the fishery to collapse. The herring catch is an important part of the North Atlantic fishing industry. In the last decade, other herring fisheries in the South Atlantic and North Pacific oceans have been expanding. At its peak in the 1930s, the California sardine industry was landing over 500,000 tonnes annually. The menhaden catch yields approximately 1 million tonnes annually, largely from the Atlantic Ocean and the Gulf of Mexico.

The huge size of the clupeoid-fish catch is not reflected in its dollar value as an economic commodity or in its significance as a source of protein in human nutrition. Nearly all anchovies, anchovetas, and menhaden and much of the herrings, sardines, and pilchards caught are reduced to fish meal to be used as an inexpensive protein supplement in livestock and poultry fodder. As fish meal, clupeoid fishes make only indirect contributions to the human diet.

Clupeoid fishes are found in shallow coastal waters and in upwelling regions. Their schooling behavior (figure 15.3) simplifies catching techniques and reduces harvesting expenses. Large purse seines (which may be 600 m long and 200 m deep; see figure 15.4 and p. 413) are used to surround and trap entire schools. Once encircled, the fishes are ladled into the ship's hold. Most clupeoid fishes are small, with an average adult length of between 15 and 25 cm. Their small size, however, is compensated for by other characteristics that enhance their economic usefulness. These fishes have fine gill rakers that enable them to feed on small organisms close to the base of marine food webs. Mature Peruvian anchovetas feed principally on chain-forming diatoms and other relatively large phytoplankton aggregations. Herring generally feed on herbivorous zooplankton (see figure 2.11).

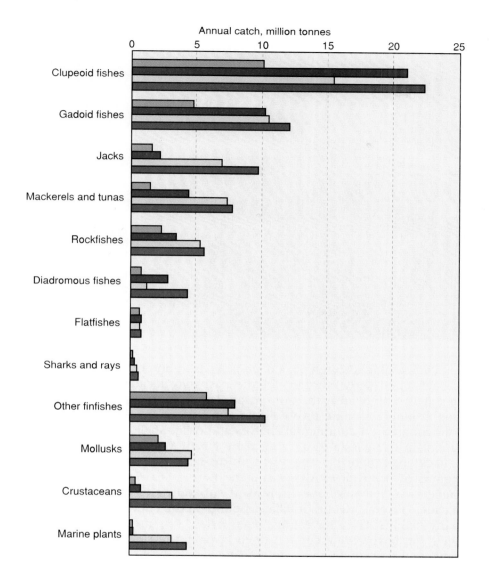

Annual catch, million tonnes

Figure 15.1
Global marine and estuarine harvest by major categories: 1960 (green), 1970 (blue), 1980 (gold), 1990 (maroon).
Data from UNFAO catch and landing statistics.

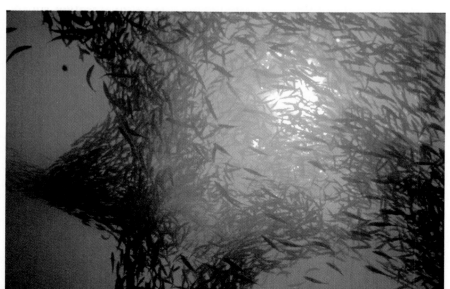

Figure 15.2
Schooling anchovy, *Engraulis*.
Photo by Tom Phillipp.

Figure 15.3

A modern tuna seiner with its purse sein surrounding a tuna school
Courtesy W. Perryman, National Marine Fisheries Service.

The combined catch of cod, pollack, hake, and other gadoid fishes has remained fairly constant for the past two decades. Gadoid fishes usually live on or near the bottom, are larger than clupeoid fishes, and feed at higher trophic levels. Fishing operations for these species are concentrated on continental shelves and other shallow areas. Cod are caught in the coastal waters of many North Atlantic nations. Alaskan pollack are caught by Japanese and Russian trawler fleets in shallow regions of the Gulf of Alaska and the Bering Sea. Pollack catches by these two nations have increased substantially in the past two decades. Several species of hakes abound on the continental shelves of many oceans. They have long been prized by international fishing fleets but have only recently gained favor in the United States, where they are usually marketed as whitefish.

Redfishes, bass, sea perch, and rockfishes also live near the bottom. No single species of this group dominates the catch statistics, yet the combined 1990 catch amounted to over 5 million tonnes. Sold fresh or frozen, most of these fishes are popular fish-market items.

Horse and jack mackerel are primarily pelagic carnivores that generally feed on smaller, anchovy-sized fishes and invertebrates. Mackerel resemble tuna in form, but they seldom exceed 30 to 40 cm in length. The fisheries for these mackerel groups are scattered in coastal and offshore waters.

Tuna are the largest of the commercial species discussed here. Tuna weighing over 100 kg are not unusual, but most range from 5 to 20 kg. Yellowfin, bigeye, albacore, and skipjack tuna account for the majority of tuna catches. Tuna are often the top carnivores in complex food webs and may be separated from the primary producers by seven or more trophic levels. These fishes are active predators of smaller, more abundant animals, especially clupeoid fishes and sauries, and are captured in or near nutrient-rich waters where prey fish abound.

Figure 15.4

Tuna being ladled out of net shown in Fig. 15.3.

Courtesy W. Perryman, National Marine Fisheries Service.

U.S. tuna fishers rely heavily on long-range purse seiners to harvest schooling species of tuna, especially skipjack and yellowfin tuna (figure 15.4). Yellowfin tuna exhibit a strong but poorly understood behavioral association with some subtropical dolphins (*Stenella*). Because dolphins are easily visible at the sea surface, purse seining of yellowfin tuna is simply a matter of setting the nets around a dolphin school with the assumption (usually correct) that a school of yellowfin tuna is just below. In 1972 alone, this practice led to the deaths of more than 300,000 dolphins, which were captured along with the tuna by the U.S. tuna-fishing fleet. Since then, tighter restrictions on both design and operation of purse seining, imposed by the 1972 U.S. Marine Mammal Protection Act, have reduced annual dolphin mortality due to purse seining to about 20,000 for the U.S. tuna fleet. Much of this apparent reduction, however, reflects a widespread move of the fleet to foreign registry, places where their actions are not subject to U.S. regulations.

To counter the adverse publicity associated with this high dolphin mortality, in 1990, several large tuna processing companies announced new policies to buy and market only "dolphin-safe" tuna, tuna caught without setting purse seines on dolphin schools. It remains to be seen whether these policies can be effectively monitored by the tuna-producing industry.

To exploit the more widely dispersed bluefin, tropical yellowfin, and bigeye tuna, laborious longline fishing methods are also used. The tuna aggregate in a narrow zone of relatively high primary productivity that straddles the equator in the eastern half of the Pacific Ocean (see figure 5.23). These large predators feed on sardines, lantern fishes and other small fishes, and crustaceans, which in turn graze on the smaller zooplankton. The longline fisheries of the equatorial Pacific Ocean have developed into a valuable oceanic fishery, with Mexico and Japan each taking about 100,000 tonnes of yellowfin, bigeye, and bluefin tunas in 1990. The longline gear consists of floating mainlines that extend as far as 100 km along the surface. Hanging from each mainline are about 2,000 equally spaced vertical lines that terminate with baited hooks. Once a longline set is in place, fishers move along the mainline, remove hooked tuna, and rebait and replace the hooks.

In recent years, many fisheries have begun using gill nets. Designed to entangle fishes in the net fabric, these large rectangular nets are either anchored to shore or allowed to drift. Although effective at entrapping fishes, they are also invisible and indiscriminant killers of nontarget species. These nets are fast becoming major causes of mortality for some species of sharks and many seabird and marine mammal populations.

Major Fishing Areas of the World Ocean

The shallow water over continental shelves and near-surface banks encourages rapid regeneration of critical nutrients in the photic zone. These nutrients are prevented from escaping into deeper water and accumulating below the upper mixed layer, where return to the surface requires a much longer time. High levels of production by phytoplankton in neritic waters are further enhanced because the resulting animal production is "crowded" into a water column usually less than 200 m deep. In contrast, organic material produced over deep-ocean basins must be shared by the many consumers thinly dispersed through several thousands of meters of water.

About 90% of the marine catch is taken from continental shelves and overlying neritic waters, a region representing less than 8% of the total oceanic area. Bottom fishes and benthic invertebrates account for about 15% of the total global catch. Halibut, flounder, sole, and other flatfishes are benthic species commonly caught in shallow waters using bottom trawls. High prices and stable markets for these fishes have resulted in extensive fishing of most known stocks. The North Atlantic stocks of many types of bottom fishes have been subject to heavy fishing pressures for most of this century, and several are seriously overfished.

Other major fishing areas are centered in regions of upwelling where abundant supplies of critical nutrients from deeper waters are returned to the photic zone. Upwelling may, depending on the locality, occur sporadically, on a seasonal basis, or continually throughout the year. (Mechanisms of Antarctic, equatorial, and coastal upwelling were described in chapter 5.)

The nutrient-rich waters surrounding the Antarctic continent sustain very high summertime primary production. Yet, with the exception of the near-defunct pelagic whaling industry, no large fishery has yet developed in Antarctic waters. Regions of coastal upwelling are most apparent along the western coasts of Africa and North and South America (see figure 5.15), but they do occur to lesser degrees along other coastlines. High yields of commercial species result from increased rates of primary production over shallow bottoms. The greatest concentrations of clupeoid fishes are found in regions of coastal upwelling: Peruvian anchoveta from the Peru Current upwelling area, pilchard from a similar area in the Benguela Current off the west coast of South Africa, and, before a drastic decline in the 1940s, sardines from the California Current. The Peruvian and Benguelan upwelling systems also support enormous populations of cormorants, pelicans, penguins, and other seabirds.

A Perspective on Sources of Seafoods

Humans are omnivores. We obtain nourishment from a tremendous variety of plants and animals. Yet the staples of the human diet can be narrowed down to three types of plants and four groups of animals. The plant staples include cereals (rice, wheat, corn, and other cereal grains), vegetables, and fruits, nuts, and berries. Beef, poultry (plus the milk and egg products of these animals), pork, and fishes provide the major share of the food to satisfy the carnivore in us. It is important to understand just how significant the present marine contribution of food is to the total human diet and how meaningful it may be in the future.

Table 15.1 compares categories of plant and animal foods from both terrestrial and oceanic production systems based on the state of technology used in the production of each food category. Plant production is separated into gathering, the casual use of untended wild plants (figure 15.5), and farming, the agricultural tending of domesticated plant species. Comparable categories for animals are used: hunting of wild animals and herding of genetically improved, controlled domesticated animals. These terms are also applied to marine food items. Only a very few marine plants are presently farmed, whereas oysters, some clams, and a few other marine animals are "herded." Table 15.1 lists the 1990 food production statistics for each of the four production categories for land and ocean. Production statistics for wild plants and wild animals grown on land can only be approximated. The remaining figures are compiled annually by the UNFAO.

Table 15.1
Human Food from Land and Ocean Production Systems, 1990

Food Type	Category of Production	Food Production ($\times 10^6$ tonnes)	
		Land	Ocean
Plants	Gathering	150	4
	Farming	5,000	1
Animals	Hunting	120	81
	Herding	680	2
Total		5,950	83
	Less That Used for Fish Meal		−27
	For Human Consumption		56 + 5

Compiled from FAO Production Yearbook and Yearbook of Fisheries Statistics *1990, Ryther 1981, and Vitousek et al. 1986.*

Figure 15.5

A kelp harvester in operation. As the harvester moves backward, the kelp is cut below the sea surface and then pulled up the stern loading ramp.
Courtesy Kelco, a division of Merck Pharmaceuticals.

The information presented in table 15.1 forces some uncomfortable, yet undeniable conclusions. The majority of the seafoods harvested are animals three or four trophic levels above the primary producers. In terrestrial agricultural systems, plants and herbivores are harvested and losses to higher trophic levels are avoided. Furthermore, marine animals consist of wild, unimproved stocks that are hunted rather than controlled and domesticated. Although technologically advanced ships, nets, and fish-finding gear may be used, little of the monetary gains have been reinvested to improve fish stocks.

Another disturbing trend in the use of living marine resources has developed in the past four decades. In 1955, 86% of the world fish

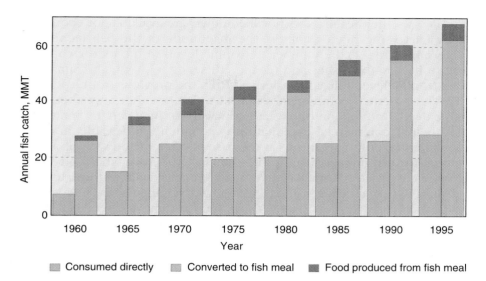

Figure 15.6

Disposition of the world fish catch, 1960–1995.
Data from FAO catch and landing statistics.

catch was consumed directly, and only 14% was reduced to fish meal for use as a protein supplement for domestic livestock. By 1970, the fish meal fraction had increased to about 40% of the world catch (figure 15.6) and has remained at 30–35% since. Assuming the entire 29 million tonnes of fishes reduced to meal in 1995 were fed to pigs and chickens with trophic efficiencies of 20%, less than 6 million additional tonnes of edible pork and poultry were produced. So the actual amount of human food derived from the 1995 fish catch was not the total catch of 92 million tonnes. Instead, it was nearer 69 million tonnes; 63 million tonnes of edible fishes and 6 million tonnes of livestock raised on fish meal derived from 29 million tonnes of fishes.

Long food webs, unsophisticated production systems, and expanding industrial uses of marine organisms all severely limit the capacity of marine food production. The marine environment just does not produce very much food; for the past several decades, only 1% of the food consumed by the world human population came from the sea. The likelihood of greatly increasing that fraction of our marine diet depends on the magnitude of still unharvested fish stocks and how we manage the future harvesting of those stocks, while simultaneously restraining future human population growth.

Our total global fish catch for the 25 years following World War II increased about 6% a year. The global human population was expanding at about 2% each year for the same period. Since 1970, however, increases in total catch size have declined to a more modest 1.2% per year, so no real gains against the food needs of our growing population have been made in the past quarter-century. Is the leveling off of the size of the global fish catch an indication that the global fishing industry is approaching some limit on further growth? Not necessarily, but this question is not easily answered. The ultimate limit on marine sources of food is established by the rate of photosynthesis by primary producers. The marine production system begins at the first trophic level with a net production of 250 to 300 billion (2.5 to 3×10^{11}) tonnes of material annually (see table 5.1).

Because neither markets nor techniques are available for economically harvesting phytoplankton, the magnitude of the fish harvest is determined by: (1) the number of trophic levels in the food web leading to harvestable fishes and (2) the efficiency with which animals at one trophic level utilize food derived from the previous trophic level.

The potential commercial value of a fish or other animal is, to a large degree, a function of its size. Animals must be a certain minimum size before commercial exploitation is economically feasible. In addition, the size of the organisms at the base of marine food webs is an essential feature in establishing the number of steps in the food web. Food webs based on smaller phytoplankton cells (such as those in subtropical gyres) generally consist of a greater number of trophic levels, as do food webs leading to large fishes such as tuna (see figure 12.18).

In general, phytoplankton cells decrease in size from greater than 100 μm in coastal and upwelled waters to less than 25 μm in the open ocean. Several moderate-sized zooplankton species such as *Euphausia pacifica,* which function as herbivores in coastal North Pacific waters, must move one step up the food web and assume a carnivorous mode of feeding in offshore waters because the phytoplankton there are too small to be captured. Other forms of oceanic zooplankton occupying the third trophic level are no more than 1 or 2 mm in length. Virtually all species of herbivorous copepods in the open ocean are preyed upon by chaetognaths, which, in turn, become food for small fishes.

Thus, in the open ocean environment, three or four trophic levels are required to produce animals only a few centimeters in length. Food webs leading to tuna, squids, and other commercially important open-ocean species consist of an average of five trophic levels, with an average number of three trophic levels for commercial species in coastal waters and one and one-half for upwelling areas. The latter number is low because many clupeoid fishes taken from upwelling areas graze directly on phytoplankton without any intermediate trophic levels.

Accurate estimates of ecological efficiencies in marine food webs are difficult to calculate. Efficiency factors are based on the growth of organisms, which is, in turn, a function of food assimilation minus waste and metabolic costs (such as respiration and locomotion). These factors vary widely between species, between individuals of the same species, and between various stages in the life cycle of a species. Young growing individuals often exhibit efficiencies as high as 30%, but their ecological efficiencies decline to nearly 0% at maturity. Thus, efficiency estimates for populations composed of a variety of age groups can only be approximated. It is even more difficult to approximate the ecological efficiencies of entire trophic levels consisting of a diverse group of animal types, each with its own peculiar age structure and growth rate. Reasonable estimates of average ecological efficiencies are 10% for the oceanic province, 15% for coastal regions, and 20% for areas of upwelling. With estimates of net phytoplankton production (presented in table 5.1 and repeated in part in table 15.2), the number of trophic levels, and the efficiency of exchange between the trophic levels for each marine production province, we can estimate the total potential production of fish from the sea (last column, table 15.2) to be about 350 million tonnes annually.

Table 15.2

Estimates of Fish Production for the Marine Production Provinces

Net annual primary production (N.A.P.P.) based on values in the right column of table 5.1 × 10 to convert carbon weight to wet (live) weight.

Province	N.A.P.P. ($\times 10^9$ Tonnes)*	No. of Trophic Levels	Trophic Efficiency (%)	Potential Fish Production ($\times 10^6$ tonnes)
Open ocean	209	5	10	2
Coastal	68	3	15	230
Upwelling	1	1.5	20	120
Total	278			352

*From Ryther 1969, Falkowski 1980, Walsh 1984, Smith and Nelson 1986, and Tett 1984.

The prediction of a potential annual production of about 350 million tonnes of animals is not equivalent to an actual harvest of that magnitude: Losses to birds, larger fishes, and other predators are significant in some fisheries. Human preferences for some species of fishes will cause other species to remain underutilized. To maintain a continuing supply of raw materials, fishing activities must allow a reasonably large fraction (generally one-half to two-thirds) of utilized fish stocks to escape and reproduce so that harvesting can continue on a sustained yield basis. Finally, some fishing practices, especially trawling and longlining, produce large and unused by-catches of noncommercial or undersized animals that are dumped overboard.

Thus, using existing fishing methods to harvest presently exploited types of seafoods, the resource potential exists to double our present global fish catch. But a twofold increase of the 1% that seafood contributes to our present diet is still a very small portion of our total food needs, and continued increases in the human population will surely offset all of those gains in future fisheries production. Two other possibilities for increasing our future seafood harvests are discussed in the following sections.

Mariculture

Mariculture, the application of farming techniques to grow, manage, and harvest marine animals and plants, may vary from simple enhancement of natural populations by releasing hatchery-reared juvenile fishes to intensive captive maintenance of species for their entire life span. As the numbers in table 15.1 indicate, farming of marine plants and animals presently contributes less than 10% to the total of marine food production.

Fish farming has been practiced for centuries in Southeast Asia, Japan, and China. Mullets and milkfish are grown in shallow estuarine ponds, where they graze on algae, detritus, and small animals. Estuaries, salt marshes, and other productive coastal habitats offer a tremendous potential for cultivating fishes in closed pond systems. Yet it is unlikely that "feedlot" production of inexpensive marine organisms will soon be a reality. Intensive mariculture activities modify the nature of coastal estuaries, and salt marshes are common locations for these activities. For each fish pond installed, a portion of the native fish and shrimp populations already contributing to previously established local fisheries will be displaced or denied access to these productive coastal waters. Mariculture development has also become a major factor in the degradation of tropical and subtropical mangrove communities.

Salmon ranching in Japan, Norway, and the United States has grown rapidly in the past two decades. Salmon smolts raised in hatcheries begin the prolonged ocean phase of their life cycle at release points and return to these points two to four years later. An adult return equal to 1 to 2% of the smolts released is now being achieved. Each year a total of about 3 billion smolts are released; at least 30 million are expected to return to their release sites for easy capture and processing. Globally, the yield from farmed salmon now rivals the catch from wild salmon stocks. Schemes to use the high seas to graze other species of pelagic fishes will remain impractical as long as the "farmer" cannot be assured that others will refrain from harvesting his or her fishes after they have grown.

For practical reasons, mariculturists must confine their activities to coastal waters. In special circumstances, polluting nutrients from sewer treatment plants or hot water from coastal power-generating stations could be used beneficially. Diverted into fish ponds, nutrients and the warm water could enhance the growth of primary producers to feed the fishes. But these are special circumstances. More commonly, mariculture programs are, and will continue to be, restricted by adverse problems of coastal pollution and habitat destruction and by competing uses for the same land.

Moving Down the Food Web

The ideal species for mariculture are plants at the first trophic level or algal grazers, detritus feeders, and other omnivores not far removed from the primary producers. Oysters, mussels, abalones, lobsters, and small amounts of red algae are grown in controlled environments. Even genetic selection for improved survival and faster growth is being employed on a limited basis. But mariculture is a costly business; it is and probably will be restricted to expensive luxury food items that generate large returns on investments. Captive-raised oysters and lobsters add little to the diets of most people on the earth and certainly nothing to the diets of those who most need it.

In theory, if we could harvest what fishes eat rather than harvesting the fishes themselves, a tenfold increase in the harvest could be realized because one trophic level and its associated energy loss would be eliminated. Rather than contemplating limits of 100 or 200 million tonnes of fishes, we could look forward to harvesting 1 or 2 billion tonnes of marine food each year. Serious problems, however, block this path to greatly increased marine harvests. Almost without exception, animals occupying lower trophic levels are smaller and more dispersed than the animals now harvested.

Although the technological developments necessary to harvest the zooplankton and smaller fishes that constitute these lower trophic levels are not insurmountable, they may not be worthwhile. These smaller, more dispersed animals are more difficult to harvest;

the additional energy needed to collect these small food items may exceed the energy gained from the additional harvest. It is likely to continue to be more efficient for us to wait until the larger animals have eaten the smaller ones.

Here is one example: A century ago, the phytoplankton crop around the Antarctic continent supported a tremendous assemblage of zooplankton, particularly the krill *Euphausia superba* (see figure 12.7 for distribution). In turn, krill fed large populations of blue, fin, and humpback whales. With these whale populations now seriously reduced, they no longer serve as intermediaries to harvest the krill for us. Fisheries scientists from several countries are test-harvesting krill with an eye to expanded future production. Preliminary estimates of an annual sustained harvest of 100 million and even 200 million tonnes have been suggested for this one species alone. The potential of such massive harvesting is a complex issue with several ramifications unrelated to food production.

A major problem inhibiting the use of krill is what to do with it once it is caught. There is little demand for fresh or frozen krill. It is nutritious, but few people will buy it in its natural form. Most plans for using krill include some sort of processing to disguise it or to convert it to an odorless, powdered protein concentrate. The protein concentrate could then be mixed with grain flour to make high-protein breads, pastas, rice cakes, tortillas, or almost any other common food item.

The energy costs involved in catching krill, processing it to a palatable form, and transporting it to markets in the Northern Hemisphere will be enormous. Thus, the likelihood of efficiently harvesting krill and other exotic organisms low in marine food webs is presently marginal at best, and needed increases in food production in the immediate future continue to rely on improving techniques for terrestrial crop production using currently available techniques to grow well-known crops already in demand by society.

The Problems of Overexploitation

Commercial and subsistence fishing represent a form of predation that has predictable effects on the prey species. When fishing of new stocks begins, initial catches are generally large and include a high proportion of large fish. Continued or increased fishing pressure tends to reduce the average size and sometimes the abundance of the stock. If the fishing effort is matched to the growth and reproductive potential of the stock, then a **maximum sustainable yield** of fishes can be caught year after year without causing major upsets in the stock abundance. Too often, though, the fishing pressure becomes much greater than the stock can withstand. Losses to fishing and natural predators together exceed recruitment of young animals, and populations decline. These declines are quickly reflected in reduced catches.

Numerous examples of overfished stocks can be found in most segments of the fishing industry. Most of the popular species of halibut, plaice, cod, ocean perch, herring, and salmon of the North Atlantic and Pacific oceans are being or already have been overexploited. So are many of the warm-water tuna stocks and most of the large whales.

Both the UNFAO and the U.S. National Academy of Sciences recently concluded that overexploitation due to the industrialization of fishing has been the major factor in endangering the continued functioning of several large-scale marine ecosystems. The two examples discussed in the following paragraphs are sufficient to demonstrate some of the problems created by overfishing activities that seem in direct conflict with the fishing industry's own best interests.

The Peruvian Anchoveta

The Peruvian anchoveta, *Engraulis ringens,* is a typical clupeoid fish similar to the anchovy shown in figure 15.2. It is a small, fast-growing filter feeder that schools in the upwelling areas of the Peru Current. The first commercial use of the anchoveta was indirect; from the time of the Incas, droppings or **guano** deposits from the nesting colonies of seabirds that fed on anchoveta have been collected and used as a major source of fertilizer. These seabirds, primarily Peruvian boobies, brown pelicans, and guanay cormorants (figure 15.7), annually converted about 4 million tonnes of anchoveta to an inexpensive fertilizer widely used by Peru's subsistence farmers.

Commercial exploitation of the Peruvian anchoveta for reduction to fish meal began in 1950. The next year, 7,000 tonnes were landed. After 1955, the growth of the fishery was explosive (figure 15.8); over 2 million tonnes were landed in 1960. By 1970, the catch of this one species had surpassed 13 million tonnes, almost one-fifth of the entire world fish harvest for that year. Nearly all the catch was taken by local fishers and reduced to fish meal and oil for export.

Figure 15.7

Guanay cormorants, *Phalocrocorax,* at their nests on an island off the coast of Peru in the 1920s.
From Murphy 1925.

Accompanying the meteoric rise in commercial anchoveta catches was a drastic drop in the number of guano birds that depended on the anchoveta for food. From 28 million in 1956, the guano bird population was reduced to 6 million during the ENSO year of 1957 (see Research in Progress: El Niño, p. 136). With upwelling blocked by the intrusion of a surface layer of warm tropical water, plankton populations were dramatically reduced, anchoveta died or moved, and the guano birds starved. After four years without an ENSO, the bird population had rebounded to 17 million, only to be hit by another ENSO in 1965. That time the bird population plummeted to 4 million, and it has not recovered.

A substantial base of biological information was collected during the development and growth of the Peruvian anchoveta fishery. With the advantages this information base provided, proper management procedures were expected to ensure a large and continuous harvest from this immense stock. The Instituto del Mar del Peru, an advisory panel of fishery experts, projected a maximum sustainable yield of approximately 9.5 million tonnes annually. At the time, the reduced bird populations were taking less than 1 million tonnes each year. Tonnage reports alone, however, are inadequate and sometimes misleading. As the fishing pressure increased over the decade of the 1960s, the average size of fishes being caught decreased, and the number needed to make a tonne increased sharply. By 1970, small anchoveta were taken with such efficiency that 95% of the juvenile fish recruited into the population were captured before their first spawning.

A glance at figure 15.8 shows that the catches of anchoveta for 1967 through 1971 exceeded the predicted maximum sustainable yield

Figure 15.8

Changes in the guano bird population and the anchoveta fish catch along the northwestern coast of South America. ENSO, El Niño–Southern Oscillation.
Adapted from Muck 1989 and annual UNFAO catch and landing statistics.

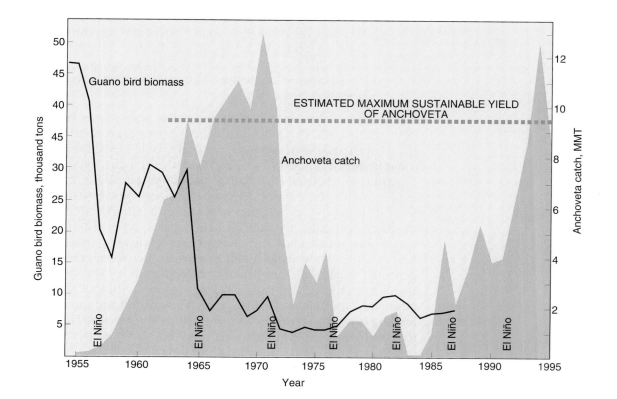

by at least 1 million tonnes each year. The industry and the regulatory agencies responsible for managing the anchoveta stocks had ample warning of what was to come. In 1972, sampling surveys indicated that the anchoveta stocks had been severely depleted and recruitment of juvenile fish was poor. As expected, the 1972 catch dropped to little more than 5 million tonnes, less than half that of the previous year. Even worse was 1973, with an estimated catch of about 2 million tonnes. Apparently, the fishing pressures of the previous decade were too much for this tremendous stock of fish, and it finally collapsed. Protection measures and greatly reduced fishing pressure during the decade following that collapse have promoted a partial recovery of anchoveta stocks, and, beginning in the late 1980s, harvests expanded once again to levels exceeding maximum sustainable yields.

The Great Whales

The history of the whaling industry has been a long and tragic one. Aboriginal hunting of coastal whales for food has occurred for several thousands years. In the eighteenth and nineteenth centuries, whaling took a new turn. Pelagic whales became major items of commerce as demand for their oil grew and whaling became a profitable commercial enterprise. Ships from a dozen nations combed the oceans for whales that could be killed with hand harpoons and lances. Right, bowhead, and gray whales were favorite targets, for they swam slowly and, once killed, they floated conveniently at the sea surface. In his famed *Moby Dick,* Herman Melville questioned the future of these great whales, faced, as they were, with

> omniscient look-outs at the mastheads of the whale-ships, now penetrating even through Behring's straits, and into the remotest secret drawers and lockers of the world; and the thousand harpoons and lances darting along all the continental coasts; the moot point is, whether Leviathan can long endure so wide a chase, and so remorseless a havoc.

By the end of the nineteenth century the gray whale, both species of right whales, and the bowhead were on the verge of extinction. Under strict international protection, the gray whale has since recovered, but the number of bowheads and right whales remains very low.

The era of modern whaling was initiated in the late nineteenth century with the cannon-fired harpoon equipped with an explosive head. The explosive harpoon was so devastatingly effective that even the large rorquals, the blue and fin whales, were taken in large numbers. These whales had previously been ignored by whalers with hand harpoons, for they were much too fast to be overtaken in sail- or oar-powered boats. The subsequent rapid decline of the whale stocks in the North Atlantic and Pacific oceans forced ambitious whalers to seek new and untouched whaling areas. They found the Antarctic, the rich feeding grounds of the largest populations of whales. The discovery of the Antarctic whale populations touched off seven decades of slaughter unparalleled in the history of whaling.

Aided by pelagic factory ships fitted with stern ramps to haul the whale carcasses aboard for processing, the kill of large rorquals rose dramatically. From 176 blue whales in 1910, the annual take climbed to

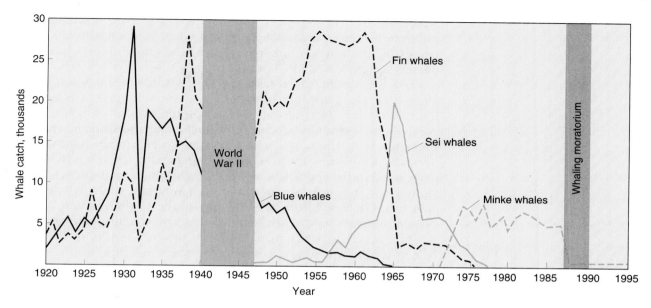

Figure 15.9

Data from UNFAO catch and landing
statistics of blue, fin, sei, and minke
whales in the Atlantic, 1920–1995.

almost 30,000 in 1931 (figure 15.9). After the peak year of 1931, blue
whales became increasingly scarce. Blue whale catches declined steadily
until they were commercially insignificant by the mid-1950s. In 1966,
only 70 blue whales were killed in the entire world ocean. Only then,
when substantial numbers could no longer be found to turn a profit, was
the hunting of blue whales banned in the Southern Hemisphere.

The trend of increasing then rapidly declining annual catches,
shown by the blue whale curve in figure 15.9, is distressingly similar
to that of the Peruvian anchoveta fishery. But the ruthless exploitation
of the great whales did not halt with the near extinction of the blue
whale. As the blue whale populations gave out, whalers switched to
the smaller, more numerous fin whales, catches of which skyrocketed
to over 25,000 whales each year for most of the 1950s. But by 1960,
the fin whale catch began to plummet, and whaling pressure was diverted to even smaller sei whales. The total sei whale population probably never exceeded 60,000. One third of the sei whale population
was killed in 1965 alone. Whaling pressure quickly pushed the catches
of this species far beyond a maximum sustainable yield. By the late
1960s, sei whales had followed their larger relatives to commercial extinction, and the whaling effort was shifted to the minke whale, an 8
to 9 m long miniature version of a blue whale (figure 7.18).

As early as 1940, whaling nations were faced with undeniable evidence that some stocks of pelagic whales were seriously overexploited. In 1948, 20 whaling nations established the International
Whaling Commission (IWC) to oversee the utilization and conservation of the world's whale resources. Unfortunately, the IWC had neither inspection nor enforcement powers. Only once during the 1963
to 1973 decade did the whaling industry actually manage to catch the
quotas established for it by the IWC. The quotas were, in effect, not
quotas at all, as there were no longer enough whales to fill them. In
1974, the IWC, under pressure from several national governments and
international conservation organizations, adopted a new set of management procedures for several geographically localized stocks of

whales. Under these procedures, all species of baleen whales in the Antarctic except the minke are classified as protected stocks, and no harvesting is allowed. Four species, the blue, gray, humpback, and right whales are protected in all oceans.

Despite the increased take of minke whales in the Antarctic since 1970, their numbers and the numbers of Antarctic penguins, seals, and fishes that rely (as did the now depleted stocks of larger baleen whales) on the Antarctic krill *Euphausia superba* have increased appreciably. With the larger whale species so efficiently removed from Antarctic food webs, these smaller, commercially less attractive species responded to the reduced competition with increased growth and maturity rates. The age at sexual maturity of Antarctic minke whales, for instance, has declined from over 15 years in the 1930s to about 7 years at present. Similar changes are seen in other krill-eating species as well.

The severe depletion of the large pristine stocks of krill-eating baleen whales led some people to the naive assumption that, without these whales, a large "surplus" of krill would exist for our harvesting. Such an assumption ignores the evidence that a new equilibrium has developed in krill-based food webs. Minke whales, birds, and seals now play larger roles than they did a century ago. This is not a unique situation, for many fish stocks depleted by excessive harvesting pressures have been replaced with other (and, from a commercial point of view, often less desirable) species. These new species assemblages, by their very existence in niches previously occupied by the exploited stocks, serve as a barrier to the recovery of those stocks.

It is presently impossible to predict the long-term fate of either the great whales or the Peruvian anchoveta. Of the two, the anchoveta has better survival and recovery potential. Anchoveta mature very rapidly, often spawning within their first year. Once mature, each female deposits 9,000 to 24,000 eggs each year for two or three years. Whether the resurgence of anchoveta stocks beginning in the late 1980s actually signals a long-term recovery or is simply part of an ongoing series of boom and bust cycles will depend on the effective implementation of a tough and intelligent management plan. However, the road to recovery for a large, slowly reproducing species such as the blue whale is fraught with unanswered questions. Can the few thousand remaining blue whales scattered over the world ocean encounter each other frequently enough to mate, reproduce, and add to their decimated numbers at a sufficient rate? Will additional blue whales meet stiff competition from other animals for food and other resources appropriated from the whales during their tragic decline? And, most important, if the population begins to increase, will humans refrain from exploiting it so that it can secure a more solid grip on survival? The reproductive resiliency of the great whales is largely unknown. They may bounce back. The gray whale did, but the right and bowhead whales still have not.

In 1982, the IWC approved a five-year moratorium, effective at the onset of the 1985–1986 whaling season, on the commercial killing of large whales. This moratorium was seen as a critical first step in assessing the impact of nearly a century of intense hunting of oceanic whale stocks. At the same time, it would give declining whale populations an opportunity to stabilize. Even after it went into effect, the moratorium was opposed by several nations with long whaling histories, especially Japan, Norway, and Iceland. These nations lobbied repeatedly for permits to take whales (particularly fin and minke whales) for scientific research purposes. When permits were denied, Japanese whaling companies repeatedly went whaling for Antarctic minkes anyway.

When the five-year moratorium against whaling expired in 1991, many IWC member nations supported its indefinite extension. Others, especially those originally opposing the moratorium, have resumed limited whaling of minke and fin whales. How the international community handles this delicate problem will be a serious test of our biological wisdom as well as our political will. With the recent increase in the size of their population, minke whales play a relatively new role as a major predator of Antarctic krill, and given the history of the whaling industry during this century, it should not be surprising to find that a nation anticipating substantial harvests of krill in the near future would strive to retain the means of controlling those species that would best compete with such a commercial venture.

The Tragedy of the Commons

Why have fishing enterprises and fishing nations repeatedly exploited the fish resources on which they depend to the point that returns on their fishing effort decline and too often disappear? Even with the maze of legal and economic considerations involved, incentives for these apparently self-defeating actions are not difficult to find. Salmon, tuna, whales, and other oceanic species are unowned resources belonging to no single nation or individual. They exist outside the jurisdictional limits of all nations and are therefore open to access by any nation. Historically, the concept of **open access** to the high seas evolved in the sixteenth and seventeenth centuries, when the right to navigate freely was more crucial than the freedom to fish. But as coastal fish stocks were depleted, fishers became increasingly dependent on distant stocks in international waters. They eventually discovered that the freedom to fish on the high seas was fundamentally different from the freedom to navigate. Unlike navigation, fishing activities remove a valuable commodity from a common resource pool (the "commons") at the expense of all, including those who do not fish.

Ideally, it is assumed that oceanic resources are unowned and open to access for all people. But with today's advancing pace of technology, some nations achieve the ability to exploit these resources more quickly than others. If the fishing activities of one nation fail to catch these unowned oceanic species, some other nation soon will. Such attitudes have led to predictable and inevitable results for species after species: increased competition between fishing nations for limited resources, duplication of effort, declining fishing efficiency, and, of course, overfishing.

Once in the net, a school of fish is no longer the property of all people; it belongs instead to those who set the net and haul the fish aboard. All people and all nations share the cost of losing the fish, the great whales, and the other marine animals that have nearly disappeared because of overfishing. Yet the short-term profits derived from overfishing are not similarly shared. This is an example of G. J. Hardin's concept of the "tragedy of the commons." The tragedy of this situation is that it best rewards those who most heavily exploit and abuse the unprotected living resources of the open ocean.

International Regulation of Fisheries

Without controls to limit the access of fishers to fish populations in international waters, concerned nations have created regulatory commissions similar to the IWC. These commissions are charged with the responsibility of governing the management and harvest of regional fish stocks. The International Commission for the Northwest Atlantic Fisheries, for instance, includes Canada, Denmark, France, Germany, Iceland, Italy, Norway, Poland, Russia, Spain, the United States, and the United Kingdom. The regulations established by this commission do not carry the weight of international law, but they are binding on member nations. Even so, they have failed to halt serious overfishing of the cod and ocean perch stocks of the northwest Atlantic.

Other commissions, particularly those regulating the halibut and salmon fisheries in the North Pacific, have been much more successful. Their success, however, is now creating new problems. Too many additional fishing boats from the United States, Canada, Korea, Japan, and Russia are being attracted to these well-managed fisheries. Under these additional fishing pressures, management problems are magnified, profits of individual fishers are diminished, and the possibilities for overexploitation are greatly enhanced. Rebuilding sustainable fisheries will require a large reduction in current fishing capacity (especially the number of fishing vessels), reduction of by-catch waste, and the development of proactive rather than reactive procedures and policies to manage fishery resources.

In the early 1970s, at the Third United Nations Conference on the Law of the Sea, the prospects for a single international convention to regulate open-ocean fish stocks seemed uncertain. The U.S. government was under pressure from its own coastal fishing interests to extend U.S. jurisdiction for fisheries out to 200 miles. They could see the distant-water fishing fleets of Russia, Japan, and Germany taking huge harvests just off their shores. The United States ended its traditional policy of opposition to extended zones of control for coastal nations in 1974. In 1976, the United States passed the Magnuson Fishery Conservation and Management Act. Under this act, the United States assumed exclusive jurisdiction over fisheries management in a zone extending 200 miles to sea. Entrance of foreign fishing vessels into this Exclusive Economic Zone (EEZ) is allowed on a permit basis only, and then the fishers are allowed to harvest only fish stocks with sustained yields greater than the harvesting capacity of U.S. fishers.

In 1982, the United Nations adopted a draft Law of the Sea (LOS) Treaty that was to go into effect 12 months after ratification. Over a decade later, ratification of the treaty was still seven nations short of the 60 needed to enter it into force. However, many nations have unilaterally adopted major elements of the LOS Treaty. The LOS Treaty includes many of the key features found in the U.S. Magnuson Fishery Conservation and Management Act, in particular, a 200-mile-wide exclusive economic zone granting coastal nations sovereign rights with respect to natural resources (including fishing), scientific research, and environmental preservation. In addition, the LOS Treaty obliges nations to prevent or control marine pollution, to promote the development

Figure 15.10

Worldwide extent of the 200-mile
exclusive economic zones
sanctioned by the United Nations.

and transfer of marine technology to developing nations, and to peace-fully settle disputes arising from the exploitation of marine resources.

The imposition of 200-mile-wide EEZs by essentially all coastal na-tions of the world dramatically changed the concept of open access for most of the world's continental shelves, coastal upwelling areas, and major fisheries lying within 200 miles of some nations' shorelines (figure 15.10). Yet neither the international LOS nor national manage-ment plans such as the U.S. Magnuson Act have resulted in reductions in fishing effort or in obvious gains in sustainable fisheries located within national EEZs.

The LOS Treaty is notably silent regarding the Antarctic upwelling area, for territorial claims on the Antarctic continent are not recog-nized. Commercial ventures there, including whale and krill harvesting, must be consistent with the goals of the 1981 convention for the Con-servation of Antarctic Marine Living Resources; namely, the "mainte-nance of the ecological relationships between harvested, dependent, and related populations of Antarctic marine living resources and the restoration of depleted populations." This was the first in a slowly grow-ing trend of national and international efforts to recognize the integrity of large marine ecosystems (LME). To date, about 50 LMEs have been identified, mostly in coastal waters, as ecological systems that respond to external stresses (such as overexploitation), environmental fluctua-tions (such as ENSO), or large-scale pollution problems.

The waters around the Antarctic continent represent the last rela-tively unspoiled large marine ecosystem on the earth. The interna-tional cooperation demonstrated so far in protecting its living re-sources has been unusual in the long history of our attempts to protect marine resources. The waters around this remote continent can continue to serve both as a laboratory for improving understand-ing of living marine systems and as a model for creating approaches to preserving those resources as a common heritage of humankind.

Summary Points

- The variety of marine organisms taken for human food is large, including finfishes, shellfishes, other invertebrates, whales, and some marine plants. Commercial fishing efforts are generally concentrated in shallow waters and upwelling regions and near ocean current or thermal boundaries.
- Presently, slightly more than 1% of the total human diet comes from the sea, and most of that tonnage is wild-caught marine animals. Some estimates of future yields suggest that the present harvest of marine food may be tripled or quadrupled.
- One of the problems that is limiting present and future marine food harvests is overexploitation of existing living marine resources. The Peruvian anchoveta and the stocks of great whales are vivid examples of commercial overexploitation followed by collapse of the stocks.
- Open access and the absence of realistic, binding regulations to govern the harvest of unowned marine resources continue to promote overexploitation with little regard to its ultimate consequences.
- Some optimistic possibilities, including mariculture and the use of krill, still exist. Yet the basic problem remains: The human population continues to increase in size. The mere existence of our enormous and ever-increasing human population will have an increasingly detrimental impact on the yield of food from the sea.

Review Questions

1. Describe some of the ways in which human efforts to increase food production on land have reduced the potential yields of food from the sea.
2. List three structural or behavioral features of anchovy, herring, and other clupeoid fishes that explain why they account for such a large portion of the total world fish catch.
3. List three specific biological reasons why the oceans, which cover over 70% of the earth's surface, produce only about 1% of the total human food supply.
4. Describe the fate of the Peruvian anchoveta fishery from the mid-1950s to the present. What impact do you think the collapse of that fishery has had on the price of market eggs in Europe? On the abundance of pelican eggs in Peru?

Challenge Questions

1. Why are most of the world's important fishing areas located in relatively shallow waters, especially along western coasts of continents? Discuss physical, chemical, biological, and economic factors in your answer.
2. Discuss the conditions which cause 99.8% of the world's plant mass to be on land even though the oceans cover over 70% of the earth's surface.
3. Discuss the factors which limit the contribution of marine sources of food (for humans) to about 1% of the total volume of food consumed by the present human population.

Books

Beddington, J. R., R. J. H. Beverton, and D. M. Lavigne. 1985. *Marine mammals and fisheries.* Boston: George Allen and Unwin.

Caddy, J. F., ed. 1989. *Marine invertebrate fisheries: Their assessment and management.* New York: Wiley-Interscience.

Crutchfield, J. A., and G. Pontecorvo. 1969. *The Pacific salmon fisheries: A study of irrational conservation.* Baltimore: Johns Hopkins Press.

Hardin, G. J. 1993. *Living within limits: Ecology, economics, and population taboos.* New York: Oxford University Press.

United Nations Food and Agriculture Organization (UNFAO). 1995. *The state of world fisheries and aquaculture.* Rome: UNFAO.

U.S. Department of Commerce. 1997. *Our living oceans: Report on the status of U.S. living marine resources.* Silver Spring, MD: U.S. Department of Commerce.

Articles

Adey, W. H. 1987. Food production in low-nutrient seas. *Bioscience* 37:340–48.

Bardach, J. 1987. Aquaculture. *Bioscience* 37:318–19.

Beddington, J. R., and R. M. May. 1982. The harvesting of interacting species in a natural ecosystem. *Scientific American* (November) 247:62–69.

Borgese, E. M. 1983. The law of the sea. *Scientific American* (March) 248:42–49.

Horn, M. H., and R. N. Gibson. 1988. Intertidal fisheries. *Scientific American* 258:64–70.

Johnson, S. W. 1994. Deposition of trawl web on an Alaska beach after implementation of MARPOL Annex V legislation. *Marine Pollution Bulletin* 28:477–81.

Koslow, J. A. 1997. Seamounts and the ecology of deep-sea fisheries. *American Scientist* 85:168–76.

Laws, R. M. 1985. The ecology of the Southern Ocean. *American Scientist* 73:26–40.

Muck, P. 1989. Major trends in the pelagic ecosystem off Peru and their implications for management. PROCOPA Contribution No. 90, pp. 386–403.

Nicol, S., and W. de la Mare. 1993. Ecosystem management and the Antarctic krill. *American Scientist* 81:36–47.

Ross, R. M., and L. B. Quetin. 1986. How productive are Antarctic krill? *Bioscience* 36:264–69.

Rudloe, J., and A. Rudloe. 1989. Shrimpers and sea turtles: A conservation impasse. *Smithsonian* 29(9):45–55.

Safina, C. 1995. The world's imperiled fish. *Scientific American* (November) 273:46–52.

Sissenwine, M. P., and A. A. Rosenberg. 1993. U.S. fisheries: Status, long-term potential yields, and stock management ideas. *Oceanus.* summer:48–54.

Taylor, R. M., P. G. O'Keefe, and C. Fitzpatrick. 1994. A snow crab, *Chionoecetes opilio* (Decapoda, Majidae), fishery collapse in Newfoundland. *Fishery Bulletin* 92:412–18.

Suggestions for Further Reading

Internet Addresses

Convention On Fishing And Conservation Of The Living Resources Of The High Seas (1958)
http://sedac.ciesin.org/entri/texts/high.seas.fishing.living. resources.1958.html

Consortium for International Earth Science Information Network (CIESIN). 1997. Environmental Treaties and Resource Indicators (ENTRI) [online]. University Center, Mich.: CIESIN. Environmental Treaties and Resource Indicators (ENTRI) Full Text
http://sedac.ciesin.org/entri

Fishery Resources: Plentiful or Diminishing?
http://www.nfi.org/fishery_resources.html

General discussion of production, regulation, growth, projections, and statistics. National Fisheries Institute
http://www.nfi.org/Welcome.html

Alaska Seafood—Natural And Wild
http://www.state.ak.us/local/akpages/COMMERCE/asmi/ fishinfo.html

"The icy waters bordering Alaska's 34,000 miles of coastline are the largest, most productive commercial fishing grounds in the world,

where about five billion pounds of seafood are harvested annually."
Alaska Seafood Marketing Institute
http://www.state.ak.us/local/akpages/COMMERCE/asmihp.html

Marine Aquaculture
http://www.cmrc.org/
Click on link for marine aquaculture. Commercial site shows how
    technology is used to develop cost-effective, environmentally safe
    marine food production methods. Caribbean Marine Research
    Center, National Undersea Research Program
    http://www.cmrc.org/

Fish: What's the Catch?
http://www.earthsave.org/fishwhat.html
Commentary on overexploitation. EarthSave International
    http://www.earthsave.org/mission.htm

Ocean Pollution

Throughout this book, an assumption has been made that pristine ocean waters are essential for maintaining healthy marine communities. This assumption is being put to a global test. Each year our growing human population generates an enormous burden of domestic and industrial wastes. Initially, these wastes may be dumped into rivers, down sewers, or up smokestacks, but ultimately most wastes make their way to the ocean. The world ocean has a large but finite capacity to assimilate these waste materials without apparent degradation of water quality. However, that capacity can be exceeded if ocean mixing processes are not sufficient to dilute or disperse wastes, creating localized water-quality problems and subsequent biological disturbances.

Municipal and industrial wastewater discharges (including urban runoff and storm drain and sewer overflows), land-based agriculture and forest harvest activities, ocean-based activities related to dredging, marine mining, drilling for and shipping of oil and gas, and direct ocean dumping are all sources of ocean pollutants. Hot water from power generation plants and nutrients from agricultural runoff and sewage outfalls (particularly phosphates and nitrogen compounds) can promote growth rates of all marine organisms, but when they are discharged into semienclosed estuaries, bays, or lagoons, they often become serious pollutants. It is paradoxical that some of the worst pollutants are the pesticide and fertilizer products used to enhance food production on land and that some of the most obvious disturbances appear in marine species harvested for human consumption.

Waste materials discharged into ocean waters can be considered pollutants if they have measurable adverse effects on natural populations (table 16.1). These pollutants include a variety of suspended solids, organic compounds, and inorganic nutrients that deplete dissolved oxygen. Other, more-persistent contaminants such as heavy metals, pesticides, radioactive wastes, and petroleum products head the list of substances that, even at low concentrations, can adversely affect the health of marine organisms and the integrity of their natural ecological relationships. Once in the ocean, persistent contaminants accumulate in marine organisms, and those accumulations are magnified as they are transferred up food chains.

Heavily polluted waters may create direct human health risks, risks associated with eating seafood contaminated with disease-producing microorganisms and toxic substances. But even at lower concentrations, marine organisms provide us with clues signaling serious physiological or habitat disturbances. Like the living resources of the sea, the ocean itself is a resource shared by most nations on the earth. It serves as a valuable and often misused global disposal site for waste substances either too plentiful or too dangerous to store on land. The

Table 16.1

A Summary of the Sources and Effects of Some Marine Pollutants

Pollutant	Sources	Effects
Particulate material	Dredged material, sewage, erosion	Smothers benthic organisms, clogs gills and filters, reduces underwater light
Dissolved nutrients	Sewage agricultural runoff	Increase phytoplankton blooms, decrease dissolved oxygen
Toxics	Pesticides, industrial wastes, oil spills, antifouling paint	Increase incidence of disease, contaminate seafood, suppress immune systems, contribute to reproductive failure
Oil	Tankers, drill sites, urban and industrial wastes	Smother organisms, clog gills, mat fur or feathers, cause anatomical and physiological abnormalities
Marine debris	Garbage, ship wastes, fishing gear	Cause physical injuries and mutilations, increase mortality

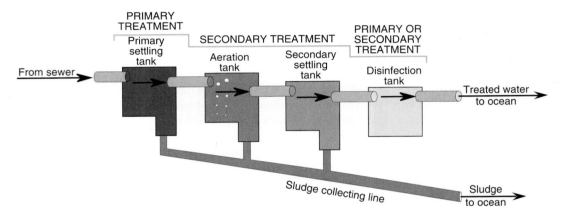

Figure 16.1

Diagram of the steps in a secondary sewage treatment plant. At each step, there is some discharge of materials to the ocean.

following discussion will describe a few substances with known serious consequences for marine life, highlighting the critical links between waste disposal, the biological integrity of marine communities, and our own health and well-being.

Sewage

A major source of coastal pollutants—human sewage—fouls bays and beaches with both toxic and nontoxic pollutants. Although billions of dollars have been invested in sewage treatment plants to treat wastewater, new and growing coastal communities have increased the amount of discharge into oceans and estuaries. The U.S. Office of Technology Assessment has identified 1,300 major industries and 600 municipal wastewater treatment plants that discharge into the coastal waters of the United States.

Typical secondary treatment is intended to separate solids and to reduce the amount of organic matter (which contributes to biochemical oxygen demand), nutrients, pathogenic bacteria, toxic pollutants, detergents, oils, and grease in wastewater. In the United States, most ocean discharges of wastewater are supposed to meet secondary treatment standards (figure 16.1), but many still do not. This problem is much greater in heavily populated developing coastal countries. In many of these countries, discharge of raw sewage and poorly treated sludge directly into estuaries and coastal ocean waters is the norm.

Figure 16.2

Dumping barged sewage sludge at
sea is now illegal in U.S. waters.
Courtesy Greenpeace.

About 15% of the 300 million tons of sewage sludge produced in
the United States each year is discharged into coastal ocean waters
through outfall pipelines or from barges. The term *sludge* describes
the mix of solids that is the end product of municipal wastewater
treatment. Sludge is not unlike detritus from marine sources in its general composition and nutritional value. It has, under ideal conditions,
some nourishment value to zooplankton and benthic detritus feeders.
But the present rate of dumping in shallow waters off the coasts of
most major urban areas is enormous, and the capacity of those areas
for accepting additional sludge is already or soon will be exceeded.

Along the northeastern U.S. coast, sewage sludge was barged to
offshore sites for dumping (figure 16.2). By 1992, when New York City
was required to cease ocean dumping, about 25 million wet tons of
sludge had been dumped at a site 160 km east of the New Jersey
coast. With a water depth of 2,400 m, sludge dumped at this site was

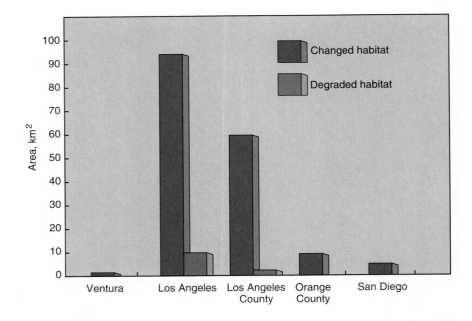

Figure 16.3

Extent of areas changed or degraded by five major sewage outfalls in the Southern California Bight, 1978–1979.
Data from Mearnes 1981.

expected to disperse and dissolve completely before reaching the bottom. When this site was visited by researchers in the submersible *Alvin* in 1989, it was apparent that appreciable amounts of sludge particles were reaching the seafloor and that these particles were contaminated with heavy metals and human sewage bacteria. The impact of these contaminants on benthic communities is still under study.

The problems faced by New York City are similar to those confronting most coastal metropolitan centers. Boston was required to cease dumping sewage sludge into its harbor in 1991, and a court-mandated treatment plant and 15-km-long outfall pipe to deliver effluents beyond the harbor have been completed. Water quality in the outer harbor has improved noticeably. Fish caught in the harbor are healthier than they were a decade ago, and the processed sludge is being converted into commercial fertilizer pellets.

In southern California, sludge containing about a quarter of a million tons of solids is discharged through 30 ocean outfalls each day. At sites near southern California outfalls, concentrations of trace metals, DDT, and PCBs are often increased severalfold in bottom sediments and in larger benthic invertebrates. Measurable changes in species diversity and biomass of benthic infauna and kelp beds can be found, but these changes depend on the rate of discharge and the degree of treatment prior to release. The impact of these heavy loads of nutrients and contaminants is only now beginning to be studied properly, and the picture they present is not simple. Increased abundance of fish and benthic invertebrates have been noted in the vicinity of some outfalls; at others, benthic communities have been noticeably degraded. DDT and PCBs have been shown to enter marine food webs and find their ways through many trophic levels to distant sites. Of the five major sewer outfalls emptying into the Southern California Bight, two have caused obvious degradation in several square kilometers around the outfall site. Together, the five outfalls have significantly changed or degraded nearly 200 km² of seafloor (figure 16.3).

Even marine communities near small population centers are not immune from the impact of sewage effluents. In Kaneohe Bay, Hawaii, two-thirds of the bay's coral reefs were destroyed, partly by the direct action of the sewage and partly by the smothering effects of large green algae whose growth had been stimulated by the increase in sewage-derived nutrients. Recent improvements in sewage treatment are promoting gradual recovery of some of the previously damaged coral reefs.

Toxic Pollutants

Many toxic substances enter the sea through sewer systems, but others originate as industrial discharges. We do not yet know how to determine the extent or fate of many toxic substances in the marine environment or how to evaluate their effects on marine life. Some of the better-known trace metals and toxic chemicals include mercury, copper, lead, and chlorinated hydrocarbons. Chlorinated hydrocarbons, synthetic chlorine-containing compounds, are created for use as pesticides or are by-products of the manufacture of plastics. They are among the most persistent and harmful of all toxic substances and include well-known products such as chlordane, lindane, heptachlor, DDT, dioxins, and PCBs (polychlorinated biphenyls).

Antifouling Paints

For centuries, boat hulls have been sheathed in copper or other metals to attempt to retard the attachment and growth of barnacles and other fouling organisms. More recently, antifouling paints using biocides made from these metals have been applied. The latest entrant in this race to find a substance to block fouling organisms is tributyltin (TBT). Paints with TBT are more effective and last longer than older copper-based antifouling paints. TBT leaches from boat hulls into harbor waters, but in amounts virtually undetectable with present analytical techniques. Yet even in these very low concentrations (a few parts per trillion), TBT harms nontarget organisms such as oysters and clams. TBT is known to deform oyster shells and to cause chronic reproductive failure in a variety of shellfish species. In 1987, the U.S. Congress passed the Antifouling Paint Control Act that classified TBT as a restricted pesticide and severely limited its use and the amount that could be used in paints. Because TBT eventually degrades to less toxic forms in the marine environment, these new restrictions have already led to decreased concentrations of tin in estuaries and bays.

DDT

By virtue of its long and widespread use, DDT and its effects on marine life have been well-documented. Dichloro-diphenyl-trichloroethane (DDT) was the first of a new class of synthetic chlorinated hydrocarbons. It became available for public use in 1945 and quickly gained international acceptance as an effective killer of most serious insect threats: house flies, lice, mosquitoes, and several crop pests. Despite these obvious benefits, the use of DDT has been banned in the United States and in some other nations. Several unfortunate characteristics have transformed DDT from a benefactor to an ecological nightmare.

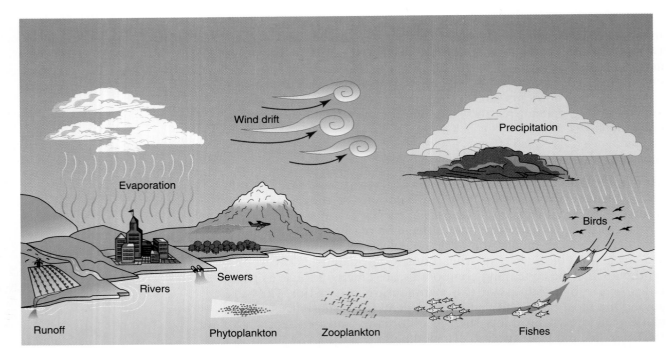

Figure 16.4

Transfer of DDT to and within marine food webs. DDT is absorbed by phytoplankton and then concentrated at each step in the food web.
Adapted from Epel and Lee 1970.

DDT is a persistent pesticide; it does not break down or lose its toxicity rapidly. Once in seawater, DDT is rapidly absorbed by suspended particles. Because it is nearly insoluble in water, measurable levels of DDT in seawater practically never occur. Yet DDT contamination from land and air has been so pervasive that it can be found in nearly all parts of the world ocean. Antarctic penguins, arctic seals, Bermuda petrels, and fishes everywhere have accumulations of DDT in their fatty tissues.

DDT is quite soluble in lipids. Fatty tissues and oil droplets concentrate the DDT absorbed on suspended particles in seawater. Phytoplankton and, to a much lesser extent, zooplankton are the initial steps in DDT's entry into marine food webs. Fishes, birds, and other predators eventually consume this plankton and its load of DDT, concentrating the toxin in their fatty tissues. At each step in the food web, further concentration, or **bioaccumulation,** occurs (figure 16.4). Eventually the DDT reaches the top carnivores.

In the 1960s, a serious DDT contamination in the Los Angeles coastal area was centered around the Los Angeles County sewer outfall at Whites Point. The source of most of the DDT was traced to a chemical plant that produced most of the world's supply of DDT. Wastes produced in the manufacture of this pesticide were washed into the sewer system at rates exceeding 100 tons a year. For comparison, the entire Mississippi River drainage system each year added only 10 tons of DDT to the waters of the Gulf of Mexico. Concentrations of DDT in marine animals occasionally exceeded the permissible limits for human consumption (5 ppm) established by the U.S. Food and Drug Administration, and some lots of canned kingfish, jack mackerel, and other species of fish were seized and could not be distributed.

Marine birds suffered the most devastating effects of DDT poisoning. As fish-eating predators, they are sometimes four or five trophic levels removed from the phytoplankton that initially absorb the DDT. The bioaccumulation of DDT that occurred at each trophic level assured these predatory birds of high DDT loads in their food. DDT and its residues block normal nerve functions in vertebrates. A few birds were found dead from extremely high concentrations of DDT. DDT also interferes with calcium deposition during the formation of eggshells. The eggshells of birds with high DDT loads were very thin and fragile. They frequently broke when laid or failed to support the weight of adult birds during incubation. The broken eggs lay in abandoned nests, mute testimony to the insidious effects of DDT.

The Brown Pelican is a common sight along both coasts of the United States and Mexico. Nesting colonies of these birds exist on islands along the warmer coasts. A decline in the number of Brown Pelicans along southern California in the late 1960s was traced to reproductive failures caused by DDT-induced eggshell thinning. In one nesting colony of 300 pelicans off the southern California coast, 12 intact eggs were laid in 1969. Of those 12, only three hatched. Some adult birds were sitting on damaged eggs. Others already had abandoned nests littered with remnants of thin-shelled and broken eggs.

DDT from the Los Angeles County sewer outfall was cited as being responsible for the local reproductive failure of the Brown Pelican and other fish-eating birds. Since the United States banned DDT in 1972, the massive doses of DDT released into California coastal waters by the Los Angeles sewer system have been reduced by 90%, and its devastating effects on seabird reproduction are slowly diminishing. Yet, DDT is still being used in other parts of the world and some is carried to the ocean each year. By itself, DDT is a threat to the continued existence of several species of marine animals. But it is only symptomatic of the greater danger posed by the many other persistent toxins of which we know even less.

Dioxins

Dioxins are another group of chlorinated compounds gaining international notoriety. The most potent dioxin is TCDD, which is toxic to birds and aquatic life at concentrations of a few parts per quadrillion. Because dioxins, like DDT, are fat soluble and stable, dioxins bioaccumulate readily and are suspected of causing cancers, developmental malformations, and immune system and reproductive difficulties.

Dioxins enter the marine environment from many sources, but most arrive in the effluent from pulp and paper manufacturing plants that chlorinate wood pulp to produce bleached paper. Trace levels of dioxins have been found in tissues of fishes collected near pulp and paper mills. Although the concentrations do not affect fishes, the bioaccumulation of dioxin may affect human health if the fishes are consumed. Some European paper-producing countries, such as Sweden, are working to reduce the amounts of dioxin discharged into the Baltic Sea by finding alternative bleaching techniques that do not use chlorine processes and by producing unbleached paper products. More studies to determine the extent and effect of dioxin in marine waters are currently under way.

Research in Progress

A Plague on Seals

In early 1988, harbor seals of the North and adjacent Baltic Sea experienced a massive and abrupt die-off. Between 17,000 and 20,000 seals mysteriously died. In some local populations, these deaths accounted for 60–70% of all harbor seals living there. Examination of the dead seals indicated that the ultimate cause of this "seal plague" was a previously unknown virus similar to those causing distemper in dogs and measles in humans. In infected seals, the virus suppresses immune system responses, leaving the individual open to secondary bacterial infections.

Massive seal die-offs are not new. Historical records indicate that massive harbor seal die-offs with symptoms like those seen in 1988 have occurred around Great Britain at least four times in the past 200 years. In 1918, harbor seal populations around Iceland succumbed in large numbers to pneumonia. Another virus-caused die-off of harbor seals occurred in New England in 1979–1980. And in 1955, more than 60% of the crabeater seals around the Antarctic died of an unidentified viral infection.

Although we know that most of these seal deaths were caused by a virus (possibly the same virus), we do not yet understand what triggers a large-scale die-off of infected individuals. Several hypotheses have been proposed, including rapidly increasing population sizes, increasing levels of environmental contaminants in food, and even increasing water temperatures. For each of these hypotheses, some evidence exists to support its claim as the factor that caused these seal populations to suddenly become susceptible to the virus in 1988. None provide a satisfactory general explanation for all of the large-scale seal die-offs observed in this century. However, experimental evidence does indicate that low concentrations of PCBs (and possibly dioxins) suppress immune system responses and interfere with embryo implantation in seals, leading to increased susceptibility to viral diseases and reduced birth rates.

For more information on this topic, see the following:
De Guise, S., et al. 1995. *Possible mechanisms of action of environmental contaminants on St. Lawrence beluga whales* (Delphinapterus leucas). Environmental Health Perspectives, *103:73–77.*
Harwood, J., and B. Grenfell. 1990. *Long-term risks of recurrent seal plagues.* Marine Pollution Bulletin *21:284–87.*

PCBs

The North Sea, surrounded as it is by many of Europe's industrialized nations, experiences some of the highest levels of marine pollution anywhere. High on the list of these pollutants are dioxins and PCBs. During the summer of 1988, more than 17,000 common and grey seals died of a viral infection that swept the Baltic, Wadden, and North seas. By the time it runs its course, this epidemic may kill as many as 80% of some North Atlantic populations of common and grey seals. Stressed seals exhibit symptoms such as lesions, encephalitis, peritonitis, osteomyelitis, and premature abortions.

The source of this epidemic has not been established with any certainty, nor has an absolute link between this viral outbreak and any specific pollutant been demonstrated. However, PCBs in the seals' food is strongly suspected because there is strong experimental evidence that indicates that low concentrations of PCBs (and possibly dioxins as well) lead to suppression of the immune systems of harbor seals. PCBs also interfere with embryo implantation in seals, leading to fewer births and lower birth weights. Seal pups born of PCB-contaminated mothers may still be confronted with higher mortality rates and suppressed immune system problems.

Figure 16.5

Oil-soaked murre after the *Exxon Valdez* oil spill, 1989.
Courtesy J. Harvey.

To examine the extent of the pollution in the Baltic Sea, the seven-nation Helsinki Convention is developing a pollution monitoring and control strategy. In the Pacific Rim nations, similar efforts are under way through the Pacific Basin Consortium on Hazardous Waste Research to identify and study pollutants present in the Pacific Ocean and to develop joint international partnerships to reduce existing and manage future marine pollutants.

Oil on Water

Torrey Canyon (1967), Santa Barbara (1969), *Amoco Cadiz* (1978), *Exxon Valdez* (1989), and *Brear* (1993)—these catastrophic oil spills engender a concern for the marine environment as no invisible contaminant can. Spilled oil floats on seawater and provides a constant reminder of its presence until it is washed ashore, sinks, or evaporates. Large volumes of oil suffocate benthic organisms by clogging their gills and filtering structures or fouling their digestive tracts. Marine birds and mammals suffer heavily as their feathers or fur become oil-soaked and matted (figure 16.5), and they lose insulation and buoyancy.

On March 22, 1989, the supertanker *Exxon Valdez* ran aground on Bligh Reef in Alaska's Prince William Sound (figure 16.6). The grounding punched holes in eight of the eleven cargo tanks and three of the seven segregated ballast tanks. The result was the largest oil spill to date in U.S. waters (242,000 barrels, or nearly 40,000,000 liters). The spill occurred in an area noted for its rich assemblages of seabirds, marine mammals, fishes, and other wildlife. It was one of the

Figure 16.6

The supertanker *Exxon Valdez* in April 1989, soon after running aground on Bligh Reef in Prince William Sound.

Courtesy J. Harvey.

most pristine stretches of coastal waters in the United States, with specially designated natural preserves such as the Kenai Fjords and Katmai National Park.

The area of the spill, because of its gravel and cobble beaches, was particularly sensitive to oil. In places, the thick, tarry crude oil penetrated more than a meter below the beach surface. High winds, waves, and currents in the days following the accident quickly spread oil over 26,000 km². The toll on wildlife was devastating. More than 33,000 dead birds were recovered, and hundreds of sea otters, seals, sea lions, and other marine mammals, as well as many thousands of commercial and noncommercial fishes, were killed. Traditional fishing and cultural activities of the native communities in Prince William Sound were halted for the year, creating enormous social and economic costs.

With the U.S. government encouraging more exploration for new seafloor oil deposits to reduce its dependency on foreign imported oil, the risks from oil spills are likely to increase substantially. To prepare for such disasters, particularly after the *Exxon Valdez* oil spill, coastal states are trying to improve oil spill contingency plans and are experimenting with oil dispersants to mitigate the environmental effects of future oil spills.

Despite the spectacular nature of major spills from tankers or offshore drilling and production platforms, more oil actually enters the marine environment in an average year in runoff from urban streets and parking lots, from leaking underground storage tanks and improperly dumped waste oil, and in bilge water from nearly every freighter, fishing vessel, and military ship afloat than does from all the oil spills worldwide in a year combined. The numerous sources and more mundane aspect of these oil pollutants make them more difficult to manage than a single dramatic spill of the same volume of oil from a wrecked tanker.

Marine Debris

Until recently, marine debris was considered to be of minor importance when compared with other pollutants. Problems caused by marine debris, however, may rival or exceed those resulting from some better known pollutants, including oil. By definition, marine debris is any manufactured object discarded in the marine environment. It may sink to the seafloor, remain suspended at mid-depths, or float at the surface and eventually be carried ashore by winds and waves.

Before the International Convention for the Prevention of Pollution from Ships (1973–1978), also known as MARPOL, and the Marine Plastics Pollution Research and Control Act of 1987 the primary source of marine debris was the massive dumping of garbage at sea by foreign and domestic merchant ships, military and fishing vessels, and recreation boats. Merchant ships generate an estimated 110,000 tonnes of marine debris yearly, and fishing vessels another 340,000 tonnes. The practice of dumping garbage at sea had been legal because no alternatives were practical. The effects have worsened because of the increase in nonbiodegradable products that float after being dumped. Recently, nonbiodegradable medical wastes in the form of used syringes and vials have been washing up on some of our beaches, apparently dumped illegally at sea.

Even though the amount of debris discarded every day is slowly being reduced with garbage compactors aboard ships and recycling or disposal opportunities in ports ashore, much plastic and other nonbiodegradable garbage continues to be discarded at sea. This debris harms marine life, damages vessels, and eventually litters beaches. Because these materials decompose very slowly, they will continue to float at the sea surface, carried by currents to harm marine life and litter our beaches for years to come.

Fishing Gear

The world's commercial fishing fleet contributes more than 100,000 tonnes of accidentally lost nets, pots, traps, and setlines and deliberately discards pieces of damaged fishing gear each year. The shift since 1940 from the use of natural fibers to virtually nonbiodegradable synthetic fibers for the construction of nets, lines, and other fishing gear has made commercial fisheries a major contributor to marine plastics pollution. The largest potential for lost and discarded fishing gear occurs in the North Pacific Ocean, where vessels of many nations operate under adverse climatic conditions. Since 1978, new gill net fisheries for squid have been developed by Japan, the Republic of Korea, and Taiwan in the central North Pacific. Over a million kilometers of gill net are now

Figure 16.7

Beached gray whale, with drift-net fabric wrapped around its tail. Six million small whales have been killed by nets since 1971.

set annually by the squid fisheries. These nets constitute a large potential source of derelict gear that, when lost, continues to drift for thousands of kilometers. Birds, mammals, and other animals that surface frequently are especially susceptible to entanglement in the floating net fabric or other plastic debris (figure 16.7). One recent survey of fur seals on the Antarctic island of South Georgia, indicated that, even on that remote island, nearly 1% of the seals were entangled in synthetic debris. Most of the seals entangled were wearing "neck collars" made of plastic strapping (59%) or fishing line and net material (29%). A large proportion of these seals exhibited signs of physical injury, and an unknown number of animals presumably died after contact with floating debris. There is some evidence, however, that MARPOL regulations are beginning to exert positive influences in reducing the volume of fishing gear intentionally discarded in international waters.

Plastics

Plastics constitute as great an environmental threat as all the other kinds of debris combined. Although plastics may break up into smaller pieces, they degrade much more slowly than most other kinds of debris, and most plastics float. Concentrations of plastics tend to be highest in the Northern Hemisphere, where vessel traffic is the heaviest, where most plastics manufacturers and fabricators are located, and where intensive recreational use is made of beaches and coastal waters.

Common plastic products include jugs, bottles, buckets, bags, sheeting, eating utensils, six-pack beverage-container yokes, life preservers, buoys, fish nets, fish net floats, fishing line, rope, and polystyrene cups and packing material. Suspension beads, the raw material

Figure 16.8

One of several posters used in campaigns to reduce marine plastic debris.

Reprinted by permission of Saltwater Productions, Anchorage, Alaska.

used by fabricators of plastic products, have become a ubiquitous component of debris. They can now be found in surface waters, in sediments, and on beaches around the world.

Even the most casual observer is sometimes overwhelmed by the startling array of plastic litter encountered on isolated beaches. A recent three-hour collection effort by more than 2,000 volunteers of SOLV (Save Oregon from Litter and Vandalism) yielded over 26 tons of plastic debris. Most of the debris was believed to have been washed ashore and not to have been left behind by beachgoers. Included in the litter were 48,898 chunks of polystyrene larger than a baseball, 2,055 bands used for strapping boxes and other kinds of cargo, 6,117 pieces of rope, 1,442 six-pack yokes, 4,787 plastic bottles and other containers, 1,097 pieces of synthetic fishing gear, 4,090 plastic bags or plastic sheets, and 5,339 plastic food utensils. The effort by SOLV proved so successful that it has become a model (figure 16.8) for annual clean-up efforts in many other coastal states.

Concluding Thoughts: Developing a Sense of Stewardship

The world ocean is a great and forgiving ecosystem, for centuries digesting the effluvia of human societies. With the human population of the world's coastal zones expected to increase by 50% by 2012, our continued use of the seas as a safe disposal site for the unwanted by-products of our civilization depends on a better understanding of the consequences of our intrusion on the workings of marine ecosystems. We are the only species on the earth capable of understanding the motives and consequences of our actions. Yet, the history of our economic involvement with marine populations has repeatedly demonstrated a serious lack of practical awareness of the fundamental disturbances our intervention causes in these natural systems. In our scramble for food, sport, and profit from the sea, we have repeatedly, and often with disastrous results, violated existing ecological relationships and invented new ones. We are positioning ourselves with increasing frequency at the tops of heavily exploited marine food webs that are contaminated with the very substances we are afraid to dispose of near our homes. It would be wise to heed the pelicans and seals as sensitive indicators of what might be in store for us if we continue our unthinking and uncaring contamination of the world ocean.

It is hoped that, as we approach the twenty-first century, we can learn to leave some old and wasteful habits behind. It will not be easy or simple, but each one of us must develop a sense of stewardship toward the world ocean and its resources that is reflected in our personal as well as our political decisions. By better understanding the fate we are shaping for the world ocean, we may yet be able to

> harmonize our civilization with the environ
> So that our children see our wisdom
> Not inherit our wastes.

Wastes and the Ocean, by Momiji

Summary Points

- Our growing human population generates an enormous burden of domestic and industrial waste that ultimately makes its way to the ocean.
- Municipal and industrial wastewater discharges (including urban runoff and storm, drain, and sewer overflows), land-based agriculture and forest harvest activities, ocean-based activities related to dredging, marine mining, drilling for and shipping of oil and gas, and direct ocean dumping are all sources of marine pollutants.
- Even at low concentrations, some waste can adversely affect the health of marine organisms and the integrity of their natural ecological relationships. Already, some habitats have been significantly changed or degraded.
- Toxic pollutants, including antifouling paints, DDT, dioxins, and PCBs, enter marine food webs at low trophic levels and at each step in the food web become more concentrated. Oil and marine debris (especially fishing gear and plastics) float on the sea surface and eventually create additional nontoxic problems on beaches, reefs, and intertidal areas.
- Each one of us must develop a sense of stewardship toward the world ocean and its resources. By caring about and understanding the workings of the world ocean, we may yet be able to minimize further degradation of an integral part of our biosphere.

1. List two types of marine pollutants that may have simultaneous benefits as well as harmful effects on the organisms they contact.
2. List two characteristics of plastic debris that make it an ever-increasing problem in the marine environment.
3. What specific and positive actions can you personally take to reduce some of the marine pollution problems described in this chapter?

Review Questions

1. Describe why fat-soluble toxins, such as DDT and dioxins, become more concentrated in tissues of animals at each higher trophic level in marine food webs. List the trophic levels leading to a North Atlantic grey seal contaminated with PCBs.
2. Why is spilled oil particularly harmful to marine birds and mammals?

Challenge Questions

Books

Clark, R. B. 1992. *Marine pollution,* 3rd ed. Oxford: Oxford University Press.

Frankel, E. G. 1995. *Ocean environmental management: A primer on the role of the oceans and how to maintain their contributions to life on Earth.* Englewood Cliffs, NJ: Prentice Hall.

Laws, E. A. 1993. *Aquatic pollution: An introductory text,* 2d ed. New York: John Wiley & Sons.

Loughlin, T. R., ed. 1994. *Impacts of the Exxon Valdez oil spill on marine mammals.* San Diego: Academic Press.

Norse, E., ed. 1993. *Global marine biological diversity: A strategy for building conservation into decision making.* Island Press.

Articles

Alexander, L. M. 1993. Large marine ecosystems. *Marine Policy* 17:186–98.

Alexander, M. 1981. Biodegradation of chemicals of environmental concern. *Science* 211:132.

Bjorndal, K. A., A. B. Bolten, and C. J. Lagueux. 1994. Ingestion of marine debris by juvenile sea turtles in coastal Florida habitats. *Marine Pollution Bulletin* 28:154–58.

Champ, M. A., and F. L. Lowenstein. 1987. The dilemma of high-technology antifouling paints. *Oceanus* 30(3):69–77.

Croxall, J. P., S. Rodwell, and I. L. Boyd. 1990. Entanglement in man-made debris of Antarctic fur seals at Bird Island, South Georgia. *Marine Mammal Science* 6:221–33.

Eberstadt, N. 1986. Population and economic growth. *Wilson Quarterly* 10:95–127.

Peakall, D. B. 1970. Pesticides and the reproduction of birds. *Scientific American* (April) 222:72–78.

Turner, M. H. 1990. Oil spill: Legal strategies block ecology communications. *Bioscience* 40:238–42.

Suggestions for Further Reading

Photo Gallery of the Exxon Valdez Grounding and Oil Spill
http://response.restoration.noaa.gov
Click on link to Photo Gallery and then click on link to Exxon Valdez Photos for series of photos related to the Exxon Valdez grounding and oil spill. One series shows oil immediately after the spill; the second series follows eight years of change at one site. Hazardous Materials Response and Assessment Division, National Ocean Service, National Oceanic and Atmospheric Administration. Also links to the Spill Information page of Hazardous Materials Response and Assessment Division of NOAA (NOAA HAZMAT), and a description of NOAA HAZMAT's involvement in the response to the Exxon Valdez spill.

Internet Addresses

Toxic Contaminants
http://www.mwra.state.ma.us/harbor/html/harbpol3.html
Contaminants found in Boston Harbor and discussion of general
    effects and durability of same. Massachusetts Water Resources
    Authority **http://www.mwra.state.ma.us/index.html**

Water Pollution
http://www.ns.ec.gc.ca/pollution/water.html
Effects of water pollution. 1997 Environment Canada. Environment
    Canada Pollution
http://www.ns.ec.gc.ca/pollution/index.html

Chronology of the North Cape Oil Spill
http://seagrant.gso.uri.edu/riseagrant/chronology.html
The North Cape oil spill—a chronology of the oil spill, damages, clean-
    up, and return to fishing. Bob Bowen and Elizabeth Gibbs, Rhode
    Island Sea Grant College Program
http://seagrant.gso.uri.edu/riseagrant/

Earthtrust's "DriftNetwork" Program
http://planet-hawaii.com/earthtrust/index.html
Several pages of discussion on the effects of driftnet use. Interspersed
    with photos. Earthtrust, Dedicated to Wildlife Conservation
    Worldwide **http://www.earthtrust.org**

Who We Are—Center for Marine Conservation
http://www.cmc-ocean/org./1_acintro.html
This page lists the organization's goals. One of many sites advocating
    good stewardship of marine resources. Center for Marine
    Conservation **http://www.cmc-ocean.org./main.html**

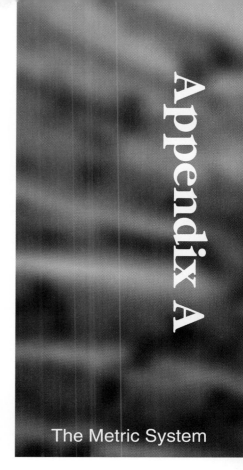
The Système International d'Unités (SI), commonly known as the metric system, is internationally accepted as the system of measure for reporting scientific and engineering data. This system is widely used because of its simplicity and ease of conversion. Unlike the English system that we commonly use, the three basic SI units for distance, volume, and mass (weight) are closely interrelated. Each basic unit is related to the others by a simple equality, making it relatively easy to convert from one unit to another (table A.1).

The basic unit of distance measure is the **meter (m)**; the meter is slightly longer than a yard. The meter, like all other basic units of the SI, is divided into smaller units by factors of 10. The next smaller distance unit is the **decimeter (dm),** or 0.1 m. Each decimeter is further divided into 10 **centimeters (cm).** Further subdivisions or multiples of the meter provide the distance units listed in table A.2.

Two types of SI units are employed to measure area. Some are simply squared distance measures, such as **square meters (m^2)** and **square centimeters (cm^2).** Other SI units are exclusively for area measure. They are listed in table A.3.

Table A.1
Common SI Prefixes

				Decimal Notation		Exponential Notation
Micro (μ)	=	one millionth	=	0.000001	=	10^{-6}
Milli (m)	=	one thousandth	=	0.001	=	10^{-3}
Centi (c)	=	one hundredth	=	0.01	=	10^{-2}
Deci (d)	=	one tenth	=	0.1	=	10^{-1}
Basic unit	=	one	=	1	=	1
Deka (dk)	=	ten	=	10	=	10
Hecto (h)	=	one hundred	=	100	=	10^2
Kilo (k)	=	one thousand	=	1,000	=	10^3
Mega	=	one million	=	1,000,000	=	10^6

Table A.2
Distance

Nannometer (nm)	=	0.000000001	meter
Micrometer (μm)	=	0.000001	meter
Millimeter (mm)	=	0.001	meter
Centimeter (cm)	=	0.01	meter
Decimeter (dm)	=	0.1	meter
Meter (m)	=	1	meter
Kilometer (km)	=	1,000	meters

Table A.3
Area

Square millimeter (mm^2)	=	0.000001	square meter
Square centimeter (cm^2)	=	0.0001	square meter
Are (a)	=	100	square meters
Hectare (ha)	=	10,000	square meters
Square kilometer (km^2)	=	1,000,000	square meters

Two separate, but related, systems of units are also used for volume measure. Some are based on the cube of distance measures, other on the **liter (l)** and its subdivisions. Some of the more common volume units are listed in table A.4.

The basis for all SI units of mass is the **gram (g).** Other mass units derived from multiples or subdivisions of the gram are listed in table A.5.

Some of the more commonly used SI-English unit equivalents are included in table A.6. Until using the SI becomes automatic, it may be helpful to learn a few of these equivalents so that you may quickly develop a mental concept of the magnitude of the units you are considering.

Table A.4
Volume

Milliliter (ml)	=	0.001	liter	=	1 cm^3
Liter (l)	=	1	cubic decimeter	=	$1{,}000 \text{ cm}^3$
Cubic centimeter (cm^3)	=	0.000001	cubic meter	=	1 ml
Cubic decimeter (dm^3)	=	0.001	cubic meter	=	1 ml

Table A.5
Mass or Weight

Milligram (mg)	=	0.001	gram
Gram (g)	=	1	gram
Kilogram (kg)	=	1,000	grams
Tonne (metric ton)	=	1,000,000	grams

Table A.6
SI-English Unit Equivalents

1 meter	=	39.37	inches
1 inch	=	2.54	centimeters
1 mile	=	1.6	kilometers
1 kilometer	=	0.62	mile
1 pound	=	453.6	grams
1 kilogram	=	2.2	pounds
1 liter	=	1.06	liquid quarts
1 liquid quart	=	0.95	liter

A List of Biologically Important Elements

Element	Symbol	Typically Forms	No. of Covalent Bonds	Ionic Charge
Hydrogen	H	Molecules and ions	1	+1
Carbon	C	Molecules	4	–
Nitrogen	N	Molecules	3	–
Oxygen	O	Molecules	2	–
Sodium	Na	Ions	–	+1
Magnesium	Mg	Ions	–	+2
Chlorine	Cl	Ions	–	–1
Potassium	K	Ions	–	+1
Calcium	Ca	Ions	–	+2

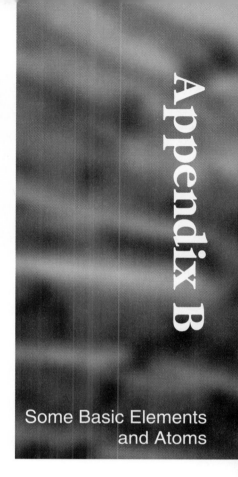

Appendix B

Some Basic Elements and Atoms

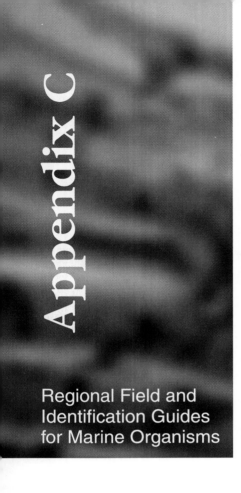

I have not personally used all of these guides and cannot vouch for their applicability or accuracy. I would appreciate hearing from you about problems with any of these or about other guides to add to this list. (e-mail address: jsumich@mail.gcccd.cc.ca.us)

Abbott, Isabella A., and E. Yale Dawson. 1978. *The seaweeds.* Dubuque, IA: Kendall/Hunt Publishing Company.

Audubon Society. 1981. *The Audubon Society field guide to North American seashore creatures.* New York: A. A. Knopf.

Booth, Ernest S. 1982. *The mammals.* Dubuque, IA: Kendall/Hunt Publishing Company.

Carcasson, R. H. 1977. *A field guide to the coral reef fishes of the Indian and West Pacific oceans.* London: Collins Publishing.

Clark, A. M., and M. E. Downey. 1992. *Starfishes of the Atlantic.* New York: Chapman and Hall.

Duncan W. H., and M. B. Duncan. 1987. *Seaside plants of the Gulf and Atlantic coasts.* Washington, D.C.: Smithsonian Institution Press.

Eschmeyer, W. N., and E. S. Herald. 1983. *A field guide to Pacific coast fishes.* Boston: Houghton Mifflin.

Gosner, K. 1978. *A field guide to the Atlantic seashore.* Peterson Field Guide Series. Boston: Houghton-Mifflin.

Gotshall, D. W. 1988. *Marine animals of Baja California.* Marina del Rey, CA: Western Marine.

Gotshall, D. W., and L. L. Laurent. 1979. *Pacific Coast subtidal marine invertebrates, a fishwatcher's guide.* Monterey, CA: Sea Challenger.

Halstead, B. W. 1988. *Poisonous and venomous marine animals of the world.* Burbank, CA: Darwin Publications.

Hinton, S. 1987. *Seashore life of southern California.* Berkeley, CA: University of California Press.

Hoese, H. D., and R. H. Moore. 1977. *Fishes of the Gulf of Mexico.* College Station, TX: Texas A & M University Press.

Human, P. 1993. *Reef coral identification: Florida, Caribbean, Bahamas, including marine plants.* New York: New World Publications.

Human, P., and N. DeLoach, eds. 1994. *Reef creature identification: Florida-Caribbean-Bahamas.* New York: New World Publications.

———. 1994. *Reef Fish Identification: Florida-Caribbean-Bahamas,* 2nd ed. New York: New World Publications.

Jahn, T. L., E. C. Bovee, and F. F. Jahn. 1978. *The protozoa.* Dubuque, IA: Kendall/Hunt Publishing Company.

Kaplan, E. H. 1984. *A field guide to southeastern and Caribbean seashores: Cape Hatteras to the Gulf Coast, Florida, and the Caribbean.* Boston: Houghton Mifflin.

Kozloff, E. N. 1987. *Marine invertebrates of the Pacific Northwest.* Seattle: University of Washington Press.

Lehmkuhl, Dennis M. 1979. *The aquatic insects.* Dubuque, IA: Kendall/Hunt Publishing Company.

Lippson, A. J., and R. L. Lippson. 1984. *Life in the Chesapeake Bay.* Baltimore: Johns Hopkins University Press.

Littler, D. S. et al. 1989. *Marine plants of the Caribbean: A field guide from Florida to Brazil.* Washington, D.C.: Smithsonian Institution Press.

Morris, B. F., and D. D. Mogelberg. 1973. *Identification manual to the pelagic Sargassum fauna.* Cambridge, MA: Bermuda Biological Station.

Ruppert, E., and R. Fox. 1988. *Seashore animals of the Southeast.* Columbia, SC: University of South Carolina Press.

Schneider, C. W., and R. B. Searles. 1991. *Seaweeds of the southeastern United States.* Durham, SC: Duke University Press.

Smith, D. L. 1977. *A guide to marine coastal plankton and marine invertebrate larvae.* Dubuque, IA: Kendall/Hunt Publishing Company.

Sterrer, W. 1986. *Marine fauna and flora of Bermuda.* New York: John Wiley and Sons.

Stokes, F. J. 1984. *Diver's and snorkeler's guide to the fishes and sea life of the Caribbean, Florida, Bahamas, and Bermuda.* Philadelphia: Academy of Natural Sciences Press.

Taylor, W. R. 1985. *Marine algae of the eastern tropical and subtropical coasts of the Americas.* Ann Arbor, MI: University of Michigan Press.

Voss, G. L. 1988. *Coral reefs of Florida.* Sarasota, FL: Pineapple Press.

This list may also be used as a guide for library acquisitions.

Books

Baker, J. D., and V. Cullen, eds. 1995. *Careers in oceanography and marine-related fields.* Virginia Beach, VA: The Oceanography Society.

Borgese, E., ed. 1979-1993. *The ocean yearbook,* vols. 1-10. Chicago: University of Chicago Press.

Cloud, P. 1987. *Biogeography and plate tectonics.* New York: Elsevier.

Cloud, P. 1988. *Oasis in space: Earth history from the beginning.* New York: W.W. Norton & Co., Inc.

Cushing, D. H., and J. J. Walsh. 1976. *The ecology of seas.* Philadelphia: W. B.. Saunders.

Duxbury, A. C., and A. B. Duxbury. 1994. *An introduction to the world's oceans.* Dubuque, IA: Wm. C. Brown Publishers.

Hardy, A. H. 1971. *The open sea: Its natural history.* Part I: "The world of plankton." Part II: "Fish and fisheries." Boston: Houghton Mifflin.

Hill, M. N, ed. 1982-83. *The sea: Ideas and observations on progress in the study of the seas,* 3 vols. New York: Interscience Publishers.

Idyll, C. P. 1976. *Abyss: The deep sea and the creatures that live in it.* New York: Thomas Y. Crowell.

Jumars, P. A. 1993. *Concepts in biological oceanography: An interdisciplinary primer.* New York: Oxford University Press.

Lalli, C. M., and T. R. Parsons. 1977. *Biological oceanography: An introduction.* Oxford, England: Butterworth Heinemann.

Levinton, J. S. 1982. *Marine ecology.* Englewood Cliffs, NJ: Prentice-Hall.

Mader, S. S. 1993. *Biology.* Dubuque, IA: Wm. C. Brown Publishers.

Marshall, N. B. 1980. *Deep-sea biology: Developments and perspectives.* New York: Garland S.T.P.M. Press.

Mills, E. L. 1989. *Biological oceanography: An early history, 1870-1960.* Ithaca, NY: Cornell University Press.

Niesen, T. M. 1982. *The marine biology coloring book.* New York: Harper and Row.

Sieburth, J. M. 1975. *Microbial seascapes: A pictorial essay on marine microorganisms and their environments.* Baltimore: University Park Press.

Steele, J. H., ed. 1973. *Marine food chains.* Edinburgh: Oliver and Boyd.

Steele, J. H. 1974. *The structure of marine ecosystems.* Cambridge, MA: Harvard University Press.

Steele, J. H., ed. 1978. *Spatial pattern in plankton communities.* New York: Plenum Press.

Valentine, J. W. 1973. *Evolutionary ecology of the marine biosphere.* Englewood Cliffs, NJ: Prentice-Hall.

Vernberg, W. B., and F. J. Vernberg. 1972. *Environmental physiology of marine animals.* New York: Springer-Verlag.

Vernberg, W., ed. 1974. *Symbiosis in the sea.* Columbia, SC: University of South Carolina Press.

Periodicals

American Naturalist
American Zoologist
Atoll Research Bulletin
Biological Bulletin
Bioscience
Bulletin of Marine Science
Copeia
Crustaceana
Deep-Sea Research
Diatom Research
Ecology
Estuaries
Estuaries and Coastal Marine Science
Fisheries Research
Fishery Bulletin
Journal of Experimental Marine Biology and Ecology
Journal of the Fisheries Research Board of Canada
Journal of Ichthyology
Journal of the Marine Biological Association, U.K.
Journal of Marine Education
Journal of Marine Research
Journal of Plankton Research
Limnology and Oceanography
Marine Biology
Marine Ecology
Marine Mammal Science
Marine Policy
Oceanography and Marine Biology, an Annual Review
Ocean Realm
Oceanus
Oikos
Pacific Science
Scientific American (frequently contains articles pertaining to marine life)
Veliger

Chapter 1

Baker, J. J., and G. E. Allen. 1981. *Matter, energy and life: An introduction to chemical concepts.* Reading, MA: Addison-Wesley.

Barkley, R. A. 1969. *Oceanographic atlas of the Pacific Ocean.* Honolulu: University of Hawaii Press.

Berner, R. A., and A. C. Lasaga. 1989. Modelling the geochemical carbon cycle. *Scientific American* 260(3):74-81.

Bogdanov, D. V. 1963. Map of the natural zones of the ocean. *Deep-Sea Research* 10:520-23.

Broecker, W. S. 1983. The ocean. *Scientific American* 249 (September) 146-60.

Selected References

General Works

Broeker, W. S., et al. 1985. Does the ocean-atmosphere system have more than one stable mode of operation? *Nature* 315:21-26.

Dietrich, G. 1980. *General oceanography.* New York: Wiley-Interscience Publishers.

Dietz, R. S., and J. C. Holden. 1970. Reconstruction of Pangaea: Breakup and dispersion of continents, Permian to present. *Journal of Geophysical Research* 75:4939-56.

Flessa, K. W. 1975. Area, continental drift and mammalian diversity. *Paleobiology* 1:189-94.

Gordon, A. L. 1986. The Southern Ocean and global climate. *Oceanus* 26(2):34-44.

Hedgpeth, J. W., ed. 1957. *Treatise on marine ecology and paleoecology.* Volume 1: Ecology. Geological Society of America. Memoir 67.

Hogg, N. 1992. The Gulf Stream and its recirculation. *Oceanus* 35(2):28-37.

Jenkyns, H. C. 1994. Early history of the oceans. *Oceanus* 36(4):49-52.

Jones, P. D., and T. M. L. Wigley. 1990. Global warming trends. *Scientific American* 263(2):84-91.

Macdonald, K. C., and B. P. Luyendyk. 1981. The crest of the East Pacific Rise. *Scientific American* 244(5):100-116.

Moore, III, B., and B. Bolin. 1986. The oceans, carbon cycle, and global climate change. *Oceanus* 29(4):16–26.

Open University. 1989. *Ocean circulation.* Oxford, England: Pergamon Press.

Pickard, G. L., and W. J. Emery, eds. 1982. *Descriptive Physical Oceanography.* New York: Pergamon Press.

Post, W. M., et al. 1990. The global carbon cycle. *American Scientist* 78(4):310–26.

Rasmusson, E. M. 1985. El Niño and variations in climate. *American Scientist* 73:168–77.

Southward, A. J. 1964. The relationship between temperature and rhythmic cirral activity in some Cirripedia considered in connection with their geographical distribution. *Helgol. wiss. Meersuntersuch.* 10:391–403.

Weller, R. A., and D. M. Farmer. 1992. Dynamics of the ocean's mixed layer. *Oceanus* 35(2):46–55.

Chapter 2

Allen, T. F. H., and T. B. Starr. 1988. *Hierarchy: Perspectives for ecological complexity.* Chicago, IL: The University of Chicago Press.

Caron, D. A. 1992. An introduction to biological oceanography. *Oceanus* 35(3):10–17.

De Duve, C. 1984. *A guided tour of the living cell.* New York: W. H. Freeman and Company.

Groves, D. I., J. S. R. Dunlop, and R. Buick. 1981. An early habitat of life. *Scientific American* 245: 64–73.

Guttman, B. S. 1976. Is "levels of organization" a useful biological concept? *Bioscience* 26:112–13.

Hardy, A. C. 1924. *The herring in relation to its animate environment,* pt. 1. The food and feeding habits of the herring. Fisheries Investigations, London, Series II, 7:1–53.

Hutchinson, G. E. 1961. The paradox of the plankton. *American Naturalist* 95:137–45.

Landry, M. R. 1976. The structure of marine ecosystems: An alternative. *Marine Biology* 35:1–7.

Levinton, J. S. 1982. *Marine ecology.* Englewood Cliffs, NJ: Prentice-Hall.

Lewin, R. A. 1982. Symbiosis and parasitism—definitions and evaluations. *Bioscience* 32:254.

Mader, S. S. 1993. *Biology.* Dubuque, IA: Wm. C. Brown Publishers.

Miller, R. S. 1967. Pattern and process in competition. *Advances in Ecological Research* 4:1–74.

Murphy, G. I. 1968. Pattern of life history and the environment. *American Naturalist* 102:391–403.

Pomeroy, L. R. 1992. The microbial food web. *Oceanus* 35(3):28–35.

Roughgarden, J. 1972. Evolution of niche width. *American Naturalist* 106:683–718.

Russell-Hunter, W. D. 1970. *Aquatic productivity.* New York: Macmillan Co.

Schoener, T. W. 1974. Resource partitioning in ecological communities. *Science* 185:27–29.

Sinclair, M. 1988. *Marine populations: An essay on population regulation and speciation.* Seattle, WA: University of Washington Press.

Southward, A. J. 1964. The relationship between temperature and rhythmic cirral activity in some Cirripedia considered in connection with their geographical distribution. *Helgol. wiss. Meersuntersuch.* 10:391–403.

Valentine, J. W. 1973. *Evolutionary ecology of the marine biosphere.* Englewood Cliffs, NJ: Prentice-Hall.

Van Valen, L. 1974. Predation and species diversity. *Journal of Theoretical Biology* 44:19–21.

Williams, G. C. 1966. Natural selection, the costs of reproduction and a refinement of Lack's principle. *American Naturalist* 100:687–90.

Wolfe, S. L. 1981. *Biology of the Cell.* Belmont, CA: Wadsworth Publishing Co.

Chapter 3

Austin, B. 1988. *Marine microbiology.* New York: Cambridge University Press.

Boatman, E. S., et al. 1987. Today's microscopy. *Bioscience* 37(6):384–94.

Bold, H. C., and M. J. Wynne. 1985. *Introduction to the algae.* Englewood Cliffs, NJ: Prentice-Hall.

Boney, A. D. 1992. *Phytoplankton.* London: E. Arnold.

Carr, N. G., and B. A. Whitton. 1983. *The biology of cyanobacteria.* Berkeley, CA: University of California Press.

Dale, B., and C. M. Yentsch. 1978. Red tide and paralytic shellfish poisoning. *Oceanus* 21 (summer):41–49.

Dawes, C. J. 1981. *Marine botany.* New York: John Wiley & Sons.

Fryxell, G. A. 1983. New evolutionary patterns in diatoms. *Bioscience* 33:92–98.

Hargraves, P. E., and F. W. French. 1983. Diatom resting spores: significance and strategies. In: *Survival strategies of the algae* by G. A. Fryxell, ed., 49–68. New York: Cambridge University Press.

Humm, H. J., and S. R. Wicks. 1980. *Introduction and guide to the marine blue-green algae.* Wiley.

Kaufman, P. B. et al. 1989. *Plants: Their biology and importance.* Addison-Wesley Educ.

Krogmann, D. W. 1981. Cyanobacteria (blue-green algae)—Their evolution and relation to other photosynthetic organisms. *Bioscience* 31:121–24.

Lembi, C. A., and J. R. Waaland, eds. 1988. *Algae and human affairs.* New York: Cambridge University Press.

Lipps, J. H. 1970. Plankton evolution. *Evolution* 24:1–22.

Margulis, L., D. Chase, and R. Guerrero. 1986. Microbial communities. *Bioscience* 36:160–70.

Marshall, H. G. 1976. Phytoplankton density along the eastern coast of U.S.A. *Marine Biology* 38:81–89.

Moll, R. A. 1977. Phytoplankton in a temperate-zone salt marsh: Net production and exchanges with coastal waters. *Marine Biology* 42:109–18.

Moore, R. E. 1977. Toxins from blue-green algae. *Bioscience* 27:797–802.

Okada, H., and A. McIntyre. 1977. Modern coccolithophores of the Pacific and North Atlantic oceans. *Micropaleontology* 23:1–55.

Paasche, E. 1968. Biology and physiology of coccolithophorids. *Annual Review of Microbiology* 22:71–86.

Peterson, M. N. A., ed. 1993. *Diversity of oceanic life: An evaluative review.* Center for Strategic and International Studies.

Pickett-Heaps, J. 1976. Cell division in eucaryotic algae. *Bioscience* 26:445–50.

Platt, T., and W. K. W. Li, eds. 1986. Photosynthetic picoplankton. *Can. Bull. Fish. Aquatic Sci.* 214:583.

Pomeroy, L. W. 1974. The ocean's food web, a changing paradigm. *Bioscience* 24:499–504.

Round, F. E., R. M. Crawford, and D. G. Mann. 1990. *The diatoms.* New York: Cambridge University Press.

Scagel, R. F., et al. 1980. *Nonvascular plants.* Belmont, CA: Wadsworth.

Smayda, T. J. 1970. The suspension and sinking of phytoplankton in the sea. *Oceanography and Marine Biology* (Annual Review) 8:353–414.

Smith, W. O., Jr., and D. M. Nelson. 1986. Importance of ice edge phytoplankton production in the southern ocean. *Bioscience* 36:251–57.

Steidinger, K. A., and K. Haddad. 1981. Biologic and hydrographic aspects of red tides. *Bioscience* 31(11):814–19.

Taylor, F. J. R., ed. 1987. *The biology of dinoflagellates.* London: Blackwell.

Vinyard, W. C. 1980. *Diatoms of North America.* Eureka, CA: Mad River Press.

Walsby, A. E. 1977. The gas vacuoles of blue-green algae. *Scientific American* (August):90–97.

Werner, D., ed. 1977. *The biology of diatoms.* Berkeley, CA: University of California Press.

Chapter 4

Abbott, I. A. 1978. *How to know the seaweeds.* Dubuque, IA: Wm. C. Brown Publishers.

Bold, H. C., and M. J. Wynne. 1985. *Introduction to the algae.* Englewood Cliffs, NJ: Prentice-Hall.

Chapman, A. R. O. 1979. *Biology of seaweeds.* Baltimore: University Park Press.

Cole, K. M., and R. G. Sheath, eds. 1990. *Biology of the red algae.* New York: Cambridge University Press.

Dawes, C. J. 1981. *Marine botany.* New York: John Wiley & Sons.

Dawson, E. Y. 1966. *Marine botany, an introduction.* New York: Holt, Rinehart and Winston.

Dring, M. J. 1982. *The biology of marine plants.* London: E. Arnold.

Duffy, J. E., and M. E. Hay. 1990. Seaweed adaptations to herbivory. *Bioscience* 40(5):368–75.

Estes, J. A., and J. F. Palmisano. 1974. Sea otters: Their role in structuring nearshore communities. *Science* 185:1058–60.

Foster, M. S. 1975. Algal succession in a *Macrocystis pyrifera* forest. *Marine Biology* 32:313–29.

Goering, J. J., and P. L. Parker. 1972. Nitrogen fixation by epiphytes on sea grasses. *Limnology and Oceanography* 17:320–23.

Jeffries, R. L. 1981. Osmotic adjustment of the response of halophytic plants to salinity. *Bioscience* 31:42–46.

King, R. J., and W. Schramm, 1976. Photosynthetic rates of benthic marine algae in relation to light intensity and seasonal variations. *Marine Biology* 37:215–22.

Koehl, M. A. R., and S. A. Wainwright. 1977. Mechanical adaptations of a giant kelp. *Limnology and Oceanography* 22:1067–71.

Lembi, C. A., and J. R. Waaland, eds. 1988. *Algae and human affairs.* New York: Cambridge University Press.

Lobban, C. S., and M. J. Winne, eds. 1982. *The biology of seaweeds.* Berkeley, CA: University of California Press.

Mann, K. H. 1972. Ecological energetics of the seaweed zone in a marine bay on the Atlantic coast of Canada. *Marine Biology* 14:199–209.

McPeak, R. H., and D. A. Glantz. 1984. Harvesting California's kelp forests. *Oceanus* 27(1):19–26.

Pettitt, J., S. Ducker, and B. Knox. 1981. Submarine pollination. *Scientific American* 244(3):134–43.

Phillips, R. C. 1978. Sea grasses and the coastal environment. *Oceanus* 21(summer):30–40.

Phleger, C. F. 1971. Effect of salinity on growth of a salt-marsh grass. *Ecology* 52:908–11.

Santelices, B. 1990. Patterns of reproduction, dispersal and recruitment in seaweeds. *Oceanography and Marine Biology* (Annual Review) 28:177–276.

Scagel, R. F., et al. 1980. *Nonvascular plants.* Belmont, CA: Wadsworth.

Schmitz, K., and C. S. Lobban. 1976. A survey of translocation in Laminariales (Phaeophyceae). *Marine Biology* 36:207–16.

Teal, J., and M. Teal. 1975. *The Sargasso Sea.* Boston: Little Brown.

Tomlinson, P. B. 1994. *The botany of mangroves.* New York: Cambridge University Press.

Chapter 5

Bainbridge, R. 1957. Size, shape and density of marine phytoplankton concentrations. *Biological Review* 32:91–115.

Baker, J. D., and W. S. Wilson. 1986. Spaceborne observations in support of earth science. *Oceanus* 29(4):76–85.

Boatman, E. S., et al. 1987. Today's microscopy. *Bioscience* 37:384–94.

Boney, A. D. 1992. *Phytoplankton.* London: E. Arnold.

Brown, O. B., et al. 1985. Phytoplankton blooming off the U.S. East Coast: A satellite description. *Science* 229:163–67.

Chisholm, S. W. 1992. What limits phytoplankton growth? *Oceanus* 35(3):36–46.

Correll, D. L. 1978. Estuarine productivity. *Bioscience* 28:646–50.

Duffy, J. E., and M. E. Hay. 1990. Seaweed adaptations to herbivory. *Bioscience* 40(5):368–75.

Falkowski, P. G., ed. 1980. *Primary productivity in the sea.* New York: Plenum Press.

Fleming, R. H. 1939. The control of diatom populations by grazing. *Journal du Conseil Permanent International pour l'Exploration de la Mer* 14:210–27.

Harris, G. P. 1986. *Phytoplankton Ecology:* Chapman and Hall.

Jenkins, W. J., and J. C. Goldman. 1985. Seasonal oxygen cycling and primary production in the Sargasso Sea. *Journal of Marine Research* 43:465–91.

King, R. J., and W. Schramm. 1976. Photosynthetic rates of benthic marine algae in relation to light intensity and seasonal variations. *Marine Biology* 37:215–22.

Landry, M. R. 1976. The structure of marine ecosystems: An alternative. *Marine Biology* 35:1–7.

Lembi, C. A., and J. R. Waaland, eds. 1988. *Algae and human affairs.* New York: Cambridge University Press.

Lobban, C. S., and M. J. Winne, eds. 1982. *The biology of seaweeds.* Berkeley, CA: University of California Press.

Malone, T. C. 1971. The relative importance of nannoplankton and net plankton as primary producers in tropical oceanic and neritic phytoplankton communities. *Limnology and Oceanography* 16:633–39.

Mann, K. H. 1973. Seaweeds: Their productivity and strategy for growth. *Science* 182:975–81.

Marshall, H. G. 1976. Phytoplankton density along the eastern coast of U.S.A. *Marine Biology* 38:81–89.

Perry, M. J. 1986. Assessing marine primary productivity from space. *Bioscience* 36:461–67.

Philander, G. 1989. El Niño and La Niña. *American Scientist* 77(5):451–59.

Pomeroy, L. R. 1974. The ocean's food web, a changing paradigm. *Bioscience* 24:449–504.

Qasim, S. Z., P. M. A. Bhattuthiri, and V. P. Devassy. 1972. The effect of intensity and quality of illumination on the photosynthesis of some tropical marine phytoplankton. *Marine Biology* 16:22–27.

Raymont, J. E. G. 1983. *Plankton and productivity in the oceans.* Volume 1: Phytoplankton. New York: Pergamon Press.

Rowan, K. S. 1989. *Photosynthetic pigments of algae.* New York: Cambridge University Press.

Russell-Hunter, W. D. 1970. *Aquatic productivity.* New York: Macmillan.

Ryther, J. H. 1969. Photosynthesis and fish production in the sea. *Science* 166:72–76.

Ryther, J. H., and C. S. Yentsch, 1957. Estimation of phytoplankton production in the ocean from chlorophyll and light data. *Limnology and Oceanography* 2:281–86.

Saffo, M. B. 1987. New light on seaweeds. *Bioscience* 37(9):654–64.

Schoen, S. B. 1987. *Langmuir circulation and small-scale patchiness of marine phytoplankton and zooplankton.* Ph.D Dissertation. Santa Barbara: University of California.

Smayda, T. J. 1970. The suspension and sinking of phytoplankton in the sea. *Oceanography and Marine Biology* (Annual Review) 8:353–414.

Smith, W. O., Jr., and D. M. Nelson. 1986. Importance of ice edge phytoplankton production in the southern ocean. *Bioscience* 36:251–57.

Steele, J. H., ed. 1973. *Marine food chains.* Edinburgh: Oliver and Boyd.

Steele, J. H. 1974. *The structure of marine ecosystems.* Cambridge, MA: Harvard University Press.

Steeman Nielsen, E. 1952. Use of radioactive carbon (C^{14}) for measuring organic production in the sea. *Journal du Conseil Permanent International pour l'Exploration de la Mer* 18:117–40.

Steeman Nielsen, E. 1975. *Marine photosynthesis.* Amsterdam: Elsevier.

Venrick, E. L., J. A. McGowan, and A. W. Mantyla. 1973. Deep maxima of photosynthetic chlorophyll in the Pacific Ocean. *Fishery Bulletin* 71:41–52.

Walsh, J. J. 1975. A spatial simulation model of the Peru upwelling ecosystem. *Deep-Sea Research* 22:201–36.

Walsh, J. J. 1984. The role of ocean biota in accelerated ecological cycles: A temporal view. *Bioscience* 34:499–507.

Whittaker, R. H., and G. E. Likens. 1973. "Carbon in the biota." In: *Carbon and the Biosphere.* Edited by G. M. Woodwell, E. V. Pecan, Technical Information Center, U.S. Atomic Energy Commission, Oak Ridge, TN.

Chapter 6

Alexander, R. M. 1979. *The invertebrates.* New York: Cambridge University Press.

Barnes, R. D. 1980. *Invertebrate zoology.* Philadelphia: Saunders College/Holt, Rinehart and Winston.

Brady, H. B. 1884. Report on the foraminifera dredged by H. M. S. *Challenger,* vol. 9, (*Zoology*): 1–814.

Brusca, R. C., and G. J. Brusca. 1990. *Invertebrates.* Sunderland, MA: Sinauer Associates.

Buzas, M. A., and S. J. Culver. 1991. Species diversity and dispersal of benthic foraminifera. *Bioscience* 41(7):483–89.

Capriulo, G. M. 1989. *Ecology of marine protozoa.* New York: Oxford University Press.

Hickman, C. P., Jr., et al. 1992. *Integrated principles of zoology.* St. Louis: Mosby College Publishing.

Levinton, J. S. 1992. The big bang of animal evolution. *Scientific American* (September):252:84–91.

Mader, S. S. 1987. *Evolution, diversity, and the environment.* Dubuque, IA: Wm. C. Brown Publishers.

McMahon, T. A., and J. T. Bonner. 1984. *On size and life.* New York: W. H. Freeman and Company.

Muscatine, L., and H. M. Lenhoff, eds. 1974. *Coelenterate biology: Reviews and new perspectives.* New York: Academic Press.

Pechenik, J. A. 1991. *Biology of the invertebrates.* Dubuque, IA: Wm. C. Brown Publishers.

Richardson, J. 1986. Brachiopods. *Scientific American* (September): 100–106.

Russell-Hunter, W. D. 1979. *A life of invertebrates.* New York: Macmillan.

Sebens, K. P. 1977. Habitat suitability, reproductive ecology and the plasticity of body size in two sea anemone populations (*Anthopleura elegantissima* and *A. xanthogrammica*). Ph.D. Dissertation. Seattle: University of Washington.

Stanley, S. M. 1975. *A theory of evolution above the species level.* Proceedings of the National Academy of Science U.S.A. 72:646–50.

Valentine, W. 1978. The evolution of multicellular plants and animals. *Scientific American* (September) 140–58.

Williams, A. B. 1984. *Shrimps, lobsters, and crabs of the Atlantic coast.* Washington, D.C.: Smithsonian Institution Press.

Willmer, P. 1990. *Invertebrate relations.* Cambridge: Cambridge University Press.

Chapter 7

Ainley, D. G., et al. 1984. *The marine ecology of birds in the Ross Sea, Antarctica.* Washington, D.C.: American Ornithologists Union.

Alderton, D. 1988. *Turtles and tortoises of the world.* New York: Facts on File, Inc.

Anderson, H. T., ed. 1969. *The biology of marine mammals.* New York: Academic Press.

Berta, A. 1994. What is a whale? *Science* 263:180–82.

Berta, A., C. E. Ray, and A. R. Wyss. 1989. Skeleton of the oldest known pinniped. *Enaliarctos mealsi. Science* 244:60–62.

Bond, C. E. 1979. *Biology of fishes.* Philadelphia: W. B. Saunders Co.

Bonner, W. N. 1982. *Seals and man: A study of interactions.* Seattle: University of Washington Press.

Compagno, L. J. V. 1988. *Sharks of the order Carcharhiniformes.* Princeton, NJ: Princeton University Press.

Croxall, J. P. 1987. *Seabirds: Feeding ecology and role in marine ecosystems.* New York: Cambridge University Press.

Dunson, W. A., ed. 1975. *The biology of sea snakes.* Baltimore: University Park Press.

Ernst, C. H., and R. W. Barbour. 1989. *Turtles of the world.* Washington, D.C.: Smithsonian Institution Press.

Gaskin, D. E. 1982. *The ecology of whales and dolphins.* Portsmouth, NH: Heinemann.

Geraci, J. R. 1978. The enigma of marine mammal strandings. *Oceanus* 21 (spring):38–47.

Herman, L. M., ed. 1980. *Cetacean behavior: Mechanisms and functions.* New York: John Wiley & Sons.

Love, J. A. 1990. *The sea otter.* London: Whittet Books.

Marshall, N. B. 1966. *The life of fishes.* New York: World Publishing Company.

Martin, A. R. 1990. *Whales and dolphins.* London: Salamander Books.

Nettleship, D. N., G. A. Sanger, and P. F. Springer, eds. 1985. *Marine birds: Their ecology and commercial fisheries relationships.* Ottawa: Canadian Wildlife Services.

Owens, D. W. 1980. The comparative reproductive physiology of sea turtles. *American Zoologist* 20:549–63.

Riedman, M. 1990. *The pinnipeds: Seals, sea lions and walruses.* Berkeley: University of California Press

Stirling, I. 1988. *Polar bears.* Ann Arbor, MI: University of Michigan Press.

Stirling, I., and D. Guravich. 1988. *Polar bears.* Ann Arbor: University of Michigan Press.

Thorson, G. 1957. Bottom communities. In Treatise on Marine Ecology and Paleoecology. *Vol. I.- Ecology,* edited by J. W. Hedgpeth. Geological Society of America. p. 461–534.

VanBlarcom, G. R., and J. A. Estes. 1988. *The community ecology of sea otters.* New York: Springer-Verlag.

Wrootton, R. J. 1990. *Ecoloty of teleost fishes.* New York: Chapman and Hall.

Würsig, B. 1988. The behavior of baleen whales. *Scientific American* 258(4):102–7.

Chapter 8

Armstrong, R. A., and R. McGehee. 1980. Competitive exclusion. *American Naturalist* 115:151–70.

Brafield, A. E. 1978. Life in sandy shores. *Studies in biology* no. 89. London: Edward Arnold.

Carson, R. L. 1979. *The edge of the sea.* Boston: Houghton Mifflin Co.

Connell, J. H. 1961. The influence of interspecific competition and other factors on the distribution of the barnacle *Chthamalus stellatus. Ecology* 42:710–23.

Dayton, P. K. 1971. Competition, disturbance, and community organization: The provision and subsequent utilization of space in a rocky intertidal community. *Ecological Monographs* 41:351–89.

Denny, M. W. 1985. Wave forces on intertidal organisms: A case study. *Limnology and Oceanography* 30:1171–87.

Eltringham, S. K. 1972. *Life in mud and sand.* New York: Crane, Russak & Co.

Epel, D. 1977. The program of fertilization. *Scientific American* (November): 129–38.

Grahame, J., and G. M. Branch. 1985. *Reproductive patterns of marine invertebrates.* Scotland: Aberdeen University Press.

Harger, J. R. E. 1972. Competitive coexistence among intertidal invertebrates. *American Scientist* 60:600–607.

Hayes, F. R. 1964. "The mud-water interface." In: *Oceanography and Marine Biology,* an Annual Review. Edited by Harold Barnes, vol. 2:122–45. New York: Hafner Press.

Hoar, W. S. 1983. *General and comparative physiology.* Englewood Cliffs, NJ: Prentice-Hall.

Lewis, J. H. 1964. *The ecology of rocky shores.* London: English Universities Press.

Little, C., and J. A. Kitching. 1996. *The biology of rocky shores.* New York: Oxford University Press.

Lubchenco, J. 1978. Plant species diversity in a marine intertidal community: Importance of herbivore food preference and algal competitive abilities. *American Naturalist* 112:23–39.

McIntyre, A. D. 1969. Ecology of marine meibenthos. *Biological Review* 44:245–90.

Menge, B. A. 1975. Brood or broadcast? The adaptive significance of different reproductive strategies in the two intertidal sea stars. *Leptasterias hexactis and Pisaster ochraceus. Marine Biology* 31:87–100.

Moore, P. G., and R. Seed. 1986. *The ecology of rock coasts.* New York: Columbia University Press.

Morse, A. N. C. 1991. How do planktonic larvae know where to settle? *American Scientist* 79:154-67.

Newell, R. C. 1979. *Biology of intertidal animals.* Faversham, Kent, U. K.: Ecological Surveys, Ltd.

Palmer, J. D. 1974. *Biological clocks in marine organisms: The control of physiological and behavioral tidal rhythms.* New York: Interscience Publishers.

Peterson, C. H. 1991. Intertidal zonation of marine invertebrates in sand and mud. *American Scientist* 79:236-48.

Reise, K. 1985. *Tidal flat ecology.* New York: Springer-Verlag.

Ricketts, C., J. Calvin, and J. W. Hedgeph. 1986. *Between Pacific tides.* Revised by D. W. Phillips. Stanford, CA: Stanford University Press.

Sanders, H. L. 1968. Marine benthic diversity: A comparative study. *American Naturalist* 102:243-82.

Scheltema, R. S. 1971. Larval dispersal as a means of genetic exchange between geographically separated populations of shallow-water benthic marine gastropods. *Biological Bulletin* 140:284-322.

Sebens, K. P. 1985. The ecology of the rocky subtidal zone. *American Scientist* 73:548-57.

Strathman, R. 1974. The spread of sibling larvae of sedentary marine invertebrates. *American Naturalist* 108:29-44.

Thorson, G. 1950. Reproduction and larval ecology of marine bottom invertebrates. *Biological Review* 25:1-45.

Thorson, G. 1961. Length of pelagic life in marine bottom invertebrates as related to larval transport by ocean currents. In: *Oceanography, AAAS,* ed. M. Sears, 455-74.

Underwood, A. J. 1974. On models for reproductive strategy in marine benthic invertebrates. *American Naturalist* 108:874-78.

Whitlatch, R. B. 1981. Patterns of resource utilization and coexistence in marine intertidal deposit-feeding communities. *Journal of Marine Research* 38:743-65.

Chapter 9

Barnes, R. S. K. 1974. Estuarine biology. *Studies in biology,* no. 49. Baltimore: University Park Press.

Bertness, M. D. 1992. The ecology of a New England salt marsh. *American Scientist* 80:260-68.

Botton, M. L., and H. H. Haskin. 1984. Distribution and feeding of the horseshoe crab. *Limulus polyphemus,* on the continental shelf off New Jersey. *Fishery Bulletin* 82(2):383-89.

Britton, J. C. 1989. *Shore ecology of the Gulf of Mexico.* Dallas: Texas Press.

Correll, D. L. 1978. Estuarine productivity. *Bioscience* 28:646-50.

Costlow, J. D., and C. G. Bookhout. 1959. The larval development of *Callinectes sapidus* Rathbun reared in the laboratory. *Biological Bulletin* 116:373-96.

Dennison, W. C., et al. 1993. Assessing water quality with submerged aquatic vegetation. *Bioscience* 43:86-93.

Durbin, A., and E. Durbin. 1974. Grazing rates of the Atlantic Nemhaden *Brevoortia tyrannus* as a function of particle size and concentration. *Marine Biology* 33:265-77.

Ernst, W. G., and J. G. Morin, eds. 1984. *The environment of the deep sea.* Englewood Cliffs, NJ: Prentice-Hall.

Garrison, D. 1976. Contribution of the net plankton and nannoplankton to the standing stocks and primary productivity in Monterey Bay, California, during the upwelling season. *Fishery Bulletin* 74:183-94.

Heinle, D. R., R. P. Harris, J. F. Ustach, and D. A. Flemer. 1977. Detritus as food for estuarine copepods. *Marine Biology* 40:341-53.

Johnson, D. R. 1985. Wind-forced dispersion of blue crab larvae in the Middle Atlantic Bight. *Continental Shelf Research* 4:425-37.

Johnson, D. R., B. S. Hester, and J. R. McConaugha. 1984. Studies of a wind mechanism influencing the recruitment of blue crabs in the Middle Atlantic Bight. *Continental Shelf Research* 3:425-37.

Kusler, J. A., W. J. Mitsch, and J. S. Larson. 1994. Wetlands. *Scientific American* (January) 270:64-70.

Lippson, J. A., and R. L. Lippson. 1984. *Life in the Chesapeake Bay.* Baltimore: Johns Hopkins University Press.

Mangelsdorf, P. C., Jr. 1967. *Salinity measurements in estuaries. In: Estuaries.* G. H. Lauff, ed. Amer. Assoc. Adv. Science Publ. no. 83:71-79.

Marshall, H. G. 1980. Seasonal phytoplankton composition in the lower Chesapeake Bay and Old Plantation Creek, Cape Charles, Virginia. *Estuaries* 3:207-16.

McLusky, D. S. 1971. *Ecology of estuaries.* London: Heinemann Educational Books, Ltd.

McRoy, C. P., and C. Helfferich. 1977. *Seagrass ecosystems.* New York: Marcel Dekker.

Miller, J. M., and M. L. Dunn. 1980. Feeding strategies and patterns of movement in juvenile estuarine fishes. In: *Estuarine perspectives.* Edited by V. S. Kennedy. New York: Academic Press.

Milliman, J. 1989. Sea levels: Past, present, and future. *Oceanus* 32(2):40-43.

Nicol, J. A. C. 1967. *The biology of marine animals.* London: Pitman and Sons.

Nichols, F., et al. 1986. Temporal dynamics of an estuary: San Francisco Bay. *Science* 231(4738):567-73.

Phillips, R. C. 1978. Seagrasses and the coastal marine environment. *Oceanus* 21:30-40.

Remane, A. 1934. Die Brackwasserfauna. *Zoologischer Anzeiger.* Supplementband 7:34-74.

Selander, R., S. Yang, R. Lewontin, and W. Johnson. 1970. Genetic variation in the horseshoe crab (*Limulus polyphemus*), a phylogenetic "relic." *Evolution* 24:402-14.

Siry, J. V. 1984. *Marshes of the ocean shore.* Austin: Texas A & M University Press.

Teal, J., and M. Teal. 1974. *Life and death of the salt marsh.* New York: Ballantine.

Tyler, M. A., and H. H. Seliger. 1978. Annual subsurface transport of a red tide dinoflagellate to its bloom area: Water circulation patterns and organism distributions in the Chesapeake Bay. *Limnology and Oceanography* 23:227-46.

Valiela, I., and J. Teal. 1979. The nitrogen budget of a salt marsh ecosystem. *Nature* 280:652-56.

Williams, A. B. 1974. The swimming crabs of the genus *Callinectes* (Decapoda: Portunidae). *Fishery Bulletin* 72:685-798.

Zedler, J., T. Winfield, and D. Mauriello. 1978. Primary productivity in a southern California estuary. *Coastal Zone* 3:649-62.

Chapter 10

Adey, W. H. 1978. Coral reef morphogenesis: A multidimensional model. *Science* 202:831-37.

Aeby, G. S. 1991. Costs and benefits of parasitism in a coral reef system. *Pacific Science* 45:85-86.

Babcock, R. C., P. L. Bull, A. J. Heyward, J. K. Oliver, C. C. Wallace, and B. L. Willis. 1986. Synchronous spawnings of 105 scleractinian coral species on the Great Barrier Reef. *Marine Biology* 90:379-394.

Barlow, G. W. 1972. The attitude of fish eye-lines in relation to body shape and to stripes and bars. *Copeia* 1:4-12.

Birkeland, C. 1989. The Faustian traits of the crown of thorns starfish. *American Scientist* 77:154-63.

Blair, S. M., T. L. McIntosh, and B. J. Mostkoff. 1994. Impacts of Hurricane Andrew on the offshore reef systems of the central and northern Dade County, Florida. *Bulletin of Marine Science* 54:961-973.

Chamberlain, J. A. 1978. Mechanical properties of coral skeleton: Compressive strength and its adaptive significance. *Paleobiology* 4:419-35.

Clifton, K. E. 1997. Mass spawning by green algae on coral reefs. *Science* 275:1116-8.

Connell, J. H. 1978. Diversity in tropical rain forests and coral reefs. *Science* 199:1302-10.

Dana, T. F. 1975. Development of contemporary Eastern Pacific coral reefs. *Marine Biology* 33:355-74.

Darwin, C. 1962. [1842]. *The structure and distribution of coral reefs*. Berkeley, CA: University of California Press.

Endean, R. 1983. *Australia's Great Barrier Reef*. New York: University of Queensland Press.

Fadlallah, Y. H. 1983. Sexual reproduction, development and larval biology in scleractinian corals: a review. Coral Reefs 2:129–50.

Falkowsky, P. G., et al. 1984. Light and the bioenergetics of a symbiotic coral. *BioScience* 34:705–9.

Falkowski, P. G., Z. Dubinsky, L. Muscatine, and L. McCloskey. 1993. Population control in symbiotic corals. *Bioscience* 43:606–11.

Fankboner, P. V. 1971. Intracellular digestion of symbiotic zooxanthellae by host amoebocytes in giant clams (Bivalvia: Tridachnidae), with a note on the nutritional role of the hypertrophied siphonal epidermis. *Biological Bulletin* 141:222–34.

Fox, D. L. 1979. *Biochromy: Natural coloration of living things*. Berkeley: University of California Press.

Goreau, T. F., N. I. Goreau, and C. M. Yonge. 1971. Reef corals: autotrophs or heterotrophs? *Biological Bulletin* 141:247–60.

Grigg, R. W. 1982. Darwin Point: A threshold for atoll formation. *Coral Reefs* 1:29–34.

Halstead, B. W., P. S. Auerbach, and D. R. Campbell. 1990. *A colour atlas of dangerous marine animals*. Boca Raton, FL: CRC Press.

Hatcher, B. G. 1990. Coral reef primary productive: a hierarchy of pattern and process. *Trends in ecology and evolution* 5:149–155.

Hawkins, J. P., C. M. Roberts, and T. Adamson. 1991. Effects of a phosphate ship grounding on a Red Sea coral reef. *Marine Pollution Bulletin* 22:538–42.

Highsmith, R. C. 1982. Reproduction by fragmentation in corals. Marine Ecology Progress Series 7:207–26.

Hutchings, P. A. 1986. Biological destruction of coral reefs: a review. *Coral Reefs* 4:239–52.

Jackson, J. B. C., and T. P. Hughes. 1985. Adaptive strategies of coral-reef invertebrates. *American Scientist* 73:265–74.

Jones, G. P. 1990. The importance of recruitment to the dynamics of a coral reef fish population. *Ecology* 71:1691–98.

Jones, O. A., and R. Endean, eds. 1976. *Biology and geology of coral reefs*. Vols I and II. New York: Academic Press.

Kaplan, E. H. 1988. *A field guide to coral reefs of the Caribbean and Florida including Bermuda and the Bahamas*. Boston: Houghton Mifflin.

Limbaugh, C. 1961. Cleaning symbiosis. *Scientific American* (August): 42–49.

Losey, G. S., Jr. 1972. The ecological importance of cleaning symbiosis. *Copeia* 4:820–33.

Mariscal, R. N. 1972. Behavior of symbiotic fishes and sea anemones. In: *Behavior of Marine Animals*. Edited by H. E. Winn and B. L. Olla. New York: Plenum Publishing.

Muscatine, L., and J. W. Porter. 1977. Reef corals: Mutualistic symbiosis adapted to nutrient-poor environments. *Bioscience* 27:454–60.

Odum, H. T., and E. P. Odum. 1955. Trophic structure and productivity of a windward coral reef community on Eniwetok Atoll. *Ecological Monographs* 25:291–320.

Porter, J. W. 1976. Autotrophy, heterotrophy, and resource partitioning in Caribbean reef-building corals. *American Naturalist* 110:731–42.

Richmond, R. H. 1985. Reversible metamorphosis in coral planula larvae. *Marine Ecology Progress Series* 22:181–5.

Sale, P. F. 1974. Mechanisms of coexistence in a guild of territorial reef fishes. *Marine Biology* 29:89–97.

Schener, P. J. 1977. Chemical communication of marine invertebrates. *Bioscience* 27:644–68.

Schuhmacher, H., and H. Zibrowius. 1985. What is hermatypic? *Coral Reefs* 4:1–9.

Scott, R. D., and H. R. Jitts. 1977. Photosynthesis of phytoplankton and zooxanthellae on a coral reef. *Marine Biology* 41:307–15.

Shlesinger, Y., and Y. Loya. 1985. Coral community reproductive patterns: Red Sea versus the Great Barrier Reef. *Science* 228:1333–35.

Sorokin, Y. I. 1972. Bacteria as food for coral reef fauna. *Oceanology* 12:169–77.

Stoddart, D. R. 1973. Coral reefs: The last two million years. *Geography* 58:313–23.

Thresher, R. E. 1984. *Reproduction in reef fishes*. Neptune City, NJ: T.H.F. Publications.

Walbran, P. D., R. A. Henderson, A. J. T. Jull, and M. J. Head. 1989. Evidence from sediments of long-term *Acanthaster planci* predation on corals of the Great Barrier Reef. *Science* 245:847–50.

Warner, R. R. 1990. Male versus female influences on mating-site determination in a coral reef fish. *Animal Behavior* 39:540–48.

Chapter 11

Arp, A. J., and J. J. Childress. 1983. Sulfide binding by the blood of the hydrothermal vent tube worm *Riftia pachyptila*. *Science* 219:295–97.

Childress, J., H. Felback, and G. Somero. 1987. Symbiosis in the deep sea. *Scientific American* 256:114–20.

Corliss, J. B., et al. 1979. Submarine thermal springs on the Galapagos Rift. *Science* 203:1073–83.

Dayton, P. K., and R. R. Hessler. 1972. Role of biological disturbance in maintaining diversity in the deep sea. *Deep Sea Research* 19:199–208.

Grassle, J. F. 1991. Deep-sea benthic biodiversity. *Bioscience* 41:464–69.

Gray, J. S. 1974. Animal-sediment relationships. *Oceanography and Marine Biology* (Annual Review) 12:223–61.

Hessler, R. R., J. D. Isaacs, and E. L. Mills. 1972. Giant amphipod from the abyssal Pacific Ocean. *Science* 175:636–37.

Higgins, R. P., and H. Thiel, eds. 1988. *Introduction to the study of meiofauna*. Washington, D.C.: Smithsonian Institution Press.

Isaacs, J. D., and R. A. Schwartzlose. 1975. Active animals of the deep-sea floor. *Scientific American* (October): 84–91.

Lutz, R. A. 1991. The biology of deep-sea vents and seeps. *Oceanus* 34(3):75–83.

Marshall, N. B. 1980. *Deep sea biology: Developments and perspectives*. New York: Garland S.T.P.M. Press.

Menzies, R. J., R. Y. George, and G. T. Rowe. 1973. *Abyssal environment and ecology of the world oceans*. New York: John Wiley.

Page, H. M., C. R. Fisher, and J. J. Childress. 1990. Role of filter-feeding in the nutritional biology of a deep-sea mussel with methanotrophic symbionts. *Marine Biology* 104:251–57.

Rex, M. A. 1973. Deep-sea species diversity: Decreased gastropod diversity at abyssal depths. *Science* 181:1051–53.

Rokop, F. J. 1974. Reproductive patterns in the deep-sea benthos. *Science* 186:743–45.

Sanders, H. L. 1968. Marine benthic diversity: A comparative study. *American Naturalist* 102:243–82.

Sanders, H. L., and R. R. Hessler. 1969. Ecology of the deep-sea benthos. *Science* 163:1419–24.

Sebens, K. P. 1985. The ecology of the rocky subtidal zone. *American Scientist* 73:548–57.

Sokolova, M. N. 1970. Weight characteristics of meiobenthos in different regions of the deep-sea trophic areas of the Pacific Ocean. *Okeanologia* (in Russian) 10:348–56.

Sverdrup, H. U.; Johnson, M. W.; and Fleming, R. H. 1942. *The oceans.-their physics, chemistry, and general biology*. Englewood Cliffs, NJ: Prentice-Hall.

Tait, R. V. 1968. *Elements of marine ecology*. London: Butterworths.

Thorson, G. 1957. "Bottom communities." In: *Treatise on marine ecology and paleoecology*, ed. J. W. Hedgepeth, Geological Society of America, Vol. 1, Ecology, 461–534.

Thorson, 1966. Some factors influencing the recruitment and establishment of marine benthic communities. *Netherlands Journal of Sea Research* 3:241–267.

Tunnicliffe, V. 1992. Hydrothermal-vent communities of the deep sea. *American Scientist* 80:336–49.

Chapter 12

Allan, J. D. 1976. Life history patterns in zooplankton. *American Naturalist* 110:165–80.

Alldredge, A. 1976. Appendicularians. *Scientific American* (July):94–102.

Alldredge, A. L., and L. P. Madin. 1982. Pelagic tunicates: Unique herbivores in the marine plankton. *Bioscience* 32:655–63.

Barham, E. G. 1966. Deep scattering layer migration and composition: Observations from a diving saucer. *Science* 151:1399–1403.

Boden, B. P., and E. M. Kampa. 1967. *The influence of natural light on the vertical migrations of an animal community in the sea.* Symposium of the Zoological Society of London 19:15–26.

Boyd, C. M. 1976. Selection of particle sizes by filter-feeding copepods: A plea for reason. *Limnology and Oceanography* 21:175–79.

Bright, T., et al. 1972. Effects of a total solar eclipse on the vertical distribution of certain oceanic zooplankters. *Limnology and Oceanography* 17:296–301.

Brinton, E. 1962. "The distribution of Pacific euphausiids." Bulletin. Scripps Institution of Oceanography 8:51–270.

Frost, B. W. 1972. Effects of size and concentration of food particles on the feeding behavior of the marine planktonic copepod *Calanus pacificus. Limnology and Oceanography* 17:805–15.

Gilmer, R. W. 1972. Free-floating mucus webs: A novel feeding adaptation for the open ocean. *Science* 176:1239–40.

Lam, R. K., and B. W. Frost. 1976. Model of copepod filtering response to changes in size and concentration of food. *Limnology and Oceanography* 21:490–500.

Mauchline, J., and L. R. Fisher. 1969. The biology of euphausiids. *Advances in Marine Biology* 7:1–454.

Porter, K. G., and J. W. Porter. 1979. Bioluminescence in marine plankton: A coevolved antipredation system. *American Naturalist* 114:458–61.

Richman, S., D. R. Heinle, and R. Huff. 1977. Grazing by adult estuarine calanoid copepods of the Chesapeake Bay. *Marine Biology* 42:69–84.

Rubenstein, D. I., and M. A. R. Koehl. 1977. The mechanisms of filter feeding: Some theoretical considerations. *American Naturalist* 111:981–94.

Russel, R. S. 1935. On the value of certain planktonic animals as indicators of water movements in the English Channel and North Sea. *Journal of the Marine Biological Association,* U.K. 20:309–32.

Russell-Hunter, W. D. 1969. *Biology of higher invertebrates.* New York: Macmillan.

Sheldon, R. W., A. Prakash, and W. H. Sutcliffe, Jr. 1972. The size distribution of particles in the ocean. *Limnology and Oceanography* 17:327–40.

Silver, M. W., A. L. Shanks, and J. D. Trent. 1978. Marine snow: Microplankton habitat and source of small-scale patchiness in pelagic populations. *Science* 201:371–73.

Smith, O. L., et al. 1971. Resource competition and an analytical model of zooplankton feeding on phytoplankton. *American Naturalist* 109:571–91.

Steele, J. H., ed. 1973. *Marine food chains.* Edinburgh: Oliver and Boyd.

Steele, J. H. 1976. "Patchiness." In: *Ecology of the Seas.* ed. D. H. Cushing and J. J. Walsh. Oxford, U.K.: Blackwell Scientific Publications.

Stoecker, D. K., and J. M. Capuzzo. 1990. Predation on protozoa: Its importance to zooplankton. *Journal of Plankton Research* 12:891–908.

Strickler, J. R. 1985. Feeding currents in calanoids: Two new hypotheses. In: *Physiological adaptations of marine animals* ed. M. S. Laverack. Symposium, *Society of Experimental Biology* 39:459–85.

Turner, J. T., P. A. Tester, and W. F. Hettler. 1985. Zooplankton feeding ecology. *Marine Biology* 90:1–8.

Vlymen, W. J. 1970. Energy expenditure of swimming copepods. *Limnology and Oceanography* 15:348–56.

Waickstead, J. H. 1976. *Marine zooplankton.* London: E. Arnold.

Zaret, T. M., and J. S. Suffern. 1976. Vertical migration in zooplankton as a predator avoidance mechanism. *Limnology and Oceanography* 21:804–13.

Chapter 13

Alexander, R. McNeill. 1988. *Elastic mechanisms in animal movement.* New York: Cambridge University Press.

Bainbridge, R. 1960. Speed and stamina in three fish. *Journal of Experimental Biology* 37:129–153.

Blake, R. W. 1983. *Fish locomotion.* New York: Cambridge University Press.

Bond, C. E. 1983. *Biology of fishes.* Philadelphia: W. B. Saunders.

Carey, F. G. 1973. Fishes with warm bodies. *Scientific American* (February):36–44.

Compagno, L. J. V. 1988. *Sharks of the order Carcharhinoformes.* Princeton: Princeton University Press.

Croxall, J. P. 1987. *Seabirds: Feeding ecology and role in marine ecosystems.* New York: Cambridge University Press.

DeLong, R. L.; and B. S. Stewart. 1991. Diving patterns of northern elephant seal bulls. *Mar. Mamm.* Sci. 7:369–384.

Denton, E. J., and J. P. Gilpin-Brown. 1973. Flotation mechanisms in modern and fossil cephalopods. *Advances in Marine Biology* 11:197–268.

Eastman, J. T., and A. L. DeVries. 1986. Antarctic fishes. *Scientific American* 255 (November):106–14.

Elsner, R., and B. Gooden. 1983. *Diving and asphyxia: A comparative study of animals and men.* New York: Cambridge University Press.

Fierstine, H. L., and V. Walters. 1968. Studies in locomotion and anatomy of scombroid fishes. *Southern California Academy of Science,* Memoirs 6:1–34.

Hoar, W. W. 1983. *General and comparative physiology.* Englewood Cliffs, NJ: Prentice Hall.

Kanwisher, J., and A. Ebling. 1957. Composition of swim bladder gas in bathypelagic fishes. *Deep-Sea Research* 4:211–17.

Keenleyside, M. H. A. 1979. *Diversity and adaptation in fish behaviour.* Springer-Verlag: New York.

Klimley, A. P. 1994. The predatory behavior of the white shark. *American Scientist* 82:122–133

Kooyman, G. 1989. *Diverse divers: Physiology and behavior.* New York: Springer-Verlag.

Kooyman, G. L. and H. T. Anderson. 1969. In: The biology of marine mammals. p. New York. Academic Press.

Kooyman, G. L., M. A. Castellini, and R. W. Davis. 1981. Physiology of diving in marine mammals. *Annual Review of Physiology* 43:343–57.

McGowan, J. A. 1972. "The nature of oceanic ecosystems." In: *The Biology of the Oceanic Pacific.* Edited by C. B. Miller. Corvallis: Oregon State University Press.

Marshall, N. B. 1966. *The life of fishes.* New York: Universe Books.

Nelson, C., and K. Johnson. 1987. *Whales and walruses as tillers of the seafloor. Scientific American* 256:112–17.

Nerini, M. 1984. A review of gray whale feeding ecology. In: *The Gray Whale, Eschrichtius robustus* (Lilljeborg, 1861). Jones, M. L.; S. L. Swartz and S. Leatherwood. Academic Press, Inc.. New York. p. 423–450.

Norris, K. S., and G. W. Harvey. 1972. "A theory for the function of the spermaceti organ of the sperm whale (Physeter catodon)." In: Animal Orientation and Navigation, pp.397–417. Washington, D.C.: National Aeronautics and Space Administration.

O'Brien, W. J., H. I. Browman, and B. I. Evans. 1990. Search strategies in foraging animals. *American Scientist* 78:152–60.

Partridge, B. L. 1982. The structure and function of fish schools. *Scientific American* 246:114–23.

Perutz, M. F. 1978. Hemoglobin structure and respiratory transport. *Scientific American* (December) 239:92–125.

Pierotti, R., and C. A. Annett. 1990. Diet and reproductive output in seabirds. *Bioscience* 40:568–74.

Pivorunas, A. 1979. The feeding mechanisms of baleen whales. *American Scientist* 67:432–40.

Sanderson, J. L., and R. Wassersug. 1990. Suspension-feeding vertebrates. *Scientific American* (January) 262:96–101.

Scholander, P. F. 1957. The wonderful net. *Scientific American* (April) 196:96–107.

Tucker, D. W. 1959. A new solution to the Atlantic eel problem. *Nature* (London) 183:495–501.

Webb, P. W. 1984. Form and function in fish swimming. *Scientific American* 251:72–82.

Zopal, W. 1987. Diving adaptations of the Weddell seal. *Scientific American* 256(6):100–105.

Chapter 14

Baker, R. C., R. Wilke, and C. H. Baltzo. 1970. The northern fur seals. *Bureau of Commercial Fisheries,* U.S. Fish and Wildlife Service, Circular 336.

Bartholomew, G. A. 1970. A model for the evolution of pinniped polygyny. *Evolution* 24:546–59.

Bonnell, M. L., and R. K. Selander. 1974. Elephant seals: Genetic variation and near extinction. *Science* 184:908–9.

Bonner, W. N. 1982. *Seals and man: A study of interactions.* Seattle: University of Washington Press.

Bonner, W. N. 1989. *Whales of the world.* New York: Facts on File, Inc.

Brown, D. H., D. K. Caldwell, and M. C. Caldwell. 1966. Observations on the behavior of wild and captive false killer whales, with notes on associated behavior of other genera of captive delphinids. *Contributions in Science,* Los Angeles County Museum 95:1–32.

Carr, A. 1965. The navigation of the green turtle. *Scientific American* (May): 79–86.

Cushing, D. H. 1968. *Fisheries biology.* Madison, Wis.: Univ. of Wisconsin Press.

Ege, V. 1939. A revision of the genus *Anguilla* Shaw, a systematic, phylogenetic, and geographical study. *Dana Reports* 3:1–256.

Elsner, R., and B. Gooden. 1983. *Diving and asphyxia: A comparative study of animals and men.* New York: Cambridge University Press.

Fish, J. F., J. L. Sumich, and G. L. Lingle. 1974. Sounds produced by the gray whale, *Eschrichtius robustus. Marine Fisheries Review* 36:38–45.

Hardin-Jones, F. R. 1968. *Fish migration.* London: Edward Arnold.

Hasler, A. D. 1966. *Underwater guideposts: Homing of salmon.* Madison, WI: University of Wisconsin Press.

Hasler, A. D., A. T. Sholz, and R. M. Horrall. 1978. Olfactory imprinting and homing in salmon. *American Scientist* 66:347–54.

Herman, L. M., ed. 1980. *Cetacean behavior: Mechanisms and functions.* New York: John Wiley & Sons.

Jumper, G. Y., Jr., and R. C. Baird. 1991. Location by olfaction: A model and application to the mating problem in the deep-sea hachetfish *Argyropelecus hemigymnus. American Naturalist* 138:1431–58.

Kalmijin, A. J. 1977. The electric and magnetic sense of sharks, skates, and rays. *Oceanus* 20:45–52.

Kellogg, W. N. 1961. *Porpoises and sonar.* Chicago: University of Chicago Press.

King, J. E. 1964. *Seals of the world.* London: British Museum of Natural History.

Klinowska, M. 1988. How brainy are cetaceans? *New Scientist.*

Koch, A. L., A. Carr, and D. W. Ehrenfeld. 1969. The problem of open-sea navigation: The migration of the green turtle to Ascension Island. *Journal of Theoretical Biology* 22:163–79.

Koehn, R. K. 1972. Genetic variation in the eel, a critique. *Marine Biology* (Berlin) 14:179–81.

Laws, R. M. 1961. Reproduction, age and growth of southern fin whales. *Discovery Reports* 31:327–486.

Lohmann, K. J. 1992. How sea turtles navigate. *Scientific American* (January) 264:100–106.

Mackintosh, N. A. 1966. The distribution of southern blue and fin whales. In: *Whales, dolphins and porpoises,* Norris, K. S. ed. Berkeley, CA: University of California Press.

Nicol, J. A. C. 1989. *The eyes of fishes.* New York: Oxford University Press.

Norris, K. S. 1968. Evolution of acoustic mechanisms in odontocete cetaceans. *Evolution and Environment* 297–324.

Norris, K. S., and G. W. Harvey. 1972. A theory for the function of the spermaceti organ of the sperm whale (*Physeter catodon*). In: *Animal orientation and navigation,* pp. 397–417. Washington, D.C.: National Aeronautics and Space Administration.

Oceanus. 1980. Special issue on sensory reception in marine organisms. 23(3).

Owens, D. W. 1980. The comparative reproductive physiology of sea turtles. *American Zoologist* 20:549–63.

Pennisi, E. 1989. Much ado about eels. *Bioscience* 39:594–98.

Pierotti, R., and C. A. Annett. 1990. Diet and reproductive output in seabirds. *Bioscience* 40:568–74.

Pike, G. C. 1962. Migration and feeding of the gray whale (*Eschrichtius gibbosus*). *Journal of the Fisheries Research Board of Canada* 19:815–38.

Rice, D. W., and A. A. Wolman. 1971. "The life history and ecology of the gray whale (*Eschrichtius robustus*)." American Society of Mammalogists, special Publication No. 3.

Robertson, D. R. 1972. Social control of sex reversal in a coral-reef fish. *Science* 177:1007–9.

Rommel, S. A., Jr., and J. D. McCleave. 1972. Oceanic electric fields: Perception by American eels? *Science* 176:1233–35.

Royce, W., L. S. Smith, and A. C. Hartt. 1968. Models of oceanic migrations of Pacific salmon and comments on guidance mechanisms. *Fishery Bulletin* 66:441–62.

Schmidt, J. 1923. Breeding places and migrations of the eel. *Nature* (London) 111:51–54.

Shapiro, D. Y. 1987. Differentiation and evolution of sex change in fishes. *Bioscience* 37:490–97.

Smith, R. J. F. 1985. *The control of fish migration.* New York: Springer-Verlag.

Tucker, D. W. 1959. A new solution to the Atlantic eel problem. *Nature* (London) 183:495–501.

Warner, R. R. 1973. Ecological and evolutionary aspects of hermaphroditism in the California sheephead. *Pimelometopon pulchrum.* Ph.D. Dissertation. San Diego: University of California.

Warner, R. R. 1984. Mating behavior and hermaphroditism in coral reef fishes. *American Scientist* 72:128–36.

Chapter 15

Adey, W. H. 1987. Food production in low-nutrient seas. *Bioscience* 37(5):340–48.

Bardach, J. 1987. Aquaculture. *Bioscience* 37(5):318–19.

Beddington, J. R., R. J. H. Beverton, and D. M. Lavigne. 1985. *Marine mammals and fisheries.* Boston: George Allen and Unwin.

Beddington, J. R., and R. M. May. 1982. The harvesting of interacting species in a natural ecosystem. *Scientific American* (November):62–69.

Bell, F. W. 1978. *Food from the sea: The economics and politics of ocean fisheries.* Denver: Westview Press.

Borgese, E. M. 1983. The law of the sea. *Scientific American* (March): 42–49.

Caddy, J. F., ed. 1989. *Marine invertebrate fisheries: Their assessment and management.* New York: Wiley-Interscience.

Cushing, D. H. 1981. *Fisheries biology.* Madison, WI: University of Wisconsin Press.

Cushing, D. H., and R. R. Dickson. 1976. The biological response in the sea to climatic changes. *Advances in Marine Biology* 14:1–122.

Emery, K. O., and C. O. Iselin. 1967. Human food from ocean and land. *Science* 157:1279–81.

F.A.O. 1995. The State of World Fisheries and Aquaculture. *United Nations Food and Agriculture Organization.* Rome.

Falkowski, P. G., ed. 1980. *Primary productivity in the sea.* New York: Plenum Press.

Food and Agricultural Organization of the United Nations. 1982. *Atlas of the living resources of the seas.* Rome: Department of Fisheries (FAO).

Food and Agricultural Organization of the United Nations. *Yearbook of fisheries statistics, catches, and landings.* Rome: Department of Fisheries (FAO).

Fye, P. M. 1982. The law of the sea. *Oceanus* 25(4):7–12.

Gulland, J. A. 1971. *The fish resources of the ocean*. Surrey, England: Fishing News (Books) Ltd.

Hardin, G. J. 1993. *Living within limits: Ecology, economics, and population taboos*. New York: Oxford University Press.

Harlan, J. R. 1976. The plants and animals that nourish man. *Scientific American* (September):89-97.

Holt, J. S. 1969. Food resources of the ocean. *Scientific American* (September):178-94.

Horn, M. H., and R. N. Gibson. 1988. Intertidal fisheries. *Scientific American* 258(1):64-70.

Johnson, S. W. 1994. Deposition of Trawl Web on an Alaska Beach after Implementation of MARPOL Annex V Legislation. *Marine Pollution Bulletin*. 28:477-481.

Laws, R. M. 1985. The ecology of the Southern Ocean. 1985. *American Scientist* 73:26-40.

Muck, P. 1989. Major Trends in the Pelagic Ecosystem off Peru and Their Implications for Management. PROCOPA Contribution # 90. 386-403.

Murphy, R. C. 1925. *Bird islands of Peru*. New York: G. P. Putnam and Sons.

Nicol, S., and W. de la Mare. 1993. Ecosystem Management and the Antarctic Krill. *American Scientist*. 81:36-47.

Robinson, M. A., and A. Crispoldi. 1975. Trends in world fisheries. *Bioscience* 18:23-9.

Ross, R. M., and L. B. Quetin. 1986. How productive are Antarctic krill? *Bioscience* 36:264-69.

Rothschild, B. J., ed. 1983. *Global fisheries: Perspectives for the 1980s*. New York: Springer-Verlag.

Rudloe, J., and A. Rudloe. 1989. Shrimpers and sea turtles: A conservation impasse. *Smithsonian* 29(9):45-55.

Ryther, J. H. 1969. Photosynthesis and fish production in the sea. *Science* 166:72-76.

Ryther, J. H., et al. 1972. Controlled eutrophication—increasing food production from the sea by recycling human wastes. *Bioscience* 22:144-52.

Ryther, J. H. 1981. Mariculture, ocean ranching, and other culture-based fisheries. *Bioscience* 31:223-30.

Safina, C. November 1995. The World's Imperiled Fish. *Scientific American*. 46-52.

Scarff, J. E. 1980. Ethical issues in whale and small cetacean management. *Environmental Ethics* 3:241-79.

Schaefer, M. B. 1970. Men, birds, and anchovies in the Peru Current—dynamic interactions. *Transactions of the American Fisheries Institute* 99:461-67.

Sissenwine M. P., and A. A. Rosenberg. 1993. U.S. Fisheries: Status, Long-Term Potential Yields, and Stock Management Ideas. *Oceanus*. Summer:48-54.

Smith, W. O., Jr., and D. M. Nelson. 1986. Importance of ice edge phytoplankton production in the southern ocean. *Bioscience* 36:251-57.

Walsh, J. J. 1984. The role of ocean biota in accelerated ecological cycles: A temporal view. *Bioscience* 34:499-507.

Vitousek, J. et al. 1986. Human appropriation of the products of photosynthesis. *Bioscience* 36:368-373.

U.S. Department of Commerce. 1997. *Our Living Oceans: Report on the Status of U.S. Living Marine Resources*. Silver Spring, MD. 150pp.

Chapter 16

Alexander, L. M. 1993. Large marine ecosystems. *Marine Policy* 17:186-98.

Alexander, M. 1981. Biodegradation of chemicals of environmental concern. *Science* 211:132.

Bjorndal, K. A., A. B. Bolten, and C. J. Lagueux. 1994. Ingestion of marine Debris by Juvenile Sea Turtles in Coastal Florida habitats. *Marine Pollution Bulletin*. 28:154-158.

Blus, L., et al. 1971. Eggshell thinning in the brown pelican: Implications of DDE. *Bioscience* 21:1213-15.

Champ, M. A., and F. L. Lowenstein. 1987. The dilemma of high-technology antifouling paints. *Oceanus* 30(3):69-77.

Charney, J. I., ed. 1982. *The new nationalism and the use of common spaces: Issues in marine pollution and the exploitation of Antarctica*. Lanham, MD: Rowman and Littlefield Publications.

Clark, R. B. 1992. *Marine pollution,* 3rd ed. Oxford University Press. Oxford, U.K.

Cox, J. L. 1972. DDT in marine plankton and fish in the California Current. *CalCOFI Reports* 16:103-11.

Croxall, J. P., S. Rodwell, and I. L. Boyd. 1990. Entanglement in man-made debris of Antarctic fur seals at Bird Island, South Georgia. *Marine Mammal Science* 6:221-33.

Eberstadt, N. 1986. Population and economic growth. *Wilson Quarterly* 10:95-127.

Epel, D., and W. L. Lee. 1970. Persistent chemicals in the marine ecosystem. *The American Biology Teacher* 207-11.

Farmingon, J. 1985. Oil pollution: A decade of monitoring. *Oceanus* 28:2-12.

Frankel, E. G. 1995. *Ocean Environmental Management: A Primer on the Role of the Oceans and How to Maintain Their Contributions to Life on Earth*. Prentice-Hall. Englewood Cliffs, NJ

Goldwater, L. J. 1971. Mercury in the environment. *Scientific American* 224 (May): 15-21.

Johnston, R., ed. 1977. *Marine pollution*. New York: Academic Press.

Laws, E. A. 1993. *Aquatic Pollution: An Introductory Text,* 2nd ed. John Wiley & Sons, Inc. New York.

Loughlin, T. R., ed. 1994. *Impacts of the Exxon Valdez Oil Spill on marine mammals*. San Diego: Academic Press.

Mearns, A. J. 1981. Effects of municipal discharges on open coastal ecosystems. In: *Marine Environmental Pollution*, ed. R. A. Geyer. Amsterdam: Elsevier.

Moriarity, F. 1983. *Ecotoxicology: The study of pollutants in ecosystems*. New York: Academic Press.

Norse, E. ed. 1993. *Global Marine Biological Diversity: A Strategy for Building Conservation into Decision Making*. Island Press.

Oceanus. 1990. Special issue on ocean disposal. 33(2).

Peakall, D. B. 1970. Pesticides and the reproduction of birds. *Scientific American* (April) 222:72-78.

Risebrough, R. W., et al. 1967. DDT residues in Pacific seabirds: A persistent insecticide in marine food chains. *Nature* 216:389-91.

Scarff, J. E. 1980. Ethical issues in whale and small cetacean management. *Environmental Ethics* 3:241-79.

Turner, M. H. 1990. Oil spill: Legal strategies block ecology communications. *Bioscience* 40:238-42.

Walsh, J. P. 1981. U.S. Policy on marine pollution. *Oceanus* 24(1):18-24.

Wilbur, R. J. 1987. Plastic in the North Atlantic. *Oceanus* 30(3):61-68.

Wurster, C. E. 1968. DDT reduces photosynthesis by marine phytoplankton. *Science* 159:1474.

Appendix

La Maraic, A. 1973. *The complete metric system with the international system of units (SI)*. Somers, New York: Abbey Books.

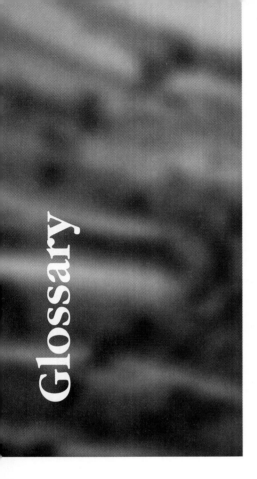

Glossary

A

aboral for radially symmetrical animals, the side of the body opposite the mouth

absorptive feeding a means of taking up dissolved food material through specialized organs or across the body wall

abyssal plains flat, sediment-covered areas in the ocean basin usually 3,000 to 5,000 m deep

accessory pigment one of several nongreen photosynthetic pigments found in marine plants that absorbs light energy from the center of the visible light spectrum and transfers it to the green pigment chlorophyll

acrorhagi nematocyst-armed defensive structures of anemones

adenosine triphosphate (ATP) a complex organic compound composed of the molecule adenosine and three phosphates, which serves in short-term energy storage and conversion in all organisms

aerobic respiration cellular respiration occurring in the presence of oxygen

air sacs lateral branches of the nasal passages of smaller toothed whales; sources of echolocation sounds

alcoholic fermentation a form of anaerobic respiration in which sugar is degraded to alcohol and CO_2 and energy are released

algal grazer an animal that consumes either large algae or thin algal films

algal ridge low, jagged coral ridge common on the windward side of coral reefs

alveoli minute air sacs in the lungs of vertebrates where gas exchange occurs

amebocyte motile cells within sponges that have the shape or properties of an amoeba

amniotic egg egg of reptiles, birds, and mammals, containing an embryo that develops within an amniotic membrane

anadromous referring to an animal (such as a salmon) that spends much of its life at sea and then returns to a freshwater stream or lake to spawn

anaerobic respiration cellular respiration occurring in the absence of oxygen

androgen in vertebrates, a male sex hormone

anoxic without oxygen

antenna an elongated sensory structure projecting from the heads of arthropods

antitropical distribution a pattern of a species' geographical distribution that extends across comparable bands of latitude in the Northern Hemisphere and the Southern Hemisphere

aorta large artery carrying blood away from the heart

aphotic zone the portion of the ocean where the absence of sunlight prohibits plant growth

apneustic breathing breathing pattern exhibited by marine mammals in which several rapid breaths alternate with a prolonged cessation of breathing

areolus a structural unit of diatom frustules

asexual reproduction reproduction by a single individual involving fission, budding, or fragmentation; does not involve the union of gametes

aspect ratio index of propulsive efficiency obtained by dividing the square of a fish's fin height by the fin area

atmosphere (ATM) a unit of pressure equal to 14.7 lb/in² and equivalent to the pressure created by a 10-m column of water

atoll a ring-shaped chain of coral reefs from which a few low islands project above the sea surface

autosome a chromosome not designated as a sex chromosome

autotroph any organism that synthesizes its organic nutrients from inorganic raw materials

auxospore the naked cell of a diatom after the frustule has been shed

B

baleen rows of comblike material that project from the outer edges of the upper jaws of filter-feeding whales

bar-built estuary a type of estuary formed behind a coastal barrier, or bar

barrier reef a coral reef separated from the shore by a lagoon

benthic division the sea bottom and the organisms that inhabit the bottom

benthos marine organisms that live in or on the sea bottom

bilateral body symmetry animal body plan with mirror-image left and right sides

bioaccumulation increasingly concentrated accumulation of substances, especially pollutants, at successively higher trophic levels in food webs

biogeochemical cycle pathway of biologically important nutrients between a living organism and its nonliving reservoir

bioluminescence production of visible light by living organisms

blackband disease a disease of reef-forming corals that results in a band of dead, blackened tissue surrounding a coral colony

blade the flattened, usually broad, leafy structure of seaweeds

blastocyst in mammals, a small ball of cells representing an early stage of embryonic development

blastula early embryonic stage of many animals, consisting of a hollow ball of cells

bleaching expulsion of pigmental zooxanthellae by reef-forming corals when stressed

bloom a dense concentration of phytoplankton that occurs in response to optimal growth conditions

bradycardia marked slowing of the heart rate during a breath-hold dive

broadcast spawners marine animals that reproduce by releasing eggs and sperm into the water

bronchi the paired ventilatory tubes of a vertebrate that branch into each lung at the lower end of the trachea

buffer a chemical substance that tends to limit changes in pH levels

buttress seaward face of a coral reef, extending from a depth of about 20 m to just below the low tide line

byssal thread a strong elastic fiber used by mussels to attach themselves to a solid substrate

C

calorie a unit of heat energy equivalent to the amount of energy required to change the temperature of 1 g of water 1°C

carnivore an animal that feeds on other animals

carpospore spores produced by the carposporophyte form of red algae

carposporophyte a generation of plants, unique to red algae, that produces carpospores

catadromous referring to fishes that migrate from fresh water to spawn in the ocean

caudal fin an enlarged fin at the posterior end of most fishes

caudal peduncle the area where the caudal fin joins the rest of a fish's body

cell wall a supportive structure that encloses the cells of most plants, bacteria, and fungi

cenosarc a thin layer of soft tissue serving to interconnect adjacent polyps of a coral

cephalization the evolutionary process of increasing specialization of the head, especially the brain and sense organs

cerata fingerlike projections along the dorsal sides of some nudibranchs

chemosynthesis bacterial synthesis of organic material from inorganic substances using chemical energy

chitin a flexible, impermeable material found in the exoskeletons of arthropods

chloride cell a specialized gill cell of bony fishes that excretes chloride

chlorophyll the green photosynthetic pigment of plants, protists, and monerans

chloroplast a subcellular structure containing chlorophyll, found in photosynthetic organisms

chromatophore surface pigment cell found in many animals that expands and contracts to produce changes in color and appearance

chromosome a subcellular structure that contains the genetic information of the cell

cilia numerous short hairlike cellular projections used for locomotion or transport

circadian rhythm a cycle of activity or behavior that recurs about once a solar day, or 24 hours

circalunadian rhythm a cycle of repeating activity each lunar day, or 24.8 hours

cleaning symbiosis a form of mutualism in which one partner picks external parasites and damaged tissue from the other partner

climax a relatively stable stage of development achieved by a community of organisms in the absence of major disturbances

clones genetically identical individuals derived from a single individual

cnidoblast cell that contains the stinging nematocyst of cnidarians

coastal plain estuary an estuary created by flooding a coastal river valley with seawater

coccolith a small calcareous plate imbedded in the cell wall of coccolithophores, a type of phytoplankton

coelom the internal body cavity of most animals, lined with a peritoneum

columella the central supporting structure of gastropod mollusk shells and of the corallites of corals

commensalism a symbiotic relationship in which the symbiont benefits without seriously affecting the host one way or another

community an assemblage of interacting populations living in a particular locale

conduction the molecular transfer of heat through a medium

consumer an organism that consumes and digests other organisms to satisfy its energy and material needs

continental boundary current a surface ocean current flowing generally north or south along a continental edge

continental drift the gradual movement of continents in response to seafloor spreading processes

continental shelf the relatively smooth underwater extension of the edge of a continent that slopes gently seaward to a depth of about 200 m

continental slope the relatively steep portion of the sea bottom between the outer edge of a continental shelf and the deep ocean basin

convection the transfer of energy by the flow or mixing of a liquid or gas

convective mixing the vertical mixing of water masses driven by wind stresses or density changes at the sea surface

corallite the calcareous skeletal cup in which a coral polyp sits

Coriolis effect the apparent change in direction of a moving object (to the left in the Southern Hemisphere and to the right in the Northern Hemisphere) due to the rotation of the earth

countercurrent an ocean current that flows directly back into another current; also, in some animals, paired blood vessels containing blood flowing in opposite directions

countershading the coloration pattern found in pelagic animals, with the upper surfaces darkly pigmented and the sides and ventral surfaces silvery or only lightly pigmented

critical depth the depth at which photosynthesis equals cell respiration

cropper a deep-sea animal in which the roles of predator and deposit feeder have merged

ctenes bands of cilia found on the body surfaces of ctenophores

cypris the final larval stage of a barnacle

cyst in diatoms, a resistant and photosynthetically inactive cell

cytoplasm the internal fluid environment of a cell

D

dark reaction that part of the photosynthetic process that, in the absence of light, utilizes the preformed high-energy molecules ATP and NADPH$_2$ to synthesize complex organic molecules

decomposer an organism that consumes and breaks down dead organic material

delayed implantation a pattern in the reproductive cycle of some mammals causing the blastocyst to remain dormant in the female's uterus for some time before implantation on the uterine wall

delta a low-lying sediment deposit often found at the mouth of a river

density ratio of the mass of a substance to its volume

deposit feeder an animal that engulfs masses of sediments and processes them through its digestive tract to extract nourishment

detritus excrement and other waste products of all types of organisms, including their remains after death

diffusion the transfer of substances along a gradient from regions of high concentrations to regions of low concentrations

diploid cell a cell that contains two of each type of chromosome characteristic of its species

diurnal tide a tidal pattern with one high tide and one low tide each lunar day

dive reflex the suite of internal responses, including bradycardia and peripheral circulation shutdown, that occurs during dives by an air-breathing vertebrate

E

ecological adaptation short-term changing responses expressed by an individual organism to its environment

ectotherm an organism whose body temperature is controlled by external heat sources

elver a juvenile eel

endoplasmic reticulum a system of folded membranes within the cytoplasm of eucaryotic cells

endotherm an animal whose body temperature is established by internal sources of heat

enzyme protein catalyst that regulates a particular chemical reaction

epifauna benthic animals that crawl about on the sea bottom or sit firmly attached to it

epipelagic zone the upper 200 m of the oceanic province

epiphyte a plant that attaches itself to other plants or animals without parasitizing them

epitheca the larger portion of a diatom frustule

estrogen a female sex hormone in vertebrates

estrous the period of highest sexual receptivity (or "heat") in some female mammals that coincides with the time of egg release by the ovary

estuary the portion of the mouth of a river in which there is substantial mixing of fresh water and seawater

eucaryotes cells characterized by an organized nucleus and other membrane-bound subcellular structures

euryhaline an organism capable of withstanding a wide range of salinities

eurythermal the ability to tolerate wide variations in environmental temperature

evolutionary adaptation the changes occurring in a population of individuals over many generations by processes of natural selection

exoskeleton an external supporting skeleton, commonly found in arthropods

external auditory canal the sound channel connecting the external and middle ears

F

fecundity the number of offspring an organism can produce in a given time span

feedback mechanisms control mechanisms in organisms and communities in which a change in a given factor either inhibits or stimulates processes controlling the production, release, or use of that factor

fertilization the fusion of two haploid gametes to produce a diploid zygote

fetch the extent of the ocean over which winds blow to create waves

finlet small median fin on the dorsal and ventral sides of the rear parts of tuna and similar fishes

fjord a deep coastal embayment caused by glacial erosion

flagellum whiplike structure used by cells for locomotion

flushing time the time required for all of the water of an estuary to be completely exchanged

food chain a diagrammatic representation of trophic relationships

food web a diagrammatic representation of the complete set of trophic relationships in the living portion of an ecosystem

form drag hydrodynamic drag on an organism caused by its cross-sectional area

frictional drag the resistance created by an animal's body surface when it moves through a fluid medium

fringing reef a large coral reef formation that closely borders the shoreline

frustule the siliceous wall of a diatom; consists of two halves

fucoxanthin a golden or brown pigment characteristic of Phaeophyta, Chrysophyta, and Pyrrophyta

G

gamete an egg or a sperm cell

gametophyte a gamete-producing haploid plant

gastrovascular cavity digestive cavity of some lower invertebrates, with a single opening for both mouth and anus

gestation period the portion of the reproductive cycle in a female mammal extending from fertilization to delivery of its offspring

gill arch in fish, the skeletal supporting structure of a gill

grana flattened saclike structures inside chloroplasts containing chlorophyll and other photosynthetic enzymes

gross primary production the total amount of photosynthesis accomplished in a given period of time

guano the droppings from seabird nesting colonies

guyot   a sunken volcanic island topped by a dead, usually flat, coral reef

gyre   the large loop of interconnected surface ocean currents within a single ocean basin, usually spanning 20° to 30° in latitude

H

hair cell   a general type of sensory cell that is equipped with sensory cilia and is sensitive to mechanical disturbance

halophyte   flowering plants that are tolerant to complete submergence in seawater

haploid cell   a cell that contains only one of each type of chromosome characteristic of its species

haptera   short, sturdy rootlike structures that form the holdfast of seaweeds

heat capacity   the measure of heat energy required to change the temperature of 1g of a substance 1°C

hemocyanin   an oxygen-binding blue pigment found in the blood of several kinds of invertebrates

hemoglobin   an oxygen-binding red blood pigment found in vertebrates and some invertebrates

herbivore   an animal adapted to feed on plants

hermaphrodite   an animal that has the sex organs of both sexes

hermatypic   a type of coral which builds massive carbonate reefs

heterotroph   an organism that is unable to synthesize its own food from inorganic substances and must utilize other organisms for nourishment

high tide   the highest level reached by the rising tide

holdfast   a structure that attaches seaweeds to the sea bottom or to other substrates

holoplankton   species of zooplankton that remain in a planktonic stage throughout their lives

homeostasis   tendency of living organisms to maintain a steady state in their internal environmental conditions, including body temperature, blood sugar level, and metabolic rate

homeotherm   an animal, such as a bird or mammal, that maintains precisely controlled internal body temperatures using its own heating and cooling mechanisms

host   one member of the host-symbiont pairing characteristic of all symbiotic relationships

hydrogen bond (H-bond)   a weak bond formed by the attractive force between the charged ends of water molecules and other charged molecules or ions

hydrostatic skeleton   the body fluid contained within some animals, against which muscles work to provide shape changes

hyperosmotic   a water medium with a higher concentration of ions than another solution and separated from it by a selectively permeable membrane

hyphae   elongated and multinucleated threadlike structures making up the body of a fungus

hypoosmotic   a water medium with a lower concentration of ions than that of another solution separated by a selectively permeable membrane

hypotheca   the smaller portion of a diatom frustule

I

infauna   animals that live within the sediment of the sea bottom

inquilinism   a protection benefit of symbiosis resulting from the proximity of a large and imposing host

interstitial animal   an animal that occupies the spaces (interstices) between sediment particles

intertidal zone   the vertical extent of the shoreline between the high and low tide lines

ion   an electrically charged atom or molecule formed by gaining or losing one or more electrons

ionic bond   in crystalline structures, an atomic bond formed between adjacent oppositely charged ions

iridocyte   a fish skin cell that contains reflecting crystals made of guanine

isohaline   having the same salinity

isosmotic   a water medium with the same concentration of ions as another solution and separated from it by a selectively permeable membrane

K

kelp   a group of large brown seaweeds

L

labyrinth organ   one of a pair of equilibrium organs in vertebrates that contains three fluid-filled semicircular canals

Langmuir cells   parallel pairs of surface ocean convection cells driven by winds

last glacial maximum (LGM)   the time that the last major continental glacial advance in the Northern Hemisphere reached its maximum extent, about 18,000 years ago

latent heat of fusion   the heat that must be extracted from a liquid to freeze it to a solid at the same temperature; for water, 80 cal/g

latent heat of vaporization   the heat energy required to convert a liquid to a gas at the same temperature; for water, 540 cal/g

latitude   the angular distance north or south of the equator of a position on the earth's surface; measured in degrees (°)

leptocephalus larva   leaf-shaped, transparent larva of some eels

light reaction   that part of the photosynthetic process that, in the presence of light, captures energy to form ATP and $NADPH_2$ to be used to synthesize complex organic molecules in the dark reaction

limiting factor   any factor necessary for the growth or health of an organism that limits further growth if in insufficient supply

littoral   the intertidal zone

longitude   the angular distance of a position on the earth's surface east or west of the Greenwich prime meridian; measured in degrees (°)

lophophore   tentacle-bearing feeding structure of ectoprocts, brachiopods, and phoronids

low tide   the lowest level reached by the falling tide

M

macrofauna   benthic animals larger than about 0.5 mm

mangal   a tropical community of mangrove plants and associated organisms

mariculture   the collective techniques applied to grow marine organisms in captive, controlled situations

maximum sustainable yield   the maximum level of fishing effort that a fish stock can withstand without experiencing major upsets in the abundance of its stock

medusa   free-swimming, or jellyfish, stage of cnidarians

meiofauna   benthic animals intermediate in size between macrofauna and microfauna

meiosis   a process of cellular division that reduces the chromosome number by half

melon   fat-filled structure in the foreheads of toothed whales

meristematic tissue   within some seaweeds, specific tissue sites where most cell division for growth occurs

meroplankton   larval forms of benthic and nektonic adults that are temporary members of the plankton community

mesoglea   the jellylike layer in the bodies of cnidarians

mesopelagic zone   the portion of the pelagic division that extends from the bottom of the epipelagic zone to about 1,000 m

metabolism   collectively, all the biochemical processes occurring in a living organism

metamere   a repeated body unit, or segment, along the long axis of some bilateral animals

microatoll   a small, flat, atoll-shaped coral structure generally found in protected coral lagoons

microbial loop   the portion of a planktonic food web composed of bacteria and other decomposers

microfauna   benthic animals smaller than about 0.1 mm

mimicry   the ability of an organism to disguise itself by assuming the behavior or appearance of another species

mitochondria   in eucaryotes, a subcellular organelle that conducts cellular respiration

mitosis   a process of cell division resulting in two descendant cells genetically identical to their parent cell

mixed semidiurnal tide   tidal pattern during a lunar day with unequal high tides and unequal low tides

molecular systematics   the study of evolutionary relationships of organisms based on comparisons of proteins and DNA fragments

mortality   the rate at which individuals of a population die

mutualism   a type of symbiotic relationship in which both the symbiont and the host benefit from the association

myoglobin   a red muscle pigment with a strong chemical affinity for oxygen (similar to that of hemoglobin)

myomere   one of a series of muscle segments along the trunk of vertebrates, especially fishes

N

nannoplankton   phytoplankton with cell sizes between 5 and 20 μm

nasal gland   salt-excreting gland in the nostrils of seabirds and reptiles

nauplius   microscopic free-swimming planktonic stage of barnacles and some crustaceans

neap tides   sets of moderate tides that alternate with spring tides and recur every two weeks

nekton   large, actively swimming marine animals

nematocyst   venomous stinging cellular organelle of cnidarians, contained within the cnidoblast

neritic province   the portion of the marine environment that overlies the continental shelves

net primary production   the measure of primary production after cell respiration is subtracted from gross primary production

neuromast   a mechanosensory cell found in vertebrates

neuston   planktonic organisms living (usually floating) at or on the sea surface

niche   the functional role of an organism as well as the suite of physical and chemical factors that limit its range of existence

nitrogen fixation   conversion by bacteria and cyanobacteria of atmospheric N_2 to other forms of nitrogen used by eucaryotic plants

non-point source   a source of pollution originating from a broad and vaguely defined source, such as fertilizers washed off fields into streams

notochord   an elongated, cartilagenous rod that forms the central skeletal support of chordate embryos

nuclear membrane   the membrane surrounding the nucleus of eucaryotic cells

nucleus   the membrane-bound central structure of eucaryotic cells that contains the chromosomes

O

oceanic province   the portion of the marine environment that overlies the deep ocean basins

oceanic reef   coral reefs associated with volcanic islands rather than continental shelves.

olfaction   ability to detect and identify chemicals dissolved in air or water by using olfactory sensory cells

open access   the concept of international law that permits free access by any nation to marine resources existing outside national jurisdictions

oral   for radially symmetrical animals, the side of the body containing the mouth

osmoregulator   an organism that regulates the salt content of its internal fluid environment

osmosis   diffusion of water across a selectively permeable membrane

osmotic conformers   organisms that tolerate large variations of internal ionic concentrations without serious damage by remaining isosmotic with the water around them

osmotic pressure   in hypoosmotic conditions, the internal fluid pressure that develops from the osmotic inflow of water

oviparity   a condition that describes the habit of releasing eggs that later hatch

ovoviviparity   an intermediate condition between viviparity and oviparity in which the eggs are incubated inside the mother until hatching

oxygen minimum zone   the ocean layer below the photic zone where dissolved oxygen concentration is lowest

ozone O_3, formed by the action of ultraviolet light acting on atmospheric O_2

P

Pangaea   the supercontinent that consisted of all the present landmasses prior to their breakup and subsequent drift to their present positions

parasitism a type of symbiotic relationship in which the parasite lives on or in the host and benefits at the expense of the host

pelagic divisim the waters of the ocean and the organisms that inhabit it

pen in squids, a thin, chitinous structure extending the length of the mantle tissue

period in ocean waves, the time required for two successive waves to pass a reference point

pH a numerical scale from 0 to 14 that is used to represent the hydrogen ion concentration of a water solution

pheromone a chemical substance used for communication between organisms of the same species

photic zone the portion of the ocean where light intensity is sufficient to accommodate plant growth

photoinhibition reduction of photsynthetic rates due to too much light

photophores an animal's light-producing organs

photosynthesis the biological synthesis of organic material from inorganic substances using light as an energy source

phototrophic referring to reliance on the photosynthetic products of symbiotic zooxanthellae to obtain nutrition; found in hermatypic corals, some sponges, and other reef invertebrates

phycobilin a type of pink or blue accessory photosynthetic pigment found in cyanobacteria and red algae

phylogenetic tree a branching diagram depicting the evolutionary relationships between major groups of organisms

physoclistous swim bladder in fish, a swim bladder lacking an air passage to the esophagus

physostomous swim bladder in fish, a swim bladder with an air passage or duct to the esophagus

phytoplankton microscopic photosynthetic members of the plankton

picoplankton the small-sized groups of phytoplankton with cells less than 2 μm in width

plankton free-floating, usually minute, organisms of the sea

planula the planktonic larval form of some corals

plasma membrane the selectively permeable outer membrane of a cell

plate tectonics the collective geological processes that move the crustal plates of the earth and cause continental drifting and seafloor spreading

pleopods abdominal paired appendages in crustaceans

pneumatic duct in fish, the connection between the esophagus and swim bladder

pneumatocyst gas-filled float present in several types of kelp plants

pneumatophore a gas-filled float used by some siphonophores to maintain buoyancy in the water

poikilotherm an organism whose body temperature varies with and is largely controlled by environmental temperatures

point source a source of pollution originating at a definable point, such as a storm drain

polar easterlies winds that blow from east to west at very high latitudes

pollen the small fertilizing structure of flowering plants that contains the male gamete

polygyny a type of social and breeding organization in which a male is dominant over and mates with several females

polyp the attached, benthic growth form of many types of cnidarians

population a group of freely interbreeding organisms of the same species

predator a carnivorous animal that feeds by killing other animals

primary producer an organism that synthesizes material by photosynthesis or chemosynthesis

primary production the synthesis of organic material from inorganic molecules

procaryotes bacteria and cyanobacteria that lack the structural complexity and defined nucleus found in eucaryotes

producer an organism, usually photosynthetic, that contributes to the net primary production of a community

pseudopodia temporary and changing extensions of the cells of some protozoans, used for locomotion

pycnocline the ocean layer, usually near the bottom of the photic zone, marked by a sharp change in density

R

radula rasping tonguelike structure of most mollusks

raphe a groove in the frustules of pennate diatoms through which cytoplasm extends for locomotion

red tide a bloom condition in which some species of dinophytes produce toxins that may cause serious mortality to other forms of marine life

reef flat the portion of a coral reef that extends behind the algal ridge to the island

rete mirabilia a complex network of capillary-sized blood vessels, usually in a countercurrent arrangement

retina the light-sensitive layer of nerve cells in the eyes of most vertebrates and some invertebrates

rhizome the horizontal underground stem of sea grasses

ridge and rise system the interconnecting chain of seafloor mountains that trace the edges of crustal plates and the sites of new oceanic crust production

S

salinity a measure of the total amount of dissolved ions in seawater

salt marsh a low-lying grass community found along tidally flooded shores of temperate-climate estuaries

saturation light intensity the light intensity that maximizes the photosynthetic rate

saxitoxin a paralytic toxin produced by dinophytes that accumulates in the butter clam (*Saxadoma*)

scavenger an animal that feeds on the dead remains of other animals and plants

school a well-defined social organization of marine animals consisting of a single species with all members of a similar size

seafloor spreading a global process in which oceanic crust moves away from ridge and rise systems where it formed

seamount an undersea volcano

selectively permeable membrane a membrane that is permeable to small molecules, usually H_2O, O_2, and CO_2, but is not permeable to larger molecules or ions

semidiurnal tides tidal patterns with two high tides and two low tides each lunar day

septa thin structures separating internal parts of organisms

sex chromosome one of a pair of chromosomes whose composition determines gender

sexual reproduction a mode of reproduction involving the production of gametes by meiosis, followed by a fusion of the gametes in fertilization

shelf break the outer edge of the continental shelf, typically 100 to 200 m deep

shelf reef a coral reef growing on the continental shelf

siphon tubelike structure of mollusks used to take in and expel water from the mantle cavity

siphuncle a central tubelike tissue connecting the chambers of shelled cephalopods such as *Nautilus*

smolt a young salmon just before it migrates downstream and out to sea

sound-scattering layer (SSL) one or more layers of midwater marine animals that reflect and scatter the sound pulses of echo sounders

species a group of organisms sharing a common ancestry, reproductive isolation from other organisms, and a capability for successful reproduction

spermaceti organ a large organ in the forehead of sperm whales that is filled with waxy spermaceti oil, a fine-quality liquid

spicules small, mineralized skeletal structures of sponges

spongin fiberous material making up the flexible skeleton of many sponges

sporangium a special plant cell or structure that produces spores

spore in plants, the single-celled haploid structure produced by sporophytes

sporophyte a spore-producing diploid plant

spring tides extremely high tides and low tides that alternate with neap tides and recur every two weeks

standing crop total amount of plant or animal material in an area at any one time

statocyst a gravitationally sensitive vesicle lined with sensory cells and containing dense bodies; found in many invertebrates

stenohaline an organism that tolerates exposure to only slight variations in salinity

stenothermal the tendency to tolerate only narrow variations in environmental temperature

stigma a structure on the female part of a flower on which pollen grains are received and germinate

stipe the flexible, stemlike structure found in the large seaweeds

stroma the part of the chloroplast containing the enzymes for the dark reaction of photosynthesis

succession the gradual replacement, through time, of one group of species in a community by other groups

surface current long-term directional flow of water at the sea surface

surface tension the mutual attraction of water molecules at the surface of a water mass that creates a flexible molecular "skin" over the water surface

suspension feeder an animal that uses a filtering device or sticky mucus to obtain plankton or detritus from the water

swim bladder gas-filled buoyancy organ of bony fishes

symbiont the beneficiary of a symbiotic relationship

symbiosis an intimate and prolonged association between two (or more) organisms in which at least one partner obtains some benefit from the relationship

T

taxon any group or category in a taxonomic system of classification; plural, taxa

taxonomy the process of classifying organisms according to their evolutionary relationships

tectonic estuary an estuary that fills in a depression created by one or more faults in the near-shore crust

temperature a relative intensity measure of the condition caused by heat

testosterone in vertebrates, a male sex hormone

thermocline the ocean layer, usually near the bottom of the photic zone, marked by a sharp change in temperature

tidal range vertical distance between high and low tides

tide a long-period wave noticeable as a periodic rise and fall of the sea surface along coastlines

trace element an element needed for normal metabolism but available only in minute amounts from the environment

trachea the windpipe of vertebrates

trade winds subtropical winds that blow from northeast to southwest in the Northern Hemisphere and from southeast to northwest in the Southern Hemisphere

trench deep area in the ocean floor, generally deeper than 6,000 m

trochophore early free-swimming, ciliated larval stage of many marine mollusks, annelid worms, ectoprocts, and brachiopods

trophic pertaining to feeding or nutrition

trophic level the position of an organism or species in a food web

trophosome an internal, symbiotic bacteria-filled organ of the giant tube worm *Riftia*

turbulence random, nonlaminar flow of a fluid

turnover rate the rate at which members of a population or community replace themselves

tympanic bulla bony case in the middle ear that encloses the sound-processing structures of mammals

U

ultraplankton phytoplankton with cells between 2 and 5 μm in width

upwelling the process that carries nutrient-rich subsurface water upward to the photic zone

urea   a nitrogen-containing waste product excreted in the urine of many vertebrates

uric acid   the main nitrogenous excretory product in birds, reptiles, some invertebrates, and insects

V

vacuole   a liquid- or food-filled cavity within a cell

veliger   early larval form of some mollusks

vena cava   the major vein returning blood to the heart of vertebrates

vertebra   one of a series of articulated bones that make up the backbone of vertebrates

vertical migration   daily or seasonal movement of small marine animals between the photic zone and midwater depths

villi   small fingerlike projections of tissue that increase surface area and improve secretion and absorption

viscosity   the resistance of water molecules to external forces that would separate them

visible light   that portion of the electromagnetic radiation to which the human eye is sensitive, usually represented as a spectrum of colors from violet to red

viviparity   a condition describing the act of giving birth to live young

W

wave   a periodic, traveling undulation of the sea surface

westerlies   winds that blow primarily from the west in the midlatitudes

X

xanthophyll   a group of yellow or golden photosynthetic pigments

Z

zoea   an early larval stage of many crustaceans

zooplankton   animal members of the plankton

zooxanthellae   symbiotic unicellular dinophytes found in corals, sea anemones, mollusks, and several other types of marine animals

zygote   the product of the fusion of two gametes to produce a diploid single cell

Index

Taxonomic

Index

Subject